T0213910

Lecture Notes in Computer Science 8451

Commenced Publication in 1973
Founding and Former Series Editors:
Gerhard Goos, Juris Hartmanis, and Jan van Leeuwen

Helmut Simonis (Ed.)

Integration of AI and OR Techniques in Constraint Programming

11th International Conference, CPAIOR 2014
Cork, Ireland, May 19-23, 2014
Proceedings

 Springer

Volume Editor

Helmut Simonis
University College Cork, Ireland
E-mail: h.simonis@4c.ucc.ie

ISSN 0302-9743   e-ISSN 1611-3349
ISBN 978-3-319-07045-2   e-ISBN 978-3-319-07046-9
DOI 10.1007/978-3-319-07046-9
Springer Cham Heidelberg New York Dordrecht London

Library of Congress Control Number: 2014938029

LNCS Sublibrary: SL 1 – Theoretical Computer Science and General Issues

Typesetting: Camera-ready by author, data conversion by Scientific Publishing Services, Chennai, India

Printed on acid-free paper

Springer is part of Springer Science+Business Media (www.springer.com)

Preface

This volume is a compilation of the research program of the 11th International Conference on the Integration of Artificial Intelligence (AI) and Operations Research (OR) Techniques in Constraint Programming (CPAIOR 2014), held in Cork, Ireland, during May 19–23, 2014.

After a successful series of five CPAIOR international workshops in Ferrara (Italy), Paderborn (Germany), Ashford (UK), Le Croisic (France), and Montreal (Canada), in 2004 CPAIOR evolved into a conference. More than 100 participants attended the first meeting held in Nice (France). In the subsequent years, CPAIOR was held in Prague (Czech Republic), Cork (Ireland), Brussels (Belgium), Paris (France), Pittsburgh (USA), Bologna (Italy), Berlin (Germany), Nantes (France), and Yorktown Heights (USA). In 2014 CPAIOR returned to Ireland.

The aim of the CPAIOR conference series is to bring together researchers from constraint programming (CP), artificial intelligence (AI), and operations research (OR) to present new techniques or applications in the intersection of these fields, as well as to provide an opportunity for researchers in one area to learn about techniques in the others. A key objective of the conference is to demonstrate how the integration of techniques from different fields can lead to highly novel and effective new methods for large and complex problems. Therefore, papers that actively combine, integrate, or contrast approaches from more than one of the areas were especially welcome. Application papers showcasing CP/AI/OR techniques on innovative and challenging applications or experience reports on such applications were also strongly encouraged.

In all, 70 long and short papers were submitted to the conference. Out of these, 33 papers were selected by the international Program Committee. Their effort to provide detailed reviews and discuss all papers in depth after author feedback to come up with a strong technical program is greatly appreciated.

The technical program of the conference was preceded by a day of workshops and a master-class. The workshops this year were selected by the workshop chair, Lars Kotthoff, and covered the following topics:

- Workshop on Multi-Stage Problems, organized by Marco Laumanns, Steven Prestwich, and Roberto Rossi.
- 5th International Workshop on Bin Packing and Placement Constraints (BPPC 2014), organized by Francois Fages and Nicolas Beldiceanu.
- e-policy Workshop, organized by Michela Milano.

The master-class was organized by Siegfried Nijssen and Lars Kotthoff on the topic of *Optimization and Machine Learning*, providing an overview of this interesting area for PhD students, academics and practitioners.

Putting together a conference requires help from many sources. I in particular want to thank the Program Committee and all other reviewers, who worked very hard in a busy period of the year. My conference co-chair, Barry O'Sullivan, was instrumental in bringing the conference back to Cork and organizing an interesting event. The help of Yuri Malitsky (publicity), Lars Kotthoff (workshops), and Barry Hurley (website) was greatly appreciated. Finally, the staff at UCC provided outstanding administrative support, without which the conference would not have been possible. Special thanks go to Linda O'Sullivan, Mary Noonan, and Caitriona Walsh.

A final acknowledgement goes to Easychair and Springer, who allowed us to put together these proceedings.

February 2014   Helmut Simonis

Organization

Program Committee

Carlos Ansótegui Gil	Universitat de Lleida, Spain
Fahiem Bacchus	University of Toronto, Canada
J. Christopher Beck	University of Toronto, Canada
Nicolas Beldiceanu	EMN, France
Hadrien Cambazard	G-SCOP, Grenoble INP, CNRS, Joseph Fourier University, France
Geoffrey Chu	Melbourne University, Australia
François Clautiaux	University of Bordeaux, France
Pierre Flener	Uppsala University, Sweden
Carla Gomes	Cornell University, USA
Tias Guns	KU Leuven, Belgium
Emmanuel Hebrard	LAAS, CNRS, France
Stefan Heinz	Zuse Institute Berlin, Germany
John Hooker	Carnegie Mellon University, USA
Frank Hutter	Freiburg University, Germany
Serdar Kadioglu	Oracle, USA
George Katsirelos	INRA, Toulouse, France
Thorsten Koch	ZIB/Matheon, Germany
Andrea Lodi	DEI, University of Bologna, Italy
Yuri Malitsky	Cork Constraint Computation Centre, Ireland
Laurent Michel	University of Connecticut, USA
Michela Milano	DEIS Università di Bologna, Italy
Nina Narodytska	University of Toronto, Canada
Yehuda Naveh	IBM, Israel
Barry O'Sullivan	4C, University College Cork, Ireland
Laurent Perron	Google France
Gilles Pesant	Ecole Polytechnique de Montreal, Canada
Thierry Petit	Ecole des Mines de Nantes/LINA, France
Claude-Guy Quimper	Université Laval, Canada
Ted Ralphs	Lehigh University, USA
Louis-Martin Rousseau	Ecole Polytechnique de Montreal, Canada
Jean-Charles Régin	Université Nice Sophia Antipolis, I3S CNRS, France
Ashish Sabharwal	IBM Research, USA
Domenico Salvagnin	University of Padoa, Italy
Pierre Schaus	n-side, Belgium
Christian Schulte	KTH Royal Institute of Technology, Sweden

Meinolf Sellmann   IBM Research, USA
Helmut Simonis   4C, University College Cork, Ireland
Michael Trick   Carnegie Mellon University, USA
Pascal Van Hentenryck   NICTA and University of Melbourne, Australia
Willem-Jan Van Hoeve   Carnegie Mellon University, USA
Petr Vilím   IBM, Czech Republic
Mark Wallace   Monash University, Australia
Tallys Yunes   University of Miami, USA

Additional Reviewers

Aharoni, Merav   Hashemi Doulabi, Seyed Hossein
Ben-Haim, Yael   Ibrahim, Mohamed
Bonfietti, Alessio   Kotthoff, Lars
Boni, Odellia   Legrain, Antoine
Borgo, Stefano   Lombardi, Michele
Carbonnel, Clément   Malapert, Arnaud
Carlsson, Mats   Mehta, Deepak
Castañeda Lozano, Roberto   Monaci, Michele
David, Philippe   Monette, Jean-Noël
de Cauwer, Milan   Papadopoulos, Alexandre
De Givry, Simon   Penz, Bernard
De Koninck, Leslie   Restrepo-Ruiz, Maria-Isabel
Derrien, Alban   Schimpf, Joachim
Fischetti, Matteo   Scott, Joseph
Fitzgerald, Tadhg   Siala, Mohamed
Fontaine, Daniel   Simonin, Gilles
Gabay, Michaël   Traversi, Emiliano
Grimes, Diarmuid   Tubertini, Paolo
Gualandi, Stefano   Xue, Yexiang
Hartert, Renaud

Table of Contents

Call-Based Dynamic Programming for the Precedence Constrained Line Traveling Salesman

Thierry Benoist[1], Antoine Jeanjean[2], and Vincent Jost[3]

[1] Innovation 24 - LocalSolver, Paris, France
tbenoist@localsolver.com
[2] Recommerce Solutions, Paris France
antoine.jeanjean@recommerce.com
[3] Grenoble-INP / UJF-Grenoble 1 / CNRS, G-SCOP UMR5272 Grenoble, France
Vincent.Jost@grenoble-inp.fr

Abstract. The Precedence Constrained Line Traveling Salesman is a variant of the Traveling Salesman Problem, where the cities to be visited lie on a line, the distance between two cities is the absolute difference between their abscissae and a partial ordering is given on the set of cities. Such a problem is encountered on linear construction schemes for instance. Using key dominance properties and lower bounds, we design a call-based dynamic program able to solve instances with up to 450 cities.

1 Introduction

The Line-TSP is a variant of the Traveling Salesman Problem (TSP) where the cities to be visited lie on a line, and the distance between two cities is the absolute difference between their abscissae. Although trivial in this pure formulation, the problem becomes interesting when side-constraints are added. For instance, [13] considers the case where each city must be visited within a certain time-window. The present paper deals with the case where a partial ordering is given on the set of cities that is to say that some precedence constraints *A must be visited before B* must be satisfied. In practice this problem is encountered on linear construction schemes when a set of partially ordered tasks (up to several hundreds) must be performed by a resource whose traveling distance must be minimized, like an excavation engine on a highway construction site for instance [8]. To the best of our knowledge, this problem was not studied in the literature before, but its NP-completeness was established as a special case in [3]. Linear structures also occur in *N-line TSP* [4,12] namely an Euclidian TSP with a limited number of of different abscissae and also in the so-called *convex hull and line TSP* [5,6].

After a formal definition of the problem we introduce a key dominance property.In section 2 we define a lower bound based on the splitting of the line in sections. Finally, we propose a call-based dynamic programming approach, where branches are pruned with our lower bound. This algorithm is experimented in section 4.

H. Simonis (Ed.): CPAIOR 2014, LNCS 8451, pp. 1–14, 2014.

1.1   Problem Definition and Notations

Definition 1. *The Precedence Constrained Line TSP (PC-Line-TSP) reads as follows.*

Given a set P of n cities, each with an abscissa X_i ($i \in [1, n]$), a partial order $\prec$ on P and an integer K; find a permutation V of P such that :

$$\sum_{i=1}^{n} \mid X_{V(i)} - X_{V(i-1)} \mid \leq K, \text{ with } X_{V(0)} = 0$$
$$\forall (i, j) \in P^2 \text{ such that } i \prec j, \ V^{-1}(i) < V^{-1}(j), \qquad (1)$$
$$\text{with } V^{-1}(i) \text{ the position of city } i \text{ in the permuation.}$$

The above sum of differences of abscissae will be referred to as the *length* of the permutation. We will denote by m the size of $\prec$ that is to say the number of ordered pairs defining this partial order. Figure 1 represents a instance of PC-Line-TSP with four cities, the Hasse diagram being represented on the left side. The dotted arrows represent a feasible solution (visiting order).

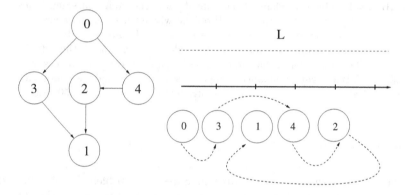

Fig. 1. A Precedence Constrained Line TSP

1.2   Properties

Definition 2. *In a solution, a* procrastination *is a city which is passed through without being visited whereas all its predecessor have already been visited.*

Definition 3. *A solution is called* dominating *(ou non-procrastinating) is it contains no procrastination. In other words it never passes through an abscissa without visiting all available cities at this abscissa (namely cities whose predecessors have already been visited).*

Lemma 1. *There exists a dominating optimal solution for each problem.*

Proof. Let V be a non dominating permutation, that is to say that there exists at least one procrastination city d. By definition there are two consecutive cities a and b in the permutation such that:

$$X_d \in [X_a, X_b[\quad or \quad X_d \in]X_b, X_a] \qquad (2)$$

and such that all predecessors of d (possibly including a) are before a in permutation V (see Figure 2), while city d is visited later, between two cities c and e. Note that c may equal b and that e may not exist (in which case d is the last city of the permutation), without affecting the validity of the reasoning below.

Let V' be the permutation obtained from V by moving city d between a and b. V' satisfies the partial order because all predecessors of d are before a in both permutations. Now we prove that the length or permutation V' is smaller or equal to the length or permutation V.

- the path $a \rightarrow d \rightarrow b$ is equal to path $a \rightarrow b$ (d being between a and b).
- the path $c \rightarrow e$ is smaller or equal to path $c \rightarrow d \rightarrow e$ since distances on a line satisfy the triangular inequality (and if e does not exists, the path $c \rightarrow d$ is merely removed).

Repeating this transformation, all procrastinations can be eliminated, while preserving or decreasing the length of the permutation, thus building a dominating permutation of equal or smaller length. Hence any problem has an optimal dominating solution. □

Fig. 2. Procrastination removal

Corollary 1. *A solution can be written as a sequence of t abscissae with $t \leq n$. This sequence (or* path*) induces a visiting order (unique but for the visiting order of cities at the same abscissa). Hence the PC-Line-TSP can be reformulated as a question: is there a sequence of abscissae (or path) whose length is smaller than K and which visits all cities ?*

Corollary 2. *The special case where cities can take only two different abscissae ($|\{X_i, i \in [0,n]\}| = 2$) is polynomial. Its optimal path alternates the two abscissae. It can be computed in linear time.*

1.3 Complexity

We have seen above that the problem is polynomial when limited to 2 abscissae. In the general case this problem was proven to be NP-hard by [3], by reduction from the Shortest Common Supersequence problem. It is NP-hard as soon as the number of number of abscissae if larger or equal to 3 (see [9]).

2 Lower Bounds

In this section we compute lower bounds for the PC-Line-TSP problem.

Trivial bound LB_0. The distance between the maximum and minium abscissae plus the distance from 0 the closest extremity is a lower bound of the traveled distance.

$$R = \max_{i \in [0,n]} X_i$$

$$L = \min_{i \in [0,n]} X_i$$

$$LB_0 = R - L + \min(|R|, |L|)$$

Definition 4. *A section $[X_l;\ X_r]$ is a pair of consecutive abscissae. The linear line is thus made of less than n disjoint sections (see Figure 4). The length of section $[X_l;\ X_r]$ is $X_r - X_l$. In what follows abscissae are indexed (from left to right) from 0 to p, with $p \leq n$. The section between abscissae $k - 1$ and k is referred to as the kth section, and its with is denoted w_k.*

In this section we define a lower bound LB_1 based on this decomposition into sections and on the polynomial 2-abscissae case evoked in corollary 2.

For each section k defined by abscissae $[X_l;\ X_r]$ we build a problem with two abscissae, setting to zero all abscissae smaller or equal to X_l and setting to $w_k = X_r - X_l$ all abscissae larger or equal to X_r. The algorithm presented below computes a lower bound to the number of times this section will be crossed.

Algorithm. The Algorithm 1 details the computation of this lower bound. The initial scan of the graph modifying abscissae is done in $O(n)$. We define $opt_l(k)$ (resp. $opt_r(k)$), the minimum number of times section k will be crossed (see Figure 3) considering that the terminal city (the last city in the permutation) lies to the left (resp. to the right) of the section.

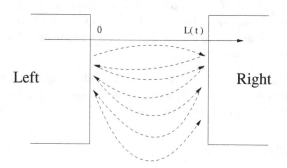

Fig. 3. Scanning a section

We denote by $B(k)$ the lower bound of the total length when the last abscissa is abscissa k ($k \in [0, p]$). It is defined by the following recursive formula:

$$
\begin{aligned}
B(0) &= \sum_{k \in [1,\ p]} opt_l(k) \\
\forall\, k &\in [1,\ p],\ B(k) = B(k-1) - opt_l(k-1) + opt_r(k-1)
\end{aligned}
\tag{3}
$$

Let F be the set of abscissae having at least one city without successor in the partial order. Since the final city of any permutation will belong to F, the following expression is a lower bound of the traveled distance of the PC-Line-TSP.

$$LB_1 = \min_{k \in F} B(k) \tag{4}$$

For each section $k \in [1,p]$, $opt_l(k)$ is computed in $O(n+m)$ with n the number of cities and m the size of the partial order. $B(0)$ is computed in $O(p)$ with $p \leq n$ and each $B(k)$ is computed in $O(1)$. Finally the complexity of the computation of this lower bound is $O(p(n+m))$. Since p is smaller than n et m is smaller than n^2, the complexity is cubic in the worst case.

Algorithm 1. BOUND-BY-SECTION

---

input : The set of abscissae, cities and the partial order
output: A lower bound of the total traveled distance

begin
  for $k \in [1,p]$ **do**
    Define the two-abscissae problem associated to section k, with starting
    city at abscissa 0
    Distance = 0
    CurrentAbscissa = 0
    Visit all available cities at CurrentAbscissa
    while *Some cities remains to be visited* **do**
      Change CurrentAbscissa to the other abscissa
      $Distance\ += w_k$
      Visit all available cities at CurrentAbscissa
    if $CurrentAbscissa = w_k$ **then**
      $opt_l(t) = Distance + w_k \; opt_r(t) = Distance$
    else
      $opt_l(t) = Distance \; opt_r(t) = Distance + w_k$
  $B(0) = \sum_{k \in [1,p]} opt_l(k)$
  for $k \in [1, p]$ **do**
    $B(k) = B(k-1) - opt_l(k-1) + opt_r(k-1)$
  return $\min_{k \in F} B(k)$

Example. Consider a PC-Line-TSP made of six cities partially ordered as in Figure 4. Here the set of possible terminal cities (cities without successors) is $\{N_1, N_4, N_5\}$. The algorithm 1 considers each of the five sections one by one. On the left of Figure 4, section 2 (between abscissae 2 and 5) is emphasized, and the two-abscissae problem associated to this section is illustrated (defined on abscissae 0 and $w_3 = 3$). The computation of $opt_l(2)$ and $opt_r(2)$ starts with the visit of city N_0 on the left side. N_1 cannot be visited yet due to its two unvisited predecessors N_2 and N_3. A first crossing of the section is needed then.

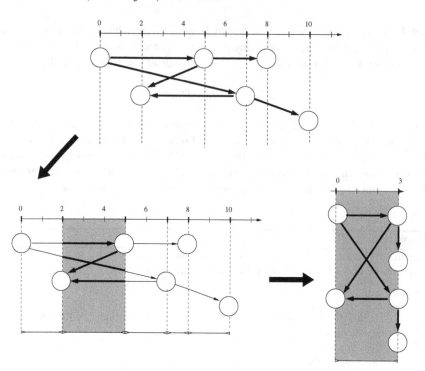

Fig. 4. A simple PC-Line-TSP instance and the two-abscissae problem attached to is second section

On the right side, N_2 and N_3 are visited, thus allowing the visit of N_4 et N_5. The section need to be crossed a second time in order to visit N_1 on the left side, thus completing the path. Finally $B(2) = 2 \times 3 = 6$.

For each section $k \in [1,5]$, $opt_l(k)$ et $opt_r(k)$ take the following values :

$$
\begin{array}{lllll}
opt_l(1) &= 2 \times 2 = 4 & et & opt_r(1) &= 1 \times 2 = 2 \\
opt_l(2) &= 2 \times 3 = 6 & et & opt_r(2) &= 3 \times 3 = 9 \\
opt_l(3) &= 2 \times 2 = 4 & et & opt_r(3) &= 3 \times 2 = 6 \\
opt_l(4) &= 2 \times 1 = 2 & et & opt_r(4) &= 1 \times 1 = 1 \\
opt_l(5) &= 2 \times 2 = 4 & et & opt_r(5) &= 1 \times 2 = 2
\end{array}
\tag{5}
$$

and we can compute the following $B(k)$:

$$
\begin{array}{lll}
B(0) &= \sum_{k \in [1,p]} opt_l(k) = 4 + 6 + 4 + 2 + 4 &= 20 \\
B(1) &= B(0) + opt_l(0) + opt_l(0) = 20 - 4 + 2 &= 18 \\
B(2) &= B(1) + opt_l(1) + opt_l(1) = 18 - 6 + 9 &= 21 \\
B(3) &= B(2) + opt_l(2) + opt_l(2) = 21 - 4 + 6 &= 23 \\
B(4) &= B(3) + opt_l(3) + opt_l(3) = 23 - 2 + 1 &= 22 \\
B(5) &= B(4) + opt_l(4) + opt_l(4) = 22 - 4 + 2 &= 20
\end{array}
\tag{6}
$$

Since $F = \{N_1, N_4, N_5\}$, the final bound is $min(B(1), B(4), B(5)) = 18$.

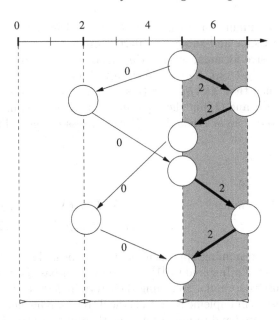

Fig. 5. Suboptimal lower bound

Counter-Example. The example on Figure 5 illustrates the non-optimality of lower bounds LB_1. There is only one possible terminal city (N_7). On section $[2, 5]$, we obtain a lower bound equal to 6 and on section $[5, 7]$ we obtain a lower bound equal to 4, hence $LB_1 = 10$. However, the optimal solution is 20 : $N_0 \ -2 \to \ N_2 \ -2 \to \ N_3 \ -3 \to \ N_1 \ -3 \to \ N_4 \ -2 \to \ N_6 \ -2 \to \ N_7 \ -3 \to \ N_5 \ -3 \to \ N_7$:

3   Exact Algorithm

3.1   Dynamic Programming

Any dominating solution to the PC-Line-TSP problem can be expressed as a sequence of left/right decisions. That is to say that each time the current abscissa has no available city, we are facing a binary choice: either go to the closest abscissa with available cities to the left or to the closest abscissa with available cities to the right. It means that a brute force enumeration of all dominating solutions has complexity $O(2^n)$, while the number of possible permutations is $n!$. We present in this section a dynamic programming approach similar to the one proposed by [7] for the classical TSP, observing that at any moment in the search, the remaining distance to be traveled only depends on the current abscissa and on the set of remaining cities. The worst case complexity of this algorithm remains $O(2^n)$, but thanks to the non-procrastination rule and to the partial ordering of cities many sets of cities cannot be encountered as a set of remaining cities, what makes this algorithm very effective in practice.

In essence, the minimum length $L_{min}(x, Q)$ for visiting a set of cities $Q \subseteq P$ starting from abscissa x and subject to partial order $\prec$ can be expressed with the following recursive formula, where $A(y)$ denotes the set of available[1] cities at abscissa y while X_R (resp. X_L) is the closest abscissa to the right of x (resp. to the left of x) such that $A(X_R) \neq \varnothing$ (resp. $A(X_L) \neq \varnothing$). Note that X_R and X_L are functions of x and Q, but these parameters are omitted in the remaining of the paper for the sake of readability. Without loss of generality we assume $A(x) = \varnothing$.

$$L_{min}(x, \varnothing) = 0$$
$$L_{min}(x, Q) = \min(\qquad X_R - x + L_{min}(X_R, Q \setminus A(X_R)),$$
$$x - X_L + L_{min}(X_L, Q \setminus A(X_L))$$

This dynamic programming approach yields the optimal solution of the PC-Line-TSP as $L_{min}(0, P)$. Inspired by [1] we design a call-based dynamic program based on this recursive formula. Following their terminology, call based dynamic programming consists in implementing a classical tree search where the optimum for each subtree is stored and re-used each time the same sub-tree is encountered (same x and same Q in our case). Compared to bottom-up dynamic programming implementation, the call-based approach allows introducing lower-bounds, upper-bounds and heuristics in order to speed up the search.

As detailed in algorithm 2, the central function becomes $L_{min}(x, Q, U)$ where U is an upper bound, and the optimum of the problem is $L_{min}(0, P, +\infty)$. As soon as a first solution is found, it is used to define upper-bounds for other branches thus excluding solutions leading to a total traveled distance larger or equal to the best found so far. The trivial lower-bound LB_0 defined in section 1 is used to eliminate such sub-optimal solutions as soon as possible. In section 4 we will also give the results obtained when using bound LB_1 instead.

The DP-labeled lines are specific to dynamic programming[2]: $storedBest[x, Q]$ is the minimum distance for visiting cities of Q when starting from abscissa x, hence before exploring a subtree (x, Q), the algorithm always check whether its minimum distance is already known (that is to say if $storedBest[x, Q]$ is defined). Similarly $storedLB[x, Q]$ is the best known lower-bound to this distance. Indeed once a subtree was vainly explored searching for a solution with a traveled distance strictly smaller than U, this information is worth storing because later in the search this subtree may be considered again with some upper bound U'. Then if $U' \leq U$ the re-exploration of this subtree is avoided. In theory the number nodes in the search tree can be larger when using lower bounds because the same sub-problem (x, Q) can be explored several time with increasing upper bounds. However the pruning effect largely compensates for this in practice. For

---

[1] These cities may be partially ordered by $\prec$ but they can all be visited if we reach abscissa y.

[2] In other words, removing the DP-labeled lines results in a classical tree search algorithm.

Algorithm 2. $L_{min}(x, Q, U)$

---

input : Current abscissa x, remaining cities Q, an upper bound U
output: The length of the best solution if $< U$, U otherwise

begin
    if $Q = \varnothing$ **then** **Return** 0

DP   **if** $storedBest[x, Q]$ **then** **Return** $storedBest[x, Q]$
DP   **if** $not(storedLB[x, Q])$ **then** $storedLB[x, Q] = LB_0(x, Q)$
DP   **if** $storedLB[x, Q] \geq U$ **then** **Return** U

    if $LB_0(x, Q) \geq U$ **then** **Return** U
    $Best = U$
    if $X_R \neq +\infty$ **then**
        $D_{right} = X_R - x$
        $Best = D_{right} + $
                $L_{min}(X_R, Q \setminus A(X_R), Best - D_{right})$

    if $X_L \neq -\infty$ **then**
        $D_{left} = x - X_L$
        $Best = D_{left} + $
                $L_{min}(X_L, Q \setminus A(X_L), Best - D_{left})$

DP   **if** $Best < U$ **then** $storedBest[x, Q] = Best$
DP   **else** $storedLB[x, Q] = U$

    Return $\min(Best, U)$

---

instance once a solution of length 35 has been found, a certain state (x, Q) might be reached after a traveling distance of 10 (visiting cities in $P \setminus Q$) hence with an upper bound of 25; but later in the search this sate may be reached after a traveling distance of 9 that is to say with an upper bound of 26, in which case this subtree must be explored again.

With $storedBest[x, Q]$ and $storedLB[x, Q]$ stored in hashtables, these values can be accessed and updated in constant time. Provided that $A(y)$ is dynamically maintained for each abscissa y (in $O(m)$ amortized complexity), X_R and X_L can be obtained in $O(p)$ (recall that m is the size of the partial order and p is the number of different abscissae). Maintaining the leftmost and rightmost abscissae in Q makes sure that LB_0 is computed in constant time. If LB_1 is used instead of LB_0, its complexity is $O(p(n + m))$ as shown in section 1.

3.2 Heuristics

In algorithm 2, the right side is systematically explored before the left side. However, different strategies can be applied. For instance a *NearestNeighbor* heuristic would consist in starting with the left side when $x - X_L < X_R - x$ (and starting with the right side otherwise). Recall that for the Euclidian TSP this simple heuristic averages less than 25% above the Held-Karp lower bound [10] and is guaranteed to remain within a $\frac{1}{2}(\log_2(N) + 1)$ ratio of the optimal solution

[11]. Alternatively an *Inertial* heuristic would consist in continuing rightwards if and only if the current abscissa was reached from the left. Finally, based on our lower bound LB_0 or LB_1, an $A\star$ heuristic consists in evaluating the lower bound on each branch and then start with the most promising one, that is to say the one with the smallest lower bound.

3.3   Dominance Rules

Lemma 2. *Let Z be a subset of P which is totally ordered by $\prec$, and let T be the subset of all cities of $P \setminus Z$ necessarily visited by a path visiting Z. If T has no successor in P' ($\nexists a \in T, b \in P', a \prec b$), then the problem restricted to cities in $P' = P \setminus T$ has the same optimal value as the initial problem.*

Proof. Clearly any solution of the initial problem is also a solution of the problem limited to P'. Inversely any solution path of the problem limited to P' is a super-sequence of the the the sequence of abscissae of set Z (ordered by $\prec$), hence the cities of T can be inserted in the solution without increasing its length, while respecting the precedences $a \prec b$ with $b \in T$. Precedences internal to P' are satisfied since these insertions do not modify the ordering of cities of P'. By hypothesis no precedence is defined from T to P'. Finally the obtained permutation has the same length as the initial solution and satisfies the partial order on P.   □

Corollary 3. *In particular, this dominance rule can be applied for any leftmost or rightmost city of P (any singleton being totally ordered by $\prec$). It means that any centrifugal connex part of the Hasse Diagram (that is made of precedences $a \prec b$ with b farther from 0 than a) can be removed from P without affecting the value of the optimal solution.*

Corollary 4. *PC-Line-TSP is fixed parameter tractable when parameterized by m.*

Proof. At least $n - m$ cities do not appear in the Hasse diagram, and thus can be removed from P without affecting the value of the optimal solution (provided that two extreme cities are kept). Consequently this reduced problem has a size limited to $m + 2$.   □

3.4   Examples

As mentioned above the worst case complexity of our dynamic programming algorithm is the same as the one of a complete scan (2^n). Before demonstrating the practical gains of dynamic programming in the next section, we exhibit below two structures for which the complexity of the dynamic program is much better.

In Figure 6, a complete scan has complexity $O(2^n)$ (one binary choice per layer), whereas dynamic programming has complexity $O(4n)$ (4 states per layer). In Figure 7 cities are distributed around a central starting city with no precedence between them, then the dynamic program explores n^2 states while complete scan remains exponential (2^n). In the latter case, the use of the dominance rule of corollary 3 reduces the number of states to 2 (all cities but two can be removed).

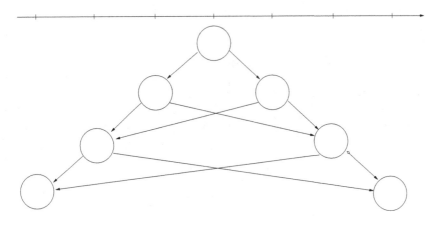

Fig. 6. Special partial orders(1)

4   Computational Results and Conclusion

4.1   Problem Instances

For generating an instance with n cities and k distinct abscissae, we start with building a partial order on $[1,n]$. For each pair (i,j) with $i < j$, a precedence $i \prec j$ is generated with probability p. Three remarks can be made on this partial order:

- It is a partial order because the generated directed graph contains no cycle (by construction all arcs are oriented toward growing integers).
- Several other Directed Acyclic Graph would represent the same partial order. We can define the density of the partial order as the number of arcs included in the transitive closure of this graph divided by the number of arcs in the total order $(n(n - 1)/2)$. A total order has a density of 100%.
- this simple method does not ensure that the generated partial orders are uniformly distributed among the set of all possible partial order. However our goal here is not to extract statistical properties but merely to generate a set of instances for comparing our algorithms. For references on the uniform generation of partial orders see [14] and [2].

Once this partial order is generated, k distinct abscissae are randomly drawn in $[0,100]$. Cities receive random abscissae from this set.

We generated 44 random instances of the PC-Line-TSP, with a number of cities (cities) from 100 to 450, a density from 5% to 85% (instances with a density of 100% were discarded) and a number of distinct abscissae from 23 to 99. We also generated 20 instances with 100 to 450 cities on 3 distinct abscissae, the first NP-Complete value (see section 1.3): any of our algorithms could solve any of these 3-abscissae instances in less than one second.

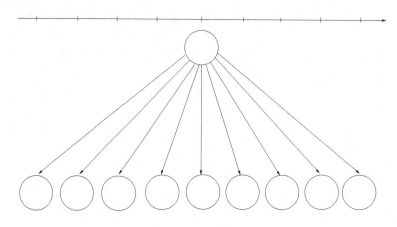

Fig. 7. Special partial orders (2)

4.2   Results

For each instance we compared 4 algorithms:

1. Our complete algorithm with bound LB_1 (bound by section defined in section 1), with our reduction algorithm enabled (corollary 3), and using heuristic *NearestNeighbor*.
2. The same algorithm as 1 using heuristic $A\star$
3. The same algorithm as 1 with bound LB_0
4. The same algorithm as 1 without our reduction algorithm

Table 1 summarizes average results obtains with these four algorithms with a time limit set to 600 seconds. Our reference algorithm found the optimal solution of 36 of the 44 instances (with both heuristics). Disabling the reduction algorithm leads to a score of 33/44 while using bound LB_0 instead of LB_1 we obtain only 27/44. In terms of completion time our second algorithm (with heuristic $A\star$) obtains the best results, with an average time of 113 seconds vs 129 seconds with heuristic *NearestNeighbor*. The impact of this heuristic is more significant when comparing the time to obtain the best solution (when both completed the search in the allocated time): with $A\star$, the time to obtain the best solution is divided by 3 on average: 31 seconds against 100 seconds.

Table 2 reports the results obtained by each algorithm on the 8 instances that could not be solved within the allocated 600 seconds. Our second algorithm (with $A\star$ heuristic) always obtains the best solution. Consequently all gaps in this table are given with respect to the results of this best algorithm. The rightmost column gives the value of our bound by sections LB_1 and the corresponding optimality gap.

As shown in table 3, the density of the partial order plays a important role in the hardness of an instance. Looking at results of algorithm 3 we see that the higher the density the easier the instance, with a number of proven optimum

Table 1. Average results on the 44 PC-Line-TSP instances (time limited to 600 seconds)

	Algorithm 1	Algorithm 2	Algorithm 3	Algorithm 4
Pruning	LB_1 + reduction	LB_1 + reduction	LB_0 + reduction	LB_1
Heuristic	NearestNeighbor	$A\star$	NearestNeighbor	NearestNeighbor
Number of proven optimum	36/44	36/44	27/44	33/44
Average time to complete	129s	113s	239s	153s
Average time to obtain best solution*	100s	31s	-	-

(*)on the 36 instances solved by 1 and 2

Table 2. Results on the 8 unsolved instances

	Algorithm 1	Algorithm 2	Algorithm 3	Algorithm 4	Lower Bound
Pruning	LB_1 + reduction	LB_1 + reduction	LB_0 + reduction	LB_1	LB_1
Heuristic	NearestNeighbor	$A\star$	NearestNeighbor	NearestNeighbor	
Pb200	**761**	761	761	761	619 (-19%)
Pb350	1904 (+3.0%)	1848	2036 (+10.2%)	2130 (+15.3%)	1563 (-15%)
Pb400A	4230 (+1.4%)	4170	4236 (+1.6%)	5118 (+22.2%)	3963 (-5%)
Pb400B	4500 (+0.1%)	4498	4500 (+0.1%)	4548 (+1.1%)	4268 (-5%)
Pb400C	3193 (+1.9%)	3133	3583 (+14.4%)	3889 (+24.1%)	2946 (-6%)
Pb400D	2345 (+5.0%)	2233	2477 (+10.9%)	2739 (+22.7%)	2171 (-3%)
Pb450A	3098 (+3.7%)	2988	3156 (+5.6%)	3688 (+23.4%)	2723 (-9%)
Pb450B	3154 (+20.3%)	2876	3460 (+9.7%)	3386 (+17.7%)	2653 (-8%)
Average gaps	+7.9%		+3.1%	+15.9%	-9%

increasing from 5 to 8. Indeed when the partial order is denser, the number of feasible permutations is smaller. The extreme case is when the density is 100% and only one permutation is allowed. Comparing algorithm 1 and 3 we see that using lower bound LB_1 pays off on all range of densities. As for the reduction algorithm its impact is higher on smaller densities, because sparse partial orders are more likely to contain centrifugal connex parts. Concerning the number of abscissae, we noticed that the number of distinct abscissae can increase the hardness of the problem or at least the complexity of our algorithms. The median number of abscissae is 60 is our benchmark. Our best algorithm (number 2) solved instances with less than 60 abscissae in 90 seconds on average (proving 21 optimum values out of 22) while instances with more than 60 abscissae are solved in 246 seconds on average (with 15 proven optimum out of 22).

Table 3. Impact of the density of the partial order (time to complete in seconds and number of proven optimum in parenthesis)

	Algorithm 1	Algorithm 2	Algorithm 3	Algorithm 4
Pruning	LB_1 + reduction	LB_1 + reduction	LB_0 + reduction	LB_1
Heuristic	NearestNeighbor	$A\star$	NearestNeighbor	NearestNeighbor
First quartile [5% to 20%]	154s (10/11)	127s (10/11)	368s (5/11)	259s (7/11)
Second quartile [20% to 48%]	245s (7/11)	198s (7/11)	344s (6/11)	277s (7/11)
Third quartile [48% to 73%]	186s (10/11)	180s (10/11)	331s (8/11)	197s (10/11)
Fourth quartile [73% to 85%]	164s (9/11)	162s (9/11)	347s (8/11)	161s (9/11)

4.3 Conclusion

Our call-based dynamic program, manages to solve to optimality instances with up to 450 cities. Examining the results we see that our lower bound LB_1, based on the splitting of the line by sections, makes possible the resolution of dense and large instances. On the other end, our domination rule dramatically improves performance on sparse instances. Finally, thanks to the call-based implementation, the algorithm can find good solutions, even for instances whose optimum could not be found within 600 seconds. In this context founding an $A\star$ heuristic on our lower bound allows finding better solutions faster.

References

1. de la Banda, M.G., Stuckey, P.J.: Dynamic programming to minimize the maximum number of open stacks. INFORMS Journal on Computing 19(4), 607–617 (2007)
2. Brightwell, G.: Models of random partial orders, pp. 53–84. Cambridge University Press (1993)
3. Charikar, M., Motwani, R., Raghavan, P., Silverstein, C.: Constrained tsp and low-power computing. In: Dehne, F., Rau-Chaplin, A., Sack, J.-R., Tamassia, R. (eds.) WADS 1997. LNCS, vol. 1272, pp. 104–115. Springer, Heidelberg (1997)
4. Cutler, M.: Efficient special case algorithms for the n-line planar traveling salesman problem. Networks 10(3), 183–195 (1980)
5. Deineko, V.G., van Dal, R., Rote, G.: The convex-hull-and-line traveling salesman problem: A solvable case. Information Processing Letters 51(3), 141–148 (1994)
6. Deineko, V.G., Woeginger, G.J.: The convex-hull-and-k-line travelling salesman problem. Inf. Process. Lett. 59(6), 295–301 (1996)
7. Held, M., Karp, R.M.: A dynamic programming approach to sequencing problems. In: Proceedings of the 1961 16th ACM National Meeting, pp. 71.201–71.204. ACM, New York (1961)
8. Jeanjean, A.: Resource scheduling optimization in mass transportation problems. In: 12th International Conference on Project Management and Scheduling, PMS 2010 (2010)
9. Jeanjean, A.: Recherche locale pour l'optimisation en variables mixtes: Méthodologie et applications industrielles. Ph.D. thesis, Laboratoire d'informatique de Polytechnique (2011)
10. Johnson, D.S., Mcgeoch, L.A.: The Traveling Salesman Problem: A Case Study in Local Optimization. John Wiley and Sons, Chichester (1997)
11. Rosenkrantz, D.J., Stearns, R.E., Lewis II, P.M.: An analysis of several heuristics for the traveling salesman problem. 6(3), 563–581 (1977)
12. Rote, G.: The n-line traveling salesman problem. Networks 22, 91–108 (1991)
13. Tsitsiklis, J.N.: Special cases of traveling salesman and repairman problems with time windows. Networks 22, 263–282 (1992)
14. Gehrlein, V.W.: On methods for generating random partial orders. Operations Research Letters 5(6), 285–291 (1986)

Stable Roommates and Constraint Programming

Patrick Prosser

School of Computing Science, University of Glasgow, Glasgow, Scotland
pat@dcs.gla.ac.uk

Abstract. In the stable roommates (SR) problem we have n agents, where each agent ranks all $n-1$ other agents. The problem is then to match agents into pairs such that no two agents prefer each other to their matched partners. A remarkably simple constraint encoding is presented that uses $O(n^2)$ binary constraints, and in which arc-consistency (the phase-1 table) is established in $O(n^3)$ time. This leads us to a specialized n-ary constraint that uses $O(n)$ additional space and establishes arc-consistency in $O(n^2)$ time. This can model stable roommates with incomplete lists (SRI), consequently it can also model stable marriage (SM) problems with complete and incomplete lists (SMI). That is, one model suffices. An empirical study is performed and it is observed that the n-ary constraint model can read in, model and output all matchings for instances with $n = 1,000$ in about 2 seconds on current hardware platforms. Enumerating all matchings is a crude solution to the egalitarian SR problem, and the empirical results suggest that although NP-hard, egalitarian SR is practically easy.

1 Introduction

In the **Stable Roommates** problem (SR) [8,7] we have an even number of agents to be matched together as couples, where each agent strictly ranks all other agents. The problem is then to match pairs of agents together such that the matching is stable, i.e. there doesn't exist a pair of agents in the matching such that $agent_i$ prefers $agent_j$ to his matched partner and $agent_j$ prefers $agent_i$ to his matched partner[1].

The **Stable Marriage** problem (SM) [4,15,5,7,16,11] is a specialized instance of stable roommates where agents have gender, such that we have two sets of agents m (men) and w (women). Each man has to be *married* to a woman and each woman to a man such that in the matching there does not exist a man m_i and a woman w_j where m_i prefers w_j to his matched partner and w_j prefers m_i to her matched partner i.e. there is no incentive for agents to divorce and elope.

Constraint programming has been applied to the stable marriage problem, probably the first efficient model being reported in 2001 [6], a 4-valued model in [12], a specialized binary constraint in [18] and an efficient n-ary constraint in [17]. This raises an obvious question: if there is an efficient constraint model for stable marriage, is there one for the more general stable roommates problem?

---

[1] For sake of brevity I assume agents are male, and hope this offends no one.

H. Simonis (Ed.): CPAIOR 2014, LNCS 8451, pp. 15–28, 2014.

In this paper I partially answer this question. I present a remarkably simple constraint model for SR, using $O(n^2)$ constraints. This model addresses SR with incomplete lists and consequently SM with incomplete lists. A more compact and computationally efficient encoding is then proposed. An empirical study is presented, comparing models and investigating the problem.

2 The Stable Roommates Problem (SR)

An example of a stable roommates instance is given in Figure 1, for $n = 10$, and this instance is taken from [7] (and we will refer to this as sr10). We have agents 1 to 10 each with a preference list, ranking the other agents. For example, $agent_1$'s first choice is for $agent_8$, then $agent_2$, followed by $agent_9$ and so on to last (9th) choice $agent_{10}$.

1 : 8 2 9 3 6 4 5 7 10	1: 8 2 3 6 4 7	(1,7) (2,3) (4,9) (5,10) (6,8)
2 : 4 3 8 9 5 1 10 6 7	2: 4 3 8 9 5 1 10 6	(1,7) (2,8) (3,5) (4,9) (6,10)
3 : 5 6 8 2 1 7 10 4 9	3: 5 6 2 1 7 10	(1,7) (2,8) (3,6) (4,9) (5,10)
4 : 10 7 9 3 1 6 2 5 8	4: 9 1 6 2	(1,4) (2,8) (3,6) (5,7) (9,10)
5 : 7 4 10 8 2 6 3 1 9	5: 7 10 8 2 6 3	(1,4) (2,9) (3,6) (5,7) (8,10)
6 : 2 8 7 3 4 10 1 5 9	6: 2 8 3 4 10 1 5 9	(1,4) (2,3) (5,7) (6,8) (9,10)
7 : 2 1 8 3 5 10 4 6 9	7: 1 8 3 5	(1,3) (2,4) (5,7) (6,8) (9,10)
8 : 10 4 2 5 6 7 1 3 9	8: 10 2 5 6 7 1	
9 : 6 7 2 5 10 3 4 8 1	9: 6 2 10 4	
10 : 3 1 6 5 2 9 8 4 7	10: 3 6 5 2 9 8	

Fig. 1. Stable roommates instance sr10 with $n = 10$ (on the left) phase-1 table (middle) and the 7 stable matchings (on the right). Instance taken from [7].

A quadratic time algorithm, essentially linear in the input size, was proposed in [8]. The algorithm has two phases. The first phase is a sequence of proposals, similar to that in the Gale Shapley algorithm [4], that results in the *phase-1 table*. The phase-1 table for sr10 is shown as the middle table in Figure 1. A sequence of *rotations* are then performed for agents with reduced preference lists that contain more than one agent. On the right hand side of Figure 1 we show the 7 stable matching that can result from this process.

3 A Simple Constraint Model

We assume that we have two dimensional integer arrays $pref$ and $rank$. Vector $pref_i$ is the preference list for $agent_i$ such that if $pref_{i,k} = j$ then $agent_j$ is $agent_i$'s k^{th} choice and $rank_{i,j} = k$. It is also assumed that if $agent_i$ finds $agent_j$ acceptable then $agent_j$ finds $agent_i$ acceptable (i.e. j appears in $pref_i$ if and only if i appears in $pref_j$). We also have the length of each agent's preference list l_i, and this allows us to model SRI instances, i.e. **Stable Roommates with Incomplete lists.**

Using sr10 as our example $pref_{3,1} = 5$ and $rank_{3,5} = 1$ ($agent_5$ is $agent_3$'s first choice) and $pref_{3,2} = 6$ and $rank_{3,6} = 2$ ($agent_6$ is $agent_3$'s second choice). Note that in sr10 $l_i = 10$ for all i, i.e. sr10 is an SR instance with complete preference lists.

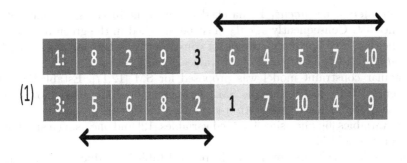

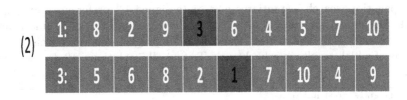

Fig. 2. A pictorial representation of the two constraints acting between $agent_1$ and $agent_3$

We have constrained integer variables a_1 to a_n, each with a domain of *ranks* $\{1..l_i+1\}$. When $a_i \leftarrow k$ this means that the corresponding agent is allocated his k^{th} choice, and that is $agent_j$ where $j = pref_{i,k}$. Furthermore, when $a_i \leftarrow l_i + 1$ the corresponding agent is matched to itself and is considered unallocated.

We can now make a declarative statement of the properties that a stable matching must have and we do this with two constraints. Given two agents, $agent_i$ and $agent_j$ who find each other acceptable, if $agent_i$ is matched to an agent he prefers less than $agent_j$ then $agent_j$ must match up with an agent that he prefers to $agent_i$ otherwise the matching will be unstable. This property must hold between every pair of agents that find each other acceptable and is expressed by constraint (2) below. Furthermore, when $agent_i$ is matched to $agent_j$ then $agent_j$ is matched to $agent_i$, and this is expressed by constraint (3).

$$\forall_{i \in [1..n]} \ a_i \in \{1..l_i + 1\} \tag{1}$$
$$\forall_{i \in [1..n]} \forall_{j \in pref_i} \ a_i > rank_{i,j} \implies a_j < rank_{j,i} \tag{2}$$
$$\forall_{i \in [1..n]} \forall_{j \in pref_i} \ a_i = rank_{i,j} \implies a_j = rank_{j,i} \tag{3}$$

This constraint is shown pictorially in Figure 2. The two constraints are shown for agents 1 and 3 in sr10. The brown box is the agent's identification number and the remaining boxes are the preference lists (a list of agents). In the top picture (1) we have the situation where $agent_1$ is matched to an agent he prefers less than $agent_3$, i.e. $agent_1$ is matched to an agent in the green part of his preference list. Consequently $agent_3$ must be matched in the green region of its preference list. The bottom picture is for constraint (3) where $agent_1$ is matched to $agent_3$, both taking the pair of red values.

A similar constraint model was proposed for SM [12,17]. Establishing arc-consistency [10,19] in that simple SM constraint model has been shown to be $O(n^3)$ although at least three $O(n^2)$ encodings have been proposed: one using boolean variables [6], one using 4-valued variables [12] and one using a specialized n-ary constraint [17].

When sr10 is made arc-consistent the phase-1 table is produced. As we can see from Figure 1 the first agent in the phase-1 table for $agent_1$ is $agent_8$ yet none of the seven solutions have a matching that contains the pair $(1, 8)$. Therefore our constraint program must backtrack, i.e. after producing the phase-1 table via propagation, search instantiates $a_1 \leftarrow 1$ (assigned 1st preference), attempts to make the model arc-consistent and fails, forcing a backtrack. To find a first solution to sr10 (a first matching) the constraint program makes 3 decisions, at least one of which results in a backtrack. To find all 7 solutions, 12 decisions are made.

```java
public class StableRoommates {

  public static void main(String[] args) throws IOException {

    BufferedReader fin = new BufferedReader(new FileReader(args[0]));
    int n = Integer.parseInt(fin.readLine());
    int[][] pref = new int[n][n];
    int[][] rank = new int[n][n];
    int[] length = new int[n];
    for (int i=0;i<n;i++){
      StringTokenizer st = new StringTokenizer(fin.readLine()," ");
      int k = 0;
      length[i] = 0;
      while (st.hasMoreTokens()){
        int j = Integer.parseInt(st.nextToken()) - 1;
        rank[i][j] = k;
        pref[i][k] = j;
        length[i] = length[i] + 1;
        k = k + 1;
      }
      rank[i][i] = k;
      pref[i][k] = i;
    }
    fin.close();
    Model model = new CPModel();
    IntegerVariable[] a = new IntegerVariable[n];
    for (int i=0;i<n;i++) a[i] = makeIntVar("a_"+ i,0,length[i],"cp:enum");
    for (int i=0;i<n;i++)
      for (int j=0;j<length[i];j++){
        int k = pref[i][j];
        model.addConstraint(implies(gt(a[i],rank[i][k]),lt(a[k],rank[k][i])));
        model.addConstraint(implies(eq(a[i],rank[i][k]),eq(a[k],rank[k][i])));
      }
    Solver solver = new CPSolver();
    solver.read(model);
    if (solver.solve().booleanValue())
      for (int i=0;i<n;i++){
        int j = pref[i][solver.getVar(a[i]).getVal()];
        if (i<j) System.out.print("("+ (i+1) +","+ (j+1) +") ");
      }
    System.out.println();
  }
}
```

Listing 1. A simple encoding for SRI, StableRoommates.java

The model was implemented in the choco constraint programming toolkit [1] using Java and the code is shown in Listing 1. The first thing to note is that everything is zero-based, such that the first agent is a_0 and the last a_{n-1} (lines 26

```
 1: 8 2 9 3 6 4 5 7
 2: 4 3 8 9 5 1 10 6
 3: 5 6 8 2 1 7 10
 4: 9 3 1 6 2
 5: 7 4 10 8 2 6 3
 6: 2 8 7 3 4 10 1 5 9
 7: 1 8 3 5
 8: 10 4 2 5 6 7 1
 9: 6 7 2 5 10 3 4
10: 3 1 6 5 2 9 8
```

Fig. 3. *Bound* phase-1 table for sr10 using bound integer variables

and 27). Lines 5 to 24 read in the problem instance, building the arrays *pref* and *rank*. To address SR with incomplete lists we add i to the end of a_i's preference list (lines 21 and 22) such that an unmatched agent is matched to itself. The constraint model is produced in lines 25 to 35 with constraint (2) posted in line 31 and constraint (3) in line 32. In lines 36 to 41 the choco toolkit searches for a first solution and prints it out.

The choco toolkit also supports bound integer variables, where only the upper and lower bounds on domains are maintained and removal of values between those bounds are performed lazily. In line 27 of Listing 1 adding the option "cp:bound" to the constructor *makeIntVarArray* changes the model so that it uses bound integer variables. When the model is made arc-consistent we then get the *bound* phase-1 table shown in Figure 3. Comparing this to Figure 1 we see that the upper and lower bounds agree with the phase-1 table but there are values between those bounds that are omitted from the enumerated domains, in particular we see that $agent_1$ has $agent_9$ in its domain yet $agent_9$ does not have $agent_1$ in its domain. Nevertheless, the constraint program maintains the desired stable roommates properties and produces the same 7 solutions as in Figure 1 and does so in less time.

The constraint model also address SRI instances (SR with incomplete lists). Figure 4 shows instance sri6, with $n = 6$. This has one stable matching $\{(1, 4), (2, 6)\}$ with agents 3 and 5 unmatched.

```
1: 2 4 5
2: 6 1 3
3: 2 4
4: 1 6 3
5: 6 1
6: 2 5 4
```

Fig. 4. SRI instance sri6. This has one solution $\{(1, 4), (2, 6)\}$.

The model also addresses stable marriage problems with complete and incomplete lists (i.e. SM and SMI). As an example consider Figure 5, a stable marriage instance with 6 men and 6 women (taken from [6]). This is shown on the left of Figure 5 with a stable matching in bold font. On the right we have the same problem represented as an SRI instance. The men are represented as agents 1 to 6 and women as agents 7 to 12. Agents 1 to 6 (the men) only find agents 7 to 12 acceptable (the women) and agents 7 to 12 (the women) find only agents 1 to 6 (the men) acceptable. To read off the SRI matching we subtract 6 from the agent matched to agents 1 to 6. Therefore our simple constraint model addresses SR, SRI, SM and SMI.

```
1 : 1 3 6 2 4 5      1 : 1 5 6 3 2 4         1 : 7 9 12 8 10 11
2 : 4 6 1 2 5 3      2 : 2 4 6 1 3 5         2 : 10 12 7 8 11 9
3 : 1 4 5 3 6 2      3 : 4 3 6 2 5 1         3 : 7 10 11 9 12 8
4 : 6 5 3 4 2 1      4 : 1 3 5 4 2 6         4 : 12 11 9 10 8 7
5 : 2 3 1 4 5 6      5 : 3 2 6 1 4 5         5 : 8 9 7 10 11 12
6 : 3 1 2 6 5 4      6 : 5 1 3 6 4 2         6 : 9 7 8 12 11 10
                                             7 : 1 5 6 3 2 4
                                             8 : 2 4 6 1 3 5
                                             9 : 4 3 6 2 5 1
                                            10 1 3 5 4 2 6
                                            11 3 2 6 1 4 5
                                            12 5 1 3 6 4 2
```

Fig. 5. Stable marriage instance sm6. On the left, the familiar SM and on the right sm6 recast as an SRI instance. Problem is taken from [6].

4   A More Efficient Model

Our constraint model can be made more computationally efficient by adopting and modifying the models in [6,12]. However, these models are bulky and quickly exhaust memory on relatively modest sized instances of SM [17]. Therefore we propose an n-ary SR constraint (SMN), similar to that proposed in [17], that can establish arc-consistency in $O(n^2)$ and takes $O(n)$ additional space (assuming we are given the arrays $pref$ and $rank$ read in on lines 5 to 24 of Listing 1). The means of reducing the computational cost is by eliminating the redundancies brought about by the arc-consistency algorithm: when a variable's domain is altered all constraints involving that variable are revised. Therefore, if a value is removed from the domain of a_i, $O(n)$ constraints will be revised. This can occur n times for an agent, and since there are n agents this results in $O(n^3)$ complexity, assuming it takes $O(1)$ time to revise a constraint as above.

With a specialized n-ary constraint we can improve upon this. We can eliminate the above redundancy by revising only the domains of agents that must be affected by a change in another variable's domain. There are five possible changes that can occur to the domain of an agent and these are:

- the upper bound of a variable decreases (Algorithm 1)
- the lower bound of a variable increases (Algorithm 2)
- a variable looses a value (Algorithm 3)
- a variable is instantiated (Algorithm 4)
- the constraint is initially posted (Algorithm 5)

Presented below are the algorithms that address these five cases and the actual choco/Java implementation (Listing 2, with imports removed for brevity). The algorithms again assume that we have constrained integer variables a_1 to a_n, each with a domain of ranks $\{1 \ldots l_i + 1\}$, and that we have the preference and rank arrays $pref$ and $rank$. In addition we require *reversible* variables lwb_i and upb_i, where lwb_i is used to store the smallest value in the domain of a_i and upb_i the largest value. By *reversible* we mean that on backtracking the values of these variables are restored. The choco toolkit provides this as class *StoredInt* (see lines 19 and 20 of Listing 2). In the complexity arguments we assume that the toolkit primitives $getMin(v)$ (get the smallest value in domain of variable v), $setMax(v, x)$ (set the upper bound of variable v's domain to be $min(max(v), x)$), $getMax(v)$ (get largest value in domain of v), $remove(v, x)$ (remove the value x from the domain of v if that value exists) and $getValue(v)$ (get the value v is instantiated to) each have a cost of $O(1)$.

deltaMin(i) (Algorithm 1). The lower bound of a_i has increased (and is now the value x, line 3). Consequently, the corresponding agent now at the top of $agent_i$'s preference list ($agent_j$ where $j = pref_{i,x}$, line 4)) can be matched to no one that he prefers less than $agent_i$ (line 5). For the corresponding agents that have been removed from $agent_i$'s preference list, and that $agent_i$ preferred to his current most preferred partner, those agents can do no worse than match up with agents that they prefer to $agent_i$ (lines 6 to 8). The new lower bound for a_i is saved in the reversible variable lwb_i. **Complexity:** This method can be called at most n times for an agent (the number of values in an agent's domain). Each time it is called the loop bound (line 6) is reduced (via line 9 on previous calls). Consequently this can reduce the maximum domain value of other agents (line 5 and line 8) at most n times. Therefore over all agents the cost of $deltaMin$ is $O(n^2)$.

deltaMax(i) (Algorithm 2). The upper bound of a_i has decreased (and now has the value x, line 3). For all corresponding agents removed from $agent_i$'s preference list we remove $agent_i$ from that agent's preference list as they can no longer be matched together (lines 4 to 6). The new upper bound is then saved in the reversible variable upb_i (line 7). **Complexity:** For an agent, this method can be called at most n times, each time with a reduced bound on the iteration in lines 4 to 6. Therefore lines 3 and 6 can be executed at most n times. Consequently the cost over all n agents is $O(n^2)$.

removeValue(i,x) (Algorithm 3). The value x has been removed from the domain of a_i consequently the corresponding agent ($agent_j$ where $j = pref_{i,x}$)

Algorithm 1. deltaMin (awakeOnInf in Listing 2).

```
1 deltaMin(int i)
2 begin
3 │  x ← getMin(a_i)
4 │  j ← pref_{i,x}
5 │  setMax(a_j, rank_{j,i})
6 │  for w ← lwb_i to x − 1 do
7 │  │   h ← pref_{i,w}
8 │  └   setMax(a_h, rank_{h,i} − 1)
9 └  lwb_i ← x
```

Algorithm 2. deltaMax (awakeOnSup in Listing 2).

```
1 deltaMax(int i)
2 begin
3 │  x ← getMax(a_i)
4 │  for y ← x + 1 to upb_i do
5 │  │   j ← pref_{i,y}
6 │  └   remove(a_j, rank_{j,i})
7 └  upb_i ← x
```

can no longer be matched to $agent_i$ (lines 3 and 4). **Complexity:** An execution is $O(1)$ cost and this can happen at most $O(n^2)$ times, i.e. n times for each of the n agents.

Algorithm 3. removeValue (awakeOnRem in Listing 2).

```
1 removeValue(int i, int x)
2 begin
3 │  j ← pref_{i,x}
4 └  remove(a_j, rank_{j,i})
```

instantiate(i) (Algorithm 4). The variable a_i has been assigned the value y (line 3) and corresponds to being matched to $agent_j$ where $j = pref_{i,y}$. All agents that $agent_i$ preferred to $agent_j$ can only be matched to agents that they prefer to $agent_i$ (lines 4 to 6). Furthermore, all agents that $agent_i$ preferred less than $agent_j$ can no longer consider $agent_i$ as a possible partner (lines 7 to 9). Finally we update the upper and lower bounds for the domain (lines 10 and 11). **Complexity:** An execution has a cost of $O(n)$ as we respond to the (at most $n − 1$) removals from the domain of the variable (lines 4 to 9). An agent can be

---

Algorithm 4. instantiate (awakeOnInst in Listing 2).

```
1  instantiate(int i)
2  begin
3  |    y ← getValue(aᵢ)
4  |    for x ← lwbᵢ to y − 1 do
5  |    |    j ← prefᵢ,ₓ
6  |    |    setMin(aⱼ, rankⱼ,ᵢ − 1)
7  |    for z ← y + 1 to upbᵢ do
8  |    |    j ← prefᵢ,z
9  |    |    remove(aⱼ, rankⱼ,ᵢ)
10 |    lwbᵢ ← y
11 |    upbᵢ ← y
```

---

instantiated with a value at most once during propagation. Consequently, over all n agents this has a cost of $O(n^2)$.

init() (Algorithm 5). This is called at the top of search, when the model is made arc-consistent by revising all the constraints. First, the upper and lower bounds for each agent are initialized (lines 2 to 4) and then propagation kicks off by making all agents consistent with respect to their most preferred partner, and this is similar to the proposal stage in [8]. **Complexity:** Line 6 is called n times and each individual call to $deltaMin(i)$ has cost $O(n)$, consequently we have an $O(n^2)$ cost in total.

Algorithm 5. init (class constructor and awake in Listing 2).

```
1  init()
2  begin
3  |    for i ← 1 to n do
4  |    |    lwbᵢ ← 1
5  |    |    upbᵢ ← lengthᵢ + 1
6  |    for i ← 1 to n do
7  |    |    deltaMin(i)
```

---

5 Empirical Study

Experiments were performed over random SR instances with complete preference lists, on a 2.4GHz Intel Xeon E5645 processor with 97 GBytes of RAM, using java version 1.6.0_26 and choco-2.1.0. We start by investigating the three models:

```java
public class SRN extends AbstractLargeIntSConstraint {

    private int n;
    private int [][] rank;
    private int [][] pref;
    private int [] length;
    private IStateInt [] upb;
    private IStateInt [] lwb;
    private IntDomainVar [] a;

    public SRN(Solver s,IntDomainVar [] a,int [][] pref,int [][] rank,int [] length){
        super(a);
        n          = a.length;
        this.a     = a;
        this.pref  = pref;
        this.rank  = rank;
        this.length = length;
        upb        = new StoredInt[n];
        lwb        = new StoredInt[n];
        for (int i=0;i<n;i++){
            upb[i]  = s.getEnvironment().makeInt(length[i]);
            lwb[i]  = s.getEnvironment().makeInt(0);
        }
    }

    public void awake() throws ContradictionException {
        for (int i=0;i<n;i++) awakeOnInf(i);
    }

    public void propagate() throws ContradictionException {}

    public void awakeOnInf(int i) throws ContradictionException {
        int x = a[i].getInf(); // best (lowest) rank for a i
        int j = pref[i][x];
        a[j].setSup(rank[j][i]);
        for (int w=lwb[i].get();w<x;w++){
            int h = pref[i][w];
            a[h].setSup(rank[h][i]-1);
        }
        lwb[i].set(x);
    }

    public void awakeOnSup(int i) throws ContradictionException {
        int x = a[i].getSup(); // worst (largest) preference for a[i]
        for (int y=x+1;y<=upb[i].get();y++){
            int j = pref[i][y];
            a[j].remVal(rank[j][i]);
        }
        upb[i].set(x);
    }

    public void awakeOnRem(int i,int x) throws ContradictionException {
        int j = pref[i][x];
        a[j].remVal(rank[j][i]);
    }

    public void awakeOnInst(int i) throws ContradictionException {
        int y = a[i].getVal();
        for (int x = lwb[i].get();x<y;x++){
            int j = pref[i][x];
            a[j].setSup(rank[j][i]-1);
        }
        for (int z=y+1;z<=upb[i].get();z++){
            int j = pref[i][z];
            a[j].remVal(rank[j][i]);
        }
        lwb[i].set(y);
        upb[i].set(y);
    }
}
```

Listing 2. SRN.java

(a) *SR*, the simple constraint model, (b) *SRB*, the simple model using bound integer variables and (c) *SRN*, the n-ary constraint model. In all cases a sample size of 100 is used, unless stated otherwise.

Figure 6 presents two scatter plots of total run time against problem size. The plot on the left is for $10 \leq n \leq 100$ and on the right $100 \leq n \leq 1,000$ (and for $n > 100$ we omit *SR*). The total run time includes time to read in the problem, build the model and then find all stable matchings for that instance. Time is measured in milliseconds. The total run times shows that *SR* does not scale beyond $n = 100$ (plot on the left) and at $n = 1,000$ (plot on the right) *SRB* typically takes 4 minutes whereas *SRN* takes 2 seconds, i.e. *SRN* is two orders of magnitude faster.

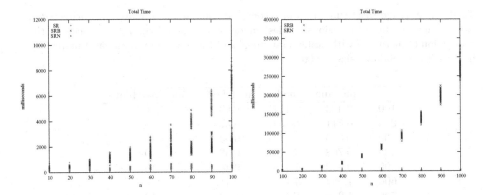

Fig. 6. Performance of the models: scatter of total time in milliseconds to find all matchings against problem size

We now investigate the problems, i.e. given n what proportion of instances have matchings? Shown in Table 1 are the proportion of instances with matchings, for $n \in \{10 \ldots 90\}$ with a sample size of 1,000. The column on the right are those reported in [8], with a sample size of 1,000 for n equal to 10 and 20, sample size 500 for $n = 30$, and sample size 200 for $40 \leq n \leq 90$.

In Table 2 we give the average total cpu time in seconds (i.e. time to read in the instance, produce the model, enumerate all solutions and output run time statistics) for $100 \leq n \leq 1,000$ using our best model (SRN). Also tabulated is the average number of nodes reported by the choco toolkit (and maximum in brackets) where a node is a decision made, and that decision might be one that leads to a failure and a backtrack. The second last column is the proportion of instances that had matchings. The last column is the maximum number of stable matchings found in an instance of size n. In all cases sample size is 100.

5.1   Discussion

Clearly (Figure 6) the n-ary encoding is orders of magnitude faster than the toolkit constraints. Although not presented, it is also more space efficient, i.e.

Table 1. Proportion of instances with solutions. Column on the right from [8].

n	Prosser	Irving
10	0.889	0.868
20	0.834	0.815
30	0.781	0.766
40	0.736	0.745
50	0.727	0.710
60	0.704	0.725
70	0.706	0.670
80	0.670	0.675
90	0.670	0.690

Table 2. Average total run times in seconds to enumerate all matchings using SRN, the average number of decisions (nodes) made by choco (maximum in brackets), the proportion of instances with stable matchings and the maximum number of matchings in an instance. Sample size is 100.

n	cpu time	nodes	matched	max matchings
100	0.423	4 (17)	0.63	9
200	0.511	6 (34)	0.52	16
300	0.645	7 (33)	0.53	16
400	0.768	7 (25)	0.38	10
500	0.950	7 (35)	0.45	16
600	1.094	7 (27)	0.41	14
700	1.290	7 (31)	0.42	12
800	1.555	8 (50)	0.44	24
900	1.786	8 (29)	0.39	12
1,000	2.046	8 (85)	0.40	40

it has a more compact model and this can be quickly constructed. Therefore it wins on two fronts: space and time.

The proportion of SR instances with matching was first investigated in [8] and later in [14] and [13]. Empirical evidence has been based on translations of the Pascal code given in the appendix of Irving's paper. Unfortunately that code has a bug and on occasion fails to find a matching when one exists. This has been observed by Stephan Mertens and independently by Ciaran McCreesh. Consequently, earlier reported results may be incorrect. The results in Table 1 use a sample size of 1,000 and might be assumed to be more accurate than those in Rob Irving's original study.

In Table 2 we have the average (and maximum) number of nodes required to find all matchings. This number is always low, and always less than 3 times the number of maximum matchings. More to the point, the model never exhibited exponential behaviour. As yet I have no explanation of why this is so, i.e. why the constraint model is so well behaved.

There are hard variants of SR. One example is egalitarian SR where a matching is to be found that minimizes the sum of the ranks, and this has been shown to be NP-hard [9]. In our constraint model an egalitarian matching is one that minimizes $\sum a_i$. Therefore we can model this problem by adding one more variable ($totalCost$), one more constraint ($totalCost = \sum a_i$) and a change from solving to minimization (line 36 of Listing 1). Naively, to find an egalitarian matching we could consider all matchings. As we see from Table 2 no instance had more than 40 matchings, no search took more than 85 nodes and the longest run time (not tabulated) was 2.6 seconds. Therefore, although NP-hard we would fail to encounter a hard instance in the problems sampled. So, (as Cheeseman, Kanefsky and Taylor famously asked [3]) where are the hard problems? As yet I do not know.

6   Conclusion

It has been demonstrated that there is a simple constraint model for the stable roommates problem. It was demonstrated that arc-consistency on this model produces the phase-1 table in $O(n^3)$ time. A backtracking search that maintains arc-consistency on each decision allows us to enumerate all matching. However, it was shown that the search process can make decisions that lead to failure. The simple model was enhanced by using bound, rather than enumerated constrained integer variables and arc-consistency delivers a *bound* phase-1 table. Nevertheless, this results in a substantial improvement in performance but the complexity of producing the phase-1 table remains $O(n^3)$. This lead to a specialized n-ary constraint with $O(n^2)$ cost for arc-consistency. Empirical study showed that this model can enumerate all matching to problems with 1,000 agents in about 2 seconds, orders of magnitude faster than the simple model.

It has also been shown that since our constraint model addresses incomplete preference lists it can also model stable marriage problems with complete and incomplete preference lists. That is, one model suffices.

Our model behaved well, never exhibiting exponential behaviour. Therefore there is work to do, to prove that the amount of backtracking is in some sense bounded by a polynomial, and this proof might be similar to that of failure-free enumeration in SM [6].

One of the first hard variants is egalitarian SR. This can be easily modeled and explored. However, it appears that it might be uninteresting. For $n \leq 1,000$ the number of matchings that need to be explored appears to be small. Furthermore, as n increases we expect that the number of instances with matchings will also fall [14,13]. Combined, this suggests that although NP-hard, egalitarian SR is easy.

All the code used in this study is available at [2].

Acknowledgements. I would like to thank Augustine Kwanashie, Ciaran McCreesh, David Manlove, Rob Irving and Ian Gent.

References

1. choco constraint programming system, http://choco.sourceforge.net/
2. Stable Roommates, http://www.dcs.gla.ac.uk/~pat/roommates/distribution
3. Cheeseman, P., Kanefsky, B., Taylor, W.M.: Where the really hard problems are, pp. 331–337. Morgan Kaufmann (1991)
4. Gale, D., Shapley, L.S.: College admissions and the stability of marriage. American Mathematical Monthly 69, 9–15 (1962)
5. Gale, D., Sotomayor, M.: Some remarks on the stable matching problem. Discrete Applied Mathematics 11, 223–232 (1985)
6. Gent, I.P., Irving, R.W., Manlove, D.F., Prosser, P., Smith, B.M.: A constraint programming approach to the stable marriage problem. In: Walsh, T. (ed.) CP 2001. LNCS, vol. 2239, pp. 225–239. Springer, Heidelberg (2001)
7. Gusfield, D., Irving, R.W.: The Stable Marriage Problem: Structure and Algorithms. The MIT Press (1989)
8. Irving, R.W.: An efficient algorithm for the "stable roommates" problem. J. Algorithms 6(4), 577–595 (1985)
9. Irving, R.W.: Optimal Stable Marriage. In: Encyclopedia of Algorithms. Springer (2008)
10. Mackworth, A.K.: Consistency in networks of relations. Artificial Intelligence 8, 99–118 (1977)
11. Manlove, D.: Algorithmics of Matching under Preferences. Theoretical Computer Science, vol. 2. World Scientific (2013)
12. Manlove, D.F., O'Malley, G.: Modelling and solving the stable marriage problem using constraint programming. In: Proceedings of the Fifth Workshop on Modelling and Solving Problems with Constraints, held at the 19th International Joint Conference on Artificial Intelligence (IJCAI 2005), pp. 10–17 (2005)
13. Mertens, S.: Random stable matchings. In: Journal of Statistical Mechanics: Theory and Experiments (2005)
14. Pittel, B., Irving, R.W.: An upper bound for the solvability of a random stable roommates instance. Random Struct. Algorithms 5(3), 465–487 (1994)
15. Roth, A.E.: The evolution of the labor market for medical interns and residents: a case study in game theory. Journal of Political Economy 92(6), 991–1016 (1984)
16. Roth, A.E., Sotomayor, M.A.O.: Two-sided matching: a study in game-theoretic modeling and analysis. Econometric Society Monographs, vol. 18. Cambridge University Press (1990)
17. Unsworth, C., Prosser, P.: An n-ary constraint for the stable marriage problem. In: Proceedings of the Fifth Workshop on Modelling and Solving Problems with Constraints, held at the 19th International Joint Conference on Artificial Intelligence (IJCAI 2005) (2005)
18. Unsworth, C., Prosser, P.: A specialised binary constraint for the stable marriage problem. In: Zucker, J.-D., Saitta, L. (eds.) SARA 2005. LNCS (LNAI), vol. 3607, pp. 218–233. Springer, Heidelberg (2005)
19. van Hentenryck, P., Deville, Y., Teng, C.-M.: A generic arc-consistency algorithm and its specializations. Artificial Intelligence 57, 291–321 (1992)

Detecting and Exploiting
Permutation Structures in MIPs

Domenico Salvagnin

DEI, University of Padova, Via Gradenigo 6b, 35131 Padova, Italy
salvagni@dei.unipd.it

Abstract. Many combinatorial optimization problems can be formulated as the search for the best possible permutation of a given set of objects, according to a given objective function. The corresponding MIP formulation is thus typically made of an assignment substructure, plus additional constraints and variables (as needed) to express the objective function. Unfortunately, the permutation structure is generally lost when the model is flattened as a mixed integer program, and state-of-the-art MIP solvers do not take full advantage of it. In the present paper we propose a heuristic procedure to detect permutation problems from their MIP formulation, and show how we can take advantage of this knowledge to speed up the solution process. Computational results on quadratic assignment and single machine scheduling problems show that the technique, when embedded in a state-of-the-art MIP solver, can indeed improve performance.

1   Introduction

Many combinatorial optimization problems can be formulated as the search for the best possible permutation of a given set of objects, according to a given objective function. Without loss of generality, in this paper we will consider problems of the form:

$$\min f(\pi) \qquad (1)$$
$$\pi \in \Pi^n \qquad (2)$$

where Π^n is the set of all permutations π of the ground set $N = \{1, \ldots, n\}$. A natural way to formulate this class of problems within a mixed-integer linear programming (MIP) paradigm is to encode the permutation π by introducing n^2 binary variables x_{ij} and $2n$ linear constraints, obtaining the so-called assignment polytope. Finally, additional artificial variables y, together with the corresponding linking constraints, are usually introduced in the model in order to properly formulate the objective function f. Note that if the objective function is linear in x, we have the so-called linear assignment polytope, which is polynomially solvable. The problem is thus reformulated as

H. Simonis (Ed.): CPAIOR 2014, LNCS 8451, pp. 29–44, 2014.
© Springer International Publishing Switzerland 2014

$$\min g(x, y) \tag{3}$$

$$\sum_{i=1}^{n} x_{ij} = 1 \quad \forall j \in N \tag{4}$$

$$\sum_{j=1}^{n} x_{ij} = 1 \quad \forall i \in N \tag{5}$$

$$Ax + By \geq b \tag{6}$$

$$x_{ij} \in \{0, 1\} \quad \forall (i, j) \in N^2 \tag{7}$$

$$y \geq 0 \tag{8}$$

It is important to note that, being a permutation problem, the structure of the formulation above is such that if all variables x are fixed, then it is always possible to compute in closed form the values of the artificial variables y, and thus to obtain a complete solution. In other words, once the permutation is known, there is no need to solve an optimization problem to compute its objective value. In particular, if it always possible to express y as a function Γ of x, such that $f(x) = g(x, \Gamma(x))$.

Many combinatorial optimization problems, as for example the quadratic assignment problem (QAP), the traveling salesman problem (TSP), and single machine scheduling, belong to the above class, although not all are always modeled in this way. For example, while the TSP is clearly a permutation problem, it is usually not formulated as such, so there is no explicit assignment substructure in the model.

Unfortunately, the permutation structure is generally lost when the model is flattened as a mixed integer program, and state-of-the-art MIP solvers do not take full advantage of it. For example, while it is always trivial to compute a feasible solution of a permutation problem, state-of-the-art MIP solvers occasionally find it very challenging to find one, even with their rich arsenal of primal heuristics. Even worse, while it is well-known that permutation problems are usually well-suited for local-search based metaheuristics, that are capable of finding near-optimal solutions with very modest computing times, MIP solvers have a very hard time in improving their first poor-quality solutions.

The main issue here is that current state-of-the-art MIP technology lacks a powerful modeling language based on global constraints, a tool which has long been standard in constraint programming [1]. As such, it is almost impossible for the modeler to pass high-level information to the solver, such as, for example, combinatorial substructures. Given the current state of things, it has become standard practice in MIP implementations to devise algorithms that basically try to *reverse-engineer* combinatorial substructures from a flat list of linear inequalities. For example, in [2] a procedure for detecting network structures was presented; such structure, when present, is then used to improve cutting plane separation. Unfortunately, while these procedures are usually cheap and effective, they are still heuristic in nature and can be fooled by the many transformations that are applied to a given MIP formulation in the preprocessing phase. Extending them to be completely preprocessing-invariant is often not done in practice for performance reasons (the resulting algorithms would be too slow), with the consequence that on some instances the substructure is not detected even if it is present.

While we cannot but share the lament for the current state of things, and encourage MIP vendors to invest into more powerful modeling interfaces to the solvers, as (partially) done, e.g., in the open solver SCIP [3], in the meantime we can still try to improve the situation for specific substructures, as is done for permutation problems in the present paper. Note that while permutation problems are usually solved with specialized codes, some challenging instances have indeed been solved with MIP technology, as for example in [4], so improving the performance of MIP solvers on this class of problems is of practical interest and can broaden the applicability of the MIP paradigm.

The outline of the paper is as follows. In Section 2 we describe a heuristic procedure to automatically detected permutation problems, while in Section 3 we show how to exploit the permutation structure to implement an efficient and general purpose primal heuristic based on local search. Two classes of permutation problems, that we used as benchmark, as described in Section 4. Computational results are given in Section 5, showing that the technique, when embedded in a of state-of-the-art MIP solver, can indeed improve performance, according to several performance measures. Conclusions and future directions of research are finally addressed in Section 6.

2   Detecting Permutation Structures

Detection of permutation problems is done in two steps:

1. in the first step we look for assignment polytopes, thus identifying the binary variables x that encode the permutation, and the corresponding assignment constraints;
2. in the second step we check that, once the variables x are fixed, the remaining variables y can indeed be computed in a straightforward way. In particular, this implies finding a topological order among variables y, such that we can compute them in one pass from left to right.

The first step is organized as a clustering algorithm. First, all constraints involving at least one non binary variable, that are not equalities and whose right-hand-size is different from 1 are removed, as they cannot be part of the assignment structure. The remaining constraints are then partitioned into clusters: each cluster can contain only constraints with the same number of variables and with pairwise disjoint support. Constraints are assigned to the first compatible cluster in a first-fit fashion; if no cluster is compatible, a new cluster is created with the current constraint in it. After the set of clusters Q has been initialized, we look for pairs (q_1, q_2) of matching clusters. Two clusters are matching if they *cover* exactly the same set of variables and each constraint in cluster q_1 intersects each constraint in cluster q_2 in exactly one variable. Intuitively, variables x are naturally double-index variables and can be arranged into a matrix,

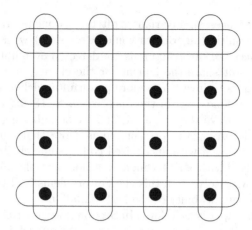

Fig. 1. Assignment structure

and the constraints in a pair of matching clusters (q_1, q_2) corresponds to the rows and columns, respectively, of this matrix, see Figure 1. Once a matching pair of cluster is found, it is removed from Q and the process continues until all pairs have been considered. Details are depicted in Algorithm 1. Note that the described detection algorithm is slightly more general than needed, since it can detect more than one assignment substructures.

The second step constructs a weighted dependency graph $G = (V, E, w)$ between the variables of the formulation, along with a topological order O on the nodes of G, and is based on row counts. At the beginning, all assignment variables x are added (in arbitrary order) as nodes to G, and to the ordered list O. All assignment constraints are removed from the model. Each constraint c_i left in the model, is assigned a count r_i, which counts all the variables in the constraint not yet in O. If some constraint ends up having a row count of zero at this step, then it means that this is not a pure permutation problem, but a constrained one, i.e., not all permutations are feasible since there are additional constraints on the assignment variables x: in this case we just abort the procedure. Then an iterative procedure begins, that loops over all constraints looking for those with $r_i = 1$. As soon as one is found, say c_i, the following steps are executed:

- let y_j be the variable left in c_i. All constraints involving y_j are considered and the singleton ones are collected in a set Y. Note that Y includes c_i.
- If $|Y| = 1$, then y_j can be computed directly from constraint c_i (plus its own bounds): y_j is added as a new node in G and to list O. In addition, edges are added to G, connecting y_j to all the other variables in c_i. Edges are weighted with the coefficients of the constraint, so that it is possible to compute the correct value of y_j given the values of the variables on which it depends. Node y_j is marked as being of type LINEAR.
- If $|Y| > 1$, then y_j depends on more than one affine expression involving other variables. If those expressions are all consistent, then y_j can be computed

---

Algorithm 1. Assignment subproblems detection.

Input: a list of constraints $C = \{c_1, \ldots, c_m\}$
Output: a list of assignment substructures $A = \{A_1, \ldots, A_k\}$
/* clustering */
1   $Q = \emptyset$;
2   **foreach** $c \in C$ **do**
3   **foreach** $q \in Q$ **do**
4   **if** c *is compatible with* c **then**
5           $q = q \cup \{c\}$;
6   **if** c *still not in a cluster* **then**
7        $Q = Q \cup \{\{c\}\}$;
/* matching clusters */
8   $A = \emptyset$;
9   **for** $q_1 \in Q$ **do**
10   **for** $q_2 \in Q$ **do**
11   **if** q_1 *and* q_2 *are matching* **then**
12         $A = A \cup \{(q_1, q_2)\}$;
13         $Q = Q \setminus \{q_1, q_2\}$;
14 **return** A

---

as either a min or a max of them (plus its own bounds). If this is the case, dummy nodes are added to G and to list O, together with the corresponding edges, in order to encode those affine expressions, similarly to the previous case. Then a new node y_j is added to G and O, with edges connecting it to the dummy nodes. Node y_j is marked as being of type MIN or MAX, depending on the direction of the inequalities. If the constraints are not consistent, then it is not possible to trivially compute y_j from the preceding variables in the topological order, meaning that this is not a permutation problem. In this case we abort the procedure.

– Variable y_j is marked as done and all row counts of constraints in which y_j appears are decreased by 1. Constraints in Y, which by definition have now a row count of 0, are removed from the model.

At each iteration at least one constraint is removed from the model, so this phase terminates in $O(m)$ iterations, where m is the number of constraints in the formulation. The procedure can be implemented quite efficiently if the constraint matrix is stored both row and column-wise.

Note that if cutting planes are added to the original formulation by the modeler, then those will end up in the dependency graph and propagate; however, since they are not needed to get a correct model, they are redundant in the graph and contribute only as a slowdown factor. For these reason, it is convenient to exploit the facilities provided by the underlying solver, if any, to mark a subset of constraints as cuts, so that they can be ignored by the graph construction algorithm (for example, the MPS and LP file formats used by CPLEX allow this extension). A similar reasoning applies to indicator constraints: if explicitly marked as such, they can be handled more efficiently by the algorithm. For example, the linear expression need not be updated if the indicator variable is false.

Example 1. Let us consider the following artificial tiny permutation problem

$$\min t$$
$$x_{11} + x_{12} = 1$$
$$x_{21} + x_{22} = 1$$
$$x_{11} + x_{21} = 1$$
$$x_{12} + x_{22} = 1$$
$$y_1 \geq 4x_{11} + 5x_{12}$$
$$y_2 \geq 3x_{21} + 2x_{22} + y_1$$
$$t \geq y_1 + 2y_2$$
$$t \geq 3y_1 + y_2$$
$$x \in \{0,1\}^4$$
$$y, w \geq 0$$

The corresponding dependency graph is depicted in Figure 2, and the ordered list is

$$O = [x_{11}, x_{12}, x_{21}, x_{22}, y_1, y_2, z_1, z_2, t]$$

□

3   Exploiting Permutation Structures

Once the permutation structure, if any, is identified, we can exploit its knowledge to improve the performance of the underlying MIP solver. A natural option, pursued in this paper, is to use the permutation structure to implement a general purpose primal heuristic, based on local-search. Alternative options, such as using the permutation structure for preprocessing and cut strengthening/separation, are possible as well, and left as future work.

The basic idea is to implement a local-search based metaheuristic, namely iterated local search (ILS) [5], using the dependency graph to explore neighborhoods and evaluate solutions. Given a permutation π, the neighborhood $\mathcal{N}(\pi)$ is defined as all permutations that can be obtained by swapping two elements of the permutation. Clearly, the neighborhood has polynomial size, containing exactly $n(n-1)/2$ permutations for each center π. The idea behind ILS is to perturb the current locally optimal solution s^* to get a new center t and call again the local search procedure from there, obtaining a new local optimum t^*. If the new solution t meets an acceptance criterion, then t is chosen as the next starting point, otherwise it is rejected and the procedure is repeated from s^*. Intuitively, ILS implements a heuristic random walk on the set of locally optimal

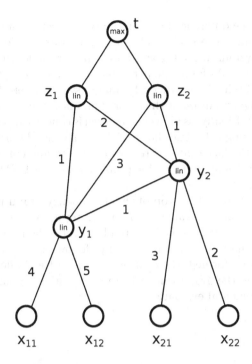

Fig. 2. Dependency graph of example problem

Algorithm 2. Basic ILS procedure

---

1 $s_0 = $ GenerateRandomSolution ();
2 $s^* = $ LocalSearch (s_0);
3 **repeat**
4    $s' = $ Perturb (s^*, history);
5    $t = $ LocalSearch (s');
6    $s^* = $ AcceptanceCriterion (s^*, t, history);
7 **until** *termination condition*;

---

solutions of a given optimization problem. A high level pseudocode for ILS is given in Algorithm 2.

Note that the perturbation mechanism and the acceptance criterion are in general dependent on the history of the system: this allows for more effective and elaborate strategies. The simplest, yet very common, acceptance criterion is to accept the new solution t if and only if its objective value is better than that of s. Other strategies include a pure random walk option, in which the new solution t is always accepted, regardless of its cost, and a simulated annealing [6,7] like acceptance criterion based on temperature, in which t is always accepted if it is an improving solution, but is also accepted with a given probability even if its objective value is worse (the probability is usually dependent on the "temperature" T of the system and on the difference between the two objective values, with slightly worsening steps being more likely). The first two strategies do not make use of the history of the system, while the third does. In our implementation, we chose the annealing criterion. As far as the perturbation step is concerned, a perturbed permutation π' is obtain from π by performing k swaps, where k is adjusted dynamically during the execution of the algorithm, and is always contained in the interval $\{k_1, \ldots, k_2\}$. Finally, as far as local search is concerned, we implemented a first-improving pivoting rule. The choice of ILS as a general purpose metaheuristic is motivated by the fact that it is relatively easy to implement and proved to be quite successful in many permutation problems, such as TSP [8], QAP [9], and scheduling problems [10].

Implementing local search on top of the dependency graph is quite straightforward. A first solution is evaluated by assigning a value to all variables x, and then computing the values of the variables y (and the intermediate expressions needed to evaluate max and min, if any) following the order in O. Then, whenever a swap is performed, the change in value of the 4 affected variables is propagated following the graph (much like in constraint propagation systems), thus achieving incremental evaluation.

4 Testbed

We considered two classes of problem that exhibit a permutation structure, namely quadratic assignment and single machine scheduling problems, as described in the next subsections.

4.1 Quadratic Assignment Problems

The NP-hard (and notoriously very difficult in practice) *Quadratic Assignment Problem* (QAP), in its Koopmans and Beckmann form [11], can be sketched as follows; see, e.g., [12] for details. We are given a complete directed graph $G = (V, A)$ with n nodes n^2 arcs, and a set of n facilities to be assigned to its nodes. In what follows, indices i, j correspond to nodes, indices u, v to facilities, $b_{ij} \geq 0$ is a given (directed) *distance* from node i to node j, and $a_{uv} \geq 0$ is a given required *flow* from facility u to facility v. By using binary variables $x_{iu} = 1$ iff facility u is assigned to node i, QAP can be stated as the following quadratic binary problem:

$$\min \sum_{i=1}^{n} \sum_{u=1}^{n} \sum_{j=1}^{n} \sum_{v=1}^{n} a_{uv} \, b_{ij} \, x_{iu} x_{jv} \tag{9}$$

$$\sum_{i=1}^{n} x_{iu} = 1 \quad \forall u \in N \tag{10}$$

$$\sum_{u=1}^{n} x_{iu} = 1 \quad \forall i \in N \tag{11}$$

$$x_{iu} \in \{0, 1\} \quad \forall (i, u) \in N^2 \tag{12}$$

Most MIP models for QAP work with additional 0-1 variables $y_{iujv} = x_{iu}x_{jv}$ that are used to linearize the quadratic objective function—the Adams-Johnson model [13] being perhaps the best-known such formulation. These kinds of models require $\Theta(n^4)$ variables and $\Theta(n^3)$ constraints, so they become huge even for medium-size instances, with unacceptable slowdowns in solving the LP relaxations during the branch-and-cut tree.

A different approach is to look for MILP models requiring just $O(n^2)$ variables and constraints. An obvious model is the MILP one credited to Kaufman and Broeckx [14] that requires the introduction of just n^2 additional (continuous) variables

$$w_{iu} = \left(\sum_{j=1}^{n} \sum_{v=1}^{n} a_{uv} b_{ij} x_{jv} \right) x_{iu} \tag{13}$$

which can be easily linearized with big-M coefficients. The corresponding MIP model reads

$$\min \sum_{i=1}^{n} \sum_{u=1}^{n} w_{iu} \tag{14}$$

$$\sum_{i=1}^{n} x_{iu} = 1 \quad \forall u \in N \tag{15}$$

$$\sum_{u=1}^{n} x_{iu} = 1 \quad \forall i \in N \tag{16}$$

$$w_{iu} \geq \sum_{j=1}^{n} \sum_{v=1}^{n} a_{uv} b_{ij} x_{jv} - M(1 - x_{iu}) \quad \forall (i, u) \in N^2 \tag{17}$$

$$x_{iu} \in \{0, 1\} \quad \forall (i, u) \in N^2 \tag{18}$$

$$w_{iu} \geq 0 \tag{19}$$

This model is known to be of little use in practice *as is* because of the big-M constraints (17). In particular, it can be proved [15] that the root-node bound is always zero. However, a much stronger formulation can be obtained by adding to (14)−(19) the (polynomial) family of cutting planes

$$w_{iu} \geq L_{iu} x_{iu} \tag{20}$$

where each L_{iu} is defined as the optimal value of the linear (and polynomially solvable) assignment problem:

$$\min \sum_{j=1}^{n} \sum_{v=1}^{n} a_{uv} b_{ij} x_{jv} \tag{21}$$

$$\sum_{j=1}^{n} x_{jv} = 1 \quad \forall v \in N \tag{22}$$

$$\sum_{v=1}^{n} x_{jv} = 1 \quad \forall j \in N \tag{23}$$

$$x_{iu} = 1 \tag{24}$$

$$x_{jv} \in \{0,1\} \quad \forall (j,v) \in N^2 \tag{25}$$

It can be shown that adding constraints (20) to the model, the resulting root relaxation bound is at least as strong as the so-called Gilmore-Lawler [16,17] bound. This lightweight model, together with the family of cutting planes (20), was used recently in [4] to solve highly symmetric QAP instances, proving to be a reasonable tradeoff between bound strength and enumeration speed.

In this paper, we considered all the instances in the standard QAPLIB [18] testbed with $n < 20$, and excluding the instances of the esc class, which are well-known to be massively symmetric. Overall, we are left with 31 instances.

As far as the structure of the problem is concerned, the quadratic assignment problem is clearly a permutation problem. Once all variables x_{iu} are assigned a value, then the value of variables w_{iu} is automatically derived by our algorithm as

$$w_{iu} = \max \left\{ \sum_{j=1}^{n} \sum_{v=1}^{n} a_{uv} b_{ij} x_{jv} - M(1 - x_{iu}), L_{iu} x_{iu}, 0 \right\} \tag{26}$$

As far as our ILS metaheuristic is concerned, evaluating a neighboring solution has cost $O(n^2)$, assuming that the constraints defining w_{iu} are dense (as is usually the case). While linear in the size of the model, this is suboptimal with respect to an ad-hoc and QAP-specific implementation, where a neighboring solution can be evaluated $O(n)$ arithmetic operations: this is the price to pay for a general purpose (and relatively simple) implementation, directly based on a linear formulation of the model. Note that in this particular case the issue could be solved by expressing constraints (17) as indicator constraints, as hinted at the end of Section 2. Indeed, whenever a variable x_{iu} flips value, we do not need to update $O(n^2)$ expressions, but only $O(n)$, since only n w_{iu} variables are nonzero in any solution, thus implementing a form of partial incremental propagation. When a non up-to-date variable w_{iu} needs to be evaluated, it is computed from scratch, which again can be done in $O(n)$ because only n x_{iu} variables are nonzero in any solution. Unfortunately, properly handling these cases complicates the implementation significantly, so we do not support it yet. Finally, we can obtain a small performance improvement by marking (20) as cuts, so that they are ignored by the algorithm.

4.2 (Weighted) Total Tardiness Minimization in Single Machine Scheduling

In single machine scheduling problems [19], we are given a set of n jobs, to be processed on a single machine, without preemption. Each job j is characterized by its processing time p_j, a due date d_j, and a nonnegative weight w_j. Different objective functions are of interest when solving single machine scheduling problems. In the present paper we will restrict to (weighted) total tardiness of the schedule.

Given a job j, its tardiness T_j is defined as

$$T_j = \max\{C_j - d_j, 0\} \tag{27}$$

where C_j is the completion time of job j. In the scheduling notation of [20], the variants considered in this paper are denoted as $1||\sum T_j$ and $1||\sum w_j T_j$, for the simple and weighted total tardiness, respectively.

For the unweighted case [19], a simple MIP model reads

$$\min \sum_{k=1}^{n} T_k \tag{28}$$

$$\sum_{j=1}^{n} x_{jk} = 1 \quad \forall k \in N \tag{29}$$

$$\sum_{k=1}^{n} x_{jk} = 1 \quad \forall j \in N \tag{30}$$

$$T_k \geq \sum_{j=1}^{n} p_j \left(\sum_{u=1}^{k} x_{ju} \right) - \sum_{j=1}^{n} d_j x_{jk} \quad \forall k \in N \tag{31}$$

$$x_{jk} \in \{0,1\} \quad \forall (j,k) \in N^2 \tag{32}$$

$$T_j \geq 0 \quad \forall j \in N \tag{33}$$

where variable $x_{jk} = 1$ iff job j is assigned position k in the processing, while T_k is the tardiness of the job in position k. In constraints (31), the first term is the cumulative processing time of the first k jobs in the sequence, while the second term is the due date of the job in position k. Note that in the model we do not explicitly keep track of the tardiness of each job by index, but only by position.

A MIP formulation for the weighted case is considerably more involved, because, differently from the unweighted case, we need to know the tardiness of each job by its index j and not just by its position k. Indeed, at least four formulations can be implemented, as surveyed in [21], with different tradeoffs between size and strength. In the present paper, we considered the one based on the assignment polytope, much in the spirit of the unweighted case. The MIP model reads

$$\min \sum_{j=1}^{n} w_j T_j \tag{34}$$

$$\sum_{j=1}^{n} x_{jk} = 1 \quad \forall k \in N \tag{35}$$

$$\sum_{k=1}^{n} x_{jk} = 1 \quad \forall j \in N \tag{36}$$

$$\gamma_1 \geq \sum_{j=1}^{n} p_j x_{j1} \tag{37}$$

$$\gamma_k \geq \gamma_{k-1} \sum_{j=1}^{n} p_j x_{jk} \quad \forall k \in N \setminus \{1\} \tag{38}$$

$$C_j \geq \gamma_k - M(1 - x_{jk}) \quad \forall (j,k) \in N^2 \tag{39}$$

$$T_j \geq C_j - d_j \quad \forall j \in N \tag{40}$$

$$x_{jk} \in \{0,1\} \quad \forall (j,k) \in N^2 \tag{41}$$

$$\gamma_k \geq 0 \quad \forall k \in N \tag{42}$$

$$C_j, T_j \geq 0 \quad \forall j \in N \tag{43}$$

As in the previous model, variable $x_{jk} = 1$ iff job j is assigned position k in the processing. In addition, γ_k is the completion time of the job in position k, C_j is the completion time of job j, and T_j is the tardiness of job j. As in the QAP case, the presence of big-M coefficients in constraints (39) makes the formulation quite weak in practice. However, by sorting the jobs by processing time before generating the model, it is possible to strengthen the model by adding a polynomial family of inequalities, which can be easily computed, see [21] for the details: we implemented this strengthened variant.

In this paper, we considered all the instances in the standard ORLIB [22] testbed with $n = 40$. Overall, there are 125 instances, and we consider both the unweighted and weighted variants (in the first case, by just ignoring the weights).

As far as the structure of the problem is concerned, with the chosen formulations the single machine scheduling problem is clearly a permutation problem. Once all variables x_{jk} are assigned a value, then we can automatically compute the value of variables T_k in the unweighted case, and of variables γ_k, C_j and T_j (in this order) in the weighted case.

As far as our ILS metaheuristic is concerned, evaluating a neighboring solution has cost $O(n)$ in the unweighted case, and $O(n^2)$ in the weighted case. Again, while linear in the size of the model, this is suboptimal in the weighted case, where an ad-hoc implementation can evaluate a neighboring solution in $O(n)$ arithmetic operations.

It is important to note that this slowdown is caused not by inefficiencies in the algorithm, but rather by the MIP formulation itself, because $O(n^2)$ linear constraints are needed to link variables γ_k and C_j. For example, if we were allowed to use nonlinear expressions, only $O(n)$ constraints of the form

$$C_j = \sum_{k=1}^{n} \gamma_k x_{jk}$$

would suffice. Note that the above constraints are essentially **element** [23] global constraints, so in the ideal case we should be able to formulate the problem using those, exploiting their presence to obtain a more efficient ILS implementation, and then let the solver automatically linearize them in order to get a standard MIP model. Thus, this is yet one more argument for implementing global constraint technology within MIP solvers.

5   Computational Experiments

We implemented our codes in C++, using IBM ILOG CPLEX 12.5.1 [24] as black box MIP solver. All tests have been performed on a cluster of identical PC, each with an Intel Xeon E3-1220 V2 CPU running at 3.10GHz and 16GB of RAM (only one CPU was used by each process). Each method was given a time limit of 10,000 seconds per instance.

We compare two variants of our state-of-the-art MIP solver on the instances described in the previous section. We denote with **detect** the full version of our code, which detects the permutation structure of the problems and applies the ILS heuristic throughout the branch-and-cut tree (implemented through the callback mechanism of CPLEX), and with **cpx** the same code with our detection algorithm disabled, in order to have a fair comparison between the two. As far as the parameters of **detect** are concerned, the detection algorithm is triggered at the end of the root node processing, and only if the integrality gap left is greater than 2%, in order to avoid wasting time in case of very easy instances. The default parameters of the ILS metaheuristic are $iterLim = 1000$ and $noImprovLim = 100$, while the perturbation interval $[k_1, k_2]$ is set to $[3, 7]$. If the integrality gap is less than 10%, then the ILS metaheuristic is run with stricter limits, namely $iterLim = 100$ and $noImprovLim = 20$. Finally, the heuristic is called every $10,000$ nodes and, if not effective for 5 times in a row, it is completely switched off for the rest of the search.

We use 4 performance measures to compare **cpx** and **detect**:

- **#solved**: number of instances solved within the time limit.
- **time**: shifted geometric mean, with a shift of 1 second, of the running time on the subset of instances solved by both methods.
- **pint**: shifted geometric mean, with a shift of 0.01, of the primal integral on the subset of instances solved by both methods (the lower the better). The primal integral [25] measures the overall behavior of the solver as far as the primal bound is concerned, and overcomes many shortcomings of traditional figures when measuring the effect of primal heuristics. Given the primal gap function $\gamma(x)$ of a feasible solution x, defined as

$$\gamma(x) = \begin{cases} 0 & \text{if } cx = \bar{z} = 0 \\ 1 & \text{if } cx \cdot \bar{z} < 0 \\ \frac{|cx - \bar{z}|}{\max(|cx|, |\bar{z}|)} & \text{otherwise.} \end{cases}$$

where $\bar{z}$ is the value of the optimal (or best known) solution, we define the primal gap pgap as a function of the running time t as

$$\text{pgap}(t) = \begin{cases} 1 & \text{if no feasible solution is known at time } t \\ \gamma(\tilde{x}) & \text{if } \tilde{x} \text{ is the incumbent at time t.} \end{cases}$$

Note that $\gamma(x)$, and thus also $\text{pgap}(t)$, is always between 0 and 1. The primal integral is defined as the integral over t of $\text{pgap}(t)$.

- gap: average final integrality gap on the subset of instances unsolved by at least one method. The integrality gap is computed as

$$\text{gap}(\bar{z}, \underline{z}) = \begin{cases} 0 & \text{if } \bar{z} = \underline{z} = 0 \\ 1 & \text{if } \bar{z} \cdot \underline{z} < 0 \\ \frac{|\bar{z} - \underline{z}|}{\max(|\bar{z}|, |\underline{z}|)} & \text{otherwise.} \end{cases}$$

where $\bar{z}$ is the value of the global primal bound and $\underline{z}$ is the value of the global dual bound. This is the integrality gap as reported by many commercial solvers, and has the advantage of always being a number between 0 and 1.

Aggregated results of the comparison between the two methods are reported in Table 1. According to the table, the two methods are approximately equivalent as far as the number of solved instances is concerned: this is not surprising, as it is well-known that in general our ability in solving MIPs is largely dominated by the dual bound, which is not affected by our method. According to many computational studies [26,27], the effect of primal heuristics on the overall solution process is approximately in the order of 10-15% on average. Still, in our testbeds, where finding good quality solutions is challenging for CPLEX, the effect of our method is a significant reduction of the overall running time, up to almost 40% on the weighted total tardiness testbed. In addition, the primal integral is also significantly reduced, dropping by a factor of 4 in the QAP

Table 1. Comparison of the two methods

testbed	method	#solved	time (s)	pint	gap
QAP	cpx	19	60.5	0.569	8.8%
	detect	20	55.5	0.113	8.3%
$1\|\| \sum T_j$	cpx	90	6.4	0.198	18.0%
	detect	88	5.3	0.128	17.6%
$1\|\| \sum w_j T_j$	cpx	40	12.8	2.140	31.8%
	detect	41	8.0	0.857	29.7%

testbed and by a factor of approximately 2 on the single scheduling instances. Finally, on the unsolved instances, the final integrality gap was also consistently reduced—although by a little amount.

6   Conclusions

We described a heuristic procedure to automatically detected permutation problems, and exploited the permutation structure to implement an efficient and general purpose primal heuristic based on local search. Computational experiments on two classes of permutation problems, namely QAP and single machine scheduling, showed that the technique, when embedded in a of state-of-the-art MIP solver, can indeed significantly improve performance.

Future research includes efficiently implementing the extensions needed to support indicator constraints, as well as devising heuristic procedures to detect other common substructures, such as, for example, linear encodings of the **element** global constraint. This would bring the general purpose ILS procedure inline with the problem specific implementations. Extensions to more general classes of permutation problem, such as rectangular or higher-dimensional assignments, could also broaden the applicability of the method.

References

1. Gent, I.P., Petrie, K.E., Puget, J.F.: Symmetry in constraint programming. In: Rossi, F., van Beek, P., Walsh, T. (eds.) Handbook of Constraint Programming, pp. 329–376. Elsevier (2006)
2. Achterberg, T., Raack, C.: The MCF-separator: detecting and exploiting multicommodity flow structures in MIPs. Mathematical Programming Computation 2(2), 125–165 (2010)
3. Achterberg, T.: SCIP: solving constraint integer programs. Mathematical Programming Computation 1(1), 1–41 (2009)
4. Fischetti, M., Monaci, M., Salvagnin, D.: Three ideas for the quadratic assignment problem. Operations Research 60(4), 954–964 (2012)
5. Lourenço, H.R., Martin, O.C., Stützle, T.: Iterated local search: Framework and applications. In: Glover, F., Kochenberger, G. (eds.) Handbook of Metaheuristics, vol. 57, pp. 321–353. Kluwer Academic Publishers (2002)
6. Kirkpatrick, S., Gelatt, C.D., Vecchi, M.P.: Optimization by simulated annealing. Science 220, 671–680 (1983)
7. Černý, V.: Thermodynamical approach to the traveling salesman problem: An efficient simulation algorithm. Journal of Optimization Theory and Applications 45(1), 41–51 (1985)
8. Johnson, D.S., McGeoch, L.A.: Experimental analysis of heuristics for the STSP. In: Gutin, G., Punnen, A. (eds.) The Traveling Salesman Problem and its Variations, pp. 369–443 (2002)
9. Stützle, T.: Iterated local search for the quadratic assignment problem. European Journal of Operational Research 174(3), 1519–1539 (2006)
10. Congram, R.K., Potts, C.N., van de Velde, S.L.: An iterated dynasearch algorithm for the single-machine total weighted tardiness scheduling problem. INFORMS Journal on Computing 14(1), 52–67 (2002)

11. Koopmans, T., Beckmann, M.: Assignment problems and the location of economic activities. Econometrica 25, 53–76 (1957)
12. Burkard, R., Dell'Amico, M., Martello, S.: Assignment Problems. SIAM (2009)
13. Adams, W., Johnson, T.: Improved linear programming-based lower bounds for the quadratic assignment problem. In: Proceedings of the DIMACS Workshop on Quadratic Assignment Problems. DIMACS Series in Discrete Mathematics and Theoretical Computer Science, vol. 16, pp. 43–75. American Mathematical Society (1994)
14. Kaufman, L., Broeckx, F.: An algorithm for the quadratic assignment problem using Benders' decomposition. European Journal of Operational Research 2, 204–211 (1978)
15. Xia, Y., Yuan, Y.: A new linearization method for quadratic assignment problem. Optimization Methods and Software 21, 803–816 (2006)
16. Gilmore, P.: Optimal and suboptimal algorithms for the quadratic assignment problem. SIAM Journal on Applied Mathematics 14, 305–313 (1962)
17. Lawler, E.: The quadratic assignment problem. Management Science 9, 586–599 (1963)
18. Burkard, R., Karisch, S., Rendl, F.: QAPLIB – A quadratic assignment problem library. European Journal of Operational Research 55, 115–119 (1991)
19. Baker, K.R., Trietsch, D.: Principles of Sequencing and Scheduling. Wiley (2009)
20. Graham, R., Lawler, E., Lenstra, J., Kan, A.R.: Optimization and approximation in deterministic sequencing and scheduling: A survey. Annals of Discrete Mathematics 5, 287–326 (1979)
21. Keha, A.B., Khowala, K., Fowler, J.W.: Mixed integer programming formulations for single machine scheduling problems. Computers & Industrial Engineering 56(1), 357–367 (2009)
22. Beasley, J.E.: OR-library: distributing test problems by electronic mail (1990)
23. Hentenryck, P.V., Carillon, J.P.: Generality versus specificity: an experience with AI and OR techniques. In: AAAI 1988 (1988)
24. IBM ILOG: CPLEX 12.5.1 User's Manual (2013)
25. Berthold, T.: Measuring the impact of primal heuristics. Operations Research Letters 41, 611–614 (2013)
26. Achterberg, T.: Constraint Integer Programming. PhD thesis, Technische Universität Berlin (2007)
27. Achterberg, T., Wunderling, R.: Mixed integer programming: Analyzing 12 years of progress. In: Facets of Combinatorial Optimization, pp. 449–481 (2013)

Solving the Quorumcast Routing Problem as a Mixed Integer Program

Quoc Trung Bui[1], Quang Dung Pham[2], and Yves Deville[1]

[1] ICTEAM, Université catholique de Louvain, Belgium
{quoc.bui,Yves.Deville}@uclouvain.be
[2] SoICT, Hanoi University of Science and Technology, Vietnam
dungpq@soict.hust.edu.vn

Abstract. The quorumcast routing problem is a generalization of multicasting which arises in many distributed applications. It consists of finding a minimum cost tree that spans the source node and at least q out of m specified nodes on a given undirected weighted graph. In this paper, we solve this problem as a mixed integer program. The experimental results show that our four approaches outperform the state of the art. A sensitivity analysis is also performed on values of q and m.

1 Introduction

Multicasting is the problem of delivering a message from a source to a given subset of nodes, called the *multicast* nodes, in a network. Suppose given an undirected graph $G = (V, E, c)$, i.e., V, E are, respectively, the set of nodes and the set of edges. Suppose further that each edge $(i, j) \in E$ is associated with a positive cost $c_{ij} \in \mathbb{R}^+$. Now, given a set of multicast nodes $S \subseteq V$, an integral value $q \leq |S|$, and a root node r (without loss the generality, we may assume that $r \in S$), the objective of the quorumcast routing problem (QRP) is to find a minimum cost tree T that spans r and at least q nodes of S [5,24,12,32,29]. QRP is NP-hard, as it reduces to the Steiner tree problem [14] when $q = |S|$. QRP appears in many distributed applications, for example, distributed synchronization and updating a replicated resource (see [5] for more details).

For solving QRP, various incomplete approaches to computing an approximation of the optimal solution have been proposed in [5,12,32,29], in which the constraint-based local search algorithm in [29] is currently the state of the art incomplete algorithm. In addition, two exact algorithms in [24,29] have been proposed for solving this problem to optimality. In [24], a partial solution is defined to be a set of sub-trees that spans the root and some multicast nodes; a partial solution is extended by adding one edge at each step until a feasible solution is constructed; a *Confined Area Pruning* scheme was introduced that allows reducing that search space. The Constraint Programming (CP) approach in [29] is currently th state of the art exact algorithm.

Contributions. In this paper, we propose four mathematical formulations for QRP and use them to solve QRP as a mixed integer program. These approaches outperform the state of the art approach based on CP. In addition, through the experimental results, we show the effect of the values q and $|S|$ on the performance of the proposed approaches.

H. Simonis (Ed.): CPAIOR 2014, LNCS 8451, pp. 45–54, 2014.

2   Mathematical Models

In this section, we propose four mathematical models for QRP. One is proposed directly on the undirected graph G, and the others are proposed on the corresponding directed graph of G that is formed by replacing each edge of G by two opposite arcs with the same cost as the original edge.

These models can exploit the properties of QRP solutions. Let T be a solution of $QRP(q, m)$ on a graph G. One can easily show that (1) all leaf nodes of T are multicast nodes [24], and (2) T spans exactly q multicast nodes.

All the models use the binary variables x_{ij} stating whether edge (i, j) is in the solution tree T. (In the undirected graph, we use the convention that $i < j$.) All the models aim at minimizing $\sum_{(i,j)\in E} c_{ij}x_{ij}$.

2.1   Natural Formulation: Model 1

In this section, we propose a formulation on the undirected graph $G = (V, E)$, called "the natural formulation." Many problems have been modeled by similar formulations [3,27,16,13,26,1]. This model introduces binary variables y_i stating whether node i is in T.

$$\sum_{i,j\in C:(i,j)\in E} x_{ij} \leq |C| - 1, \forall C \subset V, 2 \leq |C| \leq |V| - 1 \tag{1a}$$

$$\sum_{(i,j)\in E} x_{ij} + \sum_{(j,i)\in E} x_{ji} \geq y_i, \forall i \in V \tag{1b}$$

$$\sum_{(i,j)\in E} x_{ij} + \sum_{(j,i)\in E} x_{ji} \leq (|V| - 1)y_i, \forall i \in V \tag{1c}$$

$$1 + \sum_{(i,j)\in E} x_{ij} = \sum_{v\in V} y_v \tag{1d}$$

$$y_r = 1 \tag{1e}$$

$$\sum_{v\in S} y_v = q \tag{1f}$$

$$x_{ij} \in \{0, 1\}, \forall(i, j) \in E \tag{1g}$$

$$y_i \in \{0, 1\}, \forall i \in V \tag{1h}$$

In this model, the constraints (1a) are connectivity constraints (or subtour elimination constraints). The constraints (1b) and (1c) ensure that if $v \in T$, then $y_v = 1$, and if $v \notin T$, then $y_v = 0$. The constraint (1d) presents a basic property of a tree that requires the relation between the number of nodes and the number of edges. The constraints (1f) ensure that T includes exactly q multicast nodes (Property (2)). Finally, the constraint (1e) ensures that the root r is always in T. Notice that Property (1), stating that all leaf nodes are multicast nodes, could also be included, but experimental results have shown that it is useless here as well as in all subsequent models. It is therefore not considered.

In this model, there are $|E| + |V|$ variables and an exponential number of constraints.

2.2   Formulation Based on Multi-commodity Flows: Model 2

In this section, we propose a multi-commodity flow formulation on the corresponding directed graph $G' = (V, A)$. In the literature, many problems have been modeled using multi-commodity flows [16,10,7,17]. However, this problem is slightly more complex, as we do not know which multicast nodes are spanned.

This model introduces variables $y_{ij}^k \in \mathbb{R}^+$ measuring the flow, through arc $(i, j) \in A$, from the root node r to a node $k \in V \setminus \{r\}$.

$$\sum_{(r,i)\in A} (y_{ri}^k - y_{ir}^k) \le 1, \forall k \in V \tag{2a}$$

$$\sum_{k\in S, k\neq r, (r,i)\in A} (y_{ri}^k - y_{ir}^k) = q - 1 \tag{2b}$$

$$\sum_{(k,i)\in A} (y_{ki}^k - y_{ik}^k) \ge -1, \forall k \in V \tag{2c}$$

$$\sum_{k\in S, k\neq r, (k,i)\in A} (y_{ki}^k - y_{ik}^k) = -(q - 1) \tag{2d}$$

$$\sum_{(j,i)\in A} (y_{ij}^k - y_{ji}^k) = 0, \forall k \in V, i \in V \setminus \{k, r\} \tag{2e}$$

$$y_{ij}^k \le x_{ij}, \forall (i, j) \in A, \forall k \in V \cup \{r\} \tag{2f}$$

$$y_{ij}^k \ge 0, \forall (i, j) \in A, \forall k \in V \cup \{r\} \tag{2g}$$

$$x_{ij} \in \{0, 1\}, \forall (i, j) \in V \tag{2h}$$

In this model, the constraints (2f) ensure that a flow is sent along an arc only if the arc is traversed. The constraints (2b), (2d) and (2e) ensure that there exists a flow from the root r to q nodes in S (note that we assumed that r is a multicast node). The constraints (2a), (2c) and (2e) are flow conserving constraints.

In this model, there are $|A| \times |S|$ variables and a polynomial number of constraints.

2.3   Classical Formulation: Model 3

In this model, we propose a formulation, called "the classical formulation," on the corresponding directed graph $G' = (V, A)$. In the literature, many similar formulations have been proposed for problems related to finding a spanning tree [15,22,26,29].

$$x_{ir} = 0, \forall (i, r) \in A \tag{3a}$$

$$\sum_{(r,i)\in A} x_{ri} \ge 1 \tag{3b}$$

$$\sum_{u\notin C, v\in C, (u,v)\in A} x_{uv} \ge \sum_{(j,i)\in A} x_{ji}, \forall C \subset V, 2 \le |C| \le |V| - 1, \forall i \in C, r \notin C \tag{3c}$$

$$\sum_{(i,j)\in A} x_{ij} \leq 1, \forall j \in V \tag{3d}$$

$$\sum_{i\in S, i\neq r, (j,i)\in A} x_{ji} = q - 1 \tag{3e}$$

$$x_{ij} \in \{0,1\}, \forall (i,j) \in A \tag{3f}$$

In this model, the constraints (3a) indicate that there are no arcs arriving at r. The constraint (3b) ensures that there exists at least one arc leaving r. The constraints (3c) are connectivity constraints (see [11,6,19,22] for more details about the connectivity constraints). The constraint (3e) ensures that the optimal tree T' includes exactly q multicast nodes.

In this model, there are $|A|$ variables corresponding to the number of arcs in the corresponding directed graph G', and an exponential number of constraints in (3c).

2.4   Miller–Tucker–Zemlin Formulation: Model 4

In this model, we propose a formulation using Miller–Tucker–Zemlin constraints as connectivity constraints on the corresponding directed graph $G' = (V, A)$ [28,21].

This model introduces variables t_i constrained by $t_i \leq t_j$ if arc $(i,j) \in T$. These constraints prevent subtours in the solution. The variables p_i state whether or not node i is in T'.

$$p_r = 1 \tag{4a}$$

$$x_{ir} = 0, \forall (i,r) \in A \tag{4b}$$

$$\sum_{(r,i)\in A} x_{ri} \geq 1 \tag{4c}$$

$$\sum_{(i,j)\in A} x_{ij} = p_j, \forall j \in V \setminus \{r\} \tag{4d}$$

$$x_{ij} \leq p_i, \forall (i,j) \in A \tag{4e}$$

$$|V| x_{ij} + t_i + 1 \leq t_j + |V|, \forall (i,j) \in A \tag{4f}$$

$$1 + \sum_{(i,j)\in A} x_{ij} = \sum_{v\in V} p_v \tag{4g}$$

$$\sum_{i\in S} p_i = q \tag{4h}$$

$$x_{ij} \in \{0,1\}, \forall (i,j) \in A \tag{4i}$$

$$p_i \in \{0,1\}, \forall i \in V \tag{4j}$$

$$t_i \in \{1 \ldots |V|\}, \forall i \in V \tag{4k}$$

In this model, the constraints (4f) present the relative position of the nodes in the tree. They state that if $x_{ij} = 1$, then $t_i < t_j$. This prevents the solution from containing subtours. The constraints (4c) ensure that the root node r must connect to other nodes in the arborescence tree.

In this model, there are $(|A| + 2 \times |V|)$ variables and a polynomial number of constraints.

3   Solving QRP as a Mixed Integer Program

In this section, we propose four different approaches, based on the above models, to solve QRP. Model 2 and Model 4 have a polynomial numbers of variables and constraints. They can be directly used in a MIP solver (CPLEX). These approaches will be denoted *Mod2_B&B* and *Mod4_B&B*. Model 1 and Model 3 have an exponential number of constraints. The constraints are relaxed, and Branch & Cut approaches are employed.

3.1   Lazy Constraint Approach

This approach is applicable to Model 1 and Model 3, where connectivity constraints (1a) and (3c) are considered as lazy constraints. The linear programming relaxation of the initial model without lazy constraints is solved. All isolated components are identified for every feasible integral solution that is not yet feasible. If a solution to a linear programming relaxation is feasible, then there is no isolated component. To check for isolated components, we use the union-find data structure [9,18]. If there is only one component, then the solution is feasible. Otherwise, there is a cycle in some components, a lazy constraint is then added for each component as follows. For Model 3, C is the set of all nodes of the component and i is a random node in C; for Model 1, C is the set of all nodes of the component. All these lazy constraints are then added directly to the model, and the linear programming relaxation of the current model is reoptimized. This procedure is repeatedly executed until an optimal solution has been found. The two corresponding approaches will be denoted *Mod1_B&C_lazy* and *Mod3_B&C_lazy*.

3.2   Dynamic Constraint Separation Approach

This approach can be applied to Model 3, where connectivity constraints (3c) are dynamically separated [11]. This approach, denoted by *Mod3_B&C_dyn*, finds violated connectivity constraints on the support graph. Given a solution x^* to a linear programming relaxation (containing all connectivity constraints separated so far), the support graph G^* of x^* has all nodes V and edges $\{(i,j) \in E : x_{ij}^* > 0\}$[4].

This approach consists of two stages. First, the support graph is checked for isolated components not connected to the root node. For each isolated component, only constraint (3c) is added to the current model, where C is the set of all nodes in the isolated component and i is a random node in C. Second, if the support graph has only one component that includes the root node r, a maximum $r - v$-flow/minimum $r - v$-cut problem is solved for each node $v \neq r$ in the component. To solve maximum-flow/minimum-cut problems, we use code written by Skorobohatyj [30]. A maximum flow that is less than the absolute inflow to v indicates a violated connectivity constraint (3c), in which C are all nodes on the same side of the $r - v$ cut as v (of course node i in the constraint is the node v).

3.3   Preprocessing

Different reduction checks have been proposed for the minimum Steiner tree problem and others [22,19,25,31,23,2]. In the preprocessing of QRP, the following *check for useless nodes* is useful. It is performed on the undirected graph G. If a node is not a multicast node and its degree is only one, then this node (and its edge) can be removed from the graph G. If a node v is not a multicast node with exactly two neighbors u and w, then the node v and the edges (u, v) and (v, w) can be removed. If there exists an edge (u, w) with cost c_{uw}, this cost is updated to $min(c_{uw}, c_{uv} + c_{vw})$. Otherwise, an edge (u, w) is added with cost $c_{uv} + c_{vw}$. These checks can be applied iteratively until the graph remains unchanged. In practice, we limit ourselves to three iterations. Other reductions were considered, but they had only very marginal impact.

4   Computational Experiments

In [29], our approach based on CP was tested on 960 random instances, with the largest graph having 60 nodes. It was shown that the CP approach was better than the existing state of the art complete approaches. We reuse these instances. We collect these instances into a class called **C1**.

We also collect 2500 instances in a class **C2**, generated from 100 undirected graphs of 160 nodes and 25 couples $\langle q, |S| \rangle$, ranging from $\{ \langle 3, 20 \rangle$ to $\langle 119, 140 \rangle \}$. These 100 undirected graphs were extracted from 100 minimum Steiner tree instances of test set **I160** in the library SteinLib [20]. The multicast nodes were randomly chosen.

All MIP approaches were implemented in C++, using IBM Ilog Cplex Concert Technology, version 12.4. The standard Cplex cuts were automatically added. The CP approach in [29] was implemented in Comet [8]. Finally, all experiments were performed on XEN virtual machines with 1 core of a CPU Intel Core2 Quad Q6600 @2.40GHz and 1GB of RAM. The time limit for each execution of an algorithm was 30 minutes.

4.1   Comparing the Approaches

We first compare the MIP approaches as well as the CP approach from [29]. Figure 1 gives a summary of the experimental results. The columns have the following meanings: %*opt* is the percentage of instances solved to optimality within the time limit of 30 minutes; $\overline{I}$ is the average number of iterations; $\overline{N}$ is the average of the number of nodes in the branch-and-bound tree; $\overline{C}$ is the average of the number of separated constraints; $\overline{T}$ is the average computational time in seconds (on the solved instances). Figure 2 shows the evolution of the percentage of solved instances in **C2** with respect to the time limit.

It is clear that all the MIP approaches significantly outperform the CP approach. In the class **C1**, the MIP approaches are two orders of magnitude faster than CP. In the class **C2**, CP only solved two instances out of 2500, while *Mod4_B&B* solved 95.2% of the instances. In Figures 1 and 2, there is no major difference between *Mod3_B&C_lazy* and *Mod4_B&B*, nor between *Mod1_B&C_lazy* and *Mod3_B&C_dyn*. It is clear that *Mod3_B&C_lazy* and *Mod4_B&B* are the best two approaches, both in the percentage of solved instances and in the execution time. The approach *Mod2_B&B* is the worst among the MIP approaches, although it develops few nodes. The low number of nodes results from the fact that the integer relaxation version of this model is quite close to the optimal solution [17]. It is worth noting that *Mod3_B&C_dyn* has the smallest number of iterations. This mainly comes from the dynamic constraint separation approach,

Class	Approach	%opt	$\overline{I}$	$\overline{N}$	$\overline{C}$	$\overline{T}$
C1	CP	97.1	na.	na.	na.	80.47
	Mod1_B&C_lazy	100	441.7	50.43	6.58	0.57
	Mod2_B&B	100	2159	1.38	na.	1.06
	Mod3_B&C_lazy	100	398.8	77.51	81.42	0.35
	Mod3_B&C_dyn	100	308.2	3.69	158.1	1.94
	Mod4_B&B	100	506.2	55.9	na.	0.64
C2	CP	0.08	na.	na.	na.	9.74
	Mod1_B&C_lazy	78.6	92804	8507	1110	150.9
	Mod2_B&B	60.2	66716	7.05	na.	318.6
	Mod3_B&C_lazy	94.4	30938	2312	1077	97.6
	Mod3_B&C_dyn	77.2	8408	54.64	6089	153.0
	Mod4_B&B	95.2	50327	6138	na.	92.8

Fig. 1. A summary of computational results for two classes of instances

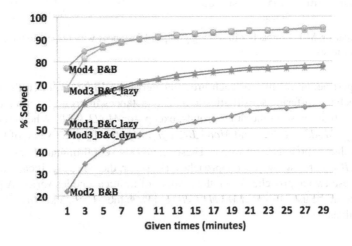

Fig. 2. Percentage of solved instances in **C2** in given times

which produces smaller and thinner branch-and-bound trees. However, the number of added constraints is much larger.

Although not reported here for reasons of space, the experimental results also showed that Property (2) of the QRP solutions (Section 2) is very useful in all the models proposed in this paper. For example, this property helps this model to solve 6.2% more instances of class **C2**.

4.2   Effect of the Values of q and $|S|$ on the Performance of the Approaches

We analyze the sensitivity of the performance to the value of q and the size of the multicast node set. We divided the instances of class **C2** into two sets of groups. In the first set, a group contains instances of the same size of multicast node set. The groups

G20 (resp. *G50, G80, G110* and *G140*) consist of all instances with $|S| = 20$ (resp. $50, 80, 110$ and 140). In the second set, a group contains 500 instances with similar $\frac{q}{|S|}$. The groups *G1* through *G5* split the instances from the smallest to the largest values of $\frac{q}{|S|}$.

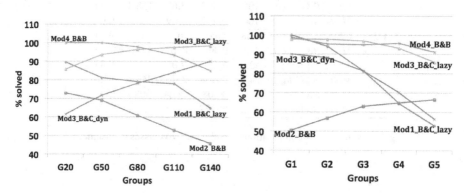

Fig. 3. Comparing MIP approaches in solving groups of instances with respect to the percentage of solved instances

The experimental results for each group are given in Figure 3. First, the different approaches have differing sensitivities to q. The instances in group *G5* are more difficult to solve than those in the other groups, except for *Mod2_B&B*. When considering the ratio $\frac{q}{|S|}$, *Mod3_B&_lazy* and *Mod3_B&_dyn* are better for high value of this ratio, while the other approaches are worse. These results also confirm that *Mod3_B&C_lazy* and *Mod4_B&B* are the two best approaches. However, there is a significant difference between these two approaches when the number of multicast nodes varies. A portfolio approach could be used to select *Mod4_B&B* for low values of $|S|$, and *Mod3_B&C_lazy* for high values.

5   Conclusion

This paper solved the quorumcast routing problem to optimality as a mixed integer program. In this paper, we proposed four mathematical formulations for QRP. We then solved QRP to optimality as a mixed integer program, introducing two constraint relaxations. The computational results showed that the MIP approaches are much more efficient than the state of the art approach (which is based on Constraint Programming). In addition, we showed that two of the approaches, *Mod3_B&C_lazy* and *Mod4_B&B*, are the two best ones. Finally, experimental results pointed out that the different approaches have different sensitivities to the parameters q and the size of the multicast node set. As future research, new separation constraints could be investigated.

Acknowledgments. The authors want to thank the anonymous reviewers for their helpful comments. This research is partially supported by the FRFC project 2.4504.10 of the Belgian FNRS and the UCLouvain Action de Recherche Concertée ICTM22C1.

References

1. Ahuja, R.K., Magnanti, T.L., Orlin, J.B.: Network flows: theory, algorithms, and applications. Prentice-Hall, Inc., Upper Saddle River (1993)
2. Cordeau, J.-F., Costa, A.M., Laporte, G.: Steiner tree problems with profits. INFOR 44, 99–115 (2006)
3. Andersen, K.A., Jørnsten, K., Lind, M.: On bicriterion minimal spanning trees: An approximation. Computers and Operations Research 23(12), 1171–1182 (1996)
4. Applegate, D.L., Bixby, R.E., Chvatal, V., Cook, W.J.: The Traveling Salesman Problem: A Computational Study. Princeton Series in Applied Mathematics. Princeton University Press, Princeton (2007)
5. Cheung, S.Y., Kumar, A.: Efficient quorumcast routing algorithms. In: Proceedings IEEE INFOCOM 1994, Networking For Global Communications, vol. 2, pp. 840–847 (June 1994)
6. Chimani, M., Kandyba, M., Ljubić, I., Mutzel, P.: Obtaining optimal k-cardinality trees fast. J. Exp. Algorithmics 14, 5:2.5–5:2.23(2010)
7. Chopra, S., Tsai, C.Y.: Polyhedral approaches for the steiner tree problem on graphs. In: Du, D.-Z., Cheng, X. (eds.) Steiner Trees in Industries, vol. 11, pp. 175–202. Kluwer Academic Publishers (2001)
8. Comet. Comet user manual, dynadec (2011), http://dynadec.com/
9. Cormen, T.H., Leiserson, C.E., Rivest, R.L., Stein, C.: Introduction to Algorithms, 3rd edn. The MIT Press (2009)
10. Requejo, A.C.C., Agra, A., Santos, E.: Formulations for the weightconstrained minimum spanning tree problem. In: AIP Conf. Proc. vol. 1281, pp. 2166–2169 (2010)
11. Drexl, M., Irnich, S.: Solving elementary shortest-path problems as mixed-integer programs. OR Spectrum, pp. 1–16 (2012)
12. Du, B., Gu, J., Tsang, D.H.K., Wang, W.: Quorumcast routing by multispace search. In: Global Telecommunications Conference on Communications: The Key to Global Prosperity, GLOBECOM 1996, vol. 2, pp. 1069–1073 (November 1996)
13. Fujie, T.: The maximum-leaf spanning tree problem: Formulations and facets. Networks 43(4), 212–223 (2004)
14. Garey, M.R., Johnson, D.S.: Computers and Intractability; A Guide to the Theory of NP-Completeness. W. H. Freeman & Co., New York (1990)
15. Goemans, M.X., Myung, Y.-S.: A catalog of steiner tree formulations. Networks 23(1), 19–28 (1993)
16. S.T. Henn.: Weight-Constrained Minimum Spanning Tree Problem. PhD thesis, University of Kaiserslautern (2007)
17. Ibrahim, M.S., Maculan, N., Minoux, M.: A strong flow-based formulation for the shortest path problem in digraphs with negative cycles. International Transactions in Operational Research 16(3), 361–369 (2009)
18. Wayne, K.: Union-find algorithms. (2008), http://www.cs.princeton.edu/r~{}s/AlgsDS07/01UnionFind.pdf
19. Koch, T., Martin, A.: Solving steiner tree problems in graphs to optimality. Networks 32, 207–232 (1998)
20. Koch, T., Martin, A., Voß, S.: SteinLib: An updated library on steiner tree problems in graphs. Technical Report ZIB-Report 00-37, Konrad-Zuse-Zentrum für Informationstechnik Berlin, Takustr. 7, Berlin (2000)
21. Kulkarni, R.V., Bhave, P.R.: Integer programming formulations of vehicle routing problems. European Journal of Operational Research 20(1), 58–67 (1985)
22. Ljubi, I., Weiskircher, R., Pferschy, U., Klau, G.W., Mutzel, P., Fischetti, M.: An algorithmic framework for the exact solution of the prize-collecting steiner tree problem. Mathematical Programming 105, 427–449 (2006)

23. Ljubi, I., Weiskircher, R., Pferschy, U., Klau, G.W., Mutzel, P., Fischetti, M.: An algorithmic framework for the exact solution of the prize-collecting steiner tree problem. In: Mathematical Progamming. Series B (2006)
24. Low, C.P.: A fast search algorithm for the quorumcast routing problem. Inf. Process. Lett. 66(2), 87–92 (1998)
25. Lucena, A., Beasley, J.E.: A branch and cut algorithm for the steiner problem in graphs. Networks 31, 39–59 (1998)
26. Lucena, A., Maculan, N., Simonetti, L.: Reformulations and solution algorithms for the maximum leaf spanning tree problem. Computational Management Science 7, 289–311 (2010)
27. Magnanti, T.L., Wolsey, L.A.: Optimal trees. In: Monma, C.L., Ball, M.O., Magnanti, T.L., Nemhauser, G.L. (eds.) Network Models. Handbooks in Operations Research and Management Science, vol. 7, ch. 9, pp. 503–615. Elsevier (1995)
28. Miller, C.E., Tucker, A.W., Zemlin, R.A.: Integer programming formulation of traveling salesman problems. J. ACM 7(4), 326–329 (1960)
29. Pham, Q., Deville, Y.: Solving the quorumcast routing problem by constraint programming. Constraints 17, 409–431 (2012)
30. G. Skorobohatyj.: Finding a minimum cut between all pairs of nodes in an undirected graph (2008),
    `http://elib.zib.de/pub/Packages/mathprog/mincut/`
    `all-pairs/index.html`
31. Uchoa, E.: Reduction tests for the prize-collecting steiner problem. Operations Research Letters 34(4), 437–444 (2006)
32. Wang, B., Hou, J.C.: An efficient QoS routing algorithm for quorumcast communication. Computer Networks Journal 44(1), 43–61 (2004)

A New MIP Model for Parallel-Batch Scheduling with Non-identical Job Sizes

Sebastian Kosch and J. Christopher Beck

Department of Mechanical & Industrial Engineering
University of Toronto, Toronto, Ontario M5S 3G8, Canada
{skosch,jcb}@mie.utoronto.ca

Abstract. Parallel-batch machine problems arise in numerous manufacturing settings from semiconductor manufacturing to printing. They have recently been addressed in constraint programming (CP) via the combination of the novel `sequenceEDD` global constraint with the existing `pack` constraint to form the current state-of-the-art approach. In this paper, we present a detailed analysis of the problem and derivation of a number of properties that are exploited in a novel mixed integer programming (MIP) model for the problem. Our empirical results demonstrate that the new model is able to outperform the CP model across a range of standard benchmark problems. Further investigation shows that the new MIP formulation improves on the existing formulation primarily by producing a much smaller model and enabling high quality primal solutions to be found very quickly.

1 Introduction

Despite the widespread application of mixed integer programming (MIP) technology to optimization problems in general and scheduling problems specifically,[1] there is a significant body of work that demonstrates the superiority of constraint programming (CP) and hybrid approaches for a number of classes of scheduling problems [1–5]. While the superiority is often a result of strong inference techniques embedded in global constraints [6–8], it is sometimes due to problem-specific implementation in the form of specialized global constraints [4] or instantiations of decomposition techniques [1–3]. The flexibility of CP and decomposition approaches which facilitates such implementations is undoubtedly positive from the perspective of solving specific problems better. However, the ability to create problem-specific approaches is in some ways in opposition to the compositionality and model-and-solve "holy grail" of CP [9]: to enable users to model and solve problems without *implementing* anything new at all.

Our overarching thesis is that, in fact, MIP technology is closer to this goal than CP, at least in the context of combinatorial optimization problems. In our investigation of this thesis, we are developing MIP models for scheduling

---

[1] For example, of the 58 papers published in the *Journal of Scheduling* in 2012, 19 use MIP, more than any other single approach.

H. Simonis (Ed.): CPAIOR 2014, LNCS 8451, pp. 55–70, 2014.

problems where the current state of the art is customized CP or hybrid approaches. Heinz et al. [10] showed that on a class of resource allocation and scheduling problems, a MIP model could be designed that was competitive with the state-of-the-art logic-based Benders decomposition. This paper represents a similar contribution in different scheduling problem: a parallel-batch processing problem which has previously been attacked by MIP, branch-and-price [11], and CP [4] with the latter representing the state of the art.

We propose a MIP model inspired by the idea of modifying a canonical feasible solution. The definition of our objective function in this novel context is not intuitive until we reason algorithmically about how constraints and assignments interact – a strategy often used in designing metaheuristics. Indeed, we suggest that the analogy between branching on independent decision variables and making moves between neighbouring schedules should be explored in more detail for a range of combinatorial problems.

In the next section we present the formal problem definition and discuss existing approaches. In Section 3 we prove a number of propositions that allow us to formally propose a novel MIP model for the problem in Section 4. Section 5 presents our empirical results, demonstrating that the performance of the new model is superior to the existing CP model, both in terms of mean time to find optimal solutions and in terms of solution quality when optimal solutions could not be found within the time limit.

2   Background

Batch machines with limited capacity exist in many manufacturing settings in forms such as ovens [12], autoclaves [4], and tanks [13]. In this paper, we tackle the problem of minimizing the maximum lateness, $L_{\max}$, in a single machine parallel batching problem where each job has an individual due date and size.

We use the following notation: a set $\mathcal{J}$ of n jobs is to be assigned to a set of n batches $\mathcal{B} = \{B_1, \ldots, B_n\}$. Batches can hold multiple jobs or remain empty. Each job j has a processing time, p_j, a size, s_j, and a due date, d_j. Jobs can be assigned to arbitrary batches, as long as the sum of the sizes of the jobs in a batch does not exceed the machine capacity, b.

The single machine processes one batch at a time. Each batch B_k has a *batch start date* S_k, a *batch processing time*, defined as the longest processing time of all jobs assigned to the batch, $P_k = \max_{j \in B_k}(p_j)$, and a *batch completion date*, which must fall before the start time of the next batch, $C_k = S_k + P_k \leq S_{k+1}$.

The lateness of a job j, L_j, is the completion time of its batch C_k less its due date d_j. The objective function is to minimize the maximum lateness over all jobs, $L_{\max} = \max_{j \in \mathcal{J}}(L_j)$. Since we are interested in the maximum lateness, only the earliest-due job in each batch matters and we define it as the *batch due date* $D_k = \min_{j \in B_k}(d_j)$.

The problem can be summarized as $1|p\text{-}batch; b < n; non\text{-}identical|L_{\max}$ [4,11], where $p\text{-}batch; b < n$ represents the resource's parallel-batch nature and its finite capacity. A version with identical job sizes was shown to be strongly NP-hard by Brucker et al. [14]; this problem, therefore, is no less difficult.

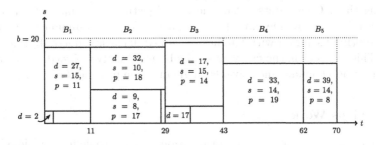

Fig. 1. An optimal solution to an example problem with eight jobs (values for s_j and p_j are not shown for the two small jobs in batches 1 and 3, respectively)

Figure 1 shows a solution to a sample problem with eight jobs and resource capacity $b = 20$. The last batch has the maximum lateness $L_5 = C_5 - D_5 = 70 - 39 = 31$.

2.1 Reference MIP Model

The problem is formally defined by MIP Model 2.1, used by Malapert et al. [4] for comparison with their CP model (see below).

$$\text{Min.} \quad L_{\max} \tag{1}$$

$$\text{s.t.} \quad \sum_{k \in \mathcal{B}} x_{jk} = 1 \qquad\qquad \forall j \in \mathcal{J} \tag{2}$$

$$\sum_{j \in \mathcal{J}} s_j x_{jk} \le b \qquad\qquad \forall k \in \mathcal{B} \tag{3}$$

$$p_j x_{jk} \le P_k \qquad\qquad \forall j \in \mathcal{J}, \forall k \in K \tag{4}$$

$$C_{k-1} + P_k = C_k \qquad\qquad \forall k \in \mathcal{B} \tag{5}$$

$$(d_{\max} - d_j)(1 - x_{jk}) + d_j \ge D_k \qquad \forall j \in \mathcal{J}, \forall k \in \mathcal{B} \tag{6}$$

$$C_k - D_k \le L_{\max} \qquad\qquad \forall k \in \mathcal{B} \tag{7}$$

$$D_{k-1} \le D_k \qquad\qquad \forall k \in \mathcal{B} \tag{8}$$

$$x_{jk} \in \{0,1\}, C_k \ge 0, P_k \ge 0, D_k \ge 0 \quad \forall j \in \mathcal{J}, \forall k \in \mathcal{B} \tag{9}$$

Model 2.1. Reference MIP model

The decision variables, x_{jk}, are binary variables where $x_{jk} = 1$ iff job j is assigned to batch k. Constraints (2) ensure that each job j is assigned to exactly one batch k. Constraints (3) ensure that no batch exceeds the machine capacity, b. Constraints (4) define each batch's processing time, P_k, as the maximum processing time of the jobs j assigned to it. Constraints (6) implement the definition of D_k while ensuring that for empty batches k, $D_k = d_{\max}$. Constraints (5) define each batch's completion time, C_k, as that of the previous batch, plus the batch's processing time. Constraints (7) define the objective value $L_{\max}$. Constraints (8) sort the batches by due date, based on a well-known dominance rule: there exists an optimal solution with batches scheduled in earliest-due-date-first order (EDD). This stems from the fact that if all jobs are already assigned, the problem reduces to a polynomial-time solvable single machine problem $(1|D_k|L_{\max})$ [15].

2.2   Previous Work

Malapert et al. [4] present a CP formulation of the problem (see Model 2.2) which relies on two global constraints: `pack` [16], which constrains the job-to-batch assignments such that no capacity limits are violated, and `sequenceEDD`, which enforces the EDD order over the batches. The implementation of the latter constraint is the main contribution of the paper and is primarily responsible for the strong performance. `sequenceEDD` includes a set of pruning rules that update the lower and upper bounds on $L_{\max}$ and on the number of batches. Based on these bounds, other assignments are then eliminated from the set of feasible assignments.

$$\text{Min.} \quad L_{\max} \tag{10}$$
$$\text{s.t.} \quad \texttt{maxOfASet}(P_k, B_k, [p_j]_k, 0) \qquad \forall k \in \mathcal{B} \tag{11}$$
$$\texttt{minOfASet}(D_k, B_k, [d_j]_k, d_{\max}) \qquad \forall k \in \mathcal{B} \tag{12}$$
$$\texttt{pack}(B_k, A_j, S_k, M, s_j) \tag{13}$$
$$\texttt{sequenceEDD}(B_k, D_k, P_k, M, L_{\max}) \tag{14}$$

Model 2.2. CP model proposed by Malapert et al.

Constraints (11) define P_k as the maximum of the set of processing times $[p_j]_k$ belonging to the jobs assigned to batch B_k, with a minimum value of 0 (note that the notation is adapted from Malapert et al. to match that in this paper). Constraints (12) define D_k as the minimum of the due dates $[d_j]_k$ associated with the set of jobs assigned to the batch B_k, with a maximum of $d_{\max}$, the largest due date among all given jobs. Constraint (13) implements the limited batch capacity b. It uses propagation rules incorporating knapsack-based reasoning, as well as a dynamic lower bound on the number of non-empty batches M [4, 16]. Note that this constraint handles the channeling between the set of jobs assigned to batch B_k, and the assigned batch index A_j for each job j. The limited capacity is enforced by setting the domain of the batch loads, S_k, to $[0, b]$. Constraint (14) ensures that the objective value $L_{\max}$ is equal to the maximum lateness of the batches scheduled according to the EDD rule.

The problem has also been addressed with a detailed branch-and-price algorithm [11], which is described in [4] as follows: each batch is a column in the column generation master problem. A solution of the master problem is a feasible sequence of batches. The objective of the subproblem is to find a batch which improves the current solution of the master problem. Malapert et al. [4] showed that their CP model was significantly faster than the branch-and-price algorithm which itself was more efficient than the reference MIP model.

Other authors have examined similar problems: Azizoglu & Webster [17] provide an exact method and a heuristic for the same problem, but minimize makespan (C_{max}) instead of L_{max}, similar to the work by Dupont and Dhaenens-Flipo [18]. Exact methods have been proposed for multi-agent variants with different objective functions by Sabouni and Jolai [19], for makespan minimization on single batch machines by Kashan et al. [20], and for makespan minimization on parallel batch machines with different release dates [21]. An extensive review of MIP models in batch processing is given by Grossmann [13].

3 Exploiting the Problem Structure

In this section, we make a series of observations about the parallel-batch scheduling problem that allow us to develop a novel MIP formulation.

3.1 The Single-EDD Schedule and Assigning Jobs to Earlier Batches

We can exploit the EDD rule to eliminate $\frac{1}{2}(n^2 - n)$ of the n^2 potential job-batch assignments a priori.

We first re-index all jobs in non-decreasing due date (and in non-decreasing processing time in case of a tie). For the remainder of this paper, consider all jobs to be indexed in this way. We then define the *single-EDD* schedule, in which each batch B_k contains the *single* job j matching its index (i.e., $x_{jk} = 1$ iff $j = k$), such that EDD is always satisfied. We refer to j as the *host job* of batch B_k, while other jobs assigned to B_k, if any, are *guest jobs*.

Lemma 1. *Consider a schedule S in which B_k is the earliest-scheduled batch such that its host job j is assigned to a later-scheduled batch B_{k+m}. In this schedule, B_k is either empty or $D_k = d_j$ (even though $j \notin B_k$).*

Proof. If B_k is non-empty, $D_k \geq d_j$: since B_k is defined as the earliest scheduled batch whose host job is scheduled later, B_k cannot host jobs due before d_j. But if $D_k > d_j$ then EDD is violated, as $D_{k+m} = d_j$. Thus, B_k is empty, or $D_k = d_j$ (due to other jobs with due date equal to d_j assigned to B_k). □

Proposition 1. *There exists an optimal solution in which job j is assigned to batch B_k, $j \leq k$.*

Proof. Consider again schedule S. Since $D_{k+m} = d_j$, EDD requires that no batch $B_q, k \leq q < k + m$, is due after d_j. By Lemma 1, we only need to consider the following two cases:

1. B_k is empty, so $P_k = 0$. Since EDD is not violated, we know that $D_q = d_j \ \forall \ B_q, k \leq q \leq k + m$. We can assign all jobs from B_{k+m} to B_k, such that $P_{k+m} = 0$. L_{max} will stay constant, as the completion time of the last-scheduled of all batches due at d_j does not change.

2. B_k is non-empty and due at $D_k = d_j$ (although $j \notin B_k$), due to at least one job g from a later-scheduled batch for which $d_g = d_j$, which is assigned to B_k. In this case, since $D_k = d_j = D_{k+m}$ and since EDD is not violated, *all* batches B_q where $k \le q \le k + m$ must be due at d_j. But then we can re-order these batches such that their respective earliest-due jobs are once again assigned to their single-EDD indices. The jobs in B_{k+m} (including j) will be assigned to B_k as a result. $L_{\max}$ is not affected by this re-assignment, as the completion time of the last-scheduled batch due at d_j does not change.

□

We thus introduce the following constraint to exclude solutions in which jobs are assigned to later batches than their single-EDD batches.

$$x_{jk} = 0 \quad \forall \{j \in \mathcal{J}, k \in \mathcal{B} | j < k\} \tag{15}$$

We can also show that in every non-empty batch B_k, the earliest-due job j must be the host job. This means that when batch B_k's host job is assigned to an earlier batch, no other jobs can be assigned to B_k; a batch that is *hostless* must be empty. This requirement rests on the following proposition.

Proposition 2. *There exists an optimal solution that has no hostless, non-empty batches.*

Proof. Consider an EDD-ordered schedule in which batch B_j is the last-scheduled batch which is hostless but non-empty: instead of its host job, only a set G of later-due guest jobs is assigned to B_j ($j \notin G$).

The earliest-due job $g \in G$ must have the same due date as batch B_{j+1}: if it is due later, EDD is violated; if it is due earlier, G is not a set of later-due guest jobs. Job g's own host batch B_g (which is hostless) cannot itself be due later than $d_g = D_{j+1}$ – this would require B_g to have guest jobs from later batches, but we defined B_k as the last batch with this property – therefore, $D_q = D_{j+1}$ for all batches $B_q, j \le q \le g$.

Then we can re-assign the guest jobs G from B_j into B_g, such that g is again host job in its own single-EDD batch. This re-assignment has no impact on $L_{\max}$ since it makes $P_j = 0$, resulting in the same completion time of the set of all batches with batch due date $D = D_{j+1}$. □

The above proposition translates to the following constraint:

$$x_{kk} \ge x_{jk} \quad \forall \{j \in \mathcal{J}, k \in \mathcal{B} | j > k\} \tag{16}$$

This observation allows us to define the due date of *all* batches to be the due date of their respective host jobs: $D_k = d_j, \forall \{j \in \mathcal{J}, k \in \mathcal{B} | j = k\}$. This rule holds even for empty batches B_k: $P_k = 0$, so $C_k = C_{k-1}$; but $D_{k-1} \le D_k$ due to this rule, so $L_{k-1} \ge L_k$ and thus L_k has no impact on $L_{\max}$.

3.2   Reformulating the Objective Function

We can formulate each batch's lateness, L_k, as its lateness in the single-EDD schedule, modified by the assignment of jobs into and out of batches $B_h, h \leq k$.

We first define $\mathcal{B}^\star \subseteq \mathcal{B}$ as the set of batches B_k which, given any EDD schedule, are the last-scheduled among all batches with due date D_k, since we can make the following observation.

Proposition 3. *Given a set of batches with equal due date in a schedule, we only need to consider the lateness of the one scheduled last.*

Proof. In an EDD ordering, the lateness of the batch scheduled last is greater than (or equal to, in the case of an empty batch) the lateness values of all other batches sharing its due date as it has the latest completion date.   □

This fact allows us to reduce the number of constraints defining $L_{\max}$, as we only need to consider batches $\mathcal{B}^\star$ as potential candidates for $L_{\max}$.

To simplify the following exposition, we define the term *move* as the reassignment of a job j from its single-EDD batch B_k to an earlier batch $B_h, h < k$, such that $x_{jk} = 0$ and $x_{jh} = 1$ and j is a guest job in B_h. Any schedule can thus be understood as a set of such moves, executed in arbitrary order starting from the single-EDD schedule. To define the objective function, we consider the change in $L_{\max}$ associated with individual moves.

Consider any EDD schedule, such as the one in Figure 2(a). Moving a job j from its single-EDD batch $B_{k=j}$ into an earlier batch B_e has the following effect:

- the lateness of all batches $B_i, i \geq k$ is reduced by p_j (Figure 2(b)),
- the lateness of all batches $B_h, h \geq e$ is increased by $\max(0, p_j - P_e)$, where P_e is the processing time of batch B_e before j is moved into it (Figure 2(c)).

In any batch, only the host job's lateness is relevant to $L_{\max}$. In other words, the lateness of batch B_k equals the lateness of job $j = k$, unless the job was moved into an earlier batch (in which case $P_k = 0$ due to Proposition 2 and $L_k = L_{k-1}$). Therefore, we can understand the lateness of batch B_k as its lateness in the single-EDD schedule, written as $L_{k,\text{single}}$, modified by the summed effect that all moves of other jobs into and out of batches $h \leq k$ have on the completion time of B_k:

$$L_k = L_{k,\text{single}} + \sum_{h \leq k} \underbrace{P_h' - p_h(2 - x_{hh})}_{T_h} \quad \forall k \in \mathcal{B}^\star \tag{17}$$

$$P_k' \geq p_j x_{jk} \qquad\qquad \forall\{j \in \mathcal{J}, k \in \mathcal{B} | j \geq k\} \tag{18}$$

$$P_k' \geq p_j \qquad\qquad \forall\{j \in \mathcal{J}, k \in \mathcal{B} | j = k\} \tag{19}$$

where $P_k' = \max(P_k, p_k) \forall k \in \mathcal{B}$ as defined in constraints (18) and (19).

For every batch $B_k \in \mathcal{B}^\star$, consider the possible scenarios for all batches $B_h, h \leq k$:

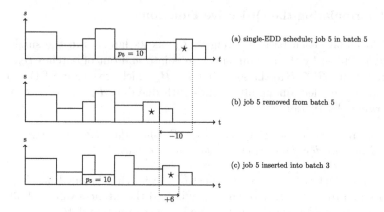

Fig. 2. Moving a job in a single-EDD schedule. Job 5 (marked "$p_5 = 10$") is moved from its single-EDD batch 5 into the earlier batch 3. This changes the lateness of job 7 (marked $\star$) from $L_{7,\text{single}}$ to $L_{7,\text{single}} - 10 + 6 = L_{7,\text{single}} - 4$.

- Batch B_h holds its host job. Then $x_{hh} = 1$ and the summand T_h evaluates to $P'_h - p_h$. If B_h has guest jobs, then $P'_h - p_h > 0$ if any of them are longer than the host job; if all guests are shorter, $P'_h = p_h$ and $T_h = 0$.
- Batch B_h is hostless and thus empty. We require $T_h = -p_h$ in accordance with Figure 2(b). To achieve this, we state in constraints (19) that P'_h never drops below the length of its host job, even when $P_h = 0$. With this in effect, the minimization objective enforces $P'_h = p_h$ and $T_h = P'_h - 2p_h = -p_h$.

Thus, we add to L_k the increase in processing time due to guests, $\max(0, P'_h - p_h)$, for every non-empty batch B_h. We subtract from L_k the host job processing time p_h for every empty batch B_h. This is congruent with Figure 2 above.

The net sum of these additions and subtractions to and from $L_{k,\text{single}}$ adjusts the lateness of batch k to its correct value given the values of x_{jk}.

Proposition 4. *Constraints* (17)–(19) *correctly define* L_k *for each batch* B_k.

Proof. By induction on k, our base case is the lateness of the first batch ($k = 1$). It is clear that the lateness of B_k is equal to its single-EDD lateness, plus $\max(P_1 - p_1, 0)$ since guest jobs may cause $P_1 > p_1$:

$$L_1 = L_{1,\text{single}} + \underbrace{P'_1 - p_1(2 - x_{11})}_{=\max(P_1 - p_1, 0)}. \tag{20}$$

Our induction hypothesis is that the proposition holds for any batch B_k:

$$L_k = L_{k,\text{single}} + \sum_{h \leq k} P'_h - p_h(2 - x_{hh}) \tag{21}$$

To show how an expression for L_{k+1} then follows, we relate L_{k+1} to L_k:

$$L_{k+1} = L_k + \underbrace{P_{k+1}}_{C_{k+1}-C_k} - \underbrace{(d_{k+1} - d_k)}_{D_{k+1}-D_k} \tag{22}$$

The difference $L_{k+1} - L_k$ can also be written in terms of single-EDD lateness values and processing time adjustments due to guests or hostlessness, all of which are expressed in known terms:

$$P_{k+1}-(d_{k+1}-d_k)=L_{k+1,\text{single}}-L_{k,\text{single}}+\begin{cases} \max(P_{k+1} - p_{k+1},0) & x_{k+1,k+1} = 1 \\ -p_{k+1} & x_{k+1,k+1} = 0 \end{cases}. \tag{23}$$

The conditional expression is equivalent to $P'_{k+1} - p_{k+1}(2 - x_{k+1,k+1})$. We can now rewrite (22) for L_{k+1} and arrive at

$$L_{k+1} = \left[L_{k,\text{single}} + \sum_{h \leq k} P'_h - p_h(2 - x_{hh}) \right]$$
$$+ L_{k+1,\text{single}} - L_{k,\text{single}} + P'_{k+1} - p_{k+1}(2 - x_{k+1,k+1}), \tag{24}$$

which, after cancelling out $L_{k,\text{single}}$ terms, becomes

$$L_{k+1} = L_{k+1,\text{single}} + \sum_{h \leq k+1} P'_h - p_h(2 - x_{hh}) \tag{25}$$

and agrees with (21). Since (25) follows from (21), and the latter is true for the base case of $k = 1$, (17) is true for all k. □

Note also that in the case of an empty batch $B_k \in \mathcal{B}^\star$, if $B_{k-1} \notin \mathcal{B}^\star$, $d_k = d_{k-1}$ and $x_{kk} = 0$, so $L_k = L_{k-1}$ as evident from (24); if $B_{k-1} \in \mathcal{B}^\star$, $d_k > d_{k-1}$, and thus $L_k < L_{k-1}$. as $d_k = d_{k-1}$ and $x_{kk} = 0$ if B_k is empty.

3.3 Additional Lazy Constraints

Lazy constraints [22] are also used in the model. Lazy constraints are constraints based on the specific problem instance. Large numbers of them are generated prior to solving, but they are not immediately used in the model. Instead, they are checked against whenever an integral solution is found, and only those that are violated are added to the LP model. In practice, only few of the lazy constraints are used in the solution process. Nevertheless, they can noticeably improve solving time in some cases.

Symmetry-Breaking Rule. This rule creates an explicit, arbitrary preference for certain solutions. Consider two schedules S_1 and S_2. Both schedules contain batches B_h and B_k, both of which are holding their respective host jobs only. Two jobs j and i are now assigned as the only guests to the two batches; furthermore $\max(p_i, p_j) \leq \min(p_h, p_k)$, $\max(s_h, s_k) + \max(s_j, s_i) \leq b$ and $\min(d_j, d_i) \geq \max(d_h, d_k)$. If $j \in B_h$ and $i \in B_k$ in schedule S_1 and vice versa in S_2, then the constraint renders S_2 infeasible.

$$2(4 - x_{hh} - x_{kk} - x_{jh} - x_{jk} - x_{ih} - x_{ik}$$
$$+ \sum_{\substack{g \\ g \neq j \\ g \neq i}} (x_{gh} + x_{gk})) \geq x_{jk} + x_{ih}$$

$$\forall \{j, i \in \mathcal{J}, \\ h, k \in \mathcal{B} \\ \mid h < k < j < i \wedge \\ [p_q \leq p_r \wedge b - s_r \geq s_q \\ \forall q \in \{j, i\}, \\ \forall r \in \{h, k\}]\} \quad (26)$$

The left-hand side of the equation evaluates to zero exactly when the above conditions are met, which in turn disallows the assignment given on the right. For all other job/batch pairings, the left side evaluates to at least two, which places no constraint on the right hand side.

This kind of symmetry-breaking rule can be extended to $m > 2$ batches, with the number of constraints growing combinatorially with m. Since it takes a constant but appreciable time to generate these constraints prior to solving, we have in our trials kept to the simplest variant shown here, and limited their use to problem instances with $n \geq 50$ jobs.

Dominance Rule on Required Assignments. A schedule is not uniquely optimal if a job j is left in its single-EDD batch although there is capacity for it in an earlier batch. This constraint can be expressed logically as: if a job j can be safely assigned to B_k without violating the capacity constraint, then j must be assigned to any earlier batch, or B_k must be empty (or both).

The left side of the above *if-then* statement is written as $(1.0 + b - s_j - \sum_{\substack{i=k \\ i \neq j}}^{n_j} s_i x_{ik})/b$, which evaluates to 1.0 or greater iff s_k plus the sizes of guest jobs in k sum to less than $b - s_j$. The constraint is written as follows:

$$2 - x_{jj} - x_{kk} \geq \left(1.0 + b - s_j - \sum_{\substack{i=k \\ i \neq j}}^{n_j} s_i x_{ik} \right) /b \quad \begin{array}{c} \forall \{j \in \mathcal{J}, k \in \mathcal{B} \\ \mid j > k \wedge p_k \geq p_j \\ \wedge s_k + s_j \leq b\} \end{array} \quad (27)$$

As with the rule above, we have found that only more difficult problems with $n \geq 50$ benefit from these constraints.

4   A New MIP Model

The full novel MIP model we are proposing is defined in Model 4.1.

Constraints (29) and (30) are uniqueness and capacity constraints: batches have to remain within capacity b, and every job can only occupy one batch. Constraints (31) and (32) define the value of P_k' for every batch k as the longest p of all jobs in k, but at least p_k. This is required in (34), which follows the explanation above. Constraints (33) ensure that no job is moved into a hostless

Min. $L_{\max}$ (28)

s.t. $\sum_k x_{jk} = 1$ $\forall j \in \mathcal{J}$ (29)

$\sum_j s_j x_{jk} \leq b$ $\forall k \in \mathcal{B}$ (30)

$P'_k \geq p_j x_{jk}$ $\forall \{j \in \mathcal{J}, k \in \mathcal{B} | j \geq k\}$ (31)

$P'_k \geq p_j$ $\forall \{j \in \mathcal{J}, k \in \mathcal{B} | j = k\}$ (32)

$x_{kk} \geq x_{jk}$ $\forall \{j \in \mathcal{J}, k \in \mathcal{B} | j > k\}$ (33)

$L_{\max} \geq L_{k,\text{single}} + \sum_{h \leq k} P'_h - p_h(2 - x_{hh})$ $\forall k \in \mathcal{B}^*$ (34)

$x_{jk} = 0$ $\forall \{j \in \mathcal{J}, k \in \mathcal{B} | j < k\}$ (35)

$(*)$ $\begin{aligned} 2(4 &- x_{hh} - x_{kk} - x_{jh} - x_{jk} - x_{ih} - x_{ik} \\ &+ \sum_{\substack{g \\ g \neq j \\ g \neq i}} (x_{gh} + x_{gk})) \geq x_{jk} + x_{ih} \end{aligned}$ $\begin{aligned} &\forall \{j, i \in \mathcal{J}, \\ &\quad h, k \in \mathcal{B} \\ &\mid h < k < j < i \wedge \\ &[p_q \leq p_r \wedge b - s_r \geq s_q \\ &\quad \forall q \in \{j, i\}, \\ &\quad \forall r \in \{h, k\}]\} \end{aligned}$ (36)

$(*)$ $2 - x_{jj} - x_{kk} \geq \left(1.0 + b - s_j - \sum_{\substack{i=k \\ i \neq j}}^{n_j} s_i x_{ik}\right) / b$ $\begin{aligned} &\forall \{j \in \mathcal{J}, k \in \mathcal{B} \\ &\mid j > k \wedge p_k \geq p_j \\ &\quad \wedge s_k + s_j \leq b\} \end{aligned}$ (37)

Model 4.1. The new MIP model. Constraints marked $(*)$ are lazy constraints.

batch, i.e. in order to move job j into batch k ($x_{jk} = 1$), job k must still be in batch k ($x_{kk} = 1$). Constraints (35) implement the requirement that jobs are only moved into earlier batches. Constraints (36) and (37) implement the additional lazy constraints described above.

5   Empirical Comparison

We empirically compared the performance of the CP model by Malapert et al. and Model 4.1. Both models were run on 120 benchmark instances as in Malapert et al. (i.e. 40 instances of each $n_j = \{20, 50, 75\}$). The benchmarks are generated as specified by Daste [11], with a capacity of $b = 10$ and values for p_j, s_j and d_j distributed as follows: $p_j = U[1, 99]$, $s_j = U[1, 10]$, and $d_j = U[0, 0.1] \cdot \tilde{C}_{\max} + U[1, 3] \cdot p_j$. $U[a, b]$ is a uniform distribution between a and b, and $\tilde{C}_{\max} = \frac{1}{bn} \cdot \left(\sum_{j=1}^{n_j} s_j \cdot \sum_{j=1}^{n_j} p_j\right)$ is an approximation of the time required to process all jobs.

Table 1. Summary of empirical results. Values are geometric means for solving time and arithmetic means for absolute gaps. No relative gaps are given due to negative lower bounds.

n_j	optimal soln. found by	instances	solving time [s]		absolute gap	
			CP	MIP	CP	MIP
20	both models	40/40	0.42	0.04	0	0
50	both models	40/40	5.67	4.16	0	0
75	both models	22/40	49.30	52.88	0	0
	CP model only	0/40	—	—	—	—
	MIP model only	13/40	> 3600	139.86	323.46	0
	neither model	5/40	> 3600	> 3600	310.40	25.00

Table 2. Average numbers of variables and constraints in reference and improved MIP models before and after processing by CPLEX's presolve routines

Mean (120 instances)	Reference MIP		Improved MIP		Reduction	
	Rows	Cols	Rows	Cols	Rows	Cols
Before presolve	7415.14	3033.81	4291.48	2882.76	−42.1%	−5.0%
After presolve	2209.34	1513.06	754.57	687.13	−65.8%	−54.6%
Reduction	−70.2%	−50.1%	−82.4%	−76.1%	—	—

The MIP benchmarks were run using CPLEX 12.5 [23] on an Intel i7 Q740 CPU (1.73 GHz) and 8 GB RAM in single-thread mode, with CPLEX parameters `Probe = Aggressive` and `MIPEmphasis = Optimality` (the latter for $n = 20$ only). The CP was implemented using the Choco solver library [24] and run on the same machine using the same problem instances.[2] Solving was aborted after a time of 3600 seconds (1 hour).

The reference MIP model solves fewer than a third of the instances within the time limit. The branch-and-price model [11] is reported to perform considerably worse than CP [4]. Therefore, neither of the two is included here.

5.1   Results

The overview in Table 1 shows aggregated results that demonstrate the performance and robustness of our new model. As shown in Figure 3, our MIP model performs better overall on instances with $n_j = 20$ and $n_j = 75$, while MIP and CP perform similarly well on intermediate problems ($n_j = 50$).

Wherever an optimal solution was not found, the improved MIP model achieved a significantly better solution quality: out of the 40 instances with $n_j = 75$, 22 were solved to optimality by both CP and MIP, 13 were solved to optimality by the MIP

---

[2] The authors would like to extend a warm thank-you to Arnaud Malapert for both providing his code and helping us run it.

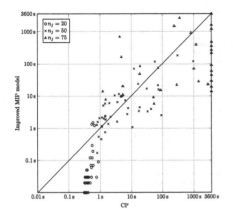

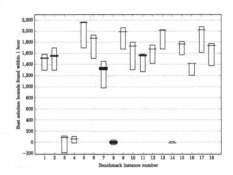

Fig. 3. Performance comparison over 120 instances, each represented by one data point. Horizontal/vertical coordinates correspond to solving time by CP model and improved MIP model, respectively. Note that 18 instances were not solved to optimality within an hour by either the CP model or both models.

Fig. 4. Comparison of solution quality for the 18 instances that were not solved to optimality within an hour by either the CP model or both models. White bars represent the LB-UB gap achieved by the CP model, black bars the LB-UB gap achieved by the improved MIP model (straight line where solved to optimality).

only, and 5 were solved by neither model within an hour. A comparison of solution quality where no optimal schedule was found confirms the robustness of the improved MIP model: as Figure 4 illustrates, the gap $(\text{UB}(L_{\max})-\text{LB}(L_{\max}))$ is consistently larger in the CP model. This means that even with very difficult problems, our model will often give near-optimal solutions more quickly than Malapert et al.'s CP model.

We further found that the lazy constraints introduced in Section 3.3 did not yield *consistent* benefits across problems; in fact, they doubled and tripled solving times for some instances. On average, however, adding the lazy constraints resulted in a speed gain on the order of about 10%, especially for larger problems $(n \geq 50)$.

6 Discussion

One likely contribution to the new model's performance is its reduced size compared to the reference MIP model: while the reference model required $3n^2 + 8n$ constraints over $n^2 + 3n$ variables, our model uses fewer than $2.5n^2 + 2.5n$ constraints over $n^2 + n$ variables.[3] In addition, CPLEX's presolve methods are more effective on our model (see Table 2), reducing its size further.

Figure 5 shows the evolution of bounds (i.e. best feasible solutions and tightest LP solutions at the nodes) for the three models over the first few seconds. It is

---

[3] The word "fewer" here arises from the problem-specific cardinality of $\mathcal{B}^\star$.

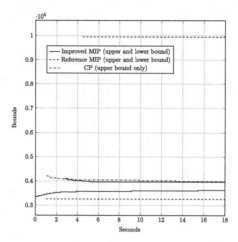

Fig. 5. Evolution of upper and lower bounds. Left cutoffs indicate the approximate mean time at which the respective bound was first found

based on the logs for the 40 instances with $n = 75$, which contain only irregular timestamps. While the new MIP model is better than the reference model on both the lower and upper bounds, the largest gain appears to derive from the latter, indicating that unlike what is commonly observed, the improved MIP model benefits not from a tighter relaxation but from being more amenable to the solver's primal heuristics.

The upper bound for the improved MIP model matches that of the CP model. CP is often able to find high quality solutions faster than MIP. However, the improved model removes that advantage on our test problems.

Making Moves in MIP Modeling. One of the novelties of the new MIP model, as well as much of its inspiration, is the consideration of *moves* from the canonical single-EDD solution. The effects of the assignment variables on the objective function can be considered discretely, allowing us to reason about them algorithmically even though they constitute a declarative model. The concept of moves is common in local search techniques including Large Neighbourhood Search (LNS) [25] where moves correspond to the removal and re-insertion of jobs from and into the schedule, similar to our reasoning in Section 3.2.

This line of reasoning presents several interesting directions for future work including (i) using the idea of moves from a canonical solution to develop MIP and CP models for other optimization problems and (ii) the derivation of dominance rules to restrict LNS moves on large problems and thereby expand the size of the neighbourhoods that can be explored.

Models vs. Global Constraints. Our results show that the novel MIP model is an improvement over previous approaches, demonstrating that at least in this case, the performance of a specialized global constraint implementation can indeed be matched and exceeded by a comparatively simple mathematical formulation. Mathematical models have the general benefit of being more readily understandable, straightforward to implement and reasonably easy to adapt to new, similar problems.

A global constraint is most valuable when it is the encapsulation of a problem structure that occurs across a number of interesting problem types. It can then be used far beyond its original context. However, with the flexibility to define arbitrary inference operations comes the temptation to develop problem-specific global constraints and to trade the ideal of re-usability for problem solving power. We believe that the collection of global constraints in CP is mature enough that most problem-specific efforts are now best placed on exploring novel ways to exploit problem structure using existing global constraints. To this end, one of our current research efforts is the development of a CP model exploiting the propositions proved in this paper without needing novel global constraints.

7 Conclusion

In this paper, we addressed an existing parallel-batch scheduling problem for which CP is the current state-of-the-art approach. Inspired by the idea of moves from a canonical solution, we proved a number of propositions allowing us to create a novel MIP model that, after presolving, is less than half the size of the previous MIP model. Empirical results demonstrated that, primarily due to the ability to find good feasible solutions quickly, the new MIP model was able to out-perform the existing CP approach over a broad range of problem instances both in terms of finding and proving optimality and in terms of finding high quality solutions when the optimal solution could not be proved.

References

[1] Hooker, J.: A hybrid method for planning and scheduling. Constraints 10, 385–401 (2005)
[2] Beck, J.C., Feng, T.K., Watson, J.P.: Combining constraint programming and local search for job-shop scheduling. INFORMS Journal on Computing 23(1), 1–14 (2011)
[3] Tran, T.T., Beck, J.C.: Logic-based benders decomposition for alternative resource scheduling with sequence-dependent setups. In: Proceedings of the Twentieth European Conference on Artificial Intelligence (ECAI 2012), pp. 774–779 (2012)
[4] Malapert, A., Guéret, C., Rousseau, L.M.: A constraint programming approach for a batch processing problem with non-identical job sizes. European Journal of Operational Research 221, 533–545 (2012)
[5] Schutt, A., Feydy, T., Stuckey, P.J., Wallace, M.: Solving RCPSP/max by lazy clause generation. Journal of Scheduling 16(3), 273–289 (2013)
[6] Baptiste, P., Le Pape, C.: Constraint propagation and decomposition techniques for highly disjunctive and highly cumulative project scheduling problems. Constraints 5(1-2), 119–139 (2000)
[7] Baptiste, P., Le Pape, C., Nuijten, W.: Constraint-based Scheduling. Kluwer Academic Publishers (2001)

[8] Vilím, P.: Edge finding filtering algorithm for discrete cumulative resources in $O(kn \log n)$. In: Gent, I.P. (ed.) CP 2009. LNCS, vol. 5732, pp. 802–816. Springer, Heidelberg (2009)

[9] Freuder, E.C.: In pursuit of the holy grail. Constraints 2, 57–61 (1997)

[10] Heinz, S., Ku, W.-Y., Beck, J.C.: Recent improvements using constraint integer programming for resource allocation and scheduling. In: Gomes, C., Sellmann, M. (eds.) CPAIOR 2013. LNCS, vol. 7874, pp. 12–27. Springer, Heidelberg (2013)

[11] Daste, D., Gueret, C., Lahlou, C.: A branch-and-price algorithm to minimize the maximum lateness on a batch processing machine. In: Proceedings of the 11th International Workshop on Project Management and Scheduling (PMS), Istanbul, Turkey, pp. 64–69 (2008)

[12] Lee, C.Y., Uzsoy, R., Martin-Vega, L.A.: Efficient algorithms for scheduling semi-conductor burn-in operations. Oper. Res. 40(4), 764–775 (1992)

[13] Grossmann, I.E.: Mixed-integer optimization techniques for the design and scheduling of batch processes. Technical Report Paper 203, Carnegie Mellon University Engineering Design Research Center and Department of Chemical Engineering (1992)

[14] Brucker, P., Gladky, A., Hoogeveen, H., Kovalyov, M.Y., Potts, C.N., Tautenhahn, T., van de Velde, S.L.: Scheduling a batching machine. Journal of Scheduling 1(1), 31–54 (1998)

[15] Pinedo, M.L.: Scheduling: Theory, Algorithms, and Systems, 2nd edn. Prentice-Hall (2003)

[16] Shaw, P.: A constraint for bin packing. In: Wallace, M. (ed.) CP 2004. LNCS, vol. 3258, pp. 648–662. Springer, Heidelberg (2004)

[17] Azizoglu, M., Webster, S.: Scheduling a batch processing machine with non-identical job sizes. International Journal of Production Research 38(10), 2173–2184 (2000)

[18] Dupont, L., Dhaenens-Flipo, C.: Minimizing the makespan on a batch machine with non-identical job sizes: An exact procedure. Computers & Operations Research 29(7), 807–819 (2002)

[19] Sabouni, M.Y., Jolai, F.: Optimal methods for batch processing problem with makespan and maximum lateness objectives. Applied Mathematical Modelling 34(2), 314–324 (2010)

[20] Kashan, A.H., Karimi, B., Ghomi, S.M.T.F.: A note on minimizing makespan on a single batch processing machine with nonidentical job sizes. Theoretical Computer Science 410(27-29), 2754–2758 (2009)

[21] Ozturk, O., Espinouse, M.L., Mascolo, M.D., Gouin, A.: Makespan minimisation on parallel batch processing machines with non-identical job sizes and release dates. International Journal of Production Research 50(20), 6022–6035 (2012)

[22] IBM ILOG: User's manual for cplex (2013)

[23] Ilog, I.: Cplex optimization suite 12.5 (2013)

[24] Choco Team: Choco: An open source java constraint programming library. version 2.1.5 (2013)

[25] Shaw, P.: Using constraint programming and local search methods to solve vehicle routing problems. In: Maher, M.J., Puget, J.-F. (eds.) CP 1998. LNCS, vol. 1520, pp. 417–431. Springer, Heidelberg (1998)

Mining (Soft-) Skypatterns Using Dynamic CSP

Willy Ugarte Rojas[1], Patrice Boizumault[1], Samir Loudni[1],
Bruno Crémilleux[1], and Alban Lepailleur[2]

[1] GREYC (CNRS UMR 6072) – University of Caen
Campus II Côte de Nacre, 14000 Caen - France
[2] CERMN (UPRES EA 4258 - FR CNRS 3038 INC3M) – University of Caen
Boulevard Becquerel, 14032 Caen Cedex - France

Abstract. Within the pattern mining area, skypatterns enable to express a user-preference point of view according to a dominance relation. In this paper, we deal with the introduction of softness in the skypattern mining problem. First, we show how softness can provide convenient patterns that would be missed otherwise. Then, thanks to Dynamic CSP, we propose a generic and efficient method to mine skypatterns as well as soft ones. Finally, we show the relevance and the effectiveness of our approach through a case study in chemoinformatics and experiments on UCI benchmarks.

1 Introduction

Discovering useful patterns from data is an important field in data mining for data analysis and is used in a wide range of applications. Many approaches have promoted the use of constraints to focus on the most promising knowledge according to a potential interest given by the final user. As the process usually produces a large number of patterns, a large effort is made to a better understanding of the fragmented information conveyed by the patterns and to produce *pattern sets* i.e. sets of patterns satisfying properties on the whole set of patterns [5]. Using the dominance relation is a recent trend in constraint-based data mining to produce useful pattern sets [19].

Skyline queries [3] enable to express a user-preference point of view according to a *dominance* relation. In a multidimensional space where a preference is defined for each dimension, a point p_i *dominates* another point p_j if p_i is better (i.e., more preferred) than p_j in at least one dimension, and p_i is not worse than p_j on every other dimension. However, while this notion of skylines has been extensively developed and researched for database applications, it has remained unused until recently for data mining purposes. Computing *skylines of patterns* from a database is clearly much harder than computing *skylines* in database applications due to the huge difference between the size of search spaces (we explain this issue in Section 5). The inherent complexity on computing skylines of patterns may explain the very few attempts in this direction.

A pioneering work [17] proposed a technique to extract skyline graphs maximizing two measures. Recently, the notion of skyline queries has been integrated into the constraint-based pattern discovery paradigm to mine skyline patterns (henceforth called *skypatterns*) [19]. As an example, a user may prefer a pattern with a high frequency, large length and a high confidence. In this case, we say that a pattern x_i dominates another pattern x_j if $freq(x_i) \geq freq(x_j)$, $size(x_i) \geq size(x_j)$, $confidence(x_i) \geq$

H. Simonis (Ed.): CPAIOR 2014, LNCS 8451, pp. 71–87, 2014.

$confidence(x_j)$ where at least one strict inequality holds. Given a set of patterns, the skypattern set contains the patterns that are not dominated by any other pattern (we formally introduce the notions in the following sections). Skypatterns are interesting for a twofold reason: they do not require any threshold on the measures and the notion of dominance provides a global interest with semantics easily understood by the user.

Nevertheless, skypatterns queries, like other kinds of queries, suffer from the stringent aspect of the constraint-based framework. Indeed, a pattern satisfies or does not satisfy the constraints. But, what about patterns that slightly miss a constraint? A pattern, close to the frontier of the dominance area, could be interesting although it is not a skypattern. In the paper, we formally introduce soft skypatterns. Note that there are very few works such as [2,21] dealing with softness into the mining process.

The contributions of this paper are the following. First, we introduce the notion of soft skypattern. Second, we propose a flexible and efficient approach to mine skypatterns as well as soft ones thanks to the Dynamic CSP (Constraint Satisfaction Problems) framework [22]. Our proposition benefits from the recent progress on cross-fertilization between data mining and Constraint Programming (CP) [4,9,7]. The common point of all these methods is to model in a declarative way pattern mining as CSP, whose resolution provides the complete set of solutions satisfying all the constraints. We show how the (soft-) skypatterns mining problem can be modeled and solved using dynamic CSPs. A major advantage of the method is to improve the mining step during the process thanks to constraints dynamically posted and stemming from the current set of candidate skypatterns. Moreover, the declarative side of the CP framework leads to a unified framework handling softness in the skypattern problem. Finally, the relevance and the effectiveness of our approach is highlighted through a case study in chemoinformatics for discovering toxicophores and experiments on UCI benchmarks.

This paper is organized as follows. Section 2 presents the context and defines skypatterns. Section 3 introduces soft skypatterns. Section 4 presents our flexible and efficient CP approach to mine skypatterns as well as soft ones. We review some related work in Section 5. Finally, Section 6 reports in depth a case study in chemoinformatics and describes experiments on UCI benchmarks.

2   The Skypattern Mining Problem

2.1   Context and Definitions

Let $\mathscr{I}$ be a set of distinct literals called *items*. An itemset (or pattern) is a non-null subset of $\mathscr{I}$. The language of itemsets corresponds to $\mathscr{L}_{\mathscr{I}} = 2^{\mathscr{I}} \setminus \emptyset$. A transactional dataset $\mathscr{T}$ is a multiset of patterns of $\mathscr{L}_{\mathscr{I}}$. Each pattern (or transaction) is a database entry. Table 1 (left side) presents a transactional dataset $\mathscr{T}$ where each transaction t_i is described by items denoted $A,\ldots,F$. The traditional example is a supermarket database in which each transaction corresponds to a customer and every item in the transaction to a product bought by the customer. An attribute (*price*) is associated to each product (see Table 1, right side).

Constraint-based pattern mining aims at extracting all patterns x of $\mathscr{L}_{\mathscr{I}}$ satisfying a query $q(x)$ (conjunction of constraints) which is usually called *theory* [12]: $Th(q) = \{x \in \mathscr{L}_{\mathscr{I}} \mid q(x) \text{ is true}\}$. A common example is the frequency measure leading to the

Table 1. Transactional dataset $\mathcal{T}$

Trans.	Items
t_1	B E F
t_2	B C D
t_3	A E F
t_4	A B C D E
t_5	B C D E
t_6	B C D E F
t_7	A B C D E F

Item	A	B	C	D	E	F
Price	30	40	10	40	70	55

minimal frequency constraint $(freq(x) \geq \theta)$. The latter provides patterns x having a number of occurrences in the dataset exceeding a given minimal threshold θ. There are other usual measures for a pattern x:

- $size(x)$ is the number of items that pattern x contains.
- $area(x) = freq(x) \times size(x)$.
- $min(x.att)$ (resp. $max(x.att)$) is the smallest (resp. highest) value of the item values of x for attribute att.
- $average(x.att)$ is the average value of the item values of x for attribute att.
- $mean(x) = (min(x.att) + max(x.att))/2$.

Considering the dataset described in Table 1, we have: $freq(BC)=5$, $size(BC)=2$ and $area(BC)=10$. Moreover, $average(BCD.price)=30$ and $mean(BCD.price)=25$.

In many applications, it is highly appropriated to look for contrasts between subsets of transactions, such as toxic and non toxic molecules in chemoinformatics (see Section 6.1). The growth-rate is a well-used contrast measure highlighting patterns whose frequency increases significantly from one subset to another [14]:

Definition 1 (Growth rate). *Let $\mathcal{T}$ be a database partitioned into two subsets $\mathcal{D}_1$ and $\mathcal{D}_2$. The growth rate of a pattern x from $\mathcal{D}_2$ to $\mathcal{D}_1$ is:*

$$m_{gr}(x) = \frac{|\mathcal{D}_2| \times freq(x, \mathcal{D}_1)}{|\mathcal{D}_1| \times freq(x, \mathcal{D}_2)}$$

The collection of patterns contains redundancy w.r.t. measures. Given a measure m, two patterns x_i and x_j are said to be equivalent if $m(x_i) = m(x_j)$. A set of equivalent patterns forms an equivalent class w.r.t. m. The largest element (i.e. the one with the highest number of items) of an equivalence class is called a **closed pattern**. More formally, a pattern x_i is closed w.r.t. m iff $\forall x_j \supsetneq x_i, m(x_j) \neq m(x_i)$. The set of closed patterns is a compact representation of the patterns (i.e we can derive all the patterns with their exact value for m from the closed ones). This definition is straightforwardly extended to a set of measures M.

2.2 Skypatterns

Skypatterns have been recently introduced by [19]. Such patterns enable to express a user-preference point of view according to a dominance relation. Given a set of patterns, the skypattern set contains the patterns that are not dominated by any other pattern.

Definition 2 (Dominance). *Given a set of measures M, a pattern x_i dominates another pattern x_j with respect to M (denoted by $x_i \succ_M x_j$), iff $\forall m \in M, m(x_i) \geq m(x_j)$ and $\exists m \in M, m(x_i) > m(x_j)$.*

Consider the example in Table 1 with $M = \{freq, area\}$. Pattern BCD dominates pattern BC because $freq(BCD) = freq(BC) = 5$ and $area(BCD) > area(BC)$. For $M = \{freq, size, average\}$, pattern BDE dominates pattern BCE because $freq(BDE) = freq(BCE) = 4$, $size(BDE) = size(BCE) = 3$ and $average(BDE.price) > average(BCE.price)$.

Definition 3 (Skypattern operator). *Given a pattern set $P \subseteq \mathcal{L}_{\mathcal{I}}$ and a set of measures M, a skypattern of P with respect to M is a pattern not dominated in P with respect to M. The skypattern operator Sky(P,M) returns all the skypatterns of P with respect to M: $Sky(P,M) = \{x_i \in P \mid \nexists x_j \in P, x_j \succ_M x_i\}$.*

The skypattern mining problem is thus to evaluate the query $Sky(\mathcal{L}_{\mathcal{I}}, M)$. For instance, from the data set in Table 1 and with $M = \{freq, size\}$, $Sky(\mathcal{L}_{\mathcal{I}}, M) = \{ABCDEF, BCDEF, ABCDE, BCDE, BCD, B, E\}$ (see Fig. 1a). The shaded area is called the *forbidden area*, as it cannot contain any skypattern. The other part is called the *dominance area*. The edge of the dominance area (bold line) marks the boundary between them.

3   The Soft Skypattern Mining Problem

This section introduces the notion of softness in the skypattern mining problem. As the skypatterns suffer from the stringent aspect of the constraint-based pattern framework, we propose to capture valuable patterns occurring in the forbidden area (that we call soft skypatterns). We define two kinds of soft skypatterns: the *edge-skypatterns* that belongs to the edge of the dominance area (see Section 3.1) and the *δ-skypatterns* that are close to this edge (see Section 3.2). The key idea is to strengthen the dominance relation in order to soften the notion of non dominated patterns.

3.1   Edge-Skypatterns

Edge-skypatterns are defined according to a dominance relation and a *Sky* operator.

Definition 4 (Strict Dominance). *Given a set of measures M, a pattern x_i strictly dominates a pattern x_j with respect to M (denoted by $x_i \gg_M x_j$), iff $\forall m \in M, m(x_i) > m(x_j)$.*

Definition 5 (Edge-skypattern operator). *Given a pattern set $P \subseteq \mathcal{L}_{\mathcal{I}}$ and a set of measures M, an edge-skypattern of P, with respect to M, is a pattern not strictly dominated in P, with respect to M. The operator Edge-Sky(P,M) returns all the edge-skypatterns of P with respect to M: $Edge\text{-}Sky(P,M) = \{x_i \in P \mid \nexists x_j \in P, x_j \gg_M x_i\}$*

Given a set of measures M, the edge-skypattern mining problem is thus to evaluate the query $Edge\text{-}Sky(P,M)$. Fig. 1a depicts the $28 = 7 + (4+8+3+4+2)$ edge-skypatterns extracted from the example in Table 1 for $M = \{freq, size\}$. Obviously, all edge-skypatterns belong to the edge of the dominance area, and seven of them are skypatterns.

Proposition 1. *For two patterns x_i and x_j, $x_i \gg_M x_j \implies x_i \succ_M x_j$. So, for a pattern set P and a set of measures M, $Sky(P,M) \subseteq Edge\text{-}Sky(P,M)$.*

3.2 δ-Skypatterns

In many cases the user is interested in patterns close to the border of the dominance area because they express a trade-off between the measures. The δ-skypatterns address this issue where δ means a percentage of relaxation allowed by the user. Let $0 < \delta \leq 1$.

Definition 6 (δ-Dominance). *Given a set of measures M, a pattern x_i δ-dominates another pattern x_j w.r.t. M (denoted by $x_i \succ^\delta_M x_j$), iff $\forall m \in M$, $(1 - \delta) \times m(x_i) > m(x_j)$.*

Definition 7 (δ-Skypattern operator). *Given a pattern set $P \subseteq \mathcal{L}_\mathcal{I}$ and a set of measures M, a δ-skypattern of P with respect to M is a pattern not δ-dominated in P with respect to M. The δ-skypattern operator δ-Sky(P,M) returns all the δ-skypatterns of P with respect to M: $\delta\text{-}Sky(P,M) = \{x_i \in P \mid \nexists x_j \in P : x_j \succ^\delta_M x_i\}$.*

The δ-skypattern mining problem is thus to evaluate the query δ-Sky(P,M). There are 38 (28+10) δ-skypatterns extracted from the example in Table 1 for $M=\{freq, size\}$ and δ=0.25. Fig. 1b only depicts the 10 δ-skypatterns that are not edge-skypatterns. Intuitively, the δ-skypatterns are close to the edge of the dominance relation, the value of δ is the maximal relative distance between a skypattern and this border.

Proposition 2. *For two patterns x_i and x_j, $x_i \succ^\delta_M x_j \implies x_i \gg_M x_j$. So, for a pattern set P and a set of measures M, Edge-Sky(P,M) $\subseteq$ δ-Sky(P,M).*

To conclude, given a pattern set $P \subseteq \mathcal{L}_\mathcal{I}$ and a set of measures M, the following inclusions hold: $Sky(P,M) \subseteq Edge\text{-}Sky(P,M) \subseteq \delta\text{-}Sky(P,M)$.

4 Mining (Soft-) Skypatterns Using Dynamic CSP

This section describes how the skypattern and the soft skypattern mining problems can be modeled and solved using Dynamic CSP [22]. The implementation was carried out in Gecode by extending the CP-based pattern extractor developed by [9]. The main idea of our of approach is to improve the mining step during the process thanks to constraints dynamically posted and stemming from the current set of the candidate skypatterns. This process stops when the forbidden area cannot be enlarged. Finally, the completeness of our approach is ensured by the completeness of the CP solver.

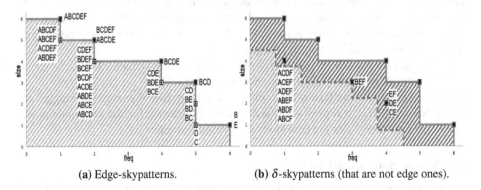

(a) Edge-skypatterns.   (b) δ-skypatterns (that are not edge ones).

Fig. 1. Soft-skypatterns extracted from the example in Table 1

4.1 Dynamic CSP

A Dynamic CSP [22] is a sequence $P_1, P_2, ..., P_n$ of CSP, each one resulting from some changes in the definition of the previous one. These changes may affect every component in the problem definition: variables, domains and constraints. *For our approach, changes are only performed by adding new constraints.* Solving such dynamic CSP involves solving a single CSP with additional constraints posted during search. Each time a new solution is found, new constraints are imposed. Such constraints will survive backtracking and state that next solutions should verify both the current set of constraints and the added ones.

4.2 Mining Skypatterns

Constraints on the dominance relation are dynamically posted during the mining process and softness is easily introduced using such constraints. Variable x will denote the (unknown) skypattern we are looking for. Changes are only performed by adding new constraints. So, we consider the sequence $P_1, P_2, ..., P_n$ of CSP where M is a set of measures, each $P_i = (\{x\}, \mathcal{L}, q_i(x))$ and:

- $q_1(x) = closed_M(x)$
- $q_{i+1}(x) = q_i(x) \wedge \phi_i(x)$ where s_i is the first solution to query $q_i(x)$

First, the constraint $closed_M(x)$ states that x must be a closed pattern w.r.t M, it allows to reduce the number of redundant patterns (see Section 2.1). Then, the constraint $\phi_i(x) \equiv \neg(s_i \succ_M x)$ states that the next solution (which is searched) will not be dominated by s_i. Using a short induction proof, we can easily argue that query $q_{i+1}(x)$ looks for a pattern x that will not be dominated by any of the patterns $s_1, s_2, ..., s_i$.

Each time the first solution s_i to query $q_i(x)$ is found, we dynamically post a new constraint $\phi_i(x)$ leading to reduce the search space. This process stops when we cannot enlarge the forbidden area (i.e. there exits n s.t. query $q_{n+1}(x)$ has no solution). For skypatterns, $\phi_i(x)$ states that $\neg(s_i \succ_M x)$ (see Definition 2):

$$\phi_i(x) \equiv \left(\bigvee_{m \in M} m(s_i) < m(x) \right) \vee \left(\bigwedge_{m \in M} m(s_i) = m(x) \right)$$

But, the n extracted patterns $s_1, s_2, ..., s_n$ are not necessarily all skypatterns. Some of them can only be "intermediate" patterns simply used to enlarge the forbidden area. A post processing step must be performed to filter all candidate patterns s_i that are not skypatterns, i.e. for which there exists s_j ($1 \le i < j \le n$) s.t. s_j dominates s_i. So mining skypatterns is achieved in a two-steps approach:

1. Compute the set $S = \{s_1, s_2, ..., s_n\}$ of candidates using Dynamic CSP.
2. Remove all patterns $s_i \in S$ that are not skypatterns.

While the number of candidates (n) could be very large (the skypattern mining problem is NP-complete), it remains reasonably-sized in practice for the experiments we conducted (see Table 2 for the case study in chemoinformatics).

4.3 Mining Soft Skypatterns

Soft skypatterns are processed exactly the same way as skypatterns. Each kind of soft skypatterns has its own constraint $\phi_i(x)$ according to its relation of dominance.

For edge-skypatterns, $\phi_i(x)$ states that $\neg(s_i \gg_M x)$ (see Definition 4):

$$\phi_i(x) \equiv \bigvee_{m \in M} m(s_i) \leq m(x)$$

For δ-skypatterns, $\phi_i(x)$ states that $\neg(s_i \succ_M^\delta x)$ (see Definition 6):

$$\phi_i(x) \equiv \bigvee_{m \in M} (1 - \delta) \times m(s_i) < m(x)$$

As previously, the n extracted patterns are not necessarily all soft skypatterns. So, a post processing is also required as for skypatterns. Once again, the number of candidates (n) remains reasonably-sized in practice for the experiments we conducted (see Table 2 for the case study in chemoinformatics and Figure 5 for UCI benchmarks).

4.4 Pattern Encoding

Let d be the 0/1 matrix where $\forall t \in \mathscr{T}, \forall i \in \mathscr{I}, (d_{t,i} = 1) \Leftrightarrow (i \in t)$. Pattern variables are set variables represented by their characteristic function with boolean variables. [4,7] model an unknown pattern x and its associated dataset $\mathscr{T}$ by introducing two sets of boolean variables: $\{X_i \mid i \in \mathscr{I}\}$ where $(X_i = 1) \Leftrightarrow (i \in x)$, and $\{T_t \mid t \in \mathscr{T}\}$ where $(T_t = 1) \Leftrightarrow (x \subseteq t)$. Each set of boolean variables aims at representing the characteristic function of the unknown pattern.

The relationship between x and $\mathscr{T}$ is modeled by posting reified constraints stating that, for each transaction $t, (T_t = 1)$ iff t is covered by x:

$$\forall t \in \mathscr{T}, (T_t = 1) \Leftrightarrow \sum_{i \in \mathscr{I}} X_i \times (1 - d_{t,i}) = 0 \tag{1}$$

4.5 Closedness Constraints

Section 2.1 recalls the definition of closed patterns satisfying closedness constraints. Let $M = \{min\}$ and $val(j)$ a function that associates an attribute value to each item j. If item i belongs to x, then its value must be greater than or equal to the min. Conversely, if this value is greater than or equal to the min, i must belong to x (if not, x would not be maximal for inclusion). So, x is a closed pattern for the measure min iff:

$$\forall i \in \mathscr{I}, (X_i = 1) \Leftrightarrow val(i) \geq min\{val(j) \mid j \in x\} \tag{2}$$

Let $M = \{freq\}$, the closedness constraint ensures that a pattern has no superset with the same frequency. So $closed_M(x)$ is modeled using Equation 1 and Equation 3.

$$\forall i \in \mathscr{I}, (X_i = 1) \Leftrightarrow \sum_{t \in \mathscr{T}} T_t \times (1 - d_{t,i}) = 0 \tag{3}$$

There are equivalences between closed patterns according to measures: the closed patterns w.r.t *mean* and *min* are the same and the closed patterns w.r.t *area, growth-rate* and *frequency* are the same [19]. The constraint $closed_M(x)$ states that x must be a closed pattern w.r.t M (the closed patterns w.r.t M gather the closed patterns w.r.t each measure of M i.e. x is closed w.r.t M iff x is closed for at least one measure $m \in M$).

5   Related Work

Computing skylines is a derivation from the maximal vector problem in computational geometry [13], the Pareto frontier [10] and multi-objective optimization. Since its redis-covery within the database community by [3], several methods have been developed for answering skyline queries [15,16,20]. These methods assume that tuples are stored in efficient tree data structures. Alternative approaches have also been proposed to help the user in selecting most significant skylines. For example, [11] measures this significance by means of the number of points dominated by a skyline.

Introducing softness for skylines. [8] have proposed thick skylines to extend the concept of skyline. A thick skyline is either a skyline point p_i, or a point p_j dominated by a skyline point p_i and such that p_j is close to p_i. In this work, the idea of softness is limited to metric semi-balls of radius $\varepsilon > 0$ centered at points p_i, where p_i are skylines.

Computing skypatterns is different from computing skylines. Skyline queries focus on the extraction of tuples of the dataset and assume that all the elements are in the dataset, while the skypattern mining task consists in extracting patterns which are elements of the frontier defined by the given measures. The skypattern problem is clearly harder because the search space for skypatterns is much larger than the search space for skylines: $O(2^{|\mathcal{I}|})$ instead of $O(|\mathcal{T}|)$ for skylines.

Computing skypatterns. [19] have proposed Aetheris, an approach taking benefit of theoretical relationships between pattern condensed representations and skypatterns. Aetheris proceeds in two steps. First, condensed representations of the whole set of patterns (i.e. closed patterns according to the considered set of measures) are extracted. Then, the operator *Sky* (see Definition 3) is applied. Nevertheless, this method can only use a crisp dominance relation. [17] deals with skyline graphs but their technique only maximizes two measures (number of vertices and edge connectivity).

CP for computing the Pareto frontier. [6] has proposed an algorithm that provides the Pareto frontier in a CSP. This algorithm is based on the concept of nogoods and uses spatial data structures (quadtrees) to arrange the set of nogoods. This approach deals for computing skylines and cannot be directly applied to skypatterns. The application is not immediate since several different patterns may correspond to a same point (they all have the same values for the considered measures). As experiments show the practical efficiency of our approach, we have considered that adding [6] to a constraint solver would require an important development time compared to the expected benefits.

6   Experimental Study

First, we report in depth a case study in chemoinformatics by performing a CPU time analysis as well as a qualitative analysis that demonstrates the usefulness and the interest of soft skypatterns (see Section 6.1). Then, using experiments on UCI benchmarks, we show and discuss the practical issues of our approach (see Section 6.2).

Aetheris *and* CP+SKY *(hard version of the skypatterns) produce exactly the same set of skypatterns.* So, the same outputs are compared Section 6.1.2 (Table 2, sky-patterns part) and Section 6.2.1 (Fig 4 and Fig 5). Up to now, there is a single work (Aetheris [19]) to extract skypatterns, no other comparison is possible on skypatterns. Finally, soft skypatterns are completely new and there is no other competitor.

All experiments were conducted on a computer running Linux operating system with a core i3 processor at 2.13 GHz and a RAM of 4 GB. Aetheris was kindly provided by A. Soulet and used in [19]. The implementation of CP+SKY was carried out in Gecode by extending the CP-based patterns extractor developed by [9].

6.1 Case Study: Discovering Toxicophores

A major issue in chemoinformatics is to establish relationships between chemicals and their activity in (eco)toxicity. Chemical fragments[1] which cause toxicity are called *toxicophores* and their discovery is at the core of prediction models in (eco)toxicity [1,18]. The aim of this study, which is part of a larger research collaboration with the CERMN Lab, is to investigate the use of softness for discovering toxicophores.

6.1.1 Experimental Protocol. The dataset is collected from the ECB web site[2]. For each chemical, the chemists associate it with hazard statement codes (HSC) in 3 categories: H400 (very toxic, CL50 $\leq$ 1 mg/L), H401 (toxic, 1 mg/L $<$ CL50 $\leq$ 10 mg/L), and H402 (harmful, 10 mg/L $<$ CL50 $\leq$ 100 mg/L). We focus on the H400 and H402 classes. The dataset $\mathscr{T}$ consists of 567 chemicals (transactions), 372 from the H400 class and 195 from the H402 class. The chemicals are encoded using 1450 frequent closed subgraphs (items) previously extracted[3] with a 1% relative frequency threshold.

In order to discover patterns as candidate toxicophores, we use both measures typically used in contrast mining [14] such as the *growth rate* (see Definition 1) since toxicophores are linked to a classification problem and measures expressing the background knowledge such as the *aromaticity* because chemists consider that this information may yield promising candidate toxicophores. Now, we describe these three measures.

- Growth rate. When a pattern has a frequency which significantly increases from the H402 class to the H400 class, then it stands a potential structural alert related to an excess of the toxicity: if a chemical has, in its structure, fragments that are related to an effect, then it is more likely to be toxic. Emerging patterns embody this natural idea by using the growth-rate measure.

- Frequency. Real-world datasets are often noisy and patterns with low frequency may be artefacts. The minimal frequency constraint ensures that a pattern is representative enough (i.e., the higher the frequency, the better is).

- Aromaticity. Chemists know that the aromaticity is a chemical property that favors toxicity since their metabolites can lead to very reactive species which can interact with biomacromolecules in a harmful way. We compute the aromaticity of a pattern as the mean of the aromaticity of its chemical fragments.

We consider four sets of measures: $M_1=\{growth\text{-}rate, freq\}$, $M_2=\{growth\text{-}rate, aromaticity\}$, $M_3=\{freq, aromaticity\}$ and $M_4=\{growth\text{-}rate, freq, aromaticity\}$. Redundancy is reduced by using closed skypatterns (see Section 4.2). For δ-skypatterns, we consider two values: $\delta = 0.1$ and $\delta = 0.2$. The extracted skypatterns and soft skypatterns are made of molecular fragments. To evaluate the presence of toxicophores, an expert analysis leads to the identification of well-known toxicophores.

[1] A fragment denotes a connected part of a chemical structure having at least one chemical bond.

[2] European Chemicals Bureau: http://echa.europa.eu/

[3] A chemical Ch contains an item A if Ch supports A, and A is a frequent subgraph of $\mathscr{T}$.

Table 2. Skypattern mining on ECB dataset

	Skypatterns				Edge-Skypatterns			δ-Skypatterns					
	CP+SKY		Aetheris		CP+Edge-SKY			CP+δ-SKY					
								$\delta = 0.1$			$\delta = 0.2$		
# of Skypatterns	# of Candidates	CPU-Time	# of closed patterns	CPU-Time	# of Edge-skypatterns	# of Candidates	CPU-Time	# of δ-skypatterns	# of Candidates	CPU-Time	# of δ-skypatterns	# of Candidates	CPU-Time
M_1	8	613	18m:34s	41,887	19m:20s	24	1,746	19m:02s	25	4,204	20m:48s	87	6,253	22m:36s
M_2	5	140	15m:32s	53,201	21m:33s	76	688	17m:51s	354	1,678	18m:14s	1,670	2,816	23m:44s
M_3	2	456	16m:45s	157,911	21m:16s	72	1,726	16m:50s	352	4,070	19m:43s	1,654	6,699	22m:25s
M_4	21	869	17m:49s	12,126	21m:40s	144	3,021	20m:27s	385	6,048	23m:36s	1,724	8,986	30m:14s

6.1.2 Performance Analysis. Table 2 reports, for each set of measures M_i: (i) the number of skypatterns that is the same for both approaches, (ii) for CP+SKY, the number of candidates (see Section 4.2) and the associated CPU-time and (iii) for Aetheris, the number of closed patterns and the associated CPU-time, (iv) the number of edge-skypatterns that are not skypatterns, the number of candidates and the required CPU-time, and (v) the number of δ-skypatterns that are not edge-skypatterns, the number of candidates and the required CPU-time. For each method, reported CPU-times include the two steps.

CP+SKY outperforms Aetheris in terms of CPU-times (see Table 2, skypatterns part). Moreover, the number of candidates generated by our approach remains small compared to the number of closed patterns computed by Aetheris. Aetheris applies the skypattern operator on the whole set of closed patterns (column 4) whereas CP+SKY applies the skypattern operator on a subset of the closed patterns (column 2). That explains why the numbers in column 2 are lower than the numbers in column 4. It shows the interest of the CP approach: thanks to the filtering of dynamically posted constraints, the search space is drastically reduced.

Finally, the number of soft skypatterns remains reasonably small. For edge skypatterns, there is a maximum of 144 patterns, while for δ-skypatterns, there is a maximum of $1,724$ patterns ($\delta = 0.2$).

6.1.3 Qualitative Analysis. In this subsection, we show that soft skypatterns enable (i) to efficiently detect well-known toxicophores emphasized by skypatterns, and (ii) to discover new and interesting toxicophores that would be missed by skypatterns.
- *Growth rate and frequency measures* (M_1). Only 8 skypatterns are found, and 3 well-known toxicophores are emphasized (see Figure 2). Two of them are aromatic compounds, namely the chlorobenzene (p_1) and the phenol rings (p_2). The third one, the organophosphorus moiety (p_3) is a component occurring in numerous pesticides. Soft skypatterns confirm the trends given by skypatterns: the chloro-substituted aromatic rings (e.g. p_4), and the organophosphorus moiety (e.g. p_5) are detected by both the edge-skypatterns and by the δ-skypatterns.
- *Growth rate and aromaticity measures* (M_2). As results for M_2 and M_3 are similar, we only report the qualitative analysis for M_2. Edge-skypatterns leads to the extraction

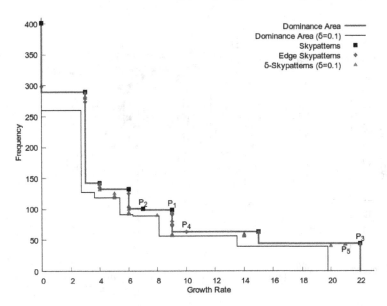

Fig. 2. Analysing the (soft-) skypatterns for M_1

of four new toxicophores: (i) nitrogen aromatic compounds: indole and benzoimidazole, (ii) S-containing aromatic compounds: benzothiophene, (iii) aromatic oxygen compounds: benzofurane, and (iv) polycyclic aromatic hydrocarbons: naphthalene. δ-skypatterns complete the list of the aromatic rings, which were not found with the skypatterns, namely biphenyl.

- Growth rate, frequency and aromaticity measures (M_4). The most interesting results are provided using M_4. Table 3 shows the ratios analysis for the (soft-) skypatterns. Col. 1 provides the name of toxicophores. Col. 2-5 give the number of (soft-) skypatterns containing one complete[4] representative fragment of each toxicophore and, between parentheses, their ratios (# of (soft-) skypatterns containing this toxicophore divided by the total # of (soft-) skypatterns, in bold at the 2nd row). Col. 6 (resp. Col. 7) gives the number of chemicals classified H400 i.e. high toxicity (resp. H402 i.e. harmful) containing at least one representative fragment of the toxicophore. Col. 8-10 show the gains provided by using soft skypatterns for discovering toxicophores (ratio soft skypatterns divided by ratio skypatterns). Bold numbers denote a gain greater than 1 and ∞ means that the toxicophore is only found by soft skypatterns.

21 skypatterns are mined (see Figure 3), and several well-known toxicophores are emphasized: the phenol ring (e_4), the chloro-substituted aromatic ring (e_3), the alkyl-substituted benzene (e_2), and the organophosphorus moiety (P_1). Besides, information dealing with nitrogen aromatic compounds are also extracted (e_1). Table 3 details the repartition of the skypatterns containing only one complete toxicophore compound, according to the toxicophores discussed above. We can observe that very few patterns are extracted.

---

[4] Patterns with only sub-fragments of a toxicophore are not taken into account.

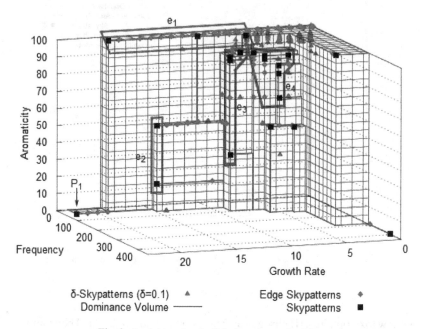

Fig. 3. Analysing the (soft-) skypatterns for M_4

Table 3. Ratio analysis of (soft-)skypattern mining

Chemical	(2)	(3)	(4)	(5)	(6)	(7)	Gain (3)	(4)	(5)
	21	**165**	**550**	**1889**					
Benzene	4 (0.19)	68 (0.41)	322 (0.59)	1373 (0.73)	63.7	18.9	**2.16**	**3.11**	**3.84**
Chlorobenzene	1 (0.05)	2 (0.01)	51 (0.09)	311 (0.16)	22.5	2.5	0.20	**1.80**	**3.20**
Phenol	1 (0.05)	11 (0.07)	32 (0.06)	302 (0.16)	25.2	3.5	**1.40**	**1.20**	**3.20**
Organophosphate **Basic**	2 (0.10)	18 (0.11)	30 (0.05)	40 (0.02)	18.0	2.5	**1.10**	0.50	0.20
Exotic		38 (0.23)	66 (0.12)	112 (0.06)	18.0	2.5	∞	∞	∞
Nitrogen aromatic rings		15 (0.09)	74 (0.13)	175 (0.09)	8.6	2.0	∞	∞	∞
Polycyclic aromatic rings		12 (0.07)	178 (0.32)	302 (0.16)	7.2	3.5	∞	∞	∞
Alkyl-substituted benzene			4 (0.02)	64 (0.12)	649 (0.34)	30.9	11.7	∞	∞	∞
Aniline			15 (0.03)	259 (0.14)	24.7	11.3		∞	∞
Alkyl-substituted aniline				157 (0.08)	12.0	7.1			∞
Chlorophenol				168 (0.09)	9.6	1.5			∞
Alkyl phenyl ether				106 (0.06)	9.9	3.0			∞
Alkyl-substituted phenol				61 (0.03)	9.6	1.5			∞
Dichlorobenzene				59 (0.03)	9.9	1.5			∞

(2) Skypatterns
(3) Edge-Skypatterns
(4) δ-Skypatterns ($\delta = 0.1$)
(5) δ-Skypatterns ($\delta = 0.2$)
(6) coverage rate on H400 (%)
(7) coverage rate on H402 (%)

Soft skypatterns enable to detect more precisely the first four toxicophores (see Table 3). For instance, 41% of edge-skypatterns extracted contain the benzene ring, against 19% for hard skypatterns (gain of 2.16: egde-skypatterns detect 2.16 times more patterns containing this fragment compared to hard ones). This gain reaches about 3.11 (resp. 3.84) for $\delta = 0.1$ (resp. 0.2). The same trends hold for chlorobenzene and phenol rings, where 16% of extracted δ-skypatterns ($\delta = 0.2$) include such fragments, against 5% in the hard case (gain of 3.20). From a chemical point of view, these fragments cover well the H400 molecules (from 18% to 63.7%), as is shown in Col. 6, thus demonstrating the toxic nature of the extracted patterns, particularly in the soft case.

Regarding the aromatic rings previously discussed (gray lines of Table 3), *several new patterns containing these toxicophores are only mined by soft skypatterns. δ-skypatterns* (with δ=0.1) allow to better discover these toxicophores compared to edge-skypatterns (average gain of about 4). *Moreover, several patterns with novel fragments of a great interest are solely detected by δ-skypatterns* (yellow lines in Table 3), particularly with δ=0.2. It is important to note that 22% of these patterns include aromatic amines (12% for aniline and 8% for substituted anilines). These two toxicophores, which cover respectively 24.7% and 12% of molecules classified H400, are very harmful to aquatic organisms. The other toxicophores are extracted by δ-skypatterns with ratios ranging from 3% to 9%.

To conclude, soft skypatterns enable to efficiently detect well-known toxicophores emphasized by skypatterns, and to discover new and interesting toxicophores that would be missed by skypatterns.

6.2 Experiments on UCI Benchmarks

Our experiments on UCI[5] benchmarks thoroughly investigate the behavior on CP+SKY and Aetheris with sets of 4 or 5 measures. We made this choice because the user often handles a limited number of measures when dealing with applications on real-world datasets (see for instance our case study in chemoinformatics in Section 6.1).

Experiments were carried out on 23 various (in terms of dimensions and density) datasets (see Col 1 of Table 4). We considered 5 measures M_6={*freq, max, area, mean, growth-rate*} and 6 sets of measures: M_6 and all the combinations of 4 measures from M_6 (noted M_1, M_2, M_3, M_4 and M_5). Measures using numeric values, like *mean*, were applied on attribute values that were randomly generated within the range $[0..1]$. For each method, reported CPU-times include the two steps.

6.2.1 Mining Skypatterns. Figure 4 shows a scatter plot of CPU-times for CP+SKY and Aetheris. Each point represents a skypattern query for a dataset: its x-value is the CPU-time the CP+SKY method took to mine it, its y-value is the CPU-time of Aetheris. We associate to each dataset a color. Moreover, we only report CPU-times for the 6 datasets requiring more than 30 seconds, either for CP+SKY or Aetheris. For both approaches, CPU times are very small and quite similar on the remaining 17 datasets.

CP+SKY outperforms Aetheris on many datasets (e.g. almost all of the points are in the left part of the plot field of Figure 4). The only exception is the dataset mushroom. This dataset, which is the largest one (both in terms of transactions and items) and with

---

[5] http://www.ics.uci.edu/ mlearn/MLRepository.html

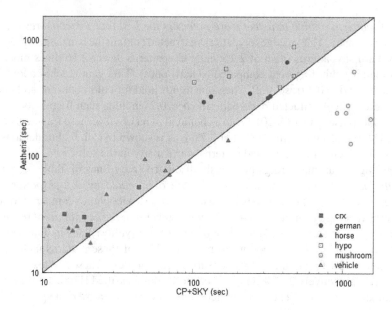

Fig. 4. Comparing CPU times on 6 UCI datasets for $M_1,\ldots,M_6$

the lowest density (around 18%), leads to the extraction of a relatively small number of closed patterns. This greatly promotes Aetheris.

Figure 5 compares, for each set of measures M_i ($1 \leq i \leq 6$), the number of closed patterns for Aetheris with the number of candidates for CP+SKY. We also report the

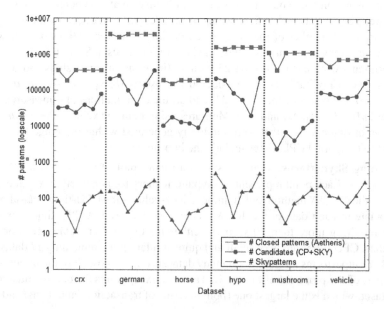

Fig. 5. Comparing # of closed patterns, candidates and skypatterns on 6 datasets

Table 4. Analysis of soft skypattern mining on UCI benchmarks for M_6

Dataset	# items	# transactions	density	CP+Edge-Sky # of Edge-skypatterns	Time (sec)	CP+δ-Sky ($\delta = 5\%$) # of δ-skypatterns	Time (sec)	CP+δ-Sky ($\delta = 10\%$) # of δ-skypatterns	Time (sec)	CP+δ-Sky ($\delta = 15\%$) # of δ-skypatterns	Time (sec)	CP+δ-Sky ($\delta = 20\%$) # of δ-skypatterns	Time (sec)
abalone	28	4,178	0.321	2,634	36	1,373	38	1,432	38	6,303	38	7,256	42
anneal	68	798	0.195	3,162	28	8,184	34	20,242	35	24,029	37	27,214	36
austral	55	690	0.272	11,714	69	34,205	70	68,855	99	69,487	102	70,652	113
breast	43	286	0.231	1,409	1	17	1	1,651	1	2,429	1	2,443	1
cleve	43	303	0.325	14,636	19	4,466	19	30,605	22	30,952	22	50,275	23
cmc	28	1,474	0.357	14,406	31	3,297	32	3,351	32	11,848	33	14,020	33
crx	59	690	0.269	29,068	134	73,627	151	73,707	159	105,344	166	165,782	167
german	76	1,000	0.276	93,087	1,157	170,169	2,614	230,457	2,995	270,435	3,439	290,654	3,483
glass	34	216	0.295	2,296	1	109	1	1,531	1	2,491	1	4,035	1
heart	38	270	0.368	15,563	15	644	16	34,136	16	42,685	18	44,114	18
hepatic	45	155	0.421	15,002	24	6,122	24	45,572	25	50,686	25	60,857	26
horse	75	300	0.235	13,068	54	39,149	60	43,073	66	55,175	68	74,275	71
hypo	47	3,163	0.389	278,625	1,343	104,147	1,387	115,126	1,402	116,654	1,463	117,089	1,487
iris	15	151	0.333	55	1	20	1	27	1	49	1	67	1
lymph	59	142	0.322	8,286	19	49,846	19	59,753	20	62,143	20	65,946	21
mushroom	119	8,124	0.193	21,639	3,241	33,757	3,328	99,852	3,336	129,383	3,407	150,965	3,614
new-thyroid	21	216	0.287	119	1	41	1	137	1	154	1	173	1
page	35	941	0.314	2,675	18	7,136	19	9,714	19	17,387	21	19,094	22
pima	26	768	0.346	1,778	5	518	5	3,439	5	4,308	5	4,358	6
tic-tac-toe	29	259	0.344	6,800	16	4,078	18	18,584	19	20,130	20	22,576	21
vehicle	58	846	0.327	76,732	687	716	689	2,457	751	3,789	782	4,369	787
wine	45	179	0.311	3,155	5	2,490	5	4,422	5	7,507	5	13,407	6
zoo	43	102	0.394	2,254	2	3,361	2	4,829	2	7,724	2	8,986	2

number of skypatterns. The number of candidates generated by our approach remains very small (some thousands) compared to the huge number of closed patterns computed by `Aetheris` (about millions). Finally, the number of skypatterns remains small.

6.2.2 Mining Soft Skypatterns. This section shows the feasibility of mining soft skypatterns on UCI Benchmarks (for these experiments, parameter δ has been set to $\{0.05, 0.1, 0.15, 0.2\}$). As our proposal is the only approach able to mine soft skypatterns, it is no longer compared with `Aetheris`. Table 4 reports, for each dataset (i) the number of edge-skypatterns that are not (hard) skypatterns, the number of candidates and the required CPU-time, (ii) for δ in $\{0.05, 0.1, 0.15, 0.2\}$ the number for δ-skypatterns that are not edge-skypatterns, the number of candidates and the required CPU-time. Even if the number of soft patterns increases with δ, our approach remains efficient: there are only 8 experiments out of 115 requiring more than 3,000 seconds.

7 Conclusion

We have introduced soft skypatterns and proposed a flexible and efficient approach to mine skypatterns as well as soft ones thanks to Dynamic CSP. The relevance and the effectiveness of our approach have been highlighted through experiments on UCI datasets and a case study in chemoinformatics.

In the future, we would like to continue to investigate where the CP approach leads to new insights into the underlying data mining problems. Thanks to CP, we would particularly like to introduce softness within other tasks such as clustering, and study the contribution of soft skypatterns for recommendation.

Acknowledgments. We would like to thank Arnaud Soulet for providing the Aetheris program and his highly valuable comments. This work is partly supported by the ANR (French Research National Agency) funded project FiCoLoFo ANR-10-BLA-0214.

References

1. Bajorath, J., Auer, J.: Emerging chemical patterns: A new methodology for molecular classification and compound selection. J. of Chemical Information and Modeling 46, 2502–2514 (2006)
2. Bistarelli, S., Bonchi, F.: Soft constraint based pattern mining. Data Knowl. Eng. 62(1), 118–137 (2007)
3. Börzönyi, S., Kossmann, D., Stocker, K.: The skyline operator. In: 17th Int. Conf. on Data Engineering, pp. 421–430. Springer (2001)
4. De Raedt, L., Guns, T., Nijssen, S.: Constraint programming for itemset mining. In: KDD 2008, pp. 204–212. ACM (2008)
5. De Raedt, L., Zimmermann, A.: Constraint-based pattern set mining. In: 7th SIAM International Conference on Data Mining. SIAM (2007)
6. Gavanelli, M.: An algorithm for multi-criteria optimization in csps. In: van Harmelen, F. (ed.) ECAI, pp. 136–140. IOS Press (2002)
7. Guns, T., Nijssen, S., De Raedt, L.: Itemset mining: A constraint programming perspective. Artif. Intell. 175(12-13), 1951–1983 (2011)
8. Jin, W., Han, J., Ester, M.: Mining thick skylines over large databases. In: Boulicaut, J.-F., Esposito, F., Giannotti, F., Pedreschi, D. (eds.) PKDD 2004. LNCS (LNAI), vol. 3202, pp. 255–266. Springer, Heidelberg (2004)
9. Khiari, M., Boizumault, P., Crémilleux, B.: Constraint programming for mining n-ary patterns. In: Cohen, D. (ed.) CP 2010. LNCS, vol. 6308, pp. 552–567. Springer, Heidelberg (2010)
10. Kung, H.T., Luccio, F., Preparata, F.P.: On finding the maxima of a set of vectors. Journal of ACM 22(4), 469–476 (1975)
11. Lin, X., Yuan, Y., Zhang, Q., Zhang, Y.: Selecting stars: The k most representative skyline operator. In: ICDE 2007, pp. 86–95 (2007)
12. Mannila, H., Toivonen, H.: Levelwise search and borders of theories in knowledge discovery. Data Mining and K. Discovery 1(3), 241–258 (1997)
13. Matousek, J.: Computing dominances in E^n. Inf. Process. Lett. 38(5), 277–278 (1991)
14. Kralj Novak, P., Lavrac, N., Webb, G.I.: Supervised descriptive rule discovery: A unifying survey of contrast set, emerging pattern and subgroup mining. Journal of Machine Learning Research 10, 377–403 (2009)
15. Papadias, D., Tao, Y., Fu, G., Seeger, B.: Progressive skyline computation in database systems. ACM Trans. Database Syst. 30(1), 41–82 (2005)
16. Papadias, D., Yiu, M., Mamoulis, N., Tao, Y.: Nearest neighbor queries in network databases. In: Encyclopedia of GIS, pp. 772–776 (2008)

17. Papadopoulos, A.N., Lyritsis, A., Manolopoulos, Y.: Skygraph: an algorithm for important subgraph discovery in relational graphs. Data Min. Knowl. Discov. 17(1), 57–76 (2008)
18. Poezevara, G., Cuissart, B., Crémilleux, B.: Extracting and summarizing the frequent emerging graph patterns from a dataset of graphs. J. Intell. Inf. Syst. 37(3), 333–353 (2011)
19. Soulet, A., Raïssi, C., Plantevit, M., Crémilleux, B.: Mining dominant patterns in the sky. In: ICDM, pp. 655–664 (2011)
20. Tan, K.-L., Eng, P.-K., Ooi, B.C.: Efficient progressive skyline computation. In: VLDB, pp. 301–310 (2001)
21. Ugarte, W., Boizumault, P., Loudni, S., Crémilleux, B.: Soft threshold constraints for pattern mining. In: Discovery Science, pp. 313–327 (2012)
22. Verfaillie, G., Jussien, N.: Constraint solving in uncertain and dynamic environments: A survey. Constraints 10(3), 253–281 (2005)

Modelling with Option Types in MiniZinc*

Christopher Mears[3], Andreas Schutt[1,2], Peter J. Stuckey[1,2], Guido Tack[1,3], Kim Marriott[1,3], and Mark Wallace[1,3]

[1] National ICT Australia (NICTA)
[2] University of Melbourne, Victoria, Australia
[3] Faculty of IT, Monash University, Australia

Abstract. Option types are a powerful abstraction that allows the concise modelling of combinatorial problems where some decisions are relevant only if other decisions are made. They have a wide variety of uses: for example in modelling optional tasks in scheduling, or exceptions to a usual rule. Option types represent objects which may or may not exist in the constraint problem being modelled, and can take an ordinary value or a special value $\top$ indicating they are *absent*. The key property of variables of option types is that if they take the value $\top$ then the constraints they appear in should act as if the variable was not in the original definition. In this paper, we explore the different ways that basic constraints can be extended to handle option types, and we show that extensions of global constraints to option types cover existing and common variants of these global constraints. We demonstrate how we have added option types to the constraint modelling language MINIZINC. Constraints over variables of option types can either be handled by transformation into regular variables without extending the requirements on underlying solvers, or they can be passed directly to solvers that support them natively.

1 Introduction

A common feature of complex combinatorial models is that some decisions are only relevant if other decisions are made. Hence some part of the model may be irrelevant dependent on decisions in another part. This is typically modelled by requiring that the "irrelevant decisions" are fixed to some default value, and none of the constraints about the irrelevant decisions are enforced. While it is certainly possible to express such models in traditional modelling languages, it is neither concise nor straightforward.

Many programming languages, including ML, Haskell and Scala, provide an *option type*, to wrap an arbitrary type with an additional "None" value to indicate that the returned value is not meaningful. This paper introduces an extension to the MINIZINC modelling language [14] that uses a similar approach to indicate that a value is irrelevant. We add support for option types in MINIZINC, that is types extended with an additional value $\top$ indicating that a decision was irrelevant. The meaning of constraints and functions on normal types is lifted to option types to create the desired behaviour that variables taking the value $\top$ do not constrain the rest of the model.

---

* NICTA is funded by the Australian Government as represented by the Department of Broadband, Communications and the Digital Economy and the Australian Research Council. The first author was sponsored by the Australian Research Council grant DP110102258.

H. Simonis (Ed.): CPAIOR 2014, LNCS 8451, pp. 88–103, 2014.

Variables of option type, or *optional variables*, are used to model objects (or relationships) that may or may not exist in the problem being modelled. Such a variable may take a value like an ordinary variable, or have an *absent* value. The most common use of optional variables in constraint programming is in modelling the start times of *optional tasks*, tasks in a scheduling problem that may or may not occur.

Example 1. In a flexible job shop scheduling problem a task t can be performed on one of k machines, with a duration d_{tl} if it runs on machine l. A common model for this is to model a single task t with start time S_t and (variable) duration D_t, as well as k optional tasks with optional start times O_{tl} and (fixed) duration d_{tl}. If task t runs on machine l then $O_{tl} = S_t$, $D_t = d_{tl}$ and $O_{tj} = \top, 1 \le j \ne l \le k$.

Optional variables are also used to model "exceptional circumstances".

Example 2. Consider an assignment problem where each of m workers should be assigned to one of n tasks, where there are too many workers for the tasks ($m > n$). A common way of modelling this in constraint programming is using the global constraint `alldifferent_except_0([w_1,...,w_m])` where worker assignments $w_i, 1 \le i \le m$ take values in the range $0..n$, where 0 represents that the worker is not assigned to any task. With optional variables this is modelled differently: each variable w_i is an optional variable with domain $1..n$ and $\top$. The $\top$ plays exactly the same role as 0 in the traditional approach, but the constraint is now simply `alldifferent([w_1,...,w_m])` applied to optional variables w_i. Lifting this constraint to work on optional variables will automatically provide the desired behaviour.

Some of the global constraints in the global constraint catalogue [1] in fact arise from the common occurrence of modelling with optional variables. We believe the global constraint catalogue could be simplified by the addition of optional variables and automatic lifting of constraints to optional variables. Optional variables can also make some forms of modelling easier even if they are not required.

Example 3. Consider an assignment problem with m workers and n tasks, but this time $n > m$. If we require that workers who work on tasks adjacent in the list are compatible, we can model this by

$$\forall 1 \le j \le n - 1. \ \forall 1 \le i, i' \le m. \ w_i = j \wedge w_{i'} = j+1 \to compatible[i, i'] \ ,$$

but this creates a very large and complex set of constraints. Instead we can model this using an `inverse` global constraint, using optional variables. Let $t_j = i$ be the optional variable representing that task j is worked on by worker i, but if no one does the task then $t_j = \top$. We can then model the above constraint using

$$inverse(w,t) \wedge \forall 1 \le j \le n-1. \ compatible[t_j, t_{j+1}] \ .$$

If $t_j = \top$ or $t_{j+1} = \top$ then the constraint $compatible[t_j, t_{j+1}]$ will automatically hold.

In the remainder of this paper we formally define option types, and how we can automatically lift the meaning of constraints and functions on ordinary types to their option type extension. We show how with option types we can concisely model common modelling idioms, and extend the expressiveness of the modelling language. The contributions of this paper are:

- A theory for extending a constraint definition over standard types to option types (Section 2.1).
- A standard approach to extending functions from standard types to option types (Section 2.2).
- An extension to MINIZINC to support option types (Section 3).
- An extension of the comprehension syntax of MINIZINC to allow iteration over variable sets and variable conditions (where clauses) using option types (Section 3.2).
- An approach to automatically translating models with option types into models without option types (Section 4). This has the advantage of providing support for the new modelling features for all solvers supporting FLATZINC without changes.
- The ability for solvers to specify their support for constraints with option types, so that models using them can make use of efficient variable implementations and propagation algorithms, where the solvers support them (Section 4.4).

2   The Logic of Option Types

Option types are defined using the type constructor opt which maps a type to an option type. If T is the set of possible values (excluding $\top$) then opt $T = T \cup \{\top\}$.

Modelling for constraint programming (CP) consists of defining relations over decision variables and functionally defined terms on these decision variables. To extend modelling to include option types, we must define how to interpret relations and functions on option types. To be useful for modelling, these definitions should have meanings that are natural to the modeller.

2.1   Lifting Constraints to Option Types

The key requirement for easy use of option types is for the modeller to clearly understand the meaning of constraints involving option types. Here we show how to automatically lift an existing constraint definition that uses standard types to be one defined on option types. The lifting is meant to reflect the implicit understanding that *"if an option type variable x takes the value $\top$ each constraint involving x should act as if x was not part of its definition."*

Projection Interpretation. Converting the implicit understanding into a formal definition, which we denote the *projection interpretation*, leads to the following:

Let $c(x_1,\ldots,x_n)$ be a constraint requiring a non-option variable of type T in the ith position, then if x_i is of type opt T, the *extended version of the constraint*, $c^{e:\{x_i\}}$ for the ith argument is

$$c^{e:\{x_i\}}(x_1,\ldots,x_n) \Leftrightarrow (x_i \neq \top \wedge c(x_1,\ldots,x_n)) \vee (x_i = \top \wedge \exists x_i.c(x_1,\ldots,x_n)) \ .$$

Thus either the variable x_i takes a non-$\top$ value and the constraint acts as usual, or the variable takes the $\top$ value and the constraint acts as if x_i was projected out.

We can extend this to lift arbitrary numbers of variables as follows. Define $c^{e:S}$ where $S \subseteq vars(c)$ as c when $S = \emptyset$ and $(c^{e:S'})^{e:\{x\}}$ when $S = S' \cup \{x\}$ otherwise. The meaning of the lifting is independent of order; e.g.:

$$c^{e:\{x\} \, e:\{y\}}(x,y,z) = c^{e:\{y\} \, e:\{x\}}(x,y,z)$$

Note that while the lifting operation does distribute over disjunction, it does not distribute over conjunction.

Example 4. When $y = \top$ the constraint $c^{e:\{y\}}(x,y,z)$ where $c(x,y,z) = x \leq y \wedge y \leq z$, is equivalent to $x \leq z$, whereas $x \leq^{e:\{y\}} y \wedge y \leq^{e:\{y\}} z$ is equivalent to *true*.

Note that pushing negations into relations is not a valid transformation with option types.

Example 5. When $y = \top$ the expression $\neg(x =^{e:\{y\}} y)$ is equivalent to *false*, while $c_2^{e:\{y\}}(x,y)$ where $c_2(x,y) \equiv x \neq y$, is equivalent to *true*.

However, this does not further restrict MiniZinc: pushing negations is already invalid for MiniZinc given the relational semantics [6] treatment of partial functions.

Since we will usually lift all arguments of a constraint to option types, we define $c^p(x_1,\ldots,x_n) = c^{e:\{x_1,\ldots,x_n\}}(x_1,\ldots,x_n)$. The lifted version of equality $x =^p y$ defines a *weak equality*, denoted $x =_w y$, which holds if x equals y or either x or y is $\top$. In order to give names to expressions of option type and support substitutivity we also need the *strong equality*, $x =_s y$ (or just $x = y$), which holds if x and y are identical, e.g. it is false if x takes value $\top$ and y takes a value different from $\top$.

Example 6. Given a constraint $O_{ij} + d_{ij} \leq S_j$ between an optional variable O_{ij} and standard variable S_j, then if $O_{ij} = \top$ the constraint is automatically satisfied. Note that it is important to disregard the possible values of the optional variable which are not $\top$ when determining the meaning. If O_{ij} had a domain of $0..10 \cup \{\top\}$ and duration $d_{ij} = 4$ then we do not want to have this constraint force $S_j \geq 4$ even when $O_{ij} = \top$. The projection interpretation does not constrain S_j since O_{ij} can take value $-\infty$ (or any value not larger than $l - 4$ where l is the lowest value in the domain of S_j).

Compression Interpretation. An alternate intuition for extending constraints to option types exists for constraints with an n-ary array argument. In this alternate interpretation, the *compression interpretation*, we treat the constraint as if $\top$ values had been removed from the array argument.

Example 7. Given a constraint `alldifferent` on array of optional variables $W = \{w_i \mid 1 \leq i \leq m\}$, then under the assumption that $w_1 = w_2 = \cdots = w_k = \top, k < m$ then the compression interpretation of $\text{alldifferent}^c([\top,\ldots,\top,w_{k+1}\ldots,w_m])$ is the constraint $\text{alldifferent}([w_{k+1},\ldots,w_m])$. Note that for `alldifferent` the compression interpretation and the projection interpretation agree, since $\text{alldifferent}^{e:W}(w)$ is equivalent to $\text{alldifferent}([w_{k+1},\ldots,w_m])$ when $w_1 = w_2 = \cdots = w_k = \top, k < m$.

While the projection and compression interpretation often agree, there are constraints where they do not.

Example 8. Consider the constraint $\texttt{sliding\_sum}(l,u,k,v)$ which requires that $l \leq v[i] + \ldots + v[i+k-1] \leq u, \forall 1 \leq i \leq n-k+1$ where n is the length of the array v. The projection interpretation allows a $\top$ value for $v[j]$ if there is some value that keeps all the sliding sums in the right range. For example $\texttt{sliding\_sum}^\text{p}(1,2,3,[1,1,\top,1,1])$ holds since it is satisfied for $v[3] = 0$, but $\texttt{sliding\_sum}^\text{p}(1,2,3,[1,1,\top,0,\top,1,1])$ does not hold (the only solution would replace each $\top$ by 0 but this fails on the middle 3). Compression eliminates the $\top$ values from the array, changing the values being summed, hence $\texttt{sliding\_sum}^\text{c}(1,2,3,[1,1,\top,1,1])$ equals $\texttt{sliding\_sum}(1,2,3,[1,1,1,1])$ and does not hold, while $\texttt{sliding\_sum}^\text{c}(1,2,3,[1,1,\top,0,\top,1,1])$ equals $\texttt{sliding\_sum}(1,2,3,[1,1,0,1,1])$, and does hold.

The compression interpretation must be modified when we have a tuple of arrays used to represent an array of tuples.

Example 9. Given a constraint $\texttt{cumulative}(s,d,r,L)$ where s are optional variables, and assuming that $s_1 = s_2 = \cdots = s_k = \top, k < n$, then the constraint under the projection interpretation is equivalent to $\texttt{cumulative}([s_{k+1},\ldots,s_n], [d_{k+1},\ldots,d_n], [r_{k+1},\ldots,r_n], L)$; that is, we treat the constraint as if the tasks whose start time is optional did not exist in the constraint. Note that this is equivalent to a compression interpretation which removes the "corresponding" values of other arrays. Under the assumption that the durations d and resources r are also optional, and assuming that $r_1 = \top, d_2 = \top, s_3 = \top, r_4 = \top, d_5 = \top, s_6 = \top$ then the projection interpretation of the constraint is equivalent to $\texttt{cumulative}([s_7,\ldots,s_n], [d_7,\ldots,d_n], [r_7,\ldots,r_n], L)$.

Given that the projection interpretation is clear for any constraint and usually agrees with the compression interpretation when that makes sense, for MINIZINC we define the projection interpretation as the meaning for lifting a constraint to option types.

2.2 Lifting Functions to Option Types

MINIZINC includes many built-in functions (and operators) and also allows users to define their own functions. Lifting existing functions from standard types to option types is also important, so that the modeller can concisely make use of functions in their models with option types. Again we want to have a clear policy so the modeller understands how functions interact with option types.

First note that mapping functions to relations and using the projection-based lifting does not give us what we want. Consider the constraint $plus(x,y,z)$ defined as $x+y=z$. Then $plus^\text{p}(i,\top,j)$ holds for all i and j (the $\top$ takes the value $j-i$), hence we have lost the *functional nature* of the expression $x+y$!

Absorption Lifting. A straightforward extension of functions to option types is to treat the absent value as an absorbing element:

$$x \oplus^{a} y \stackrel{def}{=} \begin{cases} \top & \text{if } x = \top \\ \top & \text{if } y = \top \\ x \oplus y & \text{otherwise} \end{cases}$$

The absent value can be viewed as "contagious". This definition can transform any function $\oplus : A \times B \to C$ to $\oplus^{a} : \text{opt } A \times \text{opt } B \to \text{opt } C$.

Identity Lifting. The intended meaning of the absent value is that it should be ignored wherever possible. With this in mind, binary operations of the form $\oplus : S \times S \to S$ can be lifted to $\oplus^{i} : \text{opt } S \times \text{opt } S \to \text{opt } S$ by the definition:

$$x \oplus^{i} y \stackrel{def}{=} \begin{cases} y & \text{if } x = \top \\ x & \text{if } y = \top \\ x \oplus y & \text{otherwise} \end{cases}$$

For these operations the absent value acts as the identity: for $+$ it is zero, for $\wedge$ it is *true*, and so on. When both values are absent, the result of the operation is absent. This definition is the natural conversion from the semigroup $(S, \oplus)$ to the monoid $(S \cup \{\top\}, \oplus^{i})$. Note we can still use identity lifting for operations like *min* even though its identity element $(+\infty)$ is not in the usual domain of integer variables.

Arithmetic subtraction and division are not associative, and therefore do not form semigroups. The above definition does not make sense for these operators since they do not have left-identities. However these operators do have right-identities, and so identity-lifting can be extended to this case by defining:

$$x \oplus^{r} y \stackrel{def}{=} \begin{cases} \top & \text{if } x = \top \\ x & \text{if } y = \top \\ x \oplus y & \text{otherwise} \end{cases}$$

Effectively when we lift a function then an absent value in a position where it has no identity element gives an absent result. For example $3 - \top = 3$ since $\top$ acts as 0, but $\top - 3 = \top$ since there is no identity in this position. Note that the result is absent if and only if x is absent.

Unary Operators and Functions. Unary operators and functions $\oplus : S \to S$ can be lifted naturally to $\oplus^{a} : \text{opt } S \to \text{opt } S$. Absorption lifting is the only thing that makes sense for a unary function.

Boolean Operations. Objects of type `opt bool`, optional Booleans, are not common when modelling with optional variables, since usually optionality is captured within lifted constraints (which only take two-valued truth values). But we can directly create objects of type `opt bool`. For the Boolean operations $\wedge$ and $\vee$ we use identity lifting, and negation $(\neg)$ uses absorption lifting like any other unary operator. This defines a three-valued logic where an absent value has the effect that it neither contributes to satisfying a proposition nor hinders it. For example, both $A \wedge^{i} \top$ and $A \vee^{i} \top$ are simply equivalent to A.

This is not a standard three-valued logic, for example while De Morgan's laws still hold, distribution of $\wedge$ over $\vee$, and vice versa do not. Essentially the $\top$ value acts as a context sensitive default value.

Absorption lifting of logical operators corresponds to Kleene's weak (or Bochvar's internal) three-valued logic [10,2]. Unfortunately, absorption lifting does not accord with our intuition as it makes $\top \wedge^a C$ equal to $\top$ rather than C. Note that both of the standard three-valued logics—Łukasiewicz's [18] and Kleene's (strong) three-valued logics—also conflict with our intuition, as they make $\top \wedge true$ equal to $\top$ rather than *true*.

N-ary Functions. The identity-lifted functions permit the easy definition of many useful n-ary functions over optional variables. For example:

$$\Sigma_j x_j = 0 + x_1 + x_2 + \cdots + x_n \qquad \Sigma_j^i x_j = 0 +^i x_1 +^i x_2 +^i \cdots +^i x_n$$
$$\forall j.b_j = true \wedge b_1 \wedge b_2 \wedge \cdots \wedge b_n \qquad \forall^i j.b_j = true \wedge^i b_1 \wedge^i b_2 \wedge^i \cdots \wedge^i b_n$$
$$\exists j.b_j = false \vee b_1 \vee b_2 \vee \cdots \vee b_n \qquad \exists^i j.b_j = false \vee^i b_1 \vee^i b_2 \vee^i \cdots \vee^i b_n$$

In these cases, if all variables x_j or b_j are absent then the result is the default zero, *true* or *false* value. That is, these functions have the type `array of opt` $T \rightarrow T$.

The meaning of an n-ary function where some of the arguments are $\top$ is the function with those arguments omitted, as in the compression interpretation for relations. For example, an absent value in a forall will not cause it to be *false*.

In some cases we may wish for an n-ary function of type `array of opt` $T \rightarrow$ `opt` T, where if all argument values are absent the result is also absent. This is the only choice for functions where there is no default value, such as *minimum*.

$$minimum(X) = \min(x_1, \min(x_2, \ldots \min(x_{n-1}, x_n) \ldots))$$

becomes

$$minimum^i(X) = \min^i(x_1, \min^i(x_2, \ldots \min^i(x_{n-1}, x_n) \ldots))$$

Thankfully most builtin functions in MiniZinc are either unary, binary, or n-ary functions resulting from folding binary functions. It is not necessarily obvious how to extend other functions to option types, and hence we do not propose any lifted meaning for these, although we could suggest absorption lifting as the default.

For some models it is convenient to make use of the absorption-lifted form of a function even if it has an identity-lifted form. This can make defining complex interactions of optional variables more natural.

Example 10. Consider the `span` constraint [11] on optional tasks: $span(s_0, d_0, [s_1, \ldots, s_n], [d_1, \ldots, d_n])$ where s_i are optional start times and d_i are durations. The `span` constraint ensures that task 0 starts at the earliest start time of any task $1 \le i \le n$ and ends at the last end time, and if none occurs then $s_0 = \top$. The start time constraint is captured by $s_0 =_s minimum([s_1, \ldots, s_n])$. If each of $s_1, \ldots, s_n$ are $\top$ then the *minimum* function forces $s_0 = \top$. In contrast the end time constraint is not captured by $s_0 +^i d_0 =_s maximum([s_1 +^i d_1, \ldots, s_n +^i d_n])$ since even absent start times can contribute to the *max* if the corresponding duration is large. Instead we need $s_0 +^a d_0 =_s maximum([s_1 +^a d_1, \ldots, s_n +^a d_n])$ where $+^a$ is the absorption lifted version of the $+$ function. Only the tasks that occur will contribute to the *maximum*.

3 Using Option Types in MiniZinc

Option types are included in MINIZINC with the addition of a new type constructor opt, which maps a type to an option version of the type. An ordinary constraint lifted to option types uses the projection interpretation, and a function lifted to option types uses absorption lifting, except for the binary operators where identity lifting applies (and folds over these operators).

3.1 Basic Modelling with Option Types

Two versions of equality are provided for option types: a strong equality $=_s$, denoted =, which ensures that both sides are identical, and a weak equality $=_w$, denoted ~=, which also holds if either side is $\top$. The actual value $\top$ is represented in models using _ and has polymorphic type opt $T. But we suggest avoiding using it directly, and instead provide two (polymorphic) primitive constraints defined on option types:

```
predicate occurs(var opt $T: x) = not absent(x);
predicate absent(var opt $T: x) = (x = _);
```

where $occurs(x)$ iff $x \neq \top$ and $absent(x)$ iff $x = \top$. Clearly $occurs$ is the negation of $absent$. Both are provided for clarity of modelling.

Example 11. A MINIZINC model for flexible job shop scheduling as discussed in Example 1 is shown below. Note that the optional variables are used to give start times for the version of the tasks starting on each machine. The disjunctive constraint is the version lifted for option types. The alternative constraints are directly defined on option types; they enforce the span constraint (see Example 10) as well as ensuring at most one optional task actually occurs. The redundant cumulative constraint ensures no more than k tasks run at one time. The aim is to minimize the latest end time.

```
int: horizon;                % time horizon
set of int: Time = 0..horizon;
int: n;                      % number of tasks
set of int: Task = 1..n;
int: k;                      % number of machines
set of int: Machine = 1..k;
array[Task,Machine] of int: d;
int: maxd = max([ d[t,m] | t in Task, m in Machine ]);
int: mind = min([ d[t,m] | t in Task, m in Machine ]);
array[Task] of var Time: S;
array[Task] of var mind..maxd: D;
array[Task,Machine] of var opt Time: O;
constraint forall(m in Machine)
          (disjunctive([O[t,m]|t in Task],[d[t,m]|t in Task]));
constraint forall(t in Task)(alternative(S[t],D[t],
            [O[t,m]|m in Machine],[d[t,m]|m in Machine]));
constraint cumulative(S,D,[1|i in Task],k);
solve minimize max(t in Task)(S[t] + D[t]);
```

Example 12. A MINIZINC model for the problem of assigning *m* workers to *n* tasks discussed in Example 2 is shown below. The constraint is simply an `alldifferent` over optional variables.

```
int: n;              % number of tasks
int: m;              % number of workers
array[1..m] of var opt 1..n: w; % task for each worker
constraint alldifferent(w);
solve satisfy;
```

Example 13. Similarly a MINIZINC model for the compatible worker assignment problem discussed in Example 3 is shown below.

```
int: n;              % number of tasks
int: m;              % number of workers
array[1..m,1..m] of bool: compatible;
array[1..m] of var 1..n: w;       % task for each worker
array[1..n] of var opt 1..m: t;   % worker for each task
constraint inverse(w,t);
constraint forall(j in 1..n-1)(compatible[t[j],t[j+1]]);
solve satisfy;
```

Example 14. The `span` constraint of Example 10 can be expressed as shown below, where ~+ is a new operator added to MINIZINC to encode absorption lifted addition.

```
predicate span(var opt Time:s0, var int: d0,
               array[int] of var opt Time:s, array[int] of int:d) =
   s0 = min(s) /\
   (absent(s0) -> d0 = 0) /\
   s0 ~+ d0 = max([s[i] ~+ d[i] | i in index_set(s)]);
```

The first line makes use of *strong* equality to ensure that s_0 is absent if each of the tasks in *s* is absent. The second line of the definition ensures that if the spanning task is absent, then its duration is fixed (to zero). The third line use absorption lifted + (written ~+) to constrain the end time. We can define `alternative` using `span`.

3.2 Extending Comprehension Syntax Using Option Types

A significant advantage of the addition of option types for MINIZINC is that it allows us to increase the expressiveness of comprehensions in the language. At present an array comprehension expression of the form $[f(i) \mid i$ in S where $c(i)]$ requires that S is a fixed set and $c(i)$ is a condition that does not depend on decision variables (has type/mode `par bool`). Once we include option types this can be relaxed.

Suppose $f(i)$ is of type T. We allow $c(i)$ to be dependent on decisions (type/mode `var bool`) by modifying the array comprehension to output an array of `opt` T rather than an array of T, with the interpretation that if $c(i) = false$ then $f(i)$ is replaced by $\top$.

Hence we rewrite $[f(i) \mid i \text{ in } S \text{ where } c(i)]$ as[1]

$$[\,\texttt{if}\ c(i)\ \texttt{then}\ f(i)\ \texttt{else}\ \top\ \texttt{endif} \mid i \text{ in } S\,]$$

Once we have extended the `where` clause to be decision dependent it is straightforward to also support comprehensions where the set S is itself a decision variable. The comprehension $[f(i) \mid i \text{ in } S \text{ where } c(i)]$ where S is a *set decision variable* is equivalent to $[f(i) \mid i \text{ in } \text{ub}(S) \text{ where } i \in S \wedge c(i)]$ where $\text{ub}(S)$ returns an upper bound on the set S. Note that all set variables in MINIZINC are guaranteed to have a known finite upper bound. Generating comprehensions over variable sets can be highly convenient for modelling. The constraint language ESSENCE [7] has a similar feature but only for specific functions like *sum*. In MINIZINC since comprehensions generate arrays that can be then used as arguments to any predicate or function, this is impossible without option types.

Example 15. Consider the Balanced Academic Curriculum problem [4] (problem 30 in CSPlib (www.csplib.org)). The computation of the total load $T[s]$ of a student s, taking courses $C[s]$ where course c has load $L[c]$, is usually defined in MINIZINC as

```
array[Student] of var set of Course: C;
constraint forall(s in Student)
           (T[s] = sum(c in Course)(bool2int(c in C[s])*L[c]));
```

but a more natural formulation is simply

```
array[Student] of var set of Course: C;
constraint forall(s in Student)(T[s] = sum(c in C[s])(L[c]));
```

which can be transformed automatically to

```
array[Student] of var set of Course: C;
constraint forall(s in Student)
           (T[s] = sum(c in Course)
                        (if c in C[s] then L[c] else _ endif));
```

The *sum* is now over optional integers. We can use the same syntax even for a user defined function, e.g. if we are interested in calculating average course load $A[s]$

```
constraint forall(s in Student)(A[s] = average(c in C[s])(L[c]))
   ;
function var float: average(array[int] of var opt int:x) =
   int2float(sum(x)) /
   sum([1.0 | i in index_set(x) where occurs(x[i])]);
```

This also allows us to use option types to express *open global constraints* in a closed world [9], that is global constraints that act on a set of variables which is part of the decisions made, but is bounded. For example if the set decision variable S holds the set of indices of variables x to be made all different, we can express this as

---

[1] To support this we need to extend `if-then-else-endif` in MINIZINC to allow nonparametric tests, or use the alternate translation $[\top, f(i)][\texttt{bool2int}(c(i)) + 1]$ (that is, an array lookup returning $\top$ if $c(i)$ is false and $f(i)$ otherwise).

```
constraint alldifferent([ x[i] | i in S ]);
```

which becomes

```
constraint alldifferent([ if i in S then x[i] else _ endif |
                          i in ub(S) ]);
```

creating a call to lifted `alldifferent` with exactly the right behaviour (variables whose index is not in S are ignored).

4 Implementing Option Types in MiniZinc

Changes to MINIZINC to support option types are surprisingly small. The major change is the addition of the `opt` type constructor to the language and an additional automatic coercion from type T to `opt` T, and the extension of type inference to handle the new type constructor and coercion.

The remainder of the changes are simply adding a library of lifted definitions of predicates and functions to the system. In this library a decomposition is defined for each of the FLATZINC predicates lifted to option types. Similarly the global constraints of MINIZINC are defined in a library lifted to option types.

4.1 Rewriting Option Type Variables

Critically the option type library *translates* away option types so that underlying solvers do not need to support (and indeed never see) option types. The key to translation is replacing a variable x of type `opt` T by two variables: a Boolean o_x encoding $occurs(x)$, and v_x of type T encoding the value of x if it is not $\top$. If the optional variable x does not occur then $o_x = false$ and the value v_x is left unconstrained.

These are encoded using special functions $v_x = deopt(x)$ and $o_x = occurs(x)$. These two functions are the "primitive" operations on option variables. They are used as terms to represent the encoding, and rely on common subexpression elimination to ensure that there is a unique representation for each optional variable x. These functions are extended to arrays of variables in the natural way. Only in the last stage of translation to FLATZINC are these expressions replaced by new FLATZINC variables. Finally, variables on option types are removed if they do not appear in any constraints in the resulting FLATZINC model.

4.2 Lifting Constraints

Given this encoding we can lift constraints automatically according to the projection interpretation. Existential quantification corresponds to introducing fresh variables in a `let` expression. For example, given a predicate p(array[int] of var int: x, var int: y), the MINIZINC compiler could automatically generate the lifted version

```
predicate p(array[int] of var opt int: x, var opt int: y) =
  let { array[index_set(x)] of var int: xx; constraint xx ~= x;
        var int: yy; constraint  yy ~= y; } in p(xx,yy);
```

Although this automatic lifting would be correct, we can often define better versions of lifted constraints that avoid introducing all those temporary variables.

Example 16. The lifted version on integer disequality int_ne can be defined without introducing additional variables as

```
predicate int_ne(var opt int: x, var opt int: y) =
    (occurs(x) /\ occurs(y)) -> (deopt(x) != deopt(y));
```

It simply enforces the disequality if both optional variables occur. Similar definitions exist for all primitive constraints.

4.3   Global Constraints over Option Types

Using the usual MINIZINC rewriting capabilities we can define default decompositions for global constraints over option types. In many cases these can be the same as the regular decomposition.

Example 17. The standard decomposition for alldifferent is given below rewritten for option types by adding opt to the type.

```
predicate alldifferent(array[int] of var opt int: x) =
        forall(i,j in index_set(x) where i < j)(x[i] != x[j]);
```

This gives a correct implementation since the calls to != will be replaced by int_ne over option types, as defined above in Example 16.

For certain global constraints we may be able to do better than simply reusing a standard decomposition.

Example 18. If the underlying solver supports alldifferent_except_0 we can use that to translate alldifferent on optional variables. In the new array *ox*, we shift the values of the original array *x* to be above 0 and map ⊤ for *x* variables to 0.

```
predicate alldifferent(array[int] of var opt int: x) =
  let { int: l = lb_array(x); int: u = ub_array(x);
        array[index_set(x)] of var 0..u-l+1: ox; } in
  alldifferent_except_0(ox) /\
  forall(i in index_set(x))
        ((absent(x[i]) -> ox[i] = 0) /\
         (occurs(x[i]) -> ox[i] = deopt(x[i])-l+1));
```

Example 19. A solver supporting the cumulative constraint could implement the disjunctive constraint by modelling absent tasks as using zero resources:

```
predicate disjunctive(array[int] of var opt int: s,
                      array[int] of int: d) =
    cumulative(deopt(s),d,
               [bool2int(occurs(s[i])) | i in index_set(s)],1);
```

4.4 Native Support for Option Types

To achieve better performance, solvers that natively support some option types can define predicates directly (without decomposition) and they will be passed directly to the FLATZINC model used by the solver. In order to mix the usage of option types where some functions and predicates are natively supported and some are not, we require solvers that natively support option types to natively support the primitive functions $deopt(x)$ and $occurs(x)$.

Example 20. For a solver that natively supports optional tasks we can add this declaration to its globals library to pass the disjunctive constraint directly to the solver:

```
predicate disjunctive(array[int] of var opt int: s,
                      array[int] of int: d);
```

4.5 Different Encodings

An alternative encoding of option integer type variables (see [17] for details) is to replace each optional integer variable x ranging over $b..e$ by a Boolean $o_x = occurs(x)$ (as before) and two integer variables $l_x = lower(x)$ ranging over $b..e+1$ encoding the lower bound on x if it occurs, and $u_x = upper(x)$ ranging over $b-1..e$ encoding the upper bound if it occurs. These are related by $o_x \leftrightarrow l_x = u_x$, $\neg o_x \rightarrow l_x = e+1$, $\neg o_x \rightarrow u_x = b-1$. Propagators are extended to enforce lower bounds on x as lower bounds on l_x, and never enforce a lower bound greater than e. Similarly upper bounds on x are enforced as upper bounds on u_x and never less than $b-1$. The advantage of this representation is that in a CP solver with explanation [15] there are literals encoding optional lower bounds $[l_x \geq d]$ and optional upper bounds $[u_x \leq d]$.

Example 21. Using this alternate encoding we can define the disjunctive constraint to make use of a builtin disjunctive on these variables opt3_disjunctive

```
predicate disjunctive(array[int] of var opt int: s,
                      array[int] of int: d) =
    opt3_disjunctive([occurs(s[i]) | i in index_set(s)],
                     [lower(s[i])  | i in index_set(s)],
                     [upper(s[i])  | i in index_set(s)], d);
```

5   Experiments

In this section, we show that models involving optional variables yield efficient solver-level models, comparable in performance to a manual encoding of optionality. We have extended our MINIZINC compiler to handle option types as described in Sect. 4.

As an experiment, we model the flexible job shop scheduling problem as in Example 11 and solve it using a lazy clause generation solver, which is the state of the art for flexible job shop problems as described in [17]. The alternative constraints are implemented using a decomposition into element constraints. For the disjunctive constraints, we use the three-variable encoding from Sect. 4.5 and compare it with the following simple decomposition:

Table 1. Experimental results

Instance	decomposition			global			hand-written [17]		
	objective	*runtime*	#CP	*objective*	*runtime*	#CP	*objective*	*runtime*	#CP
fattahi/mfjs1	468*	1.60s	8250	468*	1.01s	387	468*	0.97s	388
fattahi/mfjs2	446*	1.53s	10133	446*	1.00s	327	446*	1.01s	330
fattahi/mfjs3	466*	4.49s	24506	466*	3.10s	681	466*	3.37s	697
fattahi/mfjs4	554*	8.84s	24546	554*	7.24s	1024	554*	6.98s	964
fattahi/mfjs5	514*	8.67s	23399	514*	7.30s	881	514*	7.20s	900
fattahi/mfjs6	634*	41.85s	37025	634*	33.70s	3666	634*	32.30s	3474
fattahi/mfjs7	959	—	—	909	—	—	909	—	—
fattahi/mfjs8	1095	—	—	889	—	—	889	—	—
fattahi/mfjs9	1466	—	—	1123	—	—	1123	—	—
fattahi/mfjs10	1609	—	—	1395	—	—	1395	—	—

```
predicate disjunctive(array[int] of var opt int: s,
                      array[int] of int: d) =
  forall (i,j in index_set(s) where i<j)
      ( s[i] ~+ d[i] <= s[j] \/ s[j] ~+ d[j] <= s[i] );
```

Table 1 compares the results of running the `lazyfd` G12 solver [5] on the *same*
MINIZINC model using option types with two alternate definitions for `disjunctive`:
the decomposition and a global using `opt3_disjunctive` from Section 4.5. The
mapping to FLATZINC without option types is managed completely automatically. We
compare against the hand-written MiniZinc model (hand-written) used in [17], which
uses the same global constraint but with a manual encoding of the option types. Unlike [17], we use a fixed search in order to test the difference between the models. We
compare the best objective value found (a * indicates the optimal value was proven),
the runtime (or — for $> 600s$), and the number of choice points (#CP). Comparing the
results shows that the automatic three-variable `disjunctive` decomposition matches
the performance of the hand-written models from [17], the only difference being a slight
variation in the explored search tree. The models using the simple decomposition, which
can be used with any solver even if it does not support the global disjunctive constraint
with optional tasks, scale quite well at least for the smaller examples.

6   Related Work and Conclusion

Modelling dependent decisions by hand is common in constraint programming, and
arguably accounts for many models using implication to control whether constraints are
active or not. Optional tasks [11,12] were explicitly added to the modelling language
OPL to handle the common case of scheduling with optional variables. They correspond
to option tuple types (start, duration, end) rather than optional start times we use here.
When MINIZINC is extended to handle tuple types, we will be able to model optional
tasks in exactly the same manner.

Conditional constraint satisfaction problems (CCSPs) [13,8] are strongly related to
option types. A CCSP is a CSP with control on which variables are *active*; i.e., participate

in a solution. It splits the variables V into an *initial* always active set V_I, and a *possibly* active set V_P, and splits constraints into regular *compatibility constraints* and *activity constraints*, which control which variables need to take a value. Activity constraints can be of the form $c \xrightarrow{incl} v$ meaning if c holds v is active, and $c \xrightarrow{excl} v$ meaning if c holds v is inactive. A compatibility constraint is *relevant* if all of its variables are active; irrelevant constraints are not imposed. We can straightforwardly model CCSPs using option types. Variables in V_P are lifted to option types. A compatibility constraint c is *relevant* if $rel(c) \equiv \wedge_{v \in vars(c) \cap V_P} occurs(v)$, and is modelled as $rel(c) \rightarrow c$. Activity constraints are modelled directly as $rel(c) \wedge c \rightarrow occurs(v)$ for inclusion and $rel(c) \wedge c \rightarrow absent(v)$ for exclusion. Since conditional CSPs build on a traditional CSP framework they don't consider global constraints and complex Boolean and integer expressions and hence they in effect use a very simple form of constraint lifting, *relevance*, which is analogous to absorption lifting. This lifting does not give the behaviour we want for optional tasks for example.

Previous work [16] has attempted to reformulate CCSPs into ordinary CSPs by adding a "null" value to the domains of possibly active variables. This approach was necessary because the constraint solvers used in that work only supported extensional table constraints. Our approach generates implications involving arbitrary constraints, which builds on the support for reification in MINIZINC.

Recently Caballero *et al* [3] built a transformation that allows the user to extend base types of MINIZINC (`float`, `int` and `bool`) with additional values and map them to MINIZINC. This could be used to implement option base types, but the meaning of extended constraints and functions is left to the user to define. The work is very similar in flavor; there the type extension is more general, and the meaning defined by the user, while the translation to base MINIZINC is fixed. Here the extension is fixed—and a strong contribution is to define the meaning of the extension—while the translation to MINIZINC can be redefined.

Option types share some similarities to the treatment of partial functions in the relational semantics [6] adopted by MINIZINC. Partial functions introduce an undefined element $\perp$ which in the relational semantics percolates up to the nearest enclosing Boolean context, where it is treated as *false*. Option types are more complex, the extra value $\top$ percolates up to where there is an identity element, then acts like identity. A four-valued treatment of the semantics of MINIZINC would be ideal, but would not match the reality of the underlying two-valued solving technology.

Option types are a simple yet powerful addition to a modelling language. They allow concise and natural expression of circumstances where some decisions are irrelevant if other decisions are not made. Adding option types to MINIZINC turns out to be surprisingly easy, and also allows us to extend the comprehension syntax. We can use option types to recreate the state-of-the-art solution to flexible job shop scheduling problems.

References

1. Beldiceanu, N., Carlsson, M., Demassey, S., Petit, T.: Global constraint catalogue: Past, present and future. Constraints 12(1), 21–62 (2007)
2. Bochvar, D., Bergmann, M.: On a three-valued logical calculus and its application to the analysis of the paradoxes of the classical extended functional calculus. History and Philosophy of Logic 2, 87–112 (1981)

3. Caballero, R., Stuckey, P.J., Tenoria-Fornes, A.: Finite type extensions in constraint programming. In: Schrijvers, T. (ed.) PPDP 2013, pp. 217–228. ACM Press (2013)
4. Castro, C., Manzano, S.: Variable and value ordering when solving balanced academic curriculum problems (2001), http://arxiv.org/abs/cs/0110007
5. Feydy, T., Stuckey, P.J.: Lazy clause generation reengineered. In: Gent, I.P. (ed.) CP 2009. LNCS, vol. 5732, pp. 352–366. Springer, Heidelberg (2009)
6. Frisch, A.M., Stuckey, P.J.: The proper treatment of undefinedness in constraint languages. In: Gent, I.P. (ed.) CP 2009. LNCS, vol. 5732, pp. 367–382. Springer, Heidelberg (2009)
7. Frisch, A.M., Harvey, W., Jefferson, C., Hernández, B.M., Miguel, I.: Essence: A constraint language for specifying combinatorial problems. Constraints 13(3), 268–306 (2008)
8. Geller, F., Veksler, M.: Assumption-based pruning in conditional CSP. In: van Beek, P. (ed.) CP 2005. LNCS, vol. 3709, pp. 241–255. Springer, Heidelberg (2005)
9. van Hoeve, W.J., Régin, J.C.: Open constraints in a closed world. In: Beck, J.C., Smith, B.M. (eds.) CPAIOR 2006. LNCS, vol. 3990, pp. 244–257. Springer, Heidelberg (2006)
10. Kleene, S.C.: Introduction to Metamathematics. North Holland (1952)
11. Laborie, P., Rogerie, J.: Reasoning with conditional time-intervals. In: Wilson, D.C., Lane, H.C. (eds.) FLAIRS 2008, pp. 555–560. AAAI Press (2008)
12. Laborie, P., Rogerie, J., Shaw, P., Vilím, P.: Reasoning with conditional time-intervals part II: An algebraical model for resources. In: Lane, H.C., Guesgen, H.W. (eds.) FLAIRS 2009, pp. 201–206. AAAI Press (2009)
13. Mittal, S., Falkenhainer, B.: Dynamic constraint satisfaction problems. In: Proceedings of the National Conference on Artificial Intelligence (AAAI), pp. 25–32 (1990)
14. Nethercote, N., Stuckey, P.J., Becket, R., Brand, S., Duck, G.J., Tack, G.: MiniZinc: Towards a standard CP modelling language. In: Bessiere, C. (ed.) CP 2007. LNCS, vol. 4741, pp. 529–543. Springer, Heidelberg (2007)
15. Ohrimenko, O., Stuckey, P.J., Codish, M.: Propagation via lazy clause generation. Constraints 14(3), 357–391 (2009)
16. Sabin, M., Freuder, E.C., Wallace, R.J.: Greater efficiency for conditional constraint satisfaction. In: Rossi, F. (ed.) CP 2003. LNCS, vol. 2833, pp. 649–663. Springer, Heidelberg (2003)
17. Schutt, A., Feydy, T., Stuckey, P.J.: Scheduling optional tasks with explanation. In: Schulte, C. (ed.) CP 2013. LNCS, vol. 8124, pp. 628–644. Springer, Heidelberg (2013)
18. Łukasiewicz, J.: On three-valued logic. In: Borkowski, L. (ed.) Selected works by Jan Łukasiewicz, pp. 87–88. North Holland (1970)

Interactive Design of Sustainable Cities
with a Distributed Local Search Solver

Bruno Belin[1], Marc Christie[2], and Charlotte Truchet[1]

[1] University of Nantes, Laboratoire d'Informatique de Nantes Atlantique,
Nantes, France
[2] University of Rennes 1, IRISA/INRIA Rennes Bretagne Atlantique,
Campus de Beaulieu, Rennes, France

Abstract. Within the last decades, the design of more sustainable cities
has emerged as a central society issue. A city, in the early stage of its
design process, is modeled as a balanced set of urban shapes (residential,
commercial, or industrial units, together with infrastructures, schools,
parks) that need to be spatially organized following complex rules. To
assist urban planners and decision makers in this largely manual and
iterative endeavor, we propose the design of a computer-aided decision
tool which first automatically organizes urban shapes over a given empty
territory, and then offer interactive manipulators that allow the experts
to modify the spatial organization, while maintaining relations between
shapes and informing experts of the impact of their choices. We cast
the problem as a Local Search optimization in which we perform a sequence of swaps between urban shapes, starting from a random initial
assignment. We extend the algorithm with novel heuristics to improve
computational costs and propose an efficient distributed version. The
same algorithm is used for the automated and interactive stages of the
design process. The benefits of our approach are highlighted by examples
and feedbacks from experts in the domain.

1 Introduction

In a world where more than 50% of the population lives in urban areas and where
United Nations' projections mention a global urbanization rate of around 70%
for 2050 [1], crucial questions arise on how to develop conditions for a balance
between people, environment and cities. China for example, plans to annually
create twenty whole new cities from now to 2020, around one million inhabitants
each, to accommodate farmers in urban environments [2].

The process of designing whole new cities is by nature a collaborative endeavor
gathering urban planners and decision makers around a coarse-grain map of a
territory on which to place urban shapes such as centers, industries, housings,
commercial units, public equipments, etc. The number of elements as well as
their spatial layout needs to be strongly guided by a collection of rules related
to social, economic, energy, mobility and sustainability issues.

H. Simonis (Ed.): CPAIOR 2014, LNCS 8451, pp. 104–119, 2014.

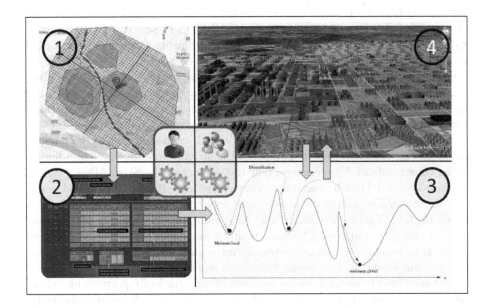

Fig. 1. A 4-stage design process of a urban environment (1) Setting contours, properties, central areas and intensity footprints (2) Computing the number of urban shapes by level of intensity (3) Automatically positioning urban shapes while enforcing constraints and favouring preferences (4) Interactively manipulating urban shapes while maintaining constraints and preferences. The last two stages are at the core of this contribution.

The urban planning community actually lacks tools to assist this initial design process, and the literature is focused on addressing the problem of predicting the evolution of urban environments. Given a current state, and a set of evolution rules (land price, employment rate, extension,...), tools such as the UrbanSim framework compute the evolution of the city using agent-based representations [3], or focus on more narrow issues [4]. In computer graphics, multiple contributions target the creation of new cities [5, 6], however primarily focusing on computational models capable of encoding the stylistic appearance of the city (eg. mimicking existing ones), rather than its functional dimensions (some aspects such as land-use are however addressed, see [7]).

In contrast, the recent work of Vanegas *etal.* [8], integrate a functional description of the city by relying of UrbanSim's evolution models, yet do not offer any interactive editing tools, and are not designed in mind for urban planners or decision makers. Finally, some design tools are available to urban planners (such as CommunityViz [9][1]), which offer automated and interactive design tools but address the problem at a very detailed level. The early stage design process of urban environments essentially remains a manual editing process.

---

[1] http://placeways.com/communityviz/

In this context, we propose a computer-aided decision tool to assist designers in their task following a 4-stage process described in Figure 1: (1) the designer sets city contours, properties, central areas and intensity footprints over a regular grid (intensity footprints are areas with a given population density); (2) a knowledge-driven process computes the number of urban shapes of each kind (housing, industry,...) per level of intensity given an expected employment rate together with country-specific values; (3) urban shapes are then automatically positioned on the regular grid while enforcing constraints and favouring preferences between urban shapes in relation to social, economic and sustainability issues; (4) the designer then manipulates the urban environment while maintaining constraints and preferences. Our tool is part of the SUSTAINS project, a national-funded French research project gathering urban planners and computer scientists. The goal of the project is to deploy a software suite for designing new cities, taking into account the city footprints, and the automated computation of energetic impacts. The suite is made for urban planners and decision makers. It will include an interactive communication tool on large tactile surfaces for public engagement. This paper focuses on stages 3 and 4.

More specifically, we address the problem of automatically positioning urban shapes on a territory, and propose a local search method in order to automatically compute realistic coarse-grain maps, which can then be interactively modified while respecting some urban constraints. Compared to the state-of-the-art, our method reduces the amount of information that the user must provide, and optimizes the land use taking into account multiple constraints and preferences. One of its key feature is the solver interactive mode, where a user can modify the solution by hand while the system maintains the constraints in interactive time. To practically address the issue of scalability, we devised a distributed version of the solving process. A video detailing results and displaying interactive modes is available here http://vimeo.com/80211470.

The paper is organized as follows. Section 2 introduces the key concepts expressed by urban planners, focusing on sustainable land-use. Section 3 describes our representation of constraints and preferences. Section 4 details our solving method, based on a distributed Local Search (LS), for generating a good initial solution and handling interaction. Section 5 presents experimental results, with a redesign of the Marne-la-Vallée city (east of Paris) which is spread over an area of 8728 hectares (87.28 km^2) with 234 644 inhabitants.

2   A Model for Early Stage Design of Sustainable Cities

The early stage design of a city consists in first selecting the number and the nature of all urban shapes composing the city and then spatially organizing them, by taking into account environmental, social, mobility and energy aspects. Several parameters have to be controlled: population density, landscape constraints, employment rate. The employment rate is calculated by dividing the number of jobs by the working-age population living in a specific area. An employment rate close to 1 corresponds to an ideal situation where each resident can access a job

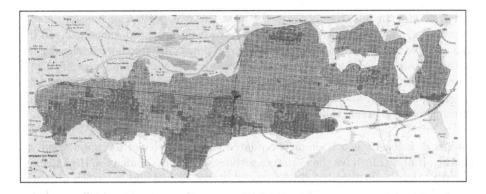

Fig. 2. Urban shapes automatically spread over the experimental area of Marne-la-Vallée city, taking into account the constraints of sustainable urban development expressed by urban planners

and housing in the area, minimizing commuting. An employment rate far from 1 has severe consequences, inducing incoming or outgoing congestions. This early stage enables decision-makers and stakeholders to agree on the broad guidelines of a preliminary project before initating a costly quantity survey.

2.1   Urban Model

In urban planning, modeling consists in simplifying the reality of the world in order to better understand how decisions and events interact. It allows to test solutions that may affect political decisions and strategies which may lead to a desirable future [10]. In this paper, the urban model proposed by urban planners is based on the core concepts that are *blocks, central areas, intensities* and *urban shapes* incorporating, in a systemic approach, major urban constraints linked to urban sustainable development.

Block. The city block whose outlines are defined by the roads is the finest level of granularity selected by urban planners. Blocks are represented in grid patterns with a mesh size of 80 meters long. It is assumed that each city block hosts one single urban shape (housing, industry, shop, school, etc).

Central Area. A very important notion for urban planners is that of central area, a structuring place which gives its name to the neighbourhood. The feeling of belonging to a neighbourhood coincides with the influence area of the place, estimated at a 300 meters radius.

Urban Shape. In our context, a urban shape represents a dominant type of land use and buildings for a single block. However, the specification of a dominant urban shape can integrate a degree of mixed use (a portion of housing, commerce, etc.). Twenty eight urban shapes have been proposed and organized in four groups: residential (detached house, town house, intermediary housing, collective housing), economic activity (industrial, craft, commercial, tertiary), infrastructure (elementary school, primary school, secondary school, high school, sports equipment), and fixed elements (roads, unbuildable zones, rivers, rough terrains, natural areas, railroads, etc).

Intensity. Intensity is a scale metric related to the density of the population in an area. It also translates the notion of activity or liveness of an area. High urban intensity represents a lively neighbourhood, dense, mixed and for which walking is the simplest way to travel for accessing all the essential urban functions. This intensity level is located in the city center, while low urban intensity is related to essentially housing areas. Six intensity levels are defined from 1 for the lowest intensity to 6 for the highest.

3 A Constraint Model of the Urban Location Problem

For a given territory, the urban model provides the number of urban shapes of each type, for each level of intensity. The objective is then to arrange the placement of shapes on the territory. We will refer to this problem as the *urban location* problem. The first task is therefore to provide a formalization of constraints and preferences related to the properties of a sustainable city. Urban planners naturally express the interactions between different urban shapes as location preferences, instead of hard constraints (for instance, manufacturing industries prefer to be next to a river and not far from a highway). We thus express these preferences into cost-functions and express the problem as an optimization problem. Yet, there also are hard constraints which strictly restricts the positions of some urban shapes (one does not place an individual house within an industrial area, for instance).

   In the end, the urban location problem encompasses a lot of constraints, some of them very specific. We distinguish in our presentation the most important, core constraints, which apply to all of the variables, and specific constraints that apply only on some areas or on some urban shapes. In the following, only a limited set of cost functions and constraints are detailed, as many of them are similar.

3.1 Grid Representation of the City

We represent the city with a regular grid where each cell corresponds to a city block. The goal is therefore to assign each cell a urban shape among the possible shapes. Some specific cells are associated with fixed elements (roads, unbuildable zones, etc), and are therefore considered as not free and left out of the problem - yet, they are kept on the map since they might interfere with the other urban forms. Furthermore, each cell is given an intensity level manually specified by the urban planners (in the early design stage). This allows for instance to represent centralities on the map (commercial areas, lively squares, etc).

For the sake of simplicity, the grid is viewed as a rectangular array of cells of size $(x \times y)$, but this representation is more symbolic than geometric in that what is important is the neighborhood and relative placement of the urban shapes, not the rectangular geometry. Each cell is a variable which value is selected among all urban shapes: we note $V_{l,c}$ the variable corresponding to the cell in line l, column c. The urban shapes are encoded into integer values: (1) Detached house, (2) Town house, (3) Intermediary housing,..., (19) Breathing space.

3.2   Core Constraints

These constraints cover the whole territory and express fundamental aspects of sustainable development.

Urban Shape Cardinality. Depending on some features of the final city (number of inhabitants, size, employment rate, etc), the urban planners are able to determine how many instances of each urban shape must appear in the city: for instance, a big city must have a certain surface of parks, a certain surface of industries, etc. In the constraint problem, this gives a hard cardinality constraint for each urban shape.

Intensity Requirement. The urban model provides a given intensity level for each urban shape, and an intensity level for each cell of the grid. Based on these elements, every assignment of urban shape to a cell must comply with the intensity correlation between urban shapes and cells. The different levels of intensity are set by the user on the map. It is a hard unary constraint.

Interaction between Urban Shapes. Each urban shape has placement preferences depending on its nature. For instance, shool units are attracted to residential units, and residential units are repelled by industrial units. We model these preferences as a function specifying the attraction (or conversely repulsion) of each urban shape for another urban shape. Between two urban shapes, possible interaction values are 0 (double attraction), 10 (single attraction), 20 (neutral), 50 (single repulsion) and 100 (double repulsion). The value of interaction decreases with the increasing distance between two cells.

The interaction preferences are expressed as a cost function. We note I the interaction matrix, which is an input of the system. $I_{p,q}$ is the interaction value between urban shape p and urban shape q. This matrix is asymmetrical, so it may happen that $I_{p,q} \neq I_{q,p}$. For a urban shape located at cell $V_{l,c}$, the interaction cost only depends on its neighbouring cells [2], namely with a set noted $\mathcal{V}_{l,c}^d$. The neighbouring cells are defined as:

---

[2] In a geometric neighbouring sense, not the neighbourhood of a LS algorithm.

$$\mathcal{V}_{l,c}^d = \{V_{i,j}, i \in [l-d, l+d], j \in [c-d, c+d]\} \setminus \{V_{l,c}\} \tag{1}$$

where: d is a parameter controlling the size. Note that this corresponds to a Moore neighborhood without its center, that is, a set of points at a bounded, non-null Chebyshev distance from the cell $V_{l,c}$. To take the distance influence into account, we consider the border of $\mathcal{V}_{l,c}^d$ noted $\bar{\mathcal{V}}_{l,c}^d$ such that:

$$\bar{\mathcal{V}}_{l,c}^d = \{V_{i,j}, (|i-l|=d) \vee (|j-c|=d)\} \tag{2}$$

Finally, for a cell $V_{l,c}$, the cost function related to our constraint is:

$$Cost_1\left(V_{l,c}\right) = \sum_{d=1}^{D} \left(\frac{\sum\limits_{v \in \bar{\mathcal{V}}_{l,c}^d} \left(I_{(V_{l,c},v)}\right)}{|\bar{\mathcal{V}}_{l,c}^d| * d^2} \right) \tag{3}$$

where: D is the maximum interaction distance (in the following, D is set to 3) to consider between two urban shapes. The cost of a cell $V_{l,c}$ includes contributions by all the rings at distance d from $V_{l,c}$, but these contributions are decreasing as d increases. Within a ring, the average contributions of the cells are divided by a correction factor of d^2, in order to obtain a similar effect as the attraction in physics.

3.3   High Level Constraints

In order to improve the sustainable aspects of our model, we add more specific constraints to improve the social equity, the preservation of environment and the economic viability.

Distance. This constraint specifies that some urban shapes must be located with a minimum distance between them. This distance is expressed as a number of cells. For example, between an individual house and a high tertiary building (R+7), we want a minimum separation distance of 4 cells. For this constraint, we use Euclidean distances that measure the distance of a straight line between two cells. We note D the distance matrix, with $D_{p,q}$ the minimum distance permitted between urban shape p and urban shape q. This matrix is symmetrical. By convention, $D_{p,q} = 0$ when there is no particular distance constraint between p and q. For a urban shape $V_{l,c}$ located at (l, c), if $\mathcal{D}_{p,q}^{Euc}$ is the Euclidean distance in number of cells separating cells p et q, the cost function related to our constraint is:

$$Cost_2\left(V_{l,c}\right) = \sum_{v \in \mathcal{V}_{l,c}^4} \left(max\left(D_{V_{l,c},v} - \mathcal{D}_{V_{l,c},v}^{Euc}, 0 \right) \right) \tag{4}$$

Critical Size Area. Industrial or artisanal areas must have at least a critical size, otherwise, the area will not be created in practice because it will not be cost-effective. The critical size is determined by urban designers for each urban shape that needs grouping (craft activity, industrial activity). There is no maximum size for a grouping. On the other hand, although it is not explicitly expressed by designers, the area must have a sufficient compact structure (a notion used by the urban designers, and intuitively meaning that circles are better than lines or flat rectangles). For these urban shapes, we thus penalize the cells which belong to too small groups, or for which the groups are not compact enough.

Accessibility. This constraint concerns only one urban shape: the breathing spaces. It specifies that, from any inhabited point of the city, a breathing space should be reacheable by walking less than fifteen minutes, which corresponds to distance of about 1.25 km (or a distance of fifteen cells). We propose two complementary versions of this constraint. First, a global version which penalizes uncovered inhabited cells in proportion to their distance to any breathing space. And second, a local version which penalizes uniformly uncovered inhabited cells. The global version is more appropriate than the second one if the number of breathing spaces to spread over the city is insufficient to cover the entire grid, but its computation time then depends on the size of the grid. In contrast, the local version can be computed locally and is relevant if the urban model provides enough breathing spaces to cover the whole area to develop.

Interspace. This is a constraint inherent to buffer spaces or public equipments. For specific urban shapes positioned near each other, it is required that they must not be contiguous (not directly touching), and be separated by a buffer space or a built equipment. The following urban shapes are considered as buffer spaces: sport equipment, breathing space and green way while urban shapes like nursery school, primary school, secondary school, high school, administrative and technical equipment are considered as built equipments. For example: between an individual house and a high collective housing (R+7), we have to position a buffer space.

Footprint. The principle is: when there is, in immediate vicinity of a secondary or high school, some particular urban shapes (individual houses or town houses, intermediary or collective housing, tertiary buildings), we must provide a place around the school, by placing, close to the building, a given number of green ways. For example, if there is around a secondary school only individual houses, town houses or intermediary housing, then we must allocate at least one green way near the school. However, if there is collective housing or tertiary buildings near the school, we have to build a bigger place around the school with at least two green ways.

Filtering. This constraint is related to central areas. Central areas are special cells marked by the user to identify the center of an urban neighbourhood (see subsection 2.1) and they can combine one, two or four cells in the same area. This constraint is used to filter the urban shapes that may occupy a central area and aid the diversity of urban shapes located on a same group. For this special cells, we favor a mix of the following urban shapes: schools, sports equipments, shops, services downtown and green ways.

4   A Solver for the Interactive Design of Sustainable Cities

The problem addressed in this paper can be viewed as a facility location problem [11], but in which all the urban elements need to be placed simultaneously. This *urban location* problem, is highly combinatorial, hence difficult to solve for a large number of cells. In addition, we need to deal with two specific requirements of the applicative context. First, the algorithm needs to scale up to the size of real-life cities, with a typical number of cells around 10000 (for a $64km^2$ city). Second, the users (urban designers and decision makers) also require to keep their hands on the system, by interactively modifying the assignment of urban elements to meet their own representations and expectations of locations.

To address both requirements, we designed a system based on two distinct solving stages. In the first step, the system computes and proposes one or several good solutions satisfying the urban constraints. And to address the issue of complexity, we developed a specific solving technique to efficiently handle the computation on multiple processors. The second step is interactive: the user modifies the map by moving single or multiple urban elements simultaneously, and the system adapts the solution by re-solving the constraints in the modified area, at a close-to-interactive frame rate, keeping the constraints satisfied as much as possible.

4.1   Initial Resolution with Adaptive Search

Our first attempt to solve this problem relied on complete CP techniques. A prototype designed in the Choco solver [12], with only some of the core constraints, failed to scale with a computation time of around half an hour for a small 16×16 map, whilst the typical size of our problems reaches 10000 variables.

Sequential Algorithm. We therefore relied on the *Adaptive Search* (AS) method which has proven its efficiency on large and various instances [13]. This meta-heuristic takes advantage of the structure of the problem in terms of constraints and variables to guide the search. Our algorithm starts from a random configuration. At this stage, we make sure that the initial assignment respects the intensity level of each cell and the given number of cells for each urban shape (this is analogous to filtering the unary constraint on the intensities and filtering the urban shapes cardinalities).

The algorithm then performs a variant of iterative repair, based on variable and constraint error information, seeking to reduce the error on the worst variable so far. The basic idea is to compute the error function for each constraint,

---

Algorithm 1. Base algorithm - initial resolution

---

1 /* Parameter: MaxRestart the number of partial resets of the algorithm
    f is the global cost function, f_i is the cost function for variable V_i
    s is the current configuration
    T is the adaptive tabu list
    j is the index of the variable with the worst cost */
2 $s \leftarrow$ random configuration ;
3 $T \leftarrow \emptyset$;
4 **while** *MaxRestart not reached* **do**
5   **while** *T is not full* **do**
6   For all i such that $V_i \notin T$, compute $f_i(s)$;
7   Select V_j a variable for which $f_j(s)$ is maximum ;
8   Compute the cost f of the configurations obtained from s by swapping V_j with another variable ;
9   Select s' the configuration for which $f(s')$ is minimum;
10   Update T by removing its oldest variable ;
11   **if** s' *can improve current solution s* **then**
12 | $s \leftarrow s'$;
        else
13         $T \leftarrow T \cup V_j$;

14 **return** s

---

then combine for each variable the errors of all the constraints in which it appears, thereby projecting constraint errors onto the relevant variables. This combination of errors is problem-dependent [13]. In our case, it is a weighted sum so that the constraints can be given different priorities.

Finally, the variable with the highest error is designated as the "culprit" and chosen for a move. In this second step, we consider the swaps involving the culprit variable and choose the best one according to f. However, there is a restriction to the set of considered swaps: we only swap two cells assigned to the same intensity level in order to satisfy the intensity constraint at any time. This swap policy ensures that the intensity constraint is kept satisfied at every iteration, as is the urban shape cardinality constraint. The algorithm also uses a short-term adaptive memory in the spirit of Tabu Search in order to prevent stagnation in local minima. Note that only the variables which do not improve f are marked tabu. When the algorithm is stuck, with all variables marked tabu, it restarts by resetting a given percentage of the variables, randomly chosen. The AS metaheuristic is described on Algorithm 1. In order to improve the speed of computations, we use a data cache for all information that require very frequent access: cost for each variable, groups, etc. We also compute all the costs incrementally, calculating only the delta induced by a swap. To efficiently distribute the AS algorithm, we also introduced two new features: a multi-candidate mode, and a multi-swap mode.

Table 1. Decomposition of the city map into four uniforms parts and distribution to the different slave processes

①② ③④	Slave	1	2	3	4	5	6	7	8	9	10
	Subarea	①	②	③	④	①;②	①;③	①;④	②;③	②;④	③;④

Multi-candidate Mode. In the AS algorithm, only one culprit variable can be candidate for a move. We introduce a *multi-candidate* mode, where all the variables with a significant cost are considered. For all of them, we consider the possible swaps as defined above. There are two reasons for this: first, the worst variable may not be the one which will achieve the best swap. Second, in a distributed mode, several candidates can be explored in parallel.

Multi-swap Mode. In addition to the multi-candidate management, we add a pool of best swaps. Instead of dealing with a single best swap for the current candidate, we store a small number of best swaps (configurable) related to each considered candidate. There is no significant time overhead because swaps must anyway be computed. This mechanism simply induces an extra memory footprint. Once the pool is filled with the best swap identified for each best candidate, we apply all the swaps of the pool in sequence, under some conditions. Typically, performing a swap may strongly change the impact of a further swap in the list. For a swap to remain valid, it must still produce a profit of a least a ratio (configurable) of its previous profit (before the first swap). In the distributed version of the algorithm, these two mechanisms (multi-candidate, multi-swap) are intended to take benefit from the work made by the different cores: each core performs an important amount of calculations, which are lost in the single-candidate, single-swap version of the algorithm. Although the moves performed in the multi modes may not be the best to apply sequentially, they still improve the cost. Applying them comes at no computational cost in the distributed version of the algorithm.

4.2   Distributed Mode

To tackle the complexity of large-scale problems, we propose to distribute the algorithm on a grid. A multi-walk parallel scheme has already been proposed for AS [14, 15], with good speed-ups on classical problems. Instead, we choose a master-slave scheme to distribute the computations. At each iteration, we parallelize the search for the best swaps [16]. We also take benefit of the geographical / geometrical layout of our problem, to assign different parts of the map to different cores in a coherent way. We detail here how the algorithm works on 10 cores, in which case the map is divided into 4 equal parts (the process is similar for a higher number of cores).

The slaves are in charge of examining the possible swaps (cost evaluation and comparison). Each slave is assigned a couple of subareas, as shown on figure 1: for instance, slave 1 investigates the swaps of cells located in subarea 1, with other cells of subarea 1. Slave 5 investigates swaps of cells in subarea 1 with cells in subarea 2. Note that the overall map is shared between each process, so that they can compute the global cost. However, each slave searches for the best possible swaps only on the subareas assigned to it. Given that the search of the best swaps does not change the state of the map, computing processes can operate in parallel without any difficulty and without changing procedures.

The slaves are synchronized with a master process that collects all the best swaps and decides which to apply. In the multi-candidate, multi-swap mode, the master process deals with the multi-candidate variables. It also collects the pool of best swaps. Finally, the master process sends the swaps that were applied to each slave.Once synchronized, the slaves can search again new swaps in parallel.

4.3   Interactive Mode

The second stage of our solving process is interactive. The purpose is to let the users have a control on the solution through interactive manipulations. In a way to smartly integrate user interactions, the idea here is to maintain the solving process active (i.e. continue swapping urban shapes) while the user is performing changes to the solution. Our interaction process is founded on the idea of having a *pool of cells* (POC). A POC represents a sub-region of the map in which urban shape assignments have already been made, and on which the solving technique presented in the previous section is applied to locally recompute a good solution. We then rely on this pool of cells to propose multiple interaction modalities, two of which are detailed in the following.

Defining Fixed vs. *Free Cells.* At any moment, the user is able to select a set of cells which he considers temporarily fixed (in the sense that the associated urban shapes are constrained to those cells). The POC is built by computing the difference between all cells of the map and the fixed cells. The solver only performs its optimization on the cells in the POC. In a similar way, the user may define a set of free cells, that will then form the POC to which the solver is applied.

Manually Moving a Urban Shape or a Set of Urban Shapes. To avoid a local change performed by the user from impacting the whole map (i.e. changing the entire solution), the direct manipulation of a set of urban shapes has been defined in a way it only locally impacts the map. To this end, a disc size is first specified by the user (see Figure 3) that defines the region in which the computations will be performed around the user's changes. Then, the users selects one or multiple urban shapes and interactively moves them around the map. The POC is then defined by all the cells within the disc, except the cells manipulated by the users (which values are fixed). Two modes of interaction are proposed. Either the computation is continuously performed as the user moves the urban shapes. In such case, the POC is reconstructed and optimized at each move. Or the computation is performed as the user drops the selected urban shapes at a new location, in which case, the POC is composed of the union between the disc area

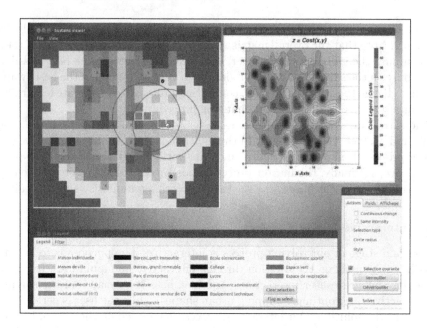

Fig. 3. Interactive mode: on the top left, a radius around the user's selected cell defines the region in which swaps will be performed to best satisfy constraints. On the top right, a heat map displays the regions in which the costs are the highest (red areas) and provides feedback on the impact of the changes w.r.t. sustainability constraints.

around the initial location of the selected urban shapes and the disc area around the final location of the dropped urban shapes. At the initial location, some cells are no more assigned, while at the final location some urban shapes need to be reassigned to cells.

These manipulation modes are furthermore supported by a visual feedback that assists the users in understanding the impact of their decisions. First, a dynamically recomputed heat map presents the regions in which the costs are the most important (see top right part of Figure 3). And second, a dynamically recomputed graphical representation displays the global change in score, together with the change in score for each constraint that is considered (distance, accessibility, critical size...). Some options of the interactive solving process are detailed in the accompanying video (see here http://vimeo.com/80211470). In particular, in situations where the user intentionally moves part of an industrial zone (on which the critical size area is specified), the solving process automatically repositions all the industrial urban shapes to reform an industrial zone of specified critical size.

Interestingly, the process is extensible to multiple users performing changes simultaneously on the same map (e.g. multiple collaborators around a large tactile device). Each user is assigned his own pool, and solvers for each pool are distributed on multiple processors. The case of intersecting pools (i.e. two users

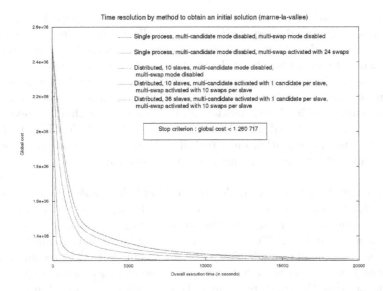

Fig. 4. Tests for different multi-candidate and multi-swap options

manipulating close regions) is easily handled by creating a unique pool being the union of the two pools.

5   Experiments

Two different experiments were performed to evaluate our system. The first evaluates the computational costs related to the initial resolution. The second, less formal, evaluates the benefits of the interactive process through an open discussion with three senior urban designers.

The application was developed in $C++$ and is based on the *EasyLocal++* framework [17]. About the accessibility constraint (section 3.3), we activated the local version that offers much better processing times on our real-life benchmark (e.g. reduced the time to find a best swap from 4 minutes to a few milliseconds on a 64x64 map). Synchronization between processes is performed by exchanging messages using the message-oriented middleware *Apache Active MQ* and the C++ library ActiveMQ-CPP.Experiments were conducted on Grid'5000, a french cluster available for research (in practice we used *granduc* in Luxembourg, with 22 nodes, each one having 2 CPU Intel 2.0GHz, 4cores/CPU, 15GB RAM). Each experiment was performed and averaged on 10 runs.

Computationnal Costs. Results presented on Fig. 4 for the city of Marne-la-Vallée (see Fig. 2) address a problem with 9038 variables (i.e. free cells), in multi-candidate and multi-swap modes. Each run was terminated either when it reached a specified timeout, or when a *good* solution was computed We define a good solution as a solution with a score lower or equal to the score of the best solution found during the distributed search with 36 slaves (that is, 1260717). The results show that the sequential process is slightly improved by activating the multi-swap mode, but the best improvements are found in the distribution scheme, which takes full benefit of the multi-candidate and multi-swap mode. In the end, considering the target score of 1260717, the speed-up between the base algorithm (multi-candidate mode disabled, multi-swap mode disabled) and its distributed version (36 slaves, multi-candidate activated with 1 candidate per slave, multi-swap activated with 10 swaps per slave) is ×11.51. This makes it possible to run our algorithm on a real-life city in less than one hour, which appears sufficient for the requirements of urban planners.

Assessment of the Interactive Mode. In the interactive mode, the user can select and change parts of the solution map, and during theses interactions, the solver is continuously running. A video accompanying this paper (available at http://vimeo.com/80211470) shows both the initial resolution and the interactive mode on a small example. As can be seen on the video, the recomputation is performed on a restricted number of cells. The computationnal cost is obviously cheaper than the initial problem and can nearly be performed in real-time. Three senior urban designers who assessed the interactive mode were very interested by the possibility to rearrange and recompute the map, while maintaining the high-level constraints such as the critical size constraint. Another important feature for the urban designers is the maintaining of intensities and centralities while rearranging the map. In practice, the user manipulations are likely to lower the score of the interaction constraint, but it is convenient not to manually deal with those structural constraints.

6   Conclusion

In this article, we have first presented means to model the *urban location* problem, an important stage of the urban planning process when designing new cities that consists in laying out urban shapes while ensuring the statisfaction of local constraints. We then have presented a solver based on a local search algorithm, the Adaptive Search, which we extended to a parallel version. The proposed system is able to deal with large-scale problems in a two-stage solving process well adapted to the need of urban designers: first, an initial resolution provides the designer with one or several good solutions. The designer can then interact with the proposed solutions, while the solver maintains the constraints active. Further research includes refining the model by considering more evolved constraints required by the designers, and intregrating it within a complete urban planning, spanning from early design to full 3D model.

Acknowledgements. Experiments presented in this paper were carried out using the Grid'5000 experimental testbed, being developed under the INRIA ALADDIN development action with support from CNRS, RENATER and several Universities as well as other funding bodies (see https://www.grid5000.fr)

References

[1] United-Nations: Urban and rural areas, world urbanization prospects: The, revision (2007)

[2] Mars, N., Hornsby, A., Foundation, D.C.: The Chinese Dream: A Society Under Construction. 010 Publishers (2008)

[3] Waddell, P.: Urbansim: Modeling urban development for land use, transportation, and environmental planning. Journal of the American Planning Association 68(3), 297–314 (2002)

[4] Dury, A., Le Ber, F., Chevrier, V.: A reactive approach for solving constraint satisfaction problems. In: Papadimitriou, C., Singh, M.P., Müller, J.P. (eds.) ATAL 1998. LNCS (LNAI), vol. 1555, pp. 397–411. Springer, Heidelberg (1999)

[5] Parish, Y.I., Müller, P.: Procedural modeling of cities. In: Proceedings of the 28th Annual Conference on Computer Graphics and Interactive Techniques, pp. 301–308. ACM (2001)

[6] Watson, B., Muller, P., Wonka, P., Sexton, C., Veryovka, O., Fuller, A.: Procedural urban modeling in practice. IEEE Computer Graphics and Applications 28(3), 18–26 (2008)

[7] Lechner, T., Ren, P., Watson, B., Brozefski, C., Wilenski, U.: Procedural modeling of urban land use. In: ACM SIGGRAPH 2006 Research posters, SIGGRAPH 2006. ACM, New York (2006)

[8] Vanegas, C.A., Aliaga, D.G., Wonka, P., Müller, P., Waddell, P., Watson, B.: Modelling the appearance and behaviour of urban spaces. In: Computer Graphics Forum, vol. 29, pp. 25–42. Wiley Online Library (2010)

[9] Kwartler, M., Bernard, R.: Communityviz: an integrated planning support system. In: Planning Support Systems (November 2001)

[10] Antoni, J.: Modéliser la ville: Formes urbaines et politiques de transport, Collection "Méthodes et approches". Economica (2010)

[11] Moujahed, S., Simonin, O., Koukam, A.: Location problems optimization by a self-organizing multiagent approach. Multiagent and Grid Systems 5(1), 59–74 (2009)

[12] Jussien, N., Rochart, G., Lorca, X.: Choco: An Open Source Java Constraint Programming Library. In: CPAIOR 2008 Workshop on Open-Source Software for Integer and Contraint Programming (CPAIOR 2008), pp. 1–10 (2008)

[13] Codognet, P., Díaz, D.: Yet another local search method for constraint solving. In: Steinhöfel, K. (ed.) SAGA 2001. LNCS, vol. 2264, pp. 73–89. Springer, Heidelberg (2001)

[14] Caniou, Y., Codognet, P., Diaz, D., Abreu, S.: Experiments in parallel constraint-based local search. In: Merz, P., Hao, J.-K. (eds.) EvoCOP 2011. LNCS, vol. 6622, pp. 96–107. Springer, Heidelberg (2011)

[15] Diaz, D., Abreu, S., Codognet, P.: Targeting the cell/be for constraint-based local search. Concurrency and Computation: Practice and Experience 24(6), 647–660 (2012)

[16] Talbi, E.: Metaheuristics: From Design to Implementation. Wiley Series on Parallel and Distributed Computing. Wiley (2009)

[17] Gaspero, L.D., Schaerf, A.: Easylocal++: An object-oriented framework for the flexible design of local-search algorithms. Softw. Pract. Exper. 33(8), 733–765 (2003)

Sliced Table Constraints:
Combining Compression and Tabular Reduction

Nebras Gharbi, Fred Hemery, Christophe Lecoutre, and Olivier Roussel

CRIL - CNRS UMR 8188,
Université Lille Nord de France, Artois,
rue de l'université, 62307 Lens cedex, France
{gharbi,hemery,lecoutre,roussel}@cril.fr

Abstract. Many industrial applications require the use of table constraints (e.g., in configuration problems), sometimes of significant size.
During the recent years, researchers have focused on reducing space and
time complexities of this type of constraint. Static and dynamic reduction based approaches have been proposed giving new compact representations of table constraints and effective filtering algorithms. In this
paper, we study the possibility of combining both static and dynamic
reduction techniques by proposing a new compressed form of table constraints based on frequent pattern detection, and exploiting it in STR
(Simple Tabular Reduction).

Introduction

Table constraints, i.e., constraints given in extension by listing the tuples of
values allowed or forbidden for a set of variables, are widely studied in constraint
programming (CP). This is because such constraints are present in many real-
world applications from areas such as design and configuration, databases, and
preferences modeling. Sometimes, table constraints provide the unique natural
or practical way for a non-expert user to express her constraints. So far, research
on table constraints has mainly focused on the development of fast algorithms to
enforce generalized arc consistency (GAC), which is a property that corresponds
to the maximum level of filtering when constraints are treated independently.
GAC algorithms for table constraints have attracted considerable interest, dating
back to GAC4 [21] and GAC-Schema [2]. Classical algorithms iterate over lists
of tuples in different ways ; e.g., see [2,19,18]. A recent AC5-based algorithm
has been proposed in [20], and has been shown efficient on table constraints of
small arity. For tables constraint of large arity, it is recognized that maintaining
dynamically the list of supports in constraint tables does pay off: these are the
variants of simple tabular reduction (STR) [23,15,16].

Table constraints are important for modeling parts of many problems, but
they admit practical boundaries because the memory space required to represent them may grow exponentially with their arity. To reduce space complexity,
researchers have focused on various forms of compression. Tries [6], Multi-valued
Decision Diagrams (MDDs) [3] and Deterministic Finite Automata (DFA) [22]

H. Simonis (Ed.): CPAIOR 2014, LNCS 8451, pp. 120–135, 2014.

are general structures used to represent table constraints in a compact way, so as to facilitate the filtering process. Cartesian product is another classical mechanism to represent compactly large sets of tuples. For instance, it has been applied successfully for handling sets of solutions [10], symmetry breaking [5,4], and learning [13,17]. So far, this form of compression has been used in two distinct GAC algorithms for table constraints: by revisiting the general GAC-schema [14] and by combining compressed tuples with STR [24]. The latter work shows how variants STR2 and STR3 can advantageously benefit from compressed tuples when the compression ratio is high.

Recently, we have proposed an original compression approach based on datamining algorithms [7], where all occurrences of the most frequent patterns in a table are replaced by their indices in a so-called patterns table. Using datamining techniques for compressing table constraints has also been studied in [11], but in a very different manner since additional variables and values are needed, and constraints are reformulated. The same authors also studied compression of SAT instances in [12]. In [7], a pattern was defined as a sequence of consecutive values, which prevented us from benefiting of optimized STR variants. In this paper, we propose to relax this condition (of consecutive values), considering any sub-tuple as a possible frequent pattern, and identifying the most frequent ones by means of data-mining techniques. Consequently, every table can be "sliced", where each slice associates a pattern μ with a sub-table containing all extensions of μ that can be found in the original table. We propose an algorithm to deal with sliced table constraints: we build it on the basis of the optimized algorithm STR2.

The paper is organized as follows. After recalling some technical background in Section 1, we present, in Section 2, a compression process for table constraints, detailing the algorithm used to obtain the new form of sliced table constraints. Next, we describe, in Section 3, an optimized algorithm to enforce GAC on sliced table constraints. Finally, after giving some experimental results in Section 4, we conclude.

1   Technical Background

A (discrete) *constraint network* (CN) N is a finite set of n variables "interconnected" by a finite set of e constraints. Each *variable x* has a *domain* which is the finite set of values that can be assigned to x. The *initial* domain of a variable x is denoted by $dom^{init}(x)$ whereas the *current* domain of x is denoted by $dom(x)$; we always have $dom(x) \subseteq dom^{init}(x)$. Each *constraint c* involves an ordered set of variables, called the *scope* of c and denoted by $scp(c)$, and is semantically defined by a *relation*, denoted by $rel(c)$, which contains the set of tuples allowed for the variables involved in c. A (positive) *table constraint c* is a constraint such that $rel(c)$ is defined explicitly by listing the tuples that are allowed by c ; an example is given below. The *arity* of a constraint c is the size of $scp(c)$, and will usually be denoted by r.

Table 1. Table constraint c on x_1, x_2, x_3, x_4, x_5

	x_1	x_2	x_3	x_4	x_5
τ_1	(c,	b,	c,	a,	c)
τ_2	(a,	a,	b,	c,	a)
τ_3	(a,	c,	b,	c,	a)
τ_4	(b,	a,	c,	b,	c)
τ_5	(b,	a,	a,	b,	b)
τ_6	(c,	c,	b,	c,	a)
τ_7	(a,	c,	a,	c,	a)

Example 1. Let c be a positive table constraint on variables x_1, x_2, x_3, x_4, x_5 with $dom(x_1) = dom(x_2) = dom(x_3) = dom(x_4) = dom(x_5) = \{a, b, c\}$. Table 1 represents the constraint c with 7 allowed tuples.

Let $X = \{x_1, \ldots, x_r\}$ be an ordered set of variables. An *instantiation* I of X is a set $\{(x_1, a_1), \ldots, (x_r, a_r)\}$ also denoted by $\{x_1 = a_1, \ldots, x_r = a_r\}$ such that $\forall i \in 1..r, a_i \in dom^{init}(x_i)$; X is denoted by $vars(I)$ and each a_i is denoted by $I[x_i]$. An instantiation I is *valid* iff $\forall(x, a) \in I, a \in dom(x)$. An r-*tuple* τ on X is a sequence of values $(a_1, \ldots, a_r)$ such that $\forall i \in 1..r, a_i \in dom^{init}(x_i)$; the individual value a_i will be denoted by $\tau[x_i]$. For simplicity, we shall use both concepts of instantiation and tuple interchangeably. For example, an r-tuple τ on $scp(c)$ is *valid* iff the underlying instantiation is valid. An r-tuple τ on $scp(c)$ is a *support* on the r-ary constraint c iff τ is a valid tuple which is allowed by c. If τ is a support on a constraint c involving a variable x and such that $\tau[x] = a$, we say that τ is a *support for* (x, a) on c. Generalized Arc Consistency (GAC) is a well-known domain-filtering consistency defined as follows:

Definition 1. *A constraint c is* generalized arc consistent *(GAC) iff $\forall x \in scp(c), \forall a \in dom(x)$, there exists at least one support for (x, a) on c. A CN N is GAC iff every constraint of N is GAC.*

Enforcing GAC is the task of removing from domains all values that have no support on a constraint. Many algorithms have been devised for establishing GAC according to the nature of the constraints. For table constraints, STR [23] is such an algorithm: it removes invalid tuples during search of supports using a sparse set data structure which separates valid tuples from invalid ones. This method of seeking supports improves search time by avoiding redundant tests on invalid tuples that have already been detected as invalid during previous GAC enforcements. STR2 [15], an optimization of STR, limits some basic operations concerning the validity of tuples and the identification of supports, through the introduction of two important sets called S^{sup} and S^{val} (described later). In the extreme best case, STR2 can be r times faster than STR.

We now introduce the concepts of pattern and sub-table that will be useful for compression.

Definition 2. *A **pattern** μ of a constraint c is an instantiation I of some variables of c. We note $scp(\mu)$ its scope, which is equal to $vars(I)$, $|\mu|$ its length,* which is equal to $|scp(\mu)|$, and $nbOcc(\mu)$ its number of occurrences in $rel(c)$, which is $|\{\tau \in rel(c) \mid \mu \subseteq \tau\}|$.

Example 2. In Table 1, $\mu_1 = \{x_1 = a, x_4 = c, x_5 = a\}$ and $\mu_2 = \{x_3 = c, x_5 = c\}$ are patterns of respective lengths 3 and 2, with $scp(\mu_1) = \{x_1, x_4, x_5\}$ and $scp(\mu_2) = \{x_3, x_5\}$.

Definition 3. *The **sub-table** T associated with a pattern μ of a constraint c is obtained by removing μ from tuples of c that contain μ and ignoring other tuples.*

$$T = \{\tau \setminus \mu \mid \tau \in rel(c) \wedge \mu \subseteq \tau\}$$

The scope of T is $scp(T) = scp(c) - scp(\mu)$

Example 3. Table 2 represents the sub-table associated with the pattern $\mu_1 = (x_1 = a, x_4 = c, x_5 = a)$ of c, described in Table 1.

Table 2. The sub-table T_1 associated with the pattern μ_1 of c

x_2	x_3
a	b
c	b
c	a

Definition 4. *An **entry** for a constraint c is a pair (μ, T) such that μ is a pattern of c and T is the sub-table associated with μ.*

Since the set of tuples represented by an entry (μ, T) represents in fact the Cartesian product of μ by T, we shall also use the notation $\mu \otimes T$ to denote a constraint entry. Notice that after the slicing process of a constraint into a set of entries, the set of tuples which are not associated with any pattern can be stored in a so called *default entry* denoted by $(\emptyset, T)$.

Example 4. The pattern $\mu = (x_1 = a, x_4 = c, x_5 = a)$ of the constraint c, depicted in Figure 2(a), appears in tuples τ_2, τ_3 and τ_7. Thus, μ and the resulting sub-table form an entry for c, as shown in Figure 2(b).

Testing the validity of classical or compressed tuples is an important operation in filtering algorithms of (compressed) table constraints. For sliced table constraints, we extend the notion of validity to constraint entries.

Definition 5. *An entry (μ, T) is valid iff at least one tuple of the Cartesian product $\mu \otimes T$ is valid. Equivalently, an entry is valid iff its pattern is valid and its sub-table contains at least one valid sub-tuple.*

Fig. 1. Example of a constraint entry

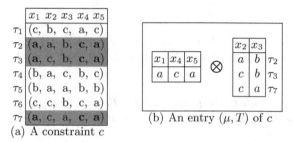

(a) A constraint c (b) An entry (μ, T) of c

2   Compression Method

Several data mining algorithms, such as Apriori [1] and FP-Growth [8] among others, can be used to identify frequent patterns. Since our objective is compression, we do not have to identify each possible frequent pattern but only the ones that are useful for compression, and specifically at most one pattern per tuple. The construction of an FP-Tree (Frequent-Pattern Tree) which is the first step in the FP-Growth algorithm is especially well suited to this goal since it identifies each long and frequent pattern. This construction requires only three scans of the table.

We briefly explain the construction of the FP-Tree in the context of table compression, using the constraint given in Table 1 as an example. Details of the general method can be found in [8,9]. The algorithm takes one parameter *minSupport* (minimum support) which is the minimal number of occurrences of a pattern that we require to consider it as frequent. In our example, we shall use *minSupport*=2 to identify patterns which occur at least twice.

In a first step, we collect the number of occurrences of each value. By abuse of terminology, we shall call frequency the number of occurrences of a value. This step requires one scan of the table. The result on our example is given in Figure 2(c). Then, in a second scan, we sort each tuple in decreasing order of frequency of values. The result is given in Figure 2(d) where the frequency of a value is given in parentheses. Values which have a frequency below the threshold *minSupport* are removed from the tuple (they are identified in bold face) because they cannot appear in a frequent pattern. Once a tuple is sorted and possibly reduced, it is inserted in the FP-Tree which is essentially a trie where each branch represents the frequent part of a tuple and each node contains the number of branches which share that node. Each edge from a parent to its child is labeled with a value. The root node does not have any label.

Figure 3(a) represents the FP-tree obtained on our running example. The first tuple inserted in the tree is the beginning of τ_1, that is $(x_1 = c, x_3 = c, x_5 = c)$. This creates the leftmost branch of the tree. Each node of this branch is given a frequency of 1. The second tuple inserted is $(x_4 = c, x_5 = a, x_1 = a, x_2 = a, x_3 = b)$ which creates the third leftmost branch in the tree (each node having a frequency of 1 at this step). When τ_3 is inserted, the new branch $(x_4 = c, x_5 = a, x_1 = a, x_2 = c, x_3 = b)$ shares its first three edges with the last branch, hence the frequency

	x_1	x_2	x_3	x_4	x_5
a	3	3	2	1	4
b	2	1	3	2	1
c	2	3	2	4	2

(c) frequencies

τ_1	(2) $x_1 = c$ (2) $x_3 = c$ (2) $x_5 = c$ (1) $\mathbf{x_2 = b}$ (1) $\mathbf{x_4 = a}$
τ_2	(4) $x_4 = c$ (4) $x_5 = a$ (3) $x_1 = a$ (3) $x_2 = a$ (3) $x_3 = b$
τ_3	(4) $x_4 = c$ (4) $x_5 = a$ (3) $x_1 = a$ (3) $x_2 = c$ (3) $x_3 = b$
τ_4	(3) $x_2 = a$ (2) $x_1 = b$ (2) $x_3 = c$ (2) $x_4 = b$ (2) $x_5 = c$
τ_5	(3) $x_2 = a$ (2) $x_1 = b$ (2) $x_3 = a$ (2) $x_4 = b$ (1) $\mathbf{x_5 = b}$
τ_6	(4) $x_4 = c$ (4) $x_5 = a$ (3) $x_2 = c$ (3) $x_3 = b$ (2) $x_1 = c$
τ_7	(4) $x_4 = c$ (4) $x_5 = a$ (3) $x_1 = a$ (3) $x_2 = c$ (2) $x_3 = a$

(d) tuples sorted according to decreasing frequencies

Fig. 2. First two steps of the compression

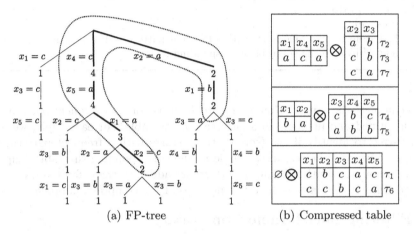

(a) FP-tree   (b) Compressed table

Fig. 3. FP-tree and compressed table

of the corresponding nodes is incremented and becomes 2. The other tuples are inserted in the same way. In the end, nodes with a frequency below the threshold *minSupport* are pruned. The remaining tree is depicted with thick lines and circled by a dashed line in Figure 3(a).

We now have to identify patterns in the FP-tree which are relevant for compression. Each node of the tree corresponds to a frequent pattern μ which can be read on the path from the root to the node. The frequency f of this pattern is given by the node itself. The savings that can be obtained by factoring this frequent pattern is $|\mu| \times (f - 1)$ values (we can save each occurrence of the pattern but one). In our example, we can see that the pattern $(x_4 = c, x_5 = a)$ can save 6 values, the pattern $(x_4 = c, x_5 = a, x_1 = a)$ can also save 6 values but the pattern $(x_4 = c, x_5 = a, x_1 = a, x_2 = c)$ can save only 4 values. Therefore, we further prune the tree by removing nodes that save less values than their parent. The leaves of the tree we obtain represent the frequent pattern used in the compression: $(x_4 = c, x_5 = a, x_1 = a)$ and $(x_2 = a, x_1 = b)$.

Algorithm 1. compress(T: table, minSupport: integer)

---

1 compute the frequency of each value of T
2 **for** $i \in 1..|T|$ **do**
3 | $\tau \leftarrow T[i]$
4 | sort τ by decreasing order of value frequency and remove values less
     | frequent than $minSupport$
5 | add τ to the FP-Tree (will update the nodes frequency)
6 | $tmp[i] \leftarrow \tau$

7 prune the tree by removing nodes which are less frequent than $minSupport$ or
   such that $|\mu| \times (f - 1)$ is smaller than for their parent
8 **for** $i \in 1..|T|$ **do**
9 | $\tau \leftarrow tmp[i]$
10 | lookup in the tree if τ starts with a frequent pattern μ. If it does not, $\mu \leftarrow \emptyset$
11 | add $T[i] \setminus \mu$ to the sub-table corresponding to μ

To complete the compression, we create an entry for each frequent pattern we
have identified and fill them in a last scan of the table. For each tuple, we use
the FP-tree to identify if the (sorted) tuple starts with a frequent pattern, in
which case we add the rest of the tuple to the corresponding sub-table. Tuples
which do not start with a frequent pattern are added to the default entry.

Algorithm 1 summarizes the different steps of the compression process.

3   Filtering Sliced Table Constraints

In order to enforce GAC on sliced table constraints, our idea is to adapt Simple
Tabular Reduction (STR), and more specifically the optimized variant STR2,
on the compressed form of this kind of constraint. As a sliced table constraint
is composed of several entries, each one composed of both a pattern and a sub-
table, the filtering process we propose acts at two distinct levels. At a high
level, the validity of each entry is checked, and at a low-level, the validity of
each pattern and each sub-tuple is checked. Remember that an entry is valid iff
both its pattern is valid and at least one tuple from its sub-table is valid (See
Definition 5). In this section, we first describe the employed data structures,
then we introduce our GAC algorithm, and finally we give an illustration.

3.1   Data Structures

A sliced table constraint c is represented by an array entries[c] of p entries.
Managing the set of valid entries, called *current*[1] entries, is performed as follows:

- entriesLimit[c] is the index of the last current entry in entries[c]. The
  elements in entries[c] at indices ranging from 1 to entriesLimit[c] are the
  current entries of c.

---

[1] Current entries correspond to valid entries at the end of the previous evocation of
the algorithm.

- removing an entry (that has become invalid) at index i is performed by a call of the form removeEntry(c, i). Such a call swaps the entries at indices i and entriesLimit$[c]$, and then decrements entriesLimit$[c]$. Note that the initial order of entries is not preserved.
- restoring a set of entries can be performed by simply changing the value of entriesLimit$[c]$.

Each entry in **entries** can be represented as a record composed of a field **pattern** and a field **subtable**. More precisely:

- the field **pattern** stores a partial instantiation μ, and can be represented in practice as a record of two arrays: one for the variables, the scope of the pattern, and the other for the values.
- the field **subtable** stores a sub-table T, and can be represented in practice as a record of two arrays: one for the variables, i.e., the scope of the sub-table T, and the other, a two-dimensional array, for the sub-tuples.

In our presentation, we shall directly handle μ and T without considering all implementation details ; for example, T will be viewed as a two-dimensional array. Managing the set of valid sub-tuples, called *current* sub-tuples, of T, is performed as follows:

- $limit[T]$ is the index of the last current sub-tuple in T. The elements in T at indices ranging from 1 to limit$[T]$ are the current sub-tuples of T.
- removing a sub-tuple (that has become invalid) at index i is performed by a call of the form removeSubtuple(T, i). Such a call swaps the sub-tuples at indices i and limit$[T]$, and then decrements limit$[T]$. Note that the initial order of sub-tuples is not preserved.
- restoring a set of sub-tuples can be performed by simply changing the value of limit$[T]$.

Note that the management of both current entries and current sub-tuples is in the spirit of STR. Also, as in [15], we introduce two sets of variables, called S^{val} and S^{sup}. The set S^{val} contains uninstantiated variables (and possibly, the last assigned variable) whose domains have been reduced since the previous invocation of the filtering algorithm on c. To set up S^{val}, we need to record the domain size of each modified variable x right after the execution of STR-slice on c: this value is recorded in lastSize$[x]$. The set S^{sup} contains uninstantiated variables (from the scope of the constraint c) whose domains contain each at least one value for which a support must be found. These two sets allow us to restrict loops on variables to relevant ones. We also use an array gacValues$[x]$ for each variable x. At any time, gacValues$[x]$ contains all values in $dom(x)$ for which a support has already been found: hence, values for a variable x without any proved support are exactly those in $dom(x) \setminus$ gacValues$[x]$. Note that the sets S^{val} and S^{sup} are initially defined with respect to the full scope of c. However, for each sub-table we also shall use local sets S^{lval} and S^{lsup} of S^{val} and S^{sup} as explained later.

3.2 Algorithm

Algorithm 2 is a filtering procedure, called STR-slice, that establishes GAC on a specified sliced table constraint c belonging to a CN N. Lines 1–10, which are exactly the same as those in Algorithm 5 of [15], allow us to initialize the sets S^{val}, S^{sup} and gacValues. Recall that S^{val} must contain the last assigned variable, denoted by lastPast(P), if it belongs to the scope of c. Lines 11–22 iterate over all current entries of c. To test the validity of an entry, we check first the validity of the pattern μ (Algorithm 3), and then, only when the pattern is valid, we check the validity of the sub-table T by scanning it (Algorithm 4). If an entry is no more valid, it is removed at line 22. Otherwise, considering the values that are present in the pattern, we have to update gacValues as well as S^{sup} when a first support for a variable is found. Lines 23–30, which are exactly the same as those in Algorithm 5 of [15], manage the reduction of domains: unsupported values are removed at line 25 and if the domain of a variable x becomes empty, an exception is thrown at line 27. Also, the set of variables X_{evt} reduced by STR-slice is computed and returned so that these "events" can be propagated to other constraints.

Algorithm 4 is an important function, called scanSubtable, of STR-slice. Its role is to iterate over all current (sub)tuples of a given sub-table, in order to collect supported values and to remove invalid tuples. Note that when this function is called, we have the guarantee that the pattern associated with the sub-table is valid (note the "and then" short-circuit operator at line 14 of Algorithm 2). The first part of the function, lines 1–10, allow us to build the local sets S^{lval} and S^{lsup} from S^{val} and S^{sup}. Such sets are obtained by intersecting S^{val} with $scp(T)$ and S^{sup} with $scp(T)$, respectively. Once the sets S^{lval} and S^{lsup} are initialized, we benefit from optimized operations concerning validity checking and support seeking, as in STR2. The second part of the function, lines 9–21, consists in iterating over all current sub-tuples of T. This is a classical STR2-like traversal of a set of tuples. Finally, line 22 returns true when there still exists at least one valid sub-tuple.

It is interesting to note the lazy synchronization performed between the global unique set S^{sup} and the specific local sets S^{lsup} (one such set per sub-table). When a variable x is identified as "fully supported", it is immediately removed from S^{sup} (see line 19 of Algorithm 2 and line 18 of Algorithm 4). Consequently, that means that the next sub-tables (entries) will benefit from such a reduction, but the information is only transmitted at initialization (lines 6–8 of Algorithm 4). On the other hand, once initialized, the global set S^{val} is never modified during the execution of STR-slice.

Backtracking Issues: In our implementation, entries and tuples can be restored by modifying the value of the limit pointers (entriesLimit[c] and limit[T] for each sub-table T of c), recorded at each search depth. Restoration is then achieved in $O(1 + p)$ (for each constraint) where p is the number of entries.

Algorithm 2. STR-slice(c: constraint)

Input : c is a sliced table constraint of the CN N to be solved
Output : the set of variables in $scp(c)$ with reduced domain

// Initialization of sets S^{val} and S^{sup}, as in STR2
1   $S^{val} \leftarrow \emptyset$
2   $S^{sup} \leftarrow \emptyset$
3   **if** lastPast$(P) \in scp(c)$ **then**
4       $\lfloor$ $S^{val} \leftarrow S^{val} \cup \{\text{lastPast}(P)\}$

5   **foreach** *variable* $x \in scp(c) \mid x \notin past(P)$ **do**
6 | gacValues$[x] \leftarrow \emptyset$
7 | $S^{sup} \leftarrow S^{sup} \cup \{x\}$
8 | **if** $|dom(x)| \neq$ lastSize$[c][x]$ **then**
9 | | $S^{val} \leftarrow S^{val} \cup \{x\}$
10 | $\lfloor$ lastSize$[c][x] \leftarrow |dom(x)|$

// Iteration over all entries of c
11   $i \leftarrow 1$
12   **while** $i \leq$ entriesLimit$[c]$ **do**
13 | $(\mu, T) \leftarrow$ entries$[c][i]$ // *ith* current entry of c
14 | **if** isValidPattern(μ) **and then** scanSubtable(T) **then**
15 | | **foreach** *variable* $x \in scp(\mu) \mid x \in S^{sup}$ **do**
16 | | | **if** $\mu[x] \notin$ gacValues$[x]$ **then**
17 | | | | gacValues$[x] \leftarrow$ gacValues$[x] \cup \{\mu[x]\}$
18 | | | | **if** $|dom(x)| = |$gacValues$[x]|$ **then**
19 | | | | $\lfloor$ $S^{sup} \leftarrow S^{sup} \setminus \{x\}$

20 | $\lfloor$ $i \leftarrow i + 1$
21 | **else**
22 | $\lfloor$ removeEntry(c, i) // entriesLimit$[c]$ decremented

// domains are now updated and X_{evt} computed, as in STR2
23   $X_{evt} \leftarrow \emptyset$
24   **foreach** *variable* $x \in S^{sup}$ **do**
25 | $dom(x) \leftarrow gacValues[x]$
26 | **if** $dom(x) = \emptyset$ **then**
27 | $\lfloor$ **throw** INCONSISTENCY
28 | $X_{evt} \leftarrow X_{evt} \cup \{x\}$
29     $\lfloor$ lastSize$[c][x] \leftarrow |dom(x)|$
30   **return** X_{evt}

Algorithm 3. isValidPattern(μ: pattern): Boolean

---

1   **foreach** *variable* $x \in scp(\mu)$ **do**
2 | **if** $\mu[x] \notin dom(x)$ **then**
3 | $\lfloor$ **return** *false*

4   **return** *true*

Algorithm 4. scanSubtable(T: sub-table): Boolean

Input   : T is a sub-table coming from an entry of the constraint c
Output : true iff there is at least one valid tuple in the sub-table T

```
// Initialization of local sets S^lval and S^lsup from S^val and S^sup
```
1 $S^{lval} \leftarrow \emptyset$
2 **foreach** *variable* $x \in S^{val}$ **do**
3 | **if** $x \in scp(T)$ **then**
4 | | $S^{lval} \leftarrow S^{lval} \cup \{x\}$

5 $S^{lsup} \leftarrow \emptyset$
6 **foreach** *variable* $x \in S^{sup}$ **do**
7 | **if** $x \in scp(T)$ **then**
8 | | $S^{lsup} \leftarrow S^{lsup} \cup \{x\}$

```
// Iteration over all (sub)tuples of T
```
9 $i \leftarrow 1$
10 **while** $i \leq$ limit$[T]$ **do**
11 | $\tau \leftarrow T[i]$ *// ith* current sub-tuple of T
12 | **if** isValidSubtuple(S^{lval}, τ) **then**
13 | | **foreach** *variable* $x \in S^{lsup}$ **do**
14 | | | **if** $\tau[x] \notin$ gacValues$[x]$ **then**
15 | | | | gacValues$[x] \leftarrow$ gacValues$[x] \cup \{\tau[x]\}$
16 | | | | **if** $|dom(x)| = |$gacValues$[x]|$ **then**
17 | | | | | $S^{lsup} \leftarrow S^{lsup} \setminus \{x\}$
18 | | | | | $S^{sup} \leftarrow S^{sup} \setminus \{x\}$

19 | | $i \leftarrow i + 1$
20 | **else**
21 | | removeSubtuple(T, i))   *// limit$[T]$* decremented

22 **return** limit$[T] > 0$

Algorithm 5. isValidSubtuple(S^{lval}: variables, τ: tuple): Boolean

---

1 **foreach** *variable* $x \in S^{lval}$ **do**
2 | **if** $\tau[x] \notin dom(x)$ **then**
3 | | **return** *false*

4 **return** *true*

However, by introducing a simple data structure, it is possible to only call the restoration procedure when necessary, limiting restoration complexity to $O(1)$ in certain cases: it suffices to register the limit pointers that need to be updated when backtracking, and this for each level. When the search algorithm backtracks, we also have to deal with the array lastSize. As mentioned in [15], we can record the content of such an array at each depth of search, so that the original state of the array can be restored upon backtracking.

As GAC-slice is a direct extension of STR2, it enforces GAC.

3.3   Illustration

Figures 4 and 5 illustrate the different steps for filtering a sliced table constraint, when STR-slice is called after an event. In Figure 4, considering that the new event is simply $x_3 \neq a$ (i.e., the removal of the value a from $dom(x_3)$), STR-slice starts checking the validity of the current entries (from 1 to entriesLimit). So, for the first entry, the validity of the pattern $\mu=\{x_1 = a, x_4 = c, x_5 = a\}$ is first checked. Since μ remains valid (our hypothesis is that the event was only $x_3 \neq a$), the sub-table of the first entry is scanned. Here, only the sub-tuple $\{x_2 = c, x_3 = a\}$ is found invalid, which modifies the value of limit for the sub-table of this first entry. After the call to STR-slice, the constraint is as in Figure 4(b).

In Figure 5, considering now that the new event is $x_3 \neq b$, we start again with the first current entry. Figuring out that the pattern is still valid, we check the validity of the associated sub-tuples. Since the sub-tuple $\{x_2 = a, x_3 = b\}$ is no more valid, it is swapped with $\{x_2 = c, x_3 = b\}$. This latter sub-tuple is then also found invalid, which sets the value of limit to 0. This is illustrated in Figure 5(a). As the sub-table of the first entry is empty, the entry is removed by swapping its position with that of last current entry. After the call to STR-slice, the constraint is as in Figure 5(b) (note that a second swap of constraint entries has been performed).

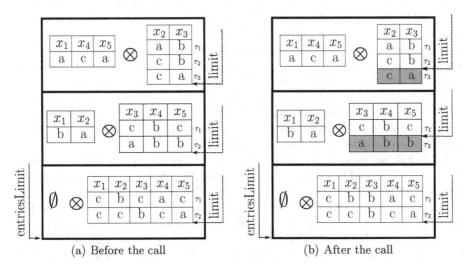

(a) Before the call   (b) After the call

Fig. 4. STR-slice called on a slice table constraint after the event $x_3 \neq a$

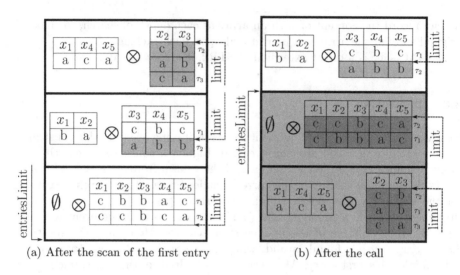

(a) After the scan of the first entry   (b) After the call

Fig. 5. From Figure 4(b), STR-slice called after the event $x_3 \neq b$

4   Experimental Results

In order to show the practical interest of our approach to represent and filter sliced table constraints, we have conducted an experimentation (with our solver AbsCon) using a cluster of bi-quad cores Xeon processors at 2.66 GHz node with 16GiB of RAM under Linux. Because STR2 and STR3 belong to the state-of-the-art GAC algorithms for table constraints, we compare the respective behaviors of STR2, STR3 and STR-slice on different series of instances[2] involving positive table constraints with arity greater than 2. STR1 is also included as a baseline.

For STR-slice, we select exclusively frequent patterns with a number of occurrences at least equal to 10% of the number of tuples in the table. This value was obtained after several experiments, similarly we chose 10 for the minimum size of sub-tables. Automatically tuning, on a specific instance, the frequency threshold for patterns and the minimum sub-table size is part of our future work.

We use MAC with the dom/ddeg variable ordering and lexico as value ordering heuristic, to solve all these instances. A time-out of $1,200$ seconds was set per instance. The two chosen heuristics guarantee that we explore the very same search tree regardless of the filtering algorithm used.

Table 3 shows mean results (CPU time in seconds) per series. For each series, the number of tested instances is given by #ins ; it corresponds to the number of instances solved by all three variants within $1,200$ seconds. Note that the mean compression ratios and CPU times (in seconds) are also given for STR-slice between parentheses. We define the compression ratio as the size of the sliced tables over the size of the initial tables, where the size of a (sliced) table is the number of values over all patterns and (sub-)tables. The results in Table 3 show that

---
[2] Available at http://www.cril.univ-artois.fr/CSC09

Table 3. Mean CPU time (in seconds) to solve instances from different series (a time-out of 1,200 seconds was set per instance) with MAC. Mean compression ratio and CPU time are given for STR-slice between parentheses.

Instance	#ins	STR1	STR2	STR3	STR-slice
a7-v24-d5-ps05	11	298.05	147.73	189.14	**115.30** (66% – 5.74)
bdd	70	44.53	**13.44**	99.21	20.35 (86% – 0.59)
crossword-ogd	43	90.05	39.35	**25.69**	29.59 (75.51% – 0.36)
crossword-uk	43	95.20	45.88	**44.33**	47.21 (88.69% – 0.18)
renault	46	19.66	14.39	**13.37**	17.20 (47.15% – 0.67)

Table 4. CPU time (in seconds) on some selected instances solved by MAC

Instance	STR1	STR2	STR3	MDD	STR-slice
a7-v24-d5-ps0.5-psh0.7-9	879	334	367	**25.5**	200 (69% – 5.41)
a7-v24-d5-ps0.5-psh0.9-6	353	195	324	**16.6**	174 (62% – 5.82)
bdd-21-2713-15-79-11	78.5	**23.5**	48.5	82.6	31.7 (88.05% – 0.28)
crossword-ogd-vg12-13	799	342	**208**	> 1,200	242 (73.46% – 0.74)
crossword-uk-vg10-13	1,173	**576**	589	> 1,200	598 (89.63% – 0.48)

STR-slice is competitive with both STR2 and STR3. Surprisingly, although the compression ratio obtained for the instances of the series *renault* is rather encouraging, the CPU time obtained for STR-slice is disappointing. We suspect that the presence of many constraints with small tables in the *renault* instances is penalizing for STR-slice because, in that case, the overhead of managing constraint entries is not counterbalanced by the small absolute spatial reduction. Table 4 presents the results obtained on some instances.

In term of space, STR3 is the variant that uses the most amount of memory (sometimes by a very large factor). STR-slice, although requiring a few additional data structures is the cheapest STR variant in term of memory (approximately as indicated by the compression ratio in Tables 3 and 4). Note that other compression approaches of the literature such as those based on MDDs [3] may outperform STR variants when the compression ratio is (very) high. This is the case, for example, on the instances of series *a7-v24-d5-ps05*. However, on other series such as *crossword*, the MDD approach can be outperformed by a large factor by the STR variants (on hard crossword instances, STR2, STR3 and STR-slice are usually about 5 times faster than MDD).

A general observation from this rather preliminary experimentation is that STR-slice is a competitor to STR2 and STR3, but a not a competitor that takes a real advantage. Several perspectives to improve this situation are developed in the conclusion.

5 Conclusion

In this paper, we have presented an original approach for filtering table constraints: it combines a new compression technique using the concept of sliced table constraints and an optimized adaptation of the tabular reduction (as in STR2). Our preliminary experimentation shows that STR-slice is a competitor to the state-of-the-art STR2 and STR3 algorithms. To make STR-slice indispensable, many further developments are necessary. First, we think that the tuning of the parameters used for guiding compression should be automatized (possibly, employing some machine learning techniques). STR-slice could then benefit from a better compression. Second, we believe that, in the rising context of big data, new constraint problems should emerge rapidly where constraints could be of (very) large arity and involve very large tables. STR-slice could advantageously handle such "huge" constraints, especially if we consider that slicing could be conducted recursively on the sub-tables (another perspective of this work). Finally, we think that the concept of sliced table constraints is interesting on its own for modeling, as certain forms of conditionality can be represented in a simple and natural way, directly with sliced table constraints.

Acknowledgments. This work has been supported by both CNRS and OSEO within the ISI project 'Pajero'.

References

1. Agrawal, R., Srikant, R.: Fast algorithms for mining association rules in large databases. In: Proceedings of the 20th International Conference on Very Large Data Bases, pp. 487–499 (1994)
2. Bessiere, C., Régin, J.: Arc consistency for general constraint networks: preliminary results. In: Proceedings of IJCAI 1997, pp. 398–404 (1997)
3. Cheng, K., Yap, R.: An MDD-based generalized arc consistency algorithm for positive and negative table constraints and some global constraints. Constraints 15(2), 265–304 (2010)
4. Fahle, T., Schamberger, S., Sellmann, M.: Symmetry breaking. In: Walsh, T. (ed.) CP 2001. LNCS, vol. 2239, pp. 93–107. Springer, Heidelberg (2001)
5. Focacci, F., Milano, M.: Global cut framework for removing symmetries. In: Walsh, T. (ed.) CP 2001. LNCS, vol. 2239, pp. 77–92. Springer, Heidelberg (2001)
6. Gent, I.P., Jefferson, C., Miguel, I., Nightingale, P.: Data structures for generalised arc consistency for extensional constraints. In: Proceedings of AAAI 2007, pp. 191–197 (2007)
7. Gharbi, N., Hemery, F., Lecoutre, C., Roussel, O.: STR et compression de contraintes tables. In: Proceedings of JFPC 2013, pp. 143–146 (2013)
8. Han, J., Pei, J., Yin, Y.: Mining frequent patterns without candidate generation. In: Proceedings of SIGMOD 2000, pp. 1–12 (2000)
9. Han, J., Pei, J., Yin, Y., Mao, R.: Mining frequent patterns without candidate generation: A frequent-pattern tree approach. Data Mining and Knowledge Discovery 8(1), 53–87 (2004)

10. Hubbe, P.D., Freuder, E.C.: An efficient cross product representation of the constraint satisfaction problem search space. In: Proceedings of AAAI 1992, pp. 421–427 (1992)
11. Jabbour, S., Sais, L., Salhi, Y.: A mining-based compression approach for constraint satisfaction problems. CoRR, abs/1305.3321 (2013)
12. Jabbour, S., Sais, L., Salhi, Y., Uno, T.: Mining-based compression approach of propositional formulae. In: Proceedings of CIKM 2013, pp. 289–298 (2013)
13. Katsirelos, G., Bacchus, F.: Generalized nogoods in CSPs. In: Proceedings of AAAI 2005, pp. 390–396 (2005)
14. Katsirelos, G., Walsh, T.: A compression algorithm for large arity extensional constraints. In: Bessière, C. (ed.) CP 2007. LNCS, vol. 4741, pp. 379–393. Springer, Heidelberg (2007)
15. Lecoutre, C.: STR2: Optimized simple tabular reduction for table constraints. Constraints 16(4), 341–371 (2011)
16. Lecoutre, C., Likitvivatanavong, C., Yap, R.: A path-optimal GAC algorithm for table constraints. In: Proceedings of ECAI 2012, pp. 510–515 (2012)
17. Lecoutre, C., Sais, L., Tabary, S., Vidal, V.: Transposition Tables for Constraint Satisfaction. In: Proceedings of AAAI 2007, pp. 243–248 (2007)
18. Lecoutre, C., Szymanek, R.: Generalized arc consistency for positive table constraints. In: Benhamou, F. (ed.) CP 2006. LNCS, vol. 4204, pp. 284–298. Springer, Heidelberg (2006)
19. Lhomme, O., Régin, J.C.: A fast arc consistency algorithm for n-ary constraints. In: Proceedings of AAAI 2005, pp. 405–410 (2005)
20. Mairy, J.-B., Van Hentenryck, P., Deville, Y.: An optimal filtering algorithm for table constraints. In: Milano, M. (ed.) CP 2012. LNCS, vol. 7514, pp. 496–511. Springer, Heidelberg (2012)
21. Mohr, R., Masini, G.: Good old discrete relaxation. In: Proceedings of ECAI 1988, pp. 651–656 (1988)
22. Pesant, G.: A regular language membership constraint for finite sequences of variables. In: Wallace, M. (ed.) CP 2004. LNCS, vol. 3258, pp. 482–495. Springer, Heidelberg (2004)
23. Ullmann, J.R.: Partition search for non-binary constraint satisfaction. Information Science 177, 3639–3678 (2007)
24. Xia, W., Yap, R.H.C.: Optimizing STR algorithms with tuple compression. In: Schulte, C. (ed.) CP 2013. LNCS, vol. 8124, pp. 724–732. Springer, Heidelberg (2013)

The PrePack Optimization Problem

Maxim Hoskins[1,2], Renaud Masson[1], Gabrielle Gauthier Melançon[1],
Jorge E. Mendoza[2], Christophe Meyer[3], and Louis-Martin Rousseau[1]

[1]CIRRELT, École Polytechnique de Montréal, Montreal, Canada
[2]Université Catholique de l'Ouest, LARIS (EA 7315), Angers, France
[3]Université du Québec à Montréal, Montreal, Canada

Abstract. The goal of packing optimization is to provide a foundation
for decisions related to inventory allocation as merchandise is brought to
warehouses and then dispatched. Major retail chains must fulfill requests
from hundreds of stores by dispatching items stored in their warehouses.
The demand for clothing items may vary to a considerable extent from
one store to the next. To take this into account, the warehouse must
pack "boxes" containing different mixes of clothing items. The number
of distinct box types has a major impact on the operating costs. Thus,
the PrePack problem consists in determining the number and contents
of the box types, as well as the allocation of boxes to stores. This paper
introduces the PrePack problem and proposes CP and MIP models and
a metaheuristic approach to address it.

1 Introduction

Major retail chains must fulfill requests from hundreds (or even thousands) of
stores by dispatching items stored in their warehouses. For instance, in the fashion industry, one must dispatch to each store an assortment of clothes of varied
colors and sizes, in such a way that the demand of the store clients is satisfied.
The items are usually shipped in boxes containing several different items.

A *box configuration* is a possible way to fill a box. The number of different
configurations used has a major impact on the operating costs. If there are many
configurations, it will be possible to satisfy the demand of the stores in a precise
manner but with elevated operating costs. If the number of configurations is
restricted, the operating costs will be low, but some items may be overstocked
or understocked at the stores.

The *PrePack* problem consists in determining the number and contents of
the box configurations, as well as the configurations to ship to each store, to
minimize overstocking and understocking. In this paper we will consider the
number of configurations to be fixed, but variants where this number is a decision
variable also exist. Moreover, the supplier will usually favor overstocking since
understocking decreases customer satisfaction. However, because overstocking is
costly for the stores they usually impose a hard limit. Furthermore, it is not
permissible to overstock an item in a store where there is no demand for it,
because the item would never be sold. Finally, the box capacities are predefined,
and all boxes must be full.

H. Simonis (Ed.): CPAIOR 2014, LNCS 8451, pp. 136–143, 2014.
© Springer International Publishing Switzerland 2014

The PrePack problem described above is of significant importance and the subject of recent patent applications [1,2,3]. However, to our knowledge, there are no current scientific publications that present an algorithm for it. The most relevant combinatorial problem, on which [1,2,3] build, is the multi-choice knapsack problem described in [4]. We formally introduce the PrePack problem and present three solution approaches. The preliminary results suggest that despite its apparent simplicity, the prepacking problem is a challenging combinatorial optimization problem that deserves to be studied more deeply.

The remainder of the paper is organized as follows. Section 2 formalizes the problem using constraint programming; Section 3 formulates the problem as a mixed integer program; and Section 4 presents hybrid metaheuristics. Section 5 discusses the results obtained by each approach, and Section 6 concludes the paper.

2   Problem Definition

A more precise description of the problem is given through the following constraint programming model. The model, which considers a fixed number of boxes where each box represents a configuration of items, uses the following indices: $i \in I$ for each item, $s \in S$ for each store, and $b \in B$ for each box configuration. From now on the sets will be implied; for example, we will write $\forall(i, s)$ instead of $\forall i \in I, s \in S$. We denote by K the set of possible box capacities (if there is no restriction on the capacities, we simply set $K = \mathbb{Z}_+$).

$$\min \text{Obj} = \sum_{i,s}(\alpha \cdot under(i, s) + \beta \cdot over(i, s)) \tag{1}$$

$$\sum_b fill(b, i) \times send(b, s) = \text{dem}(i, s) + over(i, s) - under(i, s) \ \ \forall(i, s) \tag{2}$$

$$over(i, s) \leq \text{overlimit}(i, s) \ \ \forall(i, s) \tag{3}$$

$$\sum_i fill(b, i) = capa(b) \ \ \forall(b) \tag{4}$$

$$capa(b) \in K \ \ \forall(b) \tag{5}$$

$$over(i, s), under(i, s), fill(b, i), send(b, s), capa(b) \in \mathbb{Z}_+ \ \ \forall(i, s, b) \tag{6}$$

The objective (1) is to minimize the cost of understocking (*under*) and overstocking (*over*), where α and β are the understock and overstock penalties. Constraint (2) is nonlinear and ensures that the demand (*dem*) of each item in each store is met. Here, *fill* defines the amount of each item to be packed in each box configuration and *send* indicates how many boxes of each configuration are to be sent to each store. Constraint (3) ensures that the overstocks are less than or equal to the predefined overstock allowances. Constraint (4) ensures that the boxes are full. Finally constraint (5) imposes the restrictions on the box capacities.

Although the above constraints fully define the problem, it may be helpful to add redundant constraints to reduce the solution time. In particular, we considered the following constraints:

$$\sum_b send(b, s) \le \left\lceil \frac{\sum_i \Big(\mathrm{dem}(i, s) + \mathrm{overlimit}(i, s) \Big)}{[\mathrm{min\ box\ capacity}]} \right\rceil \quad \forall(s) \qquad (7)$$

$$over(i, s) \cdot under(i, s) = 0 \quad \forall(i, s) \qquad (8)$$

$$capa(b) \ge capa(b - 1) \quad \forall(b \ge 1) \qquad (9)$$

$$\sum_i (under(i, s) + over(i, s)) \ge modulo\left(\sum_i \mathrm{dem}(i, s), 2\right) \quad \forall(s) \qquad (10)$$

Constraint (7) is interesting when the box capacities are greater than one since it limits nontrivially the number of boxes to be sent to each store. Although constraint (8) will be automatically satisfied by any optimal solution, imposing it can lead to a substantial reduction of the solution space. Note that the formulation contains several symmetries. Some of them can be removed by adding constraint (9), which ensures that the box configurations are considered in increasing order of their capacity. Finally, unlike the previous constraints, constraint (10) is valid only when the boxes are required to have an even capacity. With this constraint, if the total demand of a store is odd, then there must be some overstock or understock at that store.

3   Mixed Integer Problem Model

For this formulation we need the following variables: y_{bi} = amount of item i in box configuration b; x_{bs} = number of box configurations b shipped to store s; z_{is} = total amount of item i shipped to store s; t_{bk} = 1 if box configuration b corresponds to a box of capacity k and 0 otherwise; o_{is} = overstock of item i at store s; and u_{is} = understock of item i at store s. The overall nonlinear model is then:

$$\min Obj = \sum_{i,s} (\alpha u_{is} + \beta o_{is}) \qquad (11)$$

$$z_{is} - o_{is} + u_{is} = \mathrm{dem}(i, s) \quad \forall(i, s) \qquad (12)$$

$$z_{is} = \sum_b x_{bs} y_{bi} \quad \forall(i, s) \qquad (13)$$

$$\sum_i y_{bi} = \sum_k k \cdot t_{bk} \quad \forall(b) \qquad (14)$$

$$\sum_k t_{bk} = 1 \quad \forall(b) \qquad (15)$$

$$o_{is} \le \mathrm{overlimit}(i, s) \quad \forall(i, s) \qquad (16)$$

$$t_{bk} \in \{0, 1\}, x_{bs}, y_{bi}, z_{is} \in \mathbb{Z}_+, o_{is}, u_{is} \ge 0 \quad \forall(b, i, k, s) \qquad (17)$$

Constraints (12) and (13) ensure that the store demands are satisfied, with possible understock and overstock. Constraints (14) and (15) ensure that each box is completely filled to one of the predefined capacities. Constraints (16) restrict the overstocking, while constraint (17) defines each variable's domain.

The above formulation can be linearized by standard techniques to enable the use of standard solvers such as CPLEX. We decompose the x variables by introducing binary variables v_{bsl} such that $x_{bs} = \sum_l 2^l \cdot v_{bsl} \ \forall(b,s)$. When multiplying x_{bs} by y_{bi}, we obtain the product $w_{bisl} = v_{bsl} y_{bi}$. We replace this product by w_{bisl} and add the following constraints:

$$w_{bisl} \leq \bar{Y} v_{bsl} \ \ \forall(b,i,s,l) \tag{18}$$

$$w_{bisl} \leq y_{bi} \ \ \forall(b,i,s,l) \tag{19}$$

$$w_{bisl} \geq 0 \ \ \forall(b,i,s,l) \tag{20}$$

$$w_{bisl} \geq y_{bi} - \bar{Y}(1 - v_{bsl}) \ \ \forall(b,i,s,l) \tag{21}$$

$$z_{is} = \sum_{b,l} 2^l \cdot w_{bisl} \ \ \forall(i,s) \tag{22}$$

where $\bar{Y}$ is an upper bound on the variables y_{bi}. The linearized model is then obtained by replacing (13) by (22) and removing the x variables.

Of the constraints (7)–(10), the last one proved to be helpful when the capacities of the boxes are even, and it was introduced in the following form:

$$\sum_i (u_{is} + o_{is}) \geq modulo\left(\sum_i dem(i,s), 2\right) \qquad \forall(s) \tag{23}$$

In contrast, the symmetry-breaking constraints (9) did not help and thus were not included in the final model.

4 Hybrid Metaheuristic

We developed a two-phase hybrid metaheuristic. In the first phase, the approach uses a memetic algorithm (MA) to explore the solution space and builds a pool of *interesting* box configurations. In the second phase, the approach solves a box-to-store assignment problem, to i) choose a subset of configurations from the pool and ii) decide how many boxes of each configuration should be sent to each store.

In our MA, individuals are represented using a multi-array genotype. Each of the $|B|$ arrays in the genotype represents a box configuration, i.e., a vector of integers with $|I|$ positions. The initial population is built using three constructive heuristics: demand-driven insertion, cost-driven insertion, and random insertion. The first heuristic fills boxes with items selected based on the demand across stores. The second heuristic fills boxes based on an estimation of the impact of the resulting configuration on the objective function. The third heuristic has two steps: in the first step it randomly selects a box capacity, and in the second step it fills the selected box with a random number of items of each type.

The initial population is evolved using two evolutionary operators, namely, crossover and mutation. The crossover operator, *horizontal-vertical crossover*, recombines two parents to form four offspring. The underlying idea is to mix whole box configurations to generate a new individual and also to build new box configurations by recombining existing packing patterns. Figure 1 illustrates the operator for $|B| = 3$. The mutation operator simply sweeps each configuration in the mutating individual and swaps some of the item values. Finally, the local-search operator tries to improve an individual by changing the number of each type of item in each of the individual's configurations. During the execution of the MA, the fitness function of a solution is computed by heuristically solving the box-to-store assignment problem using a fast procedure. To control the evolutionary process, we borrowed the logic of the generational MA introduced in [5]. To support our implementation we used the Java Genetic Algorithm framework [6]. The second phase of our approach consists in solving the box-to-store

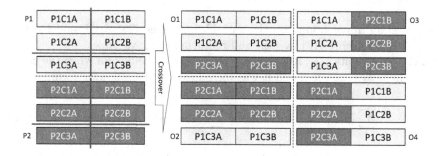

Fig. 1. Horizontal-vertical crossover example

assignment problem over a set of configurations C found during the MA. The composition of C may differ depending on the MA's parameters. To solve the assignment problem we use the following set covering model:

$$\min \sum_{i,s} \alpha u_{is} + \beta o_{is} \tag{24}$$

$$\text{s.t.} \sum_{c} a_{ic} x_{cs} = d_{is} + o_{is} - u_{is} \quad \forall (i,s) \tag{25}$$

$$\sum_{s} x_{cs} \leq \bar{X} y_c \quad \forall c \tag{26}$$

$$\sum_{c} y_c = |B| \tag{27}$$

where $c \in C$. The objective (24) is to minimize the understock u_{is} and overstock o_{is} with the parameters α and β representing the understock and overstock penalties. Constraint (25) controls the demand; a_{ic} is the number of item i present in configuration c, and x_{cs} is the number of box configurations c sent to store s. Variable y_c is 1 if configuration c is used and 0 otherwise. Constraint (26) ensures that no unused configuration is sent to a store. Constraint (27) ensures that the correct number of configurations is used to represent a solution.

To solve (24)–(27) we use either a commercial solver (CPLEX) or a large neighborhood search (LNS). Starting from a valid solution, the LNS partially destroys the solution with one operator and reconstructs it with another operator. The destruction operator selects random box configurations to remove from the solution. The reconstruction operator is based on a best-insertion algorithm. The algorithm adds to the solution the box configuration that provides the smallest increase in the value of the objective function. The objective function is determined after an assignment heuristic has assigned the box configurations to the stores. LNS repeats this process until the predetermined number of box configurations to use has been satisfied.

5   Results

We tested our approaches on variants of a real-world instance with 58 stores demanding 24 ($= 6 \times 4$) different items: T-shirts available in six different sizes and four different colors (black, blue, red, and green). Each item has a fixed overstock limit (0 or 1) for all stores but no understock limits. The available box capacities are 4, 6, 8, and 10. Finally, the overstock and understock penalties are $\beta = 1$ and $\alpha = 10$. From this instance we derived smaller instances obtained by considering only some of the colors and/or only the first ten stores. We set the maximum execution time to 15 minutes.

The results are presented in the following two tables. The second column indicates the number of box configurations used to build the solution. Columns 3 and 5 indicate the number of understocked and overstocked items across all the stores. Columns 4 and 6 indicate the time needed to prove optimality in the case of the exact models (unless the maximum time is reached, in which case we indicate the best solution found) or to find the best solution for the metaheuristic. Columns 7 and 8 indicate respectively the number of nodes in the branching tree and the resulting gap.

5.1   Exact CP and MIP Models

For the constraint programming model (CP), the results show that the model is able to quickly prove optimality when no understocking is required. This is because there exists a feasible solution for which (10) is tight. The mixed integer problem (MIP) approach performs efficiently on all the mono-color instances. However, for larger instances, there are two issues, First, it has difficulty finding good feasible solutions. For example, the LP relaxation value for the instance `BlackBluex10` coincides with the optimal value; what prevents CPLEX from solving the model is its inability to find the corresponding feasible solution. Second, it has difficulty improving the best bound; CPLEX may not be able to

Instance	No. Boxes	CP		MIP			
		Under/over stock	CPU (s)	Under/over stock	CPU (s)	No. Nodes	Gap
Black x 58	4	(0; 58)	19	(0; 58)	0.09	27	0%
Blue x 58	4	(0; 0)	118	(0; 0)	0.83	580	0%
Red x 58	4	(16; 0)	900	(16; 0)	0.42	523	0%
Green x 58	4	(0; 0)	0.5	(0; 0)	0.04	4	0%
BlackBlue x 10	7	(0; 10)	300	(7; 13)	900	88900	87.95%
BlackBlue x 58	7	(76; 110)	900	(36;106)	900	55198	87.55%
AllColor x 10	14	(33; 3)	900	(47; 19)	900	3700	98.77%
AllColor x 58	14	(401; 93)	900	-	900	0	-

improve the lower bound found at the root node. For the instance `AllColorx58`, CPLEX is not even able to process the root node (the time limit was reached as CPLEX's heuristics were trying to find a feasible solution). The results are presented with CPLEX's default parameters; some other settings have been explored without any improvement.

5.2 Hybrid Metaheuristic

The metaheuristic was run with two separate configurations: solving (24)–(27) using CPLEX over the set C made up of all the box configurations found in the individuals of the MA's final populations; and solving (24)–(27) using LNS over the set C made up of all the box configurations explored during the MA's execution (that is, a much larger set of columns). The results show that on the

Instance	Nb of boxes	CPLEX		LNS	
		under/over stock	CPU (s)	under/over stock	CPU (s)
Black x 58	4	(0; 58)	6	(0; 58)	7
Blue x 58	4	(1; 1)	7	(10; 10)	8
Red x 58	4	(58; 0)	7	(50; 0)	8
Green x 58	4	(0; 0)	7	(0; 0)	8
BlackBlue x 10	7	(4; 26)	7	(1; 11)	16
BlackBlue x 58	7	(35; 175)	43	(0; 174)	74
AllColor x 10	14	(18; 22)	49	(7; 19)	293
AllColor x 58	14	(168; 146)	273	(40; 148)	900

mono-color instances both metaheuristics find two of the four optimal solutions. On the larger instances, it is clear that the CPLEX-based model is less effective than the LNS model. This is because the latter approach has a larger solution space to explore.

6 Conclusion

We have introduced the PrePack optimization problem, a problem that does not seem to have been studied before. Our preliminary results show that this problem can be very hard, even for relatively small instances, as illustrated by the CP and MIP approaches. The hybrid metaheuristic was able to return a solution for all instances in a relatively short time. However, the results are not as good as those of the the CP and MIP approaches. This is because the quality of the results delivered by the second phase depends on the pool of generated configurations.

This problem certainly deserves further study. More valid inequalities and cuts will be necessary to improve the performance of the MIP approach. The performance of the CP model could be improved by adding effective surrogate constraints. The metaheuristic is promising because it is faster than the exact models, but the pool-generation phase needs to create more effective configurations to produce better solutions.

Acknowledgement. This paper is based on work that was initiated during the *Fifth Montreal Problem Solving Workshop* held at the *Université de Montréal* in August 2013. We would like to acknowledge financial support from the CRM and Mprime, and we thank the CRM for providing a stimulating working environment during the workshop. We are especially grateful to Eric Prescott Gagnon from JDA Software for proposing the problem and for being present throughout the week to help us understand the context.

References

1. Erie, C.W., Lee, J.S., Paske, R.T., Wilson, J.P.: Dynamic bulk packing and casing. International Business Machines Corporation, US20100049537 A1 (2010)
2. Vakhutinsky, A., Subramanian, S., Popkov, Y., Kushkuley, A.: Retail pre-pack optimizer. Oracle International Corporation, US20120284079 A1 (2012)
3. Pratt, R.W.: Computer-implemented systems and methods for pack optimization. SAS Institute, US20090271241 A1 (2009)
4. Chandra, A.K., Hirschberg, D.S., Wong, C.K.: Approximate algorithms for some generalized knapsack problems. Theoretical Computer Science 3(3), 293–304 (1976)
5. Mendoza, J.E., Medaglia, A.L., Velasco, N.: An evolutionary-based decision support system for vehicle routing: The case of a public utility. Decision Support Systems 46, 730–742 (2009)
6. Medaglia, A.L., Gutérrez, E.J.: An object-oriented framework for rapid development of genetic algorithms. Handbook of Research on Nature Inspired Computing for Economics and Management. Idea Publishing Group (2006)

An Integrated Constraint Programming Approach to Scheduling Sports Leagues with Divisional and Round-Robin Tournaments

Jeffrey Larson[1,2], Mikael Johansson[1], and Mats Carlsson[3]

[1] Automatic Control Lab, KTH, Osquldas väg 10, SE-100 44 Stockholm, Sweden
{jeffreyl,mikaelj}@kth.se
[2] Mathematics and Computer Science Division, Argonne National Laboratory,
Argonne, IL 60439, USA
jmlarson@anl.gov
[3] SICS, P.O. Box 1263, SE-164 29 Kista, Sweden
matsc@sics.se

Abstract. Previous approaches for scheduling a league with round-robin and divisional tournaments involved decomposing the problem into easier subproblems. This approach, used to schedule the top Swedish handball league Elitserien, reduces the problem complexity but can result in suboptimal schedules. This paper presents an integrated constraint programming model that allows to perform the scheduling in a single step. Particular attention is given to identifying implied and symmetry-breaking constraints that reduce the computational complexity significantly. The experimental evaluation of the integrated approach takes considerably less computational effort than the previous approach.

1 Introduction

A double round-robin tournament (DRRT), where every team plays every other team once home and once away, is one of the most common formats for a broad range of sporting events. Football leagues around the world (EPL, Serie A, La Liga, Bundesliga, Eredivisie, Allsvenskan, CONCACAF, and many others) base their season schedules on DRRTs; the format has also been used by Super Rugby, Indian Premier League (cricket), and even the Chess World Championship (where each player played every other player once as black and once as white). As the format is ubiquitous, considerable research into scheduling DRRTs efficiently and fairly has been conducted [20,21]. A comprehensive literature survey and an introduction to sports scheduling terminology can be found in [11].

There are, of course, many reasons why any league may wish to transition away from a DRRT. For example, the top Danish football league contains only 12 teams, and consists of a triple round-robin tournament to offer a sufficient number of games over the season [12]. The Swedish handball league, Elitserien, augments its traditional double round-robin format by dividing teams into two divisions, each of which hold a single round-robin tournament before a league-wide DRRT. Compared to the amount of research that has been devoted to DRRTs, these more "exotic" league formats have

H. Simonis (Ed.): CPAIOR 2014, LNCS 8451, pp. 144–158, 2014.

received limited attention in the literature (see, e.g., [21, Section 5.2] for a notable exception). This paper develops a constraint programming (CP) approach to league scheduling for the combination of divisional and round-robin play employed by Elitserien.

Since DRRT scheduling is already difficult (cf. [2] for NP-completeness results), one might be led to believe that augmentations of DRRT are even harder to address. While it is true that their complexity can make them hard to schedule, they can also allow for extra degrees of freedom or impose new constraints which make it easier to construct a schedule. For example, Elitserien requires that teams meeting three times (those in the same division) must not play at the same venue in consecutive meetings. We denote this requirement the *Alternative Venue Requirement*, or AVR for short. As we will see, this requirement ultimately makes constructing a season schedule easier.

When scheduling sport leagues, it is often common to break the problem into sub-problems that can be addressed more easily. For example, the schedule-then-break approach of Trick [20] is widely used in the scheduling of DRRTs. We adopted this approach in an earlier paper [7], where we developed a schedule for Elitserien by first constructing and enumerating a set of home-away patterns sets (HAP sets), not all of which were schedulable with respect to the AVR. A series of increasingly restrictive necessary conditions removed unschedulable HAP sets. Next, for each HAP set, tournament templates (a tournament containing generic numbers and not actual team names) were generated and ranked according to a number of factors, including carry-over effects [18], that are not easy to optimize for directly. The construction of a template was an essential step in scheduling the league, because any change in the schedule required approval from the team owners who are accustomed to working with templates. After a template was agreed upon, we used an integer programming approach to assign actual teams to the numbers in the template in a manner satisfying various constraints (e.g., venue availability or desired derby matches).

Of course, such a decomposed approach (first building a HAP set, then fixing a single template, and finally constructing a schedule) can result in a suboptimal league schedule depending on which template was chosen and the particular venue availabilities and desires that occur in a given year. Even if the league owners agree to adjust the template when the availabilities and desires become known, exploring all possibilities to find the optimal solution would result in a computationally expensive generate-and-test scheme. In contrast, this paper proposes a constraint programming approach for integrated scheduling of Elitserien that completely eliminates the intermediate steps and quickly generates a provably optimal schedule. A particular effort is made to identify and break symmetries in the problem, and substantial speed-ups are obtained over the previous approaches.

For example, for the specific requirements imposed by Elitserien for the 2013-2014 season, the CP approach implemented on a standard desktop was able to find an optimal schedule and prove its optimality in less than half a minute. Although not all leagues will allow for identical symmetry breaking tricks, we believe the approach presented in this paper is general enough to be applicable to many league formats.

Constraint programming has certainly been used for sports scheduling before. Examples of decomposed CP approaches to finding DRRT schedules include [5,19,18].

Hybridized with other methods, CP has been used to minimize travel distance in sports tournaments [1,3,10]. In [16], Régin uses a CP approach to minimizing the number of breaks (consecutive home or away games) in sports schedules. Alternative CP models of sports scheduling were discussed in [8]. However, case studies solved by *integrated* CP approaches are scarce in the literature; perhaps the problems have been assumed to be intractable without decomposition into simpler subproblems.

The outline of the paper is as follows. In Section 2, we state the requirements on the schedule to be computed. In Section 3, we present a global constraint model of the problem, and discuss the additional seasonal constraints and preferences that arise every year. In Section 4, we report the results from our computational experiments. Section 5 concludes the paper.

2   Problem Statement and Basic Tournament Properties

The requirements on Elitserien's schedule can be broadly classified into two categories: the first category (which we call *structural constraints*) addresses the schedule format and fairness in terms of breaks (consecutive home or away games), periods without games (called byes), the alternating venue requirement, and the sequence of home and away games; the second category (which we refer to as *seasonal constraints*) concerns stadium and referee availabilities, the desire to support various match-ups (such as rivalries), wishes from the media, etc. Historically, Elitserien has determined their schedule by first proposing a tournament template which addresses the structural constraints. This tournament template has numbers in place of actual teams in the schedule. Every year, the league collects information about unavailabilities and particular wishes from the clubs and assigns teams to numbers in the tournament template to form the season schedule.

Although we will develop an integrated scheduling approach which accounts for all the above concerns, it is still useful to keep the distinction between the structural and seasonal constraints. In this way, we can examine the combinatorial properties and symmetries of the tournament template generation process itself and make a direct computational comparison with the approach used in [7]. We can also clearly expose the different symmetry-breaking techniques that apply to the tournament template design and the integrated scheduling respectively.

Elitserien poses the following structural constraints on its tournament template:

C1. Both 7-team divisions must hold an SRRT to start the season.
C2. This must be followed by a DRRT between the entire league. The DRRT is organized into two SRRTs, where the second SRRT is the mirrored complement of the first: the order is reversed, home games become away games and vice versa.
C3. There must be a minimum number of breaks in the schedule.
C4. Each team has one bye during the season to occur during the divisional RRT.
C5. At no point during the season can the number of home and away games played by any team differ by more than 1.
C6. Any pair of teams must have consecutive meetings occur at different venues (AVR).
C7. Each division must have 3 pairs of complementary schedules.

Elitserien considers any consecutive home or away matches to be breaks. Therefore, a team playing away-bye-away or home-bye-home constitutes a break, as does a team ending the SRRT with the same type of game (home or away) as they start the DRRT with.

The structural constraints C1-C7 are hard constraints that do not change from year to year. However, the teams themselves might change, and every year sees different constraints on venue availabilities, match-ups, and derbies. Thus, in addition to the structural constraints, the integrated scheduling approach also accounts for the following seasonal constraints.

C8. Each division must contain a prescribed set of teams.
C9. Specific pairs of teams in each division have to be assigned complementary schedules (typically teams that come from the same city, or even share the same arena).
C10. To increase the visibility of handball, the league arranges derbies in specific periods. Elitserien derby constraints consist of a single period and a set of teams, out of which as many matches as possible should be formed. Alternatively, a single team, a single period, and a set of possible opponents are given.
C11. Venue unavailabilities have to be respected to the highest extent possible.

The league considers constraints C8-C10 hard, but allow some flexibility with constraint C11. This flexibility arises because, although a venue might be available for the target dates of a specific game round, the league allows the teams flexibility to move a game date a few days forward or backward in time. For example, a game scheduled for Saturday can be played on Friday or Sunday, depending on venue, referee, and team availabilities.

The problem that we address in this paper is how to generate a schedule that satisfies requirements C1-C10 and violates a minimum number of venue unavailabilities (requirement C11).

2.1 Basic Tournament Properties

To motivate our approach to tournament template and schedule design, we recall a basic property of Elitserien's tournament format established in [7]. The first proposition [7, Proposition 2.1] gives a lower bound on the minimum number of breaks required to schedule a league combining divisional and round-robin play.

Proposition 1. *In an n-team league (n even) with a schedule consisting of two concurrent divisional RRTs followed by two consecutive full-league RRTs, if only one bye is allowed and it must occur during the divisional RRT, any schedule must have at least $2n - 4$ breaks.*

In [7], it was also shown that it is possible to construct schedules that achieve this lower bound. This is accomplished by combining the unique divisional RRT from [4] with the fact that HAP sets must break in odd periods if they ensure that C5 is satisfied: the cumulative number of home and away games played never differs by more than 1 at any point in the season [17]. The home-away pattern for a tournament combining divisional and round-robin play can thus be constructed by combining the divisional

RRT home-away patterns in Fig. 1(a) with two copies of a full-season RRT home-away pattern in Fig. 1(b) without introducing additional breaks. To construct a given team HAP, Part I consists of one pattern from Fig. 1(a). This is followed by Part II, which consists of one pattern from Fig. 1(b). Part III is the reflected complement of the Part II pattern; i.e., if Part II ends AHH then Part III starts AAH. The schedule will also mirror this pattern. If Part II of the schedule ends with team 1 playing at team 2, Part III will start with team 2 hosting team 1.

```
       BAHAHAH       AHAHAHAHAHAHA
       HBAHAHA       AHAHAHAHAHAHH
       AHBAHAH       AHAHAHAHAHHAH
       HAHBAHA       AHAHAHAHHAHAH
       AHAHBAH       AHAHAHHAHAHAH
       HAHAHBA       AHAHHAHAHAHAH
       AHAHAHB       AHHAHAHAHAHAH
         or          HAAHAHAHAHAHA
       BHAHAHA       HAHAAHAHAHAHA
       ABHAHAH       HAHAHAAHAHAHA
       HABHAHA       HAHAHAHAAHAHA
       AHABHAH       HAHAHAHAHAAHA
       HAHABHA       HAHAHAHAHAHAA
       AHAHABH       HAHAHAHAHAHAH
       HAHAHAB
         (a)              (b)
```

Fig. 1. Left: Two HAP sets for a 7-team, no-break RRT. Right: HAP set satisfying Elitserien's requirements for a 14-team, 12-break RRT. These HAP sets are unique up to permutation of the rows.

HAP sets created as the patterns in Fig. 1 have many attractive properties. Taking the unique (up to permutation of the rows) no-break, 7-team tournament HAP set and its complement ensures that 7 teams play at home and 7 teams play away in period 8 without introducing a break. If we did not take the complement, we would have 8 teams needing to play at home without introducing a break in period 8, an impossibility. Since we are reflecting and taking the complement of Part II to schedule Part III, and breaks only occur during odd periods (to ensure the number of home and away games never differ by more than 1 at any point in the season), there are no breaks in period 9. This implies no team has a break to end the season; in other words, every team plays at home one of the last two periods of the season.

At first glance, the reflecting and taking the complement of Part II to form Part III forces teams to play the same team in periods 20 as they do in period 21 (at the opposite venue). This could be undesirable, depending on the league, but it is a non-issue for Elitserien. Period II ends before Christmas, allowing for a month-long break for Champions League competitions before Period III starts at the beginning of February.

2.2 Tournament Specific Properties

We now recall a number of properties of the unique HAP set satisfying Elitserien's requirements shown in Fig. 1. These properties will be instrumental for the development of efficient implied and symmetry-breaking constraints:

P1. In each division, one bye occurs in each period.
P2. It follows from C5 that breaks can only occur on odd periods.
P3. The three pairs of complementary schedules per division required by C7 must have breaks that are pairwise aligned, e.g.:

> HBAHAHA | HAHAHAHAHAAHA
> AHBAHAH | AHAHAHAHAHHAH
>
> HAHBAHA | HAHAAHAHAHAHA
> AHAHBAH | AHAHHAHAHAHAH
>
> HAHAHBA | HAAHAHAHAHAHA
> AHAHAHB | AHHAHAHAHAHAH

P4. Two HAPs can only be complementary if the byes occur in adjacent periods [7, Proposition 3.3]. It is apparent by visual inspection of Fig. 1 that two non-adjacent sequences are non-complementary in at least one of the periods 1 through 8.
P5. If the byes are placed as in Fig. 1, the required three pairs of complementary schedules must include teams 2, 4 and 6 of the given division.
P6. If the byes are placed as in Fig. 1, the first row of Part I is complementary to the first column for each division.

3 Constraint Model

We now describe the integrated constraint programming model in detail. We first define the variables to be used in the CP, and then the essential constraints to ensure the resultant schedule will satisfy Elitserien's structural requirements C1-C7. We next highlight some implied constraints and symmetry breaking properties that we find greatly reduce the search effort. Lastly, we model the league's seasonal constraints so we can construct the entire schedule in an integrated approach. The constraint model was encoded in MiniZinc 1.6 and executed with Gecode 3.7.0 as back-end. Full details of our experiments are given in Section 4.

3.1 Problem Variables

We first note that it is sufficient to consider only the first part of the DRRT because its second half is the mirrored complement of the first half. According to C1-C2, let $t \in 1..14$ denote a team and $p \in 1..20$ denote a period. The tournament template corresponds to the array of variables $T[t, p] \in -14..14$, where $T[t, p] < 0$ stands for team t playing away in period p, $T[t, p] > 0$ if it plays at home, and $T[t, p] = 0$ if it has a bye. The HAP set corresponds to the array of variables $H[t, p] \in \{A, B, H\}$. We also need an array $O[t, p] \in 1..14$ where $O[t, p]$ stands for the opponent of team t in period p. $O[t, p] = t$ if and only if it has a bye in that period. Finally, we need an array $B[t] \in 0..20$ which stands for the period in which the break for team t occurs, or 0 if team t has no break in its schedule.

In [6], the authors show that, if the constraint model uses opponent variables, like ours does, then an SRRT can be codified by two types of constraints, the filtering algorithms of which are crucial to performance. First, every period consists of a matching (or one-factor) of the teams. This is captured by the *symmetric alldifferent* [15] (a.k.a. *one-factor*) constraint; see (5) below. Second, the complete set of opponents for a given team i is the entire set of teams without team i. This is captured by the *alldifferent* constraint [13]; see (6) below. Alternatives to opponent variables are discussed in [8].

3.2   Structural Constraints

The T array is channeled to the O and H arrays straightforwardly:

$$T[t, p] = \begin{cases} -O[t, p], & \text{if } H[t, p] = A \\ O[t, p], & \text{if } H[t, p] = H \\ 0, & \text{if } H[t, p] = B \end{cases} , \forall t, \forall p. \qquad (1)$$

The definition of a break and the channeling between the B and H arrays are captured by the following constraint:

$$B[t] = \sum_{p \in 8..19} p \times (H[t, p] = H[t, p + 1]), \forall t. \qquad (2)$$

As mentioned above, the channeling between the O and H arrays is captured by the following constraint:

$$O[t, p] = t \Leftrightarrow H[t, p] = B, \forall t, \forall p. \qquad (3)$$

The possible HAP for any team is constrained by C3-C5 and by the fact that we know the set of sequences that must make up a HAP set satisfying Elitserien's requirements; see Fig. 1. This is easily captured by a regular expression e. The corresponding finite automaton is shown in Fig. 2, and the corresponding *regular* constraint [9] is posted on every row of H:

$$regular([H[t, p] \mid p \in 1..20], e), \forall t. \qquad (4)$$

As mentioned above, it is a fundamental RRT constraint, usually encoded with *symmetric alldifferent*, that every period consist of a matching of the teams. Unfortunately, *symmetric alldifferent* is not among the standard MiniZinc global constraints, and has no native support in Gecode. Fortunately, is is easily emulated by MiniZinc's *inverse* global constraint, which is supported by Gecode:

$$O[O[t, p], p] = t, \forall t, \forall p, \qquad (5)$$

i.e.,

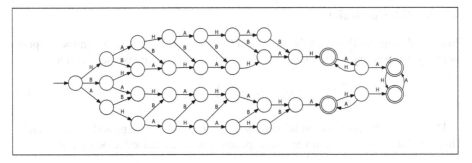

Fig. 2. The finite automaton for valid HAP. As usual, accepting states are indicated by double circles. When restricted to sequences of length 20, it accepts the combinations of rows in Fig. 1.

$$inverse([O[t,p] \mid t \in 1..14], [O[t,p] \mid t \in 1..14]), \forall p.$$

Each team must meet every other team during Part I and Part II. As mentioned above, this is another fundamental RRT constraint, easily expressed with *alldifferent*:

$$alldifferent([O[t,p] \mid p \in 1..7]) \wedge alldifferent([O[t,p] \mid p \in 8..20]), \forall t. \qquad (6)$$

Home and away must match for every team and its opponent, everywhere. This is yet another fundamental RRT constraint:

$$\begin{pmatrix} (H[t,p] = \mathtt{A} \wedge H[O[t,p],p] = \mathtt{H}) \vee \\ (H[t,p] = \mathtt{B} \wedge H[O[t,p],p] = \mathtt{B}) \vee \\ (H[t,p] = \mathtt{H} \wedge H[O[t,p],p] = \mathtt{A}) \end{pmatrix}, \forall t, \forall p. \qquad (7)$$

To encode C6, we note that it is satisfied if and only if every row of the tournament template contains distinct nonzero values. And since every team has exactly one bye, we can use *alldifferent*:

$$alldifferent([T[t,p] \mid p \in 1..20]), \forall t \qquad (8)$$

Finally, to model C7, let $i \oplus j$ denote the fact that teams i and j have complementary schedules:

$$i \oplus j \Leftrightarrow B[i] = B[j] \wedge (H[i,p] \neq H[j,p], \forall p \in 1..20).$$

Then we have:

$$\exists \{i,j,k,l,m,n\} \subset 1..7 \text{ such that } (i \oplus j) \wedge (k \oplus l) \wedge (m \oplus n) \text{ and} \\ \exists \{i,j,k,l,m,n\} \subset 8..14 \text{ such that } (i \oplus j) \wedge (k \oplus l) \wedge (m \oplus n). \qquad (9)$$

3.3 Implied Constraints

Property P3 implies that each division must have six breaks. It was determined experimentally that posting this implied constraint improves propagation (see Section 4).

$$\sum_{t\in 1..7} (B[t] > 0) = 6 \text{ and } \sum_{t\in 8..14} (B[t] > 0) = 6. \tag{10}$$

The fact that the breaks must be pairwise aligned could also be posted as a constraint. This implied constraint, however, was experimentally determined to be useless:

$$\sum_{t\in 1..7} (B[t] = p) \text{ is even and } \sum_{t\in 8..14} (B[t] = p) \text{ is even}, \forall p.$$

It's a structural property that the numbers of home and away matches must always match. Contrary to Trick's observation [21, Section 6], this implied constraint was also experimentally found to be useless:

$$\sum_{t\in 1..14} (H[t,p] = \text{A}) = \sum_{t\in 1..14} (H[t,p] = \text{H}), \forall p.$$

We have Property P1, which is useful in itself, but which is subsumed by (15), as we shall see later:

$$\begin{aligned} alldifferent([p \mid H[t,p] = \text{B}, t \in 1..7, p \in 1..7]) \wedge \\ alldifferent([p \mid H[t,p] = \text{B}, t \in 8..14, p \in 1..7]). \end{aligned} \tag{11}$$

Finally, we know from P2-P3 that out of the 14 teams, two teams must have a break each in periods 9, 11, 13, 15, 17 and 19, and two teams must have no break. This is efficiently encoded by an *global cardinality* constraint [14], and was also experimentally found to be useful:

$$\sum_{t\in 1..14} (B[t] = i) = 2, \forall i \in \{0, 9, 11, 13, 15, 17, 19\}. \tag{12}$$

3.4 Breaking Symmetries

The following constraints help remove many of the symmetries in the scheduling of Elitserien. There is an obvious home/away symmetry: given a solution, we can construct another solution by swapping home and away everywhere. This symmetry is easily broken:

$$H[1,2] = \text{H} \wedge H[2,1] = \text{A}. \tag{13}$$

A second source of symmetry is the fact that the two divisions can be swapped in the template. This symmetry can be broken by lexicographically ordering the break sequences:

$$[B[t] \mid t \in 1..7] \leq_{\text{lex}} [B[t] \mid t \in 8..14]. \tag{14}$$

There is third source of symmetry in the model: given a solution, we can construct another solution by swapping rows (teams) i and j of the same division in the arrays as well as values i and j (positive or negative) in O and T. To break this symmetry, we can fix the bye period for all teams like in Fig. 1, subsuming (11):

$$
\left(
\begin{array}{ll}
H[t,t] = \text{B} & \wedge \\
O[t,t] = t & \wedge \\
O[t,p] \in (1..7) \setminus \{t\} \quad \forall p \neq t & \wedge \\
H[t+7,t] = \text{B} & \wedge \\
O[t+7,t] = t+7 & \wedge \\
O[t+7,p] \in (8..14) \setminus \{t+7\} \quad \forall p \neq t &
\end{array}
\right), \forall t \in 1..7. \tag{15}
$$

Having fixed the bye periods in such a manner, we can use Properties P4-P5 to construct a slightly stronger version of (9) that restricts the possible pairing of complementary schedules:

$$
\begin{aligned}
(1 \oplus 2 \vee 2 \oplus 3) \wedge \\
(3 \oplus 4 \vee 4 \oplus 5) \wedge \\
(5 \oplus 6 \vee 6 \oplus 7) \wedge \\
(8 \oplus 9 \vee 9 \oplus 10) \wedge \\
(10 \oplus 11 \vee 11 \oplus 12) \wedge \\
(12 \oplus 13 \vee 13 \oplus 14)
\end{aligned} \tag{16}
$$

$$
B[t] > 0, \ \forall t \in \{2, 4, 6, 9, 11, 13\}. \tag{17}
$$

Property P6 can be posted as an implied constraint, which was determined experimentally to improve propagation, but only marginally:

$$
H[t,1] \neq H[1,t] \wedge H[t+7,1] \neq H[8,t], \forall t \in 2..7. \tag{18}
$$

Finally, alternative symmetry breaking constraints are discussed in [20].

3.5 Seasonal Constraints

The constraints described in the previous section capture the generic Elitserien structural constraints i.e., a 14-team league with a schedule consisting of (Part I) two concurrent divisional SRRTs, followed by (Part II) an RRT between all teams, and (Part III) the reverse complement of Part II. Also, for every season there are a number of extra requirements and preferences:

Team Mapping (C8). Team *names* must be substituted for team *numbers*. This requires a level of indirection in the form of another array:

$$
M[t] \in \begin{cases} 1..7 & , \text{ for teams constituting Div 1} \\ 8..14 & , \text{ for teams constituting Div 2} \end{cases}
$$

where $M[t]$ is the team number (template row) assigned to team t. The M array can be treated in two different ways, both of which are evaluated in Section 4:

1. One can let M be an array of problem variables, which allows us to keep the symmetry breaking constraint (15, 16, 17, 18).
2. One can fix M before search, but replace the same symmetry breaking constraints by (9) and (11), which make no assumptions on the placement of byes.

Complementary Schedules (C9). For reasons of e.g., venue availability, certain pairs (i, j) of teams are required to have complementary schedules:

$$M[i] \oplus M[j]$$

Derbies (C10). A derby constraint is given as a period p and a set $\mathcal{Q}$ of four teams, out of which two pairs of playing teams must be formed. Alternatively, a set $\mathcal{T}$ of three teams is given, two of which must play each other:

$$\begin{pmatrix} (O[M[i], p] = j \land O[M[j], p] = i) & \lor \\ (O[M[i], p] = k \land O[M[k], p] = i) & \lor \\ (O[M[j], p] = k \land O[M[k], p] = j) \end{pmatrix}, \text{ where } \mathcal{T} = \{i, j, k\}.$$

$$O[M[i], p] \in \{M[j] \mid j \in \mathcal{Q}\}, \forall i \in \mathcal{Q}.$$

Venue Unavailabilities (C11). These are soft constraints, turning the scheduling problem into an optimization problem. If the preferences are stated as an array:

$$N[t, p] = \begin{cases} 1 \text{ , if team } t \text{ prefers not to play at home during period } p \\ 0 \text{ , otherwise,} \end{cases}$$

then the cost function for all three parts of the schedule is:

$$\sum_{\substack{t \in 1..14, \\ p \in 1..20}} N[t, p] \times (H[M[t], p] = \text{H}) + \sum_{\substack{t \in 1..14, \\ p \in 21..33}} N[t, p] \times (H[M[t], 41 - p] = \text{A})$$

4   Experiments

The constraint model was encoded in MiniZinc 1.6 and executed with Gecode 3.7.0 as back-end on a quad core 2.8 GHz Intel Core i7-860 machine with 8MB cache per core, running Ubuntu Linux. In the parallel runs, Gecode was given 8 threads in the usual way with the command line option -p 8:

```
mzn-gecode -p 8 ...
```

The MiniZinc models and instances can be found here:

```
http://www.sics.se/~matsc/Elitserien
```

In a first experiment, we solved the generic scheduling problem, involving the structural constraints encoded by constraints (1–18) except (9) and (11), which are implied by constraints (16) and (15), respectively, but no seasonal constraints. Searching on the H variables in row-major order, followed by the O variables in row-major order, the first solution was found using one thread in 70ms and 81 failures. With the symmetry

Table 1. HAP sets distribution according to how many templates satisfying C1-C7 they admit

Range	No. HAP sets
$[10^1, 10^2)$	3
$[10^2, 10^3)$	13
$[10^3, 10^4)$	39
$[10^4, 10^5)$	41
$[10^5, 10^6)$	7
$[10^6, 10^7)$	1
Total	104

breaking described in Section 3.4, it was already known that there exist 104 distinct, feasible HAP sets. We enumerated these sets, and counted the number of solutions per set. As shown in Table 1, the solution counts vary a lot, from 34 to 2249812. This would suggest that some HAP sets can accommodate seasonal constraints much better than others.

In a second experiment, we considered the optimization problem subject to structural as well as seasonal constraints, treating the mapping M as an array of problem variables. This allowed us to keep the symmetry breaking constraints (15–18), which are very effective. Constraints (13) and (14) are however not valid in this context and were disabled.

In this experiment, we generated 20 random instances because (a) our model has only been used for one year and we wished to verify that this instance was not an especially easy case to solve, and (b) the league requested that we not divulge the true team desires for privacy reasons. The structure of the random seasonal constraints were very similar to the real ones, though:

- The partitioning of teams into divisions (C8).
- One specific pair of teams in each division to be assigned complementary schedules (C9).
- One 3-team intra-division derby set, one 3-team inter-division derby set, and one 4-team inter-division derby set (C10).
- For each team t and period p, t prefers to not play at home during period p with probability 0.054, yielding on average 25 unavailabilities (C11), which is the number of unavailabilities requested by Elitserien teams for the season that we scheduled.

We searched on the H variables in column-major order, followed by the M variables, followed by the O variables in row-major order. Fig. 3 shows the results in terms of number of instances solved to optimality as a function of elapsed time, with one curve for 1 thread and one curve for 8 threads. For 8 threads, the median solve time to optimality was 15s, the average 42s, and the standard deviation 54s. These numbers are reasonable, and show that there are no extreme outliers among our random instances. We also observe a rather uniform speed-up of 3x-4x from parallelism.

This is an improvement over the approach in [7], which first generated 80640 HAP sets satisfying C3-C5 but not necessarily schedulable, then applied necessary conditions

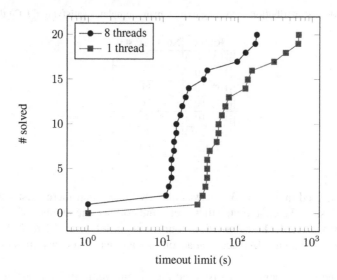

Fig. 3. Number of instances solved to optimality as function of timeout limit in seconds

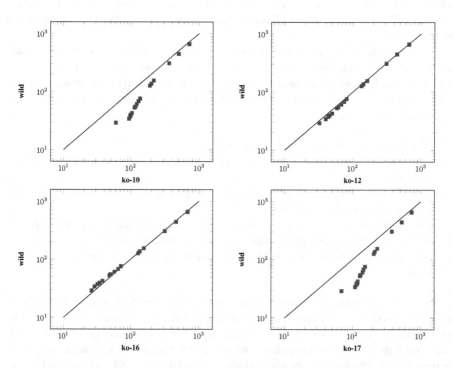

Fig. 4. Effect of knocking out a single implied constraint. Solve time to optimality in seconds, running on a single thread. The model variants are: **wild** – all constraints; **ko-10** – (10) knocked out; **ko-12** – (12) knocked out; **ko-16** – (16) knocked out and replaced by (9); **ko-17** – (17) knocked out.

for schedulability to rule out some 87% of the unschedulable HAP sets. An attempt was then made to convert the remaining HAP sets to templates by solving an integer program. The resultant templates were ranked in their carry-over effect to produce a template for the league. This template was then assigned teams with respect to the seasonal constraints C8-C11. Testing all HAP sets against the necessary conditions took nearly a day. Since the template was fixed before the seasonal constraints were available, it is likely that a suboptimal schedule was produced. Furthermore, a straightforward application of the approach in [7] to scheduling where a template does not need to be fixed a priori would clearly be inefficient: the 104 schedulable HAP sets admit 5,961,704 templates if constraints (13) and (14) are used, or 23,846,816 templates if they are not. Assuming that it takes 0.1 seconds per template to assign teams to numbers optimally, an optimistic estimate, finding the best schedule would take almost one month.

In order to gauge the effectiveness of implied and symmetry-breaking constraints (10, 12, 16, 17), we ran the same instances with the same model, except in each run, a given constraint was disabled. The results are shown as scatter plots in Fig. 4, running on a single thread for maximal runtime stability. We find that only constraints (10) and (17) had any significant impact, shortening the solve time by up to three times. An unexpected observation is that the speed-up gained from these extra constraints seems to decrease as the difficulty of the instance increases. Constraint (12) gave a marginal speedup, whereas replacing constraints (9) by (16) had a slightly detrimental effect.

In a third experiment, we considered the same optimization problem, but fixed the mapping M before search. So we had to disable constraints (15–18) and instead enable constraints (9) and (11), which make no assumptions on the placement of byes. This gave much worse results than the second experiment: on an instance that was solved with proof of optimality in 20 seconds on 8 threads in the second experiment, an optimal solution was found in about 3.5 minutes in the third experiment, with a proof of optimality obtained after 10 hours.

5 Conclusion

In this paper, we analyzed the situation where a league augments a traditional DRRT schedule by forming two divisions of teams, each of which hold an SRRT to start the season. This asymmetry (pairs of teams play three times if they are in the same division, twice otherwise) makes constructing feasible schedules an interesting problem. We highlighted the concerns of Elitserien, which we consider to be general enough to apply to many other leagues. We presented a constraint programming model for the problem, integrating the different phases that sports scheduling traditionally decomposes to, and showed a dramatic improvement over a previous approach using decomposition and integer programming. Non-decomposition approaches are rare in the sports scheduling literature. In our work, we made heavy use of implied and symmetry-breaking constraints, some of which turned out to be crucial to performance.

Acknowledgments. Jeffrey Larson and Mikael Johansson were funded in part by the Swedish Foundation for Strategic Research and the Swedish Science Council.

References

1. Benoist, T., Laburthe, F., Rottembourg, B.: Lagrange relaxation and constraint programming collaborative schemes for travelling tournament problems. In: Proceedings CPAIOR, vol. 1, pp. 15–26 (2001)
2. Briskorn, D.: Sports Leagues Scheduling. Lecture Notes in Economics and Mathematical Systems, vol. 603. Springer (2008)
3. Easton, K., Nemhauser, G.L., Trick, M.A.: The traveling tournament problem description and benchmarks. In: Walsh, T. (ed.) CP 2001. LNCS, vol. 2239, pp. 580–584. Springer, Heidelberg (2001)
4. Fronček, D., Meszka, A.: Round robin tournaments with one bye and no breaks in home-away patterns are unique. In: Multidisciplary Scheduling: Theory and Applications, pp. 331–340. MISTA, New York (July 2005) ISSN 2305-249X
5. Henz, M.: Scheduling a major college basketball conference—revisited. Operations Research 49(1), 163–168 (2001)
6. Henz, M., Müller, T., Thiel, S.: Global constraints for round robin tournament scheduling. European Journal of Operational Research 153(1), 92–101 (2004)
7. Larson, J., Johansson, M.: Constructing schedules for sports leagues with divisional and round-robin tournaments. Journal of Quantitative Analysis in Sports (to appear, 2014) doi:10.1515/jqas-2013-0090
8. Perron, L.: Alternate modeling in sport scheduling. In: van Beek, P. (ed.) CP 2005. LNCS, vol. 3709, pp. 797–801. Springer, Heidelberg (2005)
9. Pesant, G.: A regular language membership constraint for finite sequences of variables. In: Wallace, M. (ed.) CP 2004. LNCS, vol. 3258, pp. 482–495. Springer, Heidelberg (2004)
10. Rasmussen, R.V., Trick, M.A.: The timetable constrained distance minimization problem. In: Beck, J.C., Smith, B.M. (eds.) CPAIOR 2006. LNCS, vol. 3990, pp. 167–181. Springer, Heidelberg (2006)
11. Rasmussen, R.V., Trick, M.A.: Round robin scheduling - a survey. European Journal of Operational Research 188(3), 617–636 (2008)
12. Rasmussen, R.: Scheduling a triple round robin tournament for the best Danish soccer league. European Journal of Operational Research 185, 795–810 (2008)
13. Régin, J.C.: A filtering algorithm for constraints of difference in CSPs. In: 12th National Conference on Artificial Intelligence (AAAI-1994), pp. 362–367 (1994)
14. Régin, J.-C.: Generalized arc consistency for global cardinality constraint. In: Clancey, W.J., Weld, D.S. (eds.) AAAI/IAAI, vol. 1, pp. 209–215. AAAI Press / The MIT Press (1996)
15. Régin, J.-C.: The symmetric alldiff constraint. In: Dean, T. (ed.) IJCAI, pp. 420–425. Morgan Kaufmann (1999)
16. Régin, J.-C.: Minimization of the number of breaks in sports scheduling problems using constraint programming. DIMACS Series in Discrete Mathematics and Theoretical Computer Science 57, 115–130 (2001)
17. Ribeiro, C., Urrutia, S.: Scheduling the Brazilian soccer tournament with fairness and broadcast objectives. In: Burke, E.K., Rudová, H. (eds.) PATAT 2007. LNCS, vol. 3867, pp. 147–157. Springer, Heidelberg (2007)
18. Russell, R.A., Urban, T.L.: A constraint programming approach to the multiple-venue, sport-scheduling problem. Computers & Operations Research 33(7), 1895–1906 (2006)
19. Schaerf, A.: Scheduling sport tournaments using constraint logic programming. Constraints 4(1), 43–65 (1999)
20. Trick, M.A.: A schedule-then-break approach to sports timetabling. In: Burke, E., Erben, W. (eds.) PATAT 2000. LNCS, vol. 2079, pp. 242–253. Springer, Heidelberg (2001)
21. Trick, M.A.: Integer and constraint programming approaches for round-robin tournament scheduling. In: Burke, E.K., De Causmaecker, P. (eds.) PATAT 2002. LNCS, vol. 2740, pp. 63–77. Springer, Heidelberg (2003)

Local Search for a Cargo Assembly Planning Problem

Gleb Belov[1], Natashia Boland[2], Martin W.P. Savelsbergh[2],
and Peter J. Stuckey[1]

[1] Department of Computing and Information Systems
University of Melbourne, 3010 Australia
pstuckey@unimelb.edu.au
[2] School of Mathematical and Physical Sciences
University of Newcastle, Callaghan 2308, Australia

Abstract. We consider a real-world cargo assembly planning problem arising in a coal supply chain. The cargoes are built on the stockyard at a port terminal from coal delivered by trains. Then the cargoes are loaded onto vessels. Only a limited number of arriving vessels is known in advance. The goal is to minimize the average delay time of the vessels over a long planning period. We model the problem in the MiniZinc constraint programming language and design a large neighbourhood search scheme. We compare against (an extended version of) a greedy heuristic for the same problem.

Keywords: packing, scheduling, resource constraint, large neighbourhood search, constraint programming, adaptive greedy, visibility horizon.

1   Introduction

The Hunter Valley Coal Chain (HVCC) refers to the inland portion of the coal export supply chain in the Hunter Valley, New South Wales, Australia. Coal from different mines with different characteristics is 'mixed' in a stockpile at a terminal at the port to form a coal blend that meets the specifications of a customer. Once a vessel arrives at a berth at the terminal, the stockpiles with coal for the vessel are reclaimed and loaded onto the vessel. The vessel then transports the coal to its destination. The coordination of the logistics in the Hunter Valley is challenging as it is a complex system involving 14 producers operating 35 coal mines, 27 coal load points, 2 rail track owners, 4 above rail operators, 3 coal loading terminals with a total of 8 berths, and 9 vessel operators. Approximately 1700 vessels are loaded at the terminals in the Port of Newcastle each year. For more information on the HVCC see the overview presentation of the Hunter Valley Coal Chain Coordinator (HVCCC), the organization responsible for planning the coal logistics in the Hunter Valley [2].

We focus on the management of a stockyard at one of the coal loading terminals. It acts as a *cargo assembly terminal* where the coal blends assembled and stockpiled are based on the demands of the arriving ships. Our cargo assembly planning approach aims to minimize the delay of vessels, where the delay of a

H. Simonis (Ed.): CPAIOR 2014, LNCS 8451, pp. 159–175, 2014.

vessel is defined as the difference between the vessel's departure time and its earliest possible departure time, that is, the departure time in a system with infinite capacity. Minimizing the delay of vessels is used as a proxy for maximizing the throughput, i.e., the maximum number of tons of coal that can be handled per year, which is of crucial importance as the demand for coal is expected to grow substantially over the next few years. We investigate the value of information given by the *visibility horizon* — the number of future vessels whose arrival time and stockpile demands are known in advance.

The solving technology we apply is Constraint Programming (CP) using lazy clause generation (LCG) [6]. Constraint programming has been highly successful in tackling complex packing and scheduling problems [11,12]. Cargo assembly is a combined scheduling and packing problem. The specific problem is first described by Savelsbergh and Smith [9]. They propose a greedy heuristic for solving the problem and investigate some options concerning various characteristics of the problem. We present a Constraint Programming model implemented in the MiniZinc language [4]. To solve the model efficiently, we develop iterative solving methods: greedy methods to obtain initial solutions and large neighbourhood search methods [8] to improve them.

2   Cargo Assembly Planning

The starting point for this work is the model developed in [9] for stockyard planning.

The stockyard studied has four pads, A, B, C, and D, on which cargoes are assembled. Coal arrives at the terminal by train. Upon arrival at the terminal, a train dumps its contents at one of three dump stations. The coal is then transported on a conveyor to one of the pads where it is added to a stockpile by a stacker. There are six stackers, two that serve pad A, two that serve both pads B and C, and two that serve pad D. A single stockpile is built from several train loads over several days. After a stockpile is completely built, it dwells on its pad for some time (perhaps several days) until the vessel onto which it is to be loaded is available at one of the berths. A stockpile is reclaimed using a bucket-wheel reclaimer and the coal is transferred to the berth on a conveyor. The coal is then loaded onto the vessel by a shiploader. There are four reclaimers, two that serve both pads A and B, and two that serve both pads C and D. Both stackers and reclaimers travel on rails at the side of a pad. Stackers and reclaimers that serve the same pads cannot pass each other. A scheme of the stockyard is given in Figure 1.

The cargo assembly planning process involves the following steps. An incoming vessel defines a set of cargoes (different blends of coal) to be assembled and an *estimated time of arrival (ETA)*. The cargoes are assembled in the stockyard as different stockpiles. The vessel cannot arrive at berth earlier than its ETA. Once at a berth, and once all its cargoes have been assembled, the reclaiming of the stockpiles (the loading of the vessel) begins. The stockpiles are reclaimed onto the vessel in a specified order to maintain physical balancing constraints.

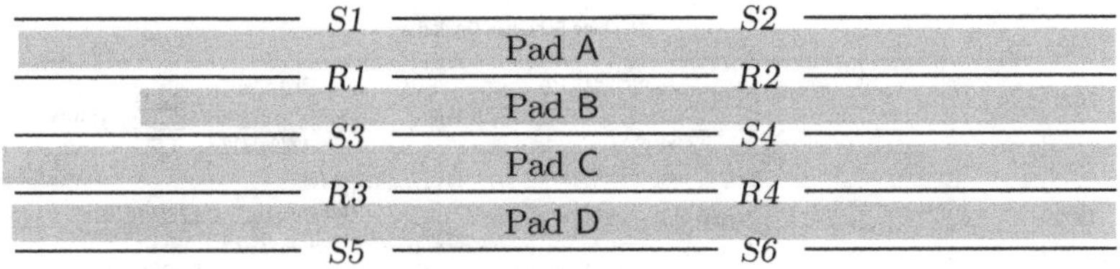

Fig. 1. A scheme of the stockyard with 4 pads, 6 stackers, and 4 reclaimers

The goal of the planning process is to maximize the throughput without causing unacceptable delays for the vessels.

When assigning each cargo of a vessel to a location in the stockyard we need to schedule the stacking and reclaiming of the stockpile taking into account limited stockyard space, stacking rates, reclaiming rates, and reclaimer movements. We model stacking and reclaiming at different levels of granularity. All reclaimer activities, e.g., the reclaimer movements along its rail track and the reclaiming of a stockpile, are modelled in time units of one minute. Stacking is modelled only at a coarse level of detail in 12 hour periods.

We assume that the time to build a stockpile is derived from the locations of the mines that contribute coal to the blend (the distance of the mines from the port). We allocate 3, 5, or 7 days to stacking of different stockpiles depending on the blend. We assume that the tonnage of the stockpile is stacked evenly over the stacking period. Since the trains that transport coal from the mines to the terminal are scheduled closer to the day of operations, this is not unreasonable. We assume that all stockpiles for a vessel are assembled on the same pad, since that leads to better results (already observed in [9]). In practice, however, there is no such restriction.

For each stockpile we need to decide a location, a stacking start time, a re-claiming start time, and which reclaimer will be used. Note that reclaiming does not have to start as soon as stacking has finished; the time between the comple-tion of stacking and the start of reclaiming is known as *dwell time*. Stockpiles cannot overlap in time and space, reclaimers can only be used on pads they serve, and reclaimers cannot cross each other on the shared track. The waiting time between the reclaiming of two consecutive stockpiles of one vessel is limited by the *continuous reclaim time limit*. The reclaiming of a stockpile, a so-called *reclaim job*, cannot be interrupted.

A cargo assembly plan can conveniently be represented using *space-time diagrams*; one space-time diagram for each of the pads in the stockyard. A space-time diagram for a pad shows for any point in time which parts of the pad are occupied by stockpiles (and thus also which parts of the pad are not occupied by stockpiles and are available for the placement of additional stockpiles) and the locations of the reclaimers serving that pad. Every pad is rectangular; how-ever its width is much smaller than its length and each stockpile is spread across the entire width. Thus, we model pads as one-dimensional entities. The loca-tion of a stockpile can be characterized by the position of its lowest end called its

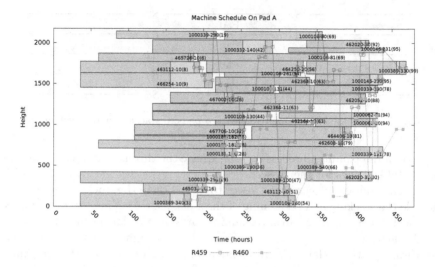

Fig. 2. A space-time diagram of pad A showing also reclaimer movements. Reclaimer R459 has to be after R460 on the pad. Both reclaimers also have jobs on pad B.

height. A stockpile occupies space on the pad for a certain amount of time. This time can be divided into three distinct parts: the *stacking part,* i.e., the time during which the stockpile is being built; the *dwell part,* i.e., the time between the end of stacking and the start of reclaiming; and a *reclaiming part,* i.e., the time during which the stockpile is reclaimed and loaded on a waiting vessel at a berth. Thus, each stockpile can be represented in a space-time diagram by a three-part rectangle as shown in Figure 2.

2.1 The Basic Constraint Programming Model

We present the model of the cargo assembly problem below; the structure corresponds directly to the implementation in MiniZinc [5]. The unit for time parameters is minutes, and for space parameters is meters. In addition, stacking start times are restricted to be multiples of 12 hours.

Parameter sets

S	—	set of stockpiles of all vessels, ordered by vessels' ETAs and reclaim sequence of each vessel's stockpiles
V	—	set of vessels, ordered by ETAs

Parameters

v_s	—	vessel for stockpile $s \in S$
eta_v	—	estimated time of arrival of vessel $v \in V$, minutes
$d_s^S \in \{4320, 7200, 10080\}$	—	stacking duration of stockpile $s \in S$, minutes
d_s^R	—	reclaiming duration of stockpile $s \in S$, minutes
l_s	—	length of stockpile $s \in S$, meters

$(H_1, \ldots, H_4) = (2142, 1905, 2174, 2156)$ —   pad lengths, meters
$\text{speed}^R = 30$ —   travel speed of a reclaimer, meters / minute
$\text{tonn}_s^{\text{daily}}$ —   daily stacking tonnage of stockpile $s \in S$, tonnes
$\text{tonn}^{\text{DIT}} = 537{,}600$ —   daily inbound throughput (total daily stacking capacity), tonnes
$\text{tonn}_k^{\text{SS}} = 288{,}000$ —   daily capacity of stacker stream $k \in \{1, 2, 3\}$, tonnes

Decisions

$p_v \in \{1, \ldots, 4\}$ —   pad on which the stockpiles of vessel $v \in V$ are assembled
$h_s \in \{0, \ldots, H_{p_{v_s}} - l_s\}$ —   position of stockpile $s \in S$ (of its 'closest to pad start' boundary) on the pad
$t_s^S \in \{0, 720, \ldots\}$ —   stacking start time of stockpile $s \in S$
$r_s \in \{1, \ldots, 4\}$ —   reclaimer used to reclaim stockpile $s \in S$
$t_s^R \in \{\text{eta}_{v_s}, \text{eta}_{v_s} + 1, \ldots\}$ —   reclaiming start time of stockpile $s \in S$

Constraints. Reclaiming of a stockpile cannot start before its vessel's ETA:

$$t_s^R \geq \text{eta}_{v_s}, \qquad \forall s \in S$$

Stacking of a stockpile starts no more than 10 days before its vessel's ETA:

$$t_s^S \geq \text{eta}_{v_s} - 14400, \qquad \forall s \in S$$

Stacking of a stockpile has to complete before reclaiming can start:

$$t_s^S + d_s^S \leq t_s^R, \qquad \forall s \in S$$

The reclaim order of the stockpiles of a vessel has to be respected:

$$t_s^R + d_s^R \leq t_{s+1}^R, \qquad \forall s \in S \text{ where } v_s = v_{s+1}$$

The continuous reclaim time limit of 5 hours has to be respected:

$$t_{s+1}^R - 300 \leq t_s^R + d_s^R, \qquad \forall s \in S \text{ where } v_s = v_{s+1}$$

A stockpile has to fit on the pad it is assigned to:

$$0 \leq h_s \leq H_{p_{v_s}} - l_s, \qquad \forall s \in S$$

Stockpiles cannot overlap in space and time:

$$p_{v_s} \neq p_{v_t} \vee h_s + l_s \leq h_t \vee h_t + l_t \leq h_s \vee t_s^R + d_s^R \leq t_t^S \vee t_t^R + d_t^R \leq t_s^S,$$
$$\forall s < t \in S$$

Reclaimers can only reclaim stockpiles from the pads they serve:

$$p_{v_s} \leq 2 \Leftrightarrow r_s \leq 2, \qquad \forall s \in S$$

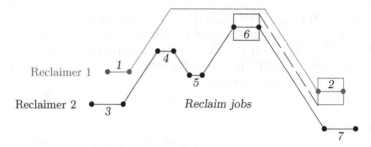

Fig. 3. A schematic example of space (vertical)-time (horizontal) location of Reclaimers 1 and 2 with some reclaim jobs. Reclaimer 2 has to stay spatially before Reclaimer 1.

If two stockpiles $s < t$ are reclaimed by the same reclaimer, then the time between the end of reclaiming the first and the start of reclaiming the second should be enough for the reclaimer to move from the middle of the first to the middle of the second:

$$r_s \neq r_t \lor \max\{(t_t^R - t_s^R - d_s^R), (t_s^R - t_t^R - d_t^R)\} \, \text{speed}^R \geq \left| h_s + \frac{l_s}{2} - h_t - \frac{l_t}{2} \right|$$

To avoid clashing, at any point in time, the position of Reclaimer 2 should be before the position of Reclaimer 1 and the position of Reclaimer 4 should be before the position of Reclaimer 3. An example of the position of Reclaimers 1 and 2 in space and time is given in Figure 3 (see also Figure 2). Because Job 3 is spatially before Job 1, there is no concern for a clash. However, since Job 6 is spatially after Job 2, we have to ensure that there is enough time for the reclaimers to get out of each other's way. The slope of the dashed line corresponds to the reclaimer's travel speed (speed^R), so we see that the time between the end of Job 6 and the start of Job 2 has to be at least $(h_6 + l_6 - h_2)/\text{speed}^R$.

We model clash avoidance by a disjunction: for any two stockpiles $s \neq t$, one of the following conditions must be met: either $(r_s \geq 3 \land r_t \leq 2)$, in which case r_s and r_t serve different pads; or $r_s < r_t$, in which case r_s does not have to be before r_t; or $h_s + l_s \leq h_t$, in which case stockpile s is before stockpile t; or, finally, enough time between the reclaim jobs exists for the reclaimers to get out of each other's way:

$$\max\{(t_t^R - t_s^R - d_s^R), (t_s^R - t_t^R - d_t^R)\} \, \text{speed}^R \geq h_s + l_s - h_t$$
$$\lor \, r_s < r_t \lor (r_s \geq 3 \land r_t \leq 2) \lor h_s + l_s \leq h_t, \quad \forall s \neq t \in S$$

Redundant cumulatives on pad space usage improved efficiency. They require derived variables l_s^p giving the 'pad length of stockpile s on pad p':

$$l_s^p = \begin{cases} l_s, & \text{if } p_{v_s} = p, \\ 0, & \text{otherwise}, \end{cases} \quad \forall s \in S, \ p \in \{1, \dots, 4\}$$

$$\texttt{cumulative}(t^S, t^R + d^R - t^S, l^p, H_p), \quad p \in \{1, \dots, 4\}$$

The stacking capacity is constrained day-wise. If a stockpile is stacked on day d and the stacking is not finished before the end of d, the full daily tonnage of that stockpile is accounted for using derived variables

$$t^{S1} = \lfloor t^S/1440 \rfloor, \quad d^{S1} = \lfloor d^S/1440 \rfloor$$

The daily stacking capacity cannot be exceeded:

$$\texttt{cumulative}(t^{S1}, d^{S1}, \text{tonn}^{\text{daily}}, \text{tonn}^{\text{DIT}})$$

The capacity of stacker stream k (a set of two stackers serving the same pads) is constrained similar to pad space usage:

$$\text{tonn}_{ks}^{\text{daily}} = \begin{cases} \text{tonn}_s^{\text{daily}}, & \text{if } (p_{v_s}, k) \in \{(1,1), (2,2), (3,2), (4,3)\} \\ 0, & \text{otherwise,} \end{cases} \quad \forall s, k$$

$$\texttt{cumulative}(t^{S1}, d^{S1}, \text{tonn}_k^{\text{daily}}, \text{tonn}_k^{\text{SS}}), \quad k \in \{1, 2, 3\}$$

The maximum number of simultaneously berthed ships is 4. We introduce derived variables for vessels' berth arrivals and use a decomposed cumulative:

$$t_v^{\text{Berth}} = t_{s^{\text{first}}(v)}^R, \quad s^{\text{first}}(v) = \min\{s | v_s = v\}, \quad \forall v \in V$$

$$\text{card}(\{u \in V \mid u \neq v, \ t_u^{\text{Berth}} \leq t_v^{\text{Berth}} \wedge t_v^{\text{Berth}} < t_u^{\text{Depart}}\}) \leq 3, \quad \forall v \in V$$

Objective Function. The objective is to minimize the sum of vessel delays. To define vessel delays, we introduce the derived variables t_v^{Depart} for vessel departure times:

$$\text{depEarliest}_v = \text{eta}_v + \sum_{s|v_s=v} d_s^R, \qquad\qquad \forall v \in V$$

$$t_v^{\text{Depart}} = t_{s^{\text{last}}(v)}^R + d_{s^{\text{last}}(v)}^R, \qquad s^{\text{last}}(v) = \max\{s|v_s = v\}, \quad \forall v \in V$$

$$\text{delay}_v = t_v^{\text{Depart}} - \text{depEarliest}_v, \qquad\qquad \forall v \in V$$

$$\textbf{objective} = \sum_v \text{delay}_v \qquad\qquad (1)$$

2.2 Solver Search Strategy

Many Constraint Programming models benefit from a custom search strategy for the solver. Similar to packing problems [1], we found it advantageous to separate branching decisions by groups of variables. We start with the most important variables — departure times of the ships (equivalently, delays). Then we fix reclaim starts, pads, reclaimers, stack starts, and pad positions. For most of the variables, we use the dichotomous strategy `indomain_split` for value selection, which divides the current domain of a variable in half and tries first to find a solution in the lower half. However, pads are assigned randomly, and reclaimers are assigned preferring lower numbers for odd vessels and higher numbers for even vessels. Pad positions are preferred so as to be closer to the native side of the chosen reclaimer, which corresponds to the idea of opportunity costs in [9].

In the greedy and LNS heuristics described next, some of the variables are fixed and the model optimizes only the remaining variables. For those free variables, we apply the search strategy described above.

2.3   A Greedy Search Heuristic with Constraint Programming

It is difficult to obtain even feasible solutions for large instances in a reasonable amount of time. Moreover, even for smaller instances, if a feasible solution is found, it is usually bad. Therefore, we apply a divide-and-conquer strategy which schedules vessels by groups (e.g., solve vessels 1–5, then vessels 6–10, then vessels 11–15, etc.). For each group, we allow the solver to run for a limited amount of time, and, if feasible solutions are found, take the best of these, or, if no feasible solution is found, we reduce the number of vessels in the group and retry. We refer to this scheme as the *extending horizon* (EH) heuristic. This heuristic is generalized in Section 2.5.

2.4   Large Neighbourhood Search

After obtaining a feasible solution, we try to improve it by re-optimizing subsets of variables while others are fixed to their current values, a *large neighbourhood search* approach [8]. We can apply this improvement approach to both complete solutions (*global LNS*) or only for the current visibility horizon (see Section 2.5). The free variables used in the large neighbourhood search are the decision variables associated with certain stockpiles.

Neighbourhood Construction Methods. We consider a number of methods for choosing which stockpile groups to re-optimize (the *neighbourhoods*):

   Spatial. Groups of stockpiles located close to each other on one pad, measured in terms of their space-time location.
   Time-based (finish). Groups of stockpiles on at most two pads with similar reclaim end times.
   Time-based (ETA). Groups of stockpiles on at most two pads belonging to vessels with similar estimated arrival times.

Examples of a spatial and a time-based neighbourhood are given in Figure 4.

First, we randomly decide which of the three types of neighbourhood to use. Next, we construct all neighbourhoods of the selected type. Finally, we randomly select one neighbourhood for resolving.

Spatial neighbourhoods are constructed as follows. In order to obtain many different neighbourhoods, every stockpile seeds a neighbourhood containing only that stockpile. Then all neighbourhoods are expanded. Iteratively, for each neighbourhood, and for each direction right, up, left, and down, independently, we add

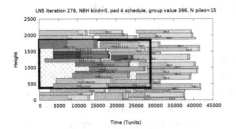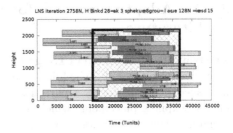

Fig. 4. Examples of LNS neighbourhoods: spatial (left) and time-based (right)

the stockpile on the same pad that is first met by the sweep line going in that direction, after the sweep line has touched the smallest enclosing rectangle of the stockpiles currently in the neighbourhood. We then add all stockpiles contained in the new smallest enclosing rectangle. We continue as long as there are neighbourhoods containing fewer than the target number of stockpiles.

Time-based neighbourhoods are constructed as follows. Stockpiles are sorted by their reclaim end time or by the ETA of the vessels they belong to. For each pair of pads, we collect all maximal stockpile subsequences of the sorted sequence of up to a target length, with stockpiles allocated to these pads.

Having constructed all neighbourhoods of the chosen type, we randomly select one neighborhood of the set. The probability of selecting a given neighborhood is proportional to its *neighborhood value*: if the *last, but not all* stockpiles of a vessel is in the neighborhood, then add the vessel's delay; instead, if *all stockpiles* of a vessel are in the neighborhood, then add 3 times the vessel's delay.

We denote the iterative large neighbourhood search method by $\text{LNS}(k_{\max}, n_{\max}, \delta)$, where for at most $k_{\max}$ iterations, we re-optimize neighborhoods of up to $n_{\max}$ stockpiles chosen using the principles outlined above, requiring that the total delay decreases at least by δ minutes in each iteration. The objective is again to minimize the total delay (1).

2.5 Limited Visibility Horizon

In the real world, only a limited number of vessels is known in advance. We model this as follows: the current *visibility horizon* is N vessels. We obtain a schedule for the N vessels and fix the decisions for the first F vessels. Then we schedule vessels $F + 1, \ldots, F + N$ (making the next F vessels visible) and so on. Let us denote this approach by VH N/F. Our default visibility horizon setting is VH 15/5, with the schedule for each visibility horizon of 15 vessels obtained using EH from Section 2.3 and then (possibly) improved by $\text{LNS}(30, 15, 12)$, i.e., 30 LNS iterations with up to 15 stockpiles in a neighbourhood, requiring a total delay improvement of at least 12 minutes. We used only time-based neighbourhoods in this case, because for small horizons, spatial neighbourhoods on one pad are too small. (Note that the special case VH 5/5 without LNS is equivalent to the heuristic EH.)

3   An Adaptive Scheme for a Heuristic from the Literature

The truncated tree search (TTS) greedy heuristic [9] processes vessels according to a given sequence. It schedules a vessel's stockpiles taking the vessel's delay into account. It performs a partial lookahead by considering *opportunity costs* of a stockpile's placement, which are related to the remaining flexibility of a reclaimer. However, it does not explicitly take later vessels into account; thus, the visibility horizon of the heuristic is one vessel. The heuristic may perform backtracking of its choices if the continuous reclaim time limit cannot be satisfied.

The default version of TTS processes vessels in their ETA order. We propose an adaptive framework for this greedy algorithm. This framework might well be used with the Constraint Programming heuristic from Section 2.3, but the latter is slower. Below we present the adaptive framework, then highlight some modelling differences between CP and TTS.

3.1   Two-Phase Adaptive Greedy Heuristic (AG)

The TTS greedy heuristic processes vessels in a given order. We propose an adaptive scheme consisting of two phases. In the first phase, we iteratively adapt the vessel order, based on vessels' delays in the generated solutions. In the second phase, earlier generated orders are randomized. Our motivation to add the randomization phase was to compare the adaptation principle to pure randomization.

For the first phase, the idea is to prioritize vessels with large delays. We introduce vessels' "weights" which are initialized to the ETAs. In each iteration, the vessels are fed to TTS in order of non-decreasing weights. Based on the generated solution, the weights are updated to an average of previous values and ETA minus a randomized delay; etc. We tried several variants of this principle and the one that seemed best is shown in Figure 5, Phase 1. The variable "oldWFactor" is the factor of old weights when averaging them with new values, starting from iteration 1 of Phase 1.

In the second phase, we randomize the orderings obtained in Phase 1. Each iteration in Phase 1 generated a vessel order $\iota = (v_1, \ldots, v_{|V|})$. Let $\mathcal{O} = (\iota_1, \ldots, \iota_k)$ be the list of orders generated in Phase 1 in non-decreasing order of TTS solution value. We select an order with index k_0 from $\mathcal{O}$ using a truncated geometric distribution with parameter $p = p_1$, TGD(p), which has the following probabilities for indexes $\{1, \ldots, k\}$:

$$P[1] = p + (1-p)^k, \quad P[2] = p(p-1), \quad P[3] = p(p-1)^2, \ldots, \quad P[k] = p(p-1)^{k-1}$$

The rationale behind this distribution is to respect the ranking of obtained solutions. A similar order randomization principle was used, e.g., in [3]. Then we modify the selected order ι_{k_0}: vessels are extracted from it, again using the truncated geometric distribution with parameter $p = p_2$, and are added to the end of the new order $\tilde{\iota}$. Then TTS is executed with $\tilde{\iota}$ and $\tilde{\iota}$ is inserted into $\mathcal{O}$ in the position corresponding to its objective value. We denote the algorithm by

```
Algorithm AG(k₁, k₂)
INPUT: Instance with V set of vessels; k₁, k₂ parameters
FUNCTION rnd(a, b) returns a pseudo-random number uniformly distributed in [a, b)
Initialize weights: 𝒲ᵥ = etaᵥ,   v ∈ V
for k = 0, k̄₁                                                    [PHASE 1]
    Sort vessels by non-decreasing values of 𝒲ᵥ,
        giving vessels' permutation ι = (v₁, ..., v_{|V|})
    Run TTS Greedy on ι
    Add ι to the sorted list 𝒪
    Set oldWFactor = rnd(0.125, 1)                      // "Value of history"
    Set 𝒟ᵥ to be the delay of vessel v ∈ V
    Let 𝒲ᵥ = oldWFactor · (𝒲ᵥ + (etaᵥ − rnd(0, 1) · 𝒟ᵥ)),   v ∈ V
end for
for k = 1, k̄₂                                                    [PHASE 2]
    Select an ordering ι from 𝒪 according to TGD(0.5)
    Create new ordering ι̃ from ι,
        extracting each new vessel according to TGD(0.85)
    Run TTS Greedy with the vessel order ι̃
    Add the new ordering ι̃ to the sorted list 𝒪
end for
```

Fig. 5. The adaptive scheme for the greedy heuristic

$AG(k_1, k_2)$, where k_1, k_2 are the number of iterations in Phases 1 and 2, respectively. Note that $AG(k_1, 0)$ is a pure Phase 1 method, while $AG(0, k_2)$ is a pure randomization method starting from the ETA order.

3.2 Differences between the Approaches

The model used by both methods is essentially identical, but there are small technical differences: the CP model uses discrete time and space and tonnages (minutes, meters, and tons), and discretizes possible stacking start times to be 12 hours apart. The discretized stacking start times reduce the search space, and may diminish solution quality, but seem reasonable given the coarse granularity of the stacking constraints imposed. The greedy method does not implement the berth constraints. If we remove them from the CP model it is solved slower, but the delay is hardly affected, so we always include them.

4   Experiments

After describing the experimental set-up, we illustrate the test data. We present numerical results starting with the value of information represented by the visibility horizon. Using the model from Section 2.1 we compare the Constraint Programming approach to the TTS heuristic and the adaptive scheme from Section 3.

The Constraint Programming models in the MiniZinc language were created by a master program written in C++, which was compiled in GNU C++.

The adaptive framework for the TTS heuristic and the TTS heuristic itself were implemented in C++ too. The MiniZinc models were processed by the finite-domain solver Opturion CPX 1.0.2 [7] which worked single-threaded on an Intel® Core™ i7-2600 CPU @ 3.40GHz under Kubuntu 13.04 Linux.

The Lazy Clause Generation [6] technology seems to be essential for our approach because our efforts to use another CP solver, Gecode 4.2.0 [10], failed even for 5-vessel subproblems. Packing problems are highly combinatorial, and this is where learning is the most advantageous. Moreover, some other LCG solvers than CPX did not work well, since this problem relies on lazy literal creation.

The solution of a MiniZinc model works in 2 phases. At first, it is *flattened*, i.e., translated into a simpler language FlatZinc. Then the actual solver is called on the flattened model. Time limits were imposed only on the second phase; in particular, we allowed at most 60 seconds in the EH heuristic and 30 seconds in an LNS iteration, see Section 2 for their details. However, reported times contain also the flattening which took a few seconds per model on average.

In EH and LNS, when writing the models with fixed subsets of the variables, we tried to omit as many irrelevant constraints as possible. In particular, this helped reduce the flattening time. For that, we imposed an upper bound of 200 hours on the maximal delay of any vessel (in the solutions, this bound was never achieved, see Figure 6 for example).

The default visibility horizon setting for our experiment, see Section 2.5, is VH 15/5: 15 vessels visible, they are approximately solved by EH and (possibly) improved by LNS(30, 15, 12); then the first 5 vessels are fixed, etc. Given the above time limits on an EH or LNS iteration, this takes less than 20 minutes to process each current visibility horizon and has shown to be usually much less because many LNS subproblems are proved infeasible rather quickly.

Our **test data** is the same as in [9]. It is historical data with compressed time to put extra pressure on the system. It has the following key properties:

- 358 vessels in the data file, sorted by their ETAs.
- One to three stockpiles per vessel, on average 1.4.
- The average interarrival time is 292 minutes.
- All ETAs are moved so that the first ETA = 10080 (7 days, to accommodate the longest build time).
- Optimizing vessel subsequences of 100 or up to 200 vessels, starting from vessels 1, 21, 41, ..., 181.

Figure 6 illustrates the test data giving the delay profile in a solution for all 358 vessels. The solution is obtained with the default visibility horizon setting VH 15/5. The most difficult subsequences seem to be the vessel groups 1..100 and 200..270.

4.1   Initial Solutions

First we look at basic methods to obtain schedules for longer sequences of vessels. This is the EH heuristic from Section 2.3 and the TTS Greedy described in

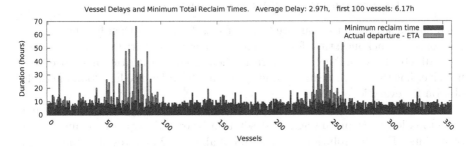

Fig. 6. Vessel delay profile in a solution of the instance 1..358

Table 1. Solutions for the 100- and up to 200-vessel instances, obtained with EH, TTS, ALL, and VH 15/5

	100 vessels								Up to 200 vessels							
	EH		TTS		ALL		VH 15/5		EH		TTS		ALL		VH 15/5	
	obj	t	obj	t	obj	t	obj	t	obj	t	obj	t	obj	t	obj	t
1st																
1	11.77	71	9.87	73	13.31	275	6.17	1509	6.15	170	5.09	90	7.06	356	3.19	1934
21	7.01	69	6.11	33	9.46	275	4.19	1758	3.75	142	3.25	68	5.08	352	2.23	2101
41	2.54	46	1.68	12	2.93	271	1.31	702	2.02	175	1.60	62	2.62	348	1.26	1465
61	0.64	42	0.61	18	0.98	273	0.51	214	3.59	252	3.25	60	5.39	351	2.63	1719
81	0.46	35	0.39	18	0.54	272	0.32	236	3.81	139	3.40	310	5.73	352	2.71	2084
101	0.33	29	0.23	7	0.52	272	0.19	202	3.39	140	3.21	62	5.14	352	1.91	2444
121	0.40	27	0.38	8	0.54	272	0.26	169	4.79	108	4.23	46	4.45	360	2.33	1815
141	2.82	154	1.44	20	2.59	273	1.35	612	4.72	220	3.26	47	4.45	353	2.45	2184
161	5.13	43	5.26	11	7.68	273	3.67	2031	3.53	101	3.26	42	5.15	352	2.25	2350
181	5.45	35	5.16	10	8.23	273	3.84	1438	3.13	70	2.93	33	4.72	328	2.17	1519
Mean	3.65	55	3.11	21	4.68	273	2.18	887	3.89	152	3.35	82	4.98	350	2.31	1961

Section 3, which fit into the visibility horizon schemes VH 5/5 and VH 1/1, respectively. We compare them to an approach to construct schedules in a single MiniZinc model (method "ALL") and to the standard visibility horizon setting VH 15/5, Section 2.5. The results are given in Table 1 for the 100-vessel and 200-vessel instances.

Method "ALL", obtaining feasible solutions for the whole 100-vessel and 200-vessel instances in a single run of the solver, became possible after a modification of the default search strategy from Section 2.2. This did not produce better results however, so we present its results only as a motivation for iterative methods for initial construction and improvement.

The default solver search strategy proved best for the iterative methods EH and LNS. But feasible solutions of complete instances *in a single model* only appeared possible with a modification. Let us call the strategy from Section 2.2 LAYERSEARCH$(1,\ldots,|V|)$ because we start with all vessels' departure times, continue with reclaim times, pads, etc. The alternative strategy can be expressed as

LAYERSEARCH(1,...,5); LAYERSEARCH(6,...,10); ...

which means: search for departure times of vessels 1,...,5; then for the reclaim times of their stockpiles; then for their pad numbers; ... departure times of vessels 6,...,10; etc. It is similar to the iterative heuristic EH with the difference that the solver has the complete model and (presumably) takes the first found feasible solution for every 5 vessels.

We had to increase the time limit per solver call: 4 minutes. But the flattening phase took longer than finding a first solution (there are a quadratic number of constraints). Feasible solutions were found in about 1–2 minutes after flattening. We also tried running the solver for longer but this did not lead to better results: the solver enumerates near the leaves of the search tree, which is not efficient in this case. Switching to the solver's default strategy after 300 seconds (search annotation cpx_warm_start [7]) gave better solutions, comparable with the EH heuristic.

In Table 1 we see that the solutions obtained by the "ALL" method are inferior to EH. Thus, for all further tests we used strategy LAYERSEARCH(1,..., $|V|$) from Section 2.2. Further, EH is inferior to TTS, both in quality and running time. This proves the efficiency of the opportunity costs in TTS and suggests using TTS for initial solutions. However, TTS runs on original real-valued data and we could not use its solutions in LNS because the latter works on rounded data which usually has small constraint violations for TTS solutions. A workaround would be to use the rounded data in TTS but given the majority of running time spent in LNS, and for simplicity we stayed with EH to obtain starting solutions. The results for VH 15/5 where LNS worked on every visibility horizon, support this choice.

4.2 Visibility Horizons

In this subsection, we look at the impact of varying the visibility horizon settings (Section 2.5), including the complete horizon (all vessels visible). More specifically, we compare $N = 1, 4, 6, 10, 15, 25$, or ∞ visible vessels and various numbers F of vessels to be fixed after the current horizon is scheduled. For $N = \infty$, we can apply a *global* solution method. Using Constraint Programming, we obtain an initial solution and try to improve it by LNS, denoted by *global LNS*, because it operates on the whole instance. Using Adaptive Greedy (Section 3), we also operate on complete schedules.

To illustrate the behaviour of global methods, we pick the difficult instance with vessels 1..100, cf. Figure 6. A graphical illustration of the progress over time of the global methods AG(130,0) and VH 15/5 + LNS(500, 15, 12) is given in Figure 7.

To investigate the value of various visibility horizons, for limited horizons, we applied the same settings as the standard one (Section 2.5): an initial schedule for the current horizon is obtained with EH and then improved with LNS(30, 15, 12).

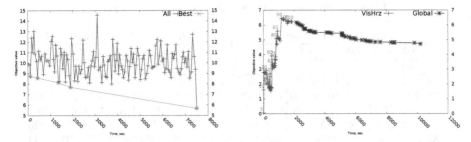

Fig. 7. Progress of the objective value in AG(130,0) (left) and VH 15/5 + LNS(500, 15, 12) (right), vessels 1..100

Table 2. Visibility horizon trade-off: all 100-vessel instances

N/F	1/1	4/2	6/3	10/5	15/7	15/1	25/12	25/5	**15/5+GLNS**
Delay, h	5.36	3.51	3.16	2.55	2.28	2.16	2.29	1.96	**1.73**
%Δ	210%	103%	83%	48%	32%	25%	33%	13%	
Time, s	114	188	202	267	916	2896	1823	3354	**4236**
%Δ	-97%	-96%	-95%	-94%	-78%	-32%	-57%	-21%	

Table 2 gives the average results for all 100-vessel instances. On average, the global Constraint Programming approach (500 LNS iterations) gives the best results, but VH 25/5 is close. Moreover, the setting VH 15/1 which invests significant effort by fixing only one vessel in a horizon, is slightly better than VH 25/12, which shows that with a smaller horizon, more computational effort can be fruitful.

The visibility horizon setting 1/1 produces the worst solutions. The TTS heuristic of Section 3 also uses this visibility horizon, but produces better results, see Table 3. The reason is probably the more sophisticated search strategy in TTS, which minimizes 'opportunity costs' related to reclaimer flexibility. At present, it is impossible to implement this complex search strategy in MiniZinc, the search sublanguage would need significant extension to do so.

4.3 Comparison of Constraint Programming and Adaptive Greedy

To compare the Constraint Programming and the AG approaches, we select the following methods: **VH 15/5** Visibility horizon 15/5; **VH 15/5+G** Visibility horizon 15/5, followed by global LNS 500; **VH 25/5** Visibility horizon 25/5; **AG$_1$** TTS Greedy, one iteration on the ETA order; **AG$_{500/500}$** Adaptive greedy, 500 iterations in both phases; **AG$_{1000/0}$** Adaptive greedy, 1000 iterations in Phase I only. The results for the 100-vessel instances are in Table 3. The pure-random configuration of the Adaptive Greedy, **AG$_{0/1000}$**, showed inferior performance, and its results are not given.

Table 3. 100 vessels: VH and LNS vs. (adaptive) greedy

	Constraint Programming						TTS Greedy and Adaptive Greedy						
	VH 15/5		VH 15/5*+G			VH 25/5		AG_1		$AG_{500/500}$			$AG_{1000/0}$		
1st	obj	t	obj	t	iter	obj	t	obj	t	obj	t	iter	obj	t	iter
1	6.17	1509	4.72	10204	338	5.44	7507	9.87	73	5.67	9992	130	5.67	7500	130
21	4.19	1758	3.17	10529	494	3.30	6626	6.11	33	3.63	10433	333	3.25	23051	991
41	1.31	702	1.24	922	0	1.24	3256	1.68	12	1.00	11253	667	1.02	12660	903
61	0.51	214	0.50	279	0	0.51	912	0.61	18	0.54	2713	126	0.54	2769	126
81	0.32	236	0.32	299	0	0.32	1266	0.39	18	0.34	11972	696	0.36	5794	344
101	0.19	202	0.19	285	2	0.18	943	0.23	7	0.21	7764	771	0.22	15	1
121	0.26	169	0.26	258	3	0.26	971	0.38	8	0.28	3589	525	0.29	201	28
141	1.35	612	0.73	4883	469	0.90	1364	1.44	20	0.80	4895	255	0.76	12969	652
161	3.67	2031	2.50	8172	369	3.51	4241	5.26	11	3.24	12574	845	3.78	4166	284
181	3.84	1438	3.64	6525	311	3.89	6450	5.16	10	3.83	6818	422	3.65	13843	809
Mean	2.18	887	**1.73**	4236	199	1.96	3354	3.11	21	**1.95**	8200	477	**1.95**	8297	427

* For limited visibility horizons, LNS(20,12,12) was applied

5 Conclusions

We consider a complex problem involving scheduling and allocation of cargo assembly in a stockyard, loading of cargoes onto vessels, and vessel scheduling. We designed a Constraint Programming (CP) approach to construct feasible solutions and improve them by Large Neighbourhood Search (LNS).

Investigation of various visibility horizon settings has shown that larger numbers of known arriving vessels lead to better results. In particular, the visibility horizon of 25 vessels provides solutions close to the best found. The new approach was compared to an existing greedy heuristic. The latter works with a visibility horizon of one vessel and, under this setting, produces better feasible solutions in less time. The reason is probably the sophisticated search strategy which cannot be implemented in the chosen CP approach at the moment. To make the comparison fairer, an adaptive iterative scheme was proposed for this greedy heuristic, which resulted in a similar performance to LNS.

Overall the CP approach using visibility horizons and LNS generated the best overall solutions in less time than the adaptive greedy approach. A significant advantage of the CP approach is that it is easy to include additional constraints, which we have done in work not reported here for space reasons.

Acknowledgments. The research presented here is supported by ARC linkage grant LP110200524. We would like to thank the strategic planning team at HVCCC for many insightful and helpful suggestions, Andreas Schutt for hints on efficient modelling in MiniZinc, as well as to Opturion for providing their version of the CPX solver under an academic license.

References

1. Clautiaux, F., Jouglet, A., Carlier, J., Moukrim, A.: A new constraint programming approach for the orthogonal packing problem. Computers & Operations Research 35(3), 944–959 (2008)
2. HVCCC: Hunter valley coal chain — overview presentation (2013), http://www.hvccc.com.au/
3. Lesh, N., Mitzenmacher, M.: BubbleSearch: A simple heuristic for improving priority-based greedy algorithms. Information Processing Letters 97(4), 161–169 (2006)
4. Marriott, K., Stuckey, P.J.: A MiniZinc tutorial (2012), http://www.minizinc.org/
5. Nethercote, N., Stuckey, P.J., Becket, R., Brand, S., Duck, G.J., Tack, G.: MiniZinc: Towards a standard CP modelling language. In: Bessière, C. (ed.) CP 2007. LNCS, vol. 4741, pp. 529–543. Springer, Heidelberg (2007)
6. Ohrimenko, O., Stuckey, P., Codish, M.: Propagation via lazy clause generation. Constraints 14(3), 357–391 (2009)
7. Opturion Pty Ltd: Opturion CPX user's guide: version 1.0.2 (2013), http://www.opturion.com
8. Pisinger, D., Ropke, S.: Large neighborhood search. In: Gendreau, M., Potvin, J.-Y. (eds.) Handbook of Metaheuristics. International Series in Operations Research & Management Science, vol. 146, pp. 399–419. Springer, US (2010)
9. Savelsbergh, M., Smith, O.: Cargo assembly planning. Tech. rep. University of Newcastle (2013) (accepted)
10. Schulte, C., Tack, G., Lagerkvist, M.Z.: Modeling and programming with Gecode (2013), http://www.gecode.org
11. Schutt, A., Feydy, T., Stuckey, P.J., Wallace, M.G.: Solving RCPSP/max by lazy clause generation. Journal of Scheduling 16(3), 273–289 (2013)
12. Schutt, A., Stuckey, P.J., Verden, A.R.: Optimal carpet cutting. In: Lee, J. (ed.) CP 2011. LNCS, vol. 6876, pp. 69–84. Springer, Heidelberg (2011)

A Logic Based Benders' Approach to the Concrete Delivery Problem

Joris Kinable[1,2,3] and Michael Trick[1]

[1] Tepper School of Business, Carnegie Mellon University,
5000 Forbes Ave Pittsburgh, PA 15213, USA
jkinable@andrew.cmu.edu, trick@cmu.edu
[2] ORSTAT - KU Leuven, Naamsestraat 69, 3000 Leuven, Belgium
[3] KU Leuven, Department of Computer Science, CODeS & iMinds-ITEC,
Gebr. De Smetstraat 1, 9000 Gent, Belgium

Abstract. This work presents an exact Logic Based Benders' decomposition for the Concrete Delivery Problem (CDP). The CDP is a complex, real world optimization problem involving the allocation and distribution of concrete to construction sites. The key scheduling issue for the CDP is the need for successive deliveries to a site to be sufficiently close in time. We decompose the CDP into a master problem and a subproblem. Based on a number of problem characteristics such as the availability of vehicles, geographical orientation of the customers and production centers, as well as the customers' demand for concrete, the master problem allocates concrete to customers. Next, the subproblem attempts to construct a feasible schedule, meeting all the routing and scheduling constraints. Infeasibilities in the schedule are communicated back to the master problem via a number of combinatorial inequalities (Benders' cuts). The master problem is solved through a Mixed Integer Programming approach, whereas the subproblem is solved via a Constraint Programming model and a dedicated scheduling heuristic. Experiments are conducted on a large number of problem instances, and compared against other exact methods presented in related literature. This algorithm is capable of solving a number of previously unsolved benchmark instances to optimality and can improve the bounds for many other instances.

Keywords: Vehicle Routing, Scheduling, Logic Based Benders' Decomposition, Integer and Constraint Programming.

1 Introduction

Many of today's real world optimization challenges do not involve just a single problem, but often comprise a multitude of interconnected problems. Although many of these optimization problems can be formulated in a single Mixed Integer (MIP) or Constraint Programming (CP) model, solving them to optimality is only possible for moderately small problem instances. The dependencies between the subproblems produce an excessive number of conditional constraints (big-M constraints), which have a significant impact on the quality of the model.

H. Simonis (Ed.): CPAIOR 2014, LNCS 8451, pp. 176–192, 2014.

Furthermore, MIP and CP solvers often make poor branching decisions, simply because the solvers are unaware of the underlying problem structure. Therefore, a major challenge lies in the design of efficient decomposition procedures for these problems.

In this work, a logic based Benders' decomposition for the Concrete Delivery Problem (CDP) is presented. The CDP, recently presented by Kinable et al. (2013), comprises the allocation and distribution of concrete to customers, under a number of routing and scheduling constraints, while maximizing the amount of concrete delivered. Concrete is transported from production centers to the customer's construction sites by a set of heterogeneous vehicles. Often, multiple deliveries for the same customer are required as the customer's demand exceeds the capacity of a single truck. Consequently, delivery schedules for different trucks need to be synchronized as deliveries for the same customer may not overlap in time. Furthermore, successive deliveries must not differ in time too much since the concrete from an early delivery must still be liquid when a second arrives.

The logic based Benders' procedure presented in this paper decomposes the CDP into a master problem and a subproblem. The master problem (MP) allocates concrete to customers, while taking a number of resource restrictions into consideration. The subproblem (SP) attempts to find a feasible delivery schedule for the concrete trucks. Whenever such a schedule does not exist, a feasibility cut is generated and added to the master problem. The master problem and subproblem are solved iteratively, until a provable optimal (and feasible) schedule is obtained.

In Kinable et al. (2013) several solution approaches for the CDP were investigated, including two exact approaches based on Mixed Integer and Constraint Programming, as well as a number of heuristic approaches. The best performance was obtained with a hybrid approach, using a dedicated scheduling heuristic, and a CP model to improve the heuristic solutions. Although good results were reported, the approach provided little insight as to the quality of the solutions. Moreover, alternative approaches to compute bounds on the optimal solution, including a Linear Programming approach could not close the optimality gap for most instances. The approach presented in this paper addresses these issues, as bounds on the optimal solution are available through the master problem.

The CDP bears strong resemblance to a number of routing and scheduling problems, including the Pickup-and-Delivery problem with Time-Windows and Split Deliveries and the Parallel Machine Scheduling Problem with Time Lags. Although the Benders' decomposition in this work is discussed in the context of CDP, we must note that the techniques presented are not uniquely confined to this application.

The remainder of this paper is structured as follows. First in Section 2, the CDP is defined in detail. Section 3 provides a literature review. Next, Section 4 presents the Benders' decomposition, defining the master and subproblem in more detail, as well as their interaction. Experiments are conducted in Section 5, thereby comparing the Benders' decomposition against other exact methods previously appeared in literature. Finally, Section 6 offers the conclusions.

Table 1. Parameters defining the CDP

Parameter	Description
P	Set of concrete production sites
C	Set of construction sites, also denoted as customers. $|C| = n$
V	$V = C \cup \{0\} \cup \{n+1\}$
$0, n+1$	resp. the start and end depots of the trucks.
K	Set of trucks
q_i	Requested amount of concrete by customer $i \in C$
l_k	Capacity of truck $k \in K$
p_k	Time required to empty truck $k \in K$
a_i, b_i	Time window during which the concrete for customer i may be delivered.
t_{ij}	Time to travel from i to j, $i, j \in V \cup P$
γ	Maximum time lag between consecutive deliveries.

2 Problem Outline

In the CDP as defined by Kinable et al. (2013), concrete has to be transported
from production sites P to a set of construction sites C. The transport is conducted by a fleet of vehicles K. Each vehicle $k \in K$ has a capacity l_k (measured in
tons of concrete) and each customer $i \in C$ has a demand q_i which usually exceeds
the capacity of a single truck. Concrete for a given customer $i \in C$ may only be
delivered within a time window $[a_i, b_i]$, and deliveries from multiple trucks to a
customer may not overlap in time. To ensure proper bonding of the concrete, the
time between two consecutive deliveries for the same customer may not exceed
γ. The time p_k, $k \in K$, required to perform a single delivery is truck dependent.
Deliveries may not be preempted and trucks are always filled to their maximum
capacity. Furthermore, the payload of a single truck may not be shared amongst
multiple customers; whenever the capacity of a truck exceeds the customer's remaining demand, the excess amount is considered waste. A customer $i \in C$ is
satisfied if at least q_i tons of concrete have been delivered. The objective of the
problem is to maximize the total demand of the satisfied customers.

The routing part of the CDP problem is similar to many other pickup and
delivery problems. Initially, the trucks are all stationed at a starting depot.
Trucks first have to drive from the starting depot to some production site to
load concrete. After loading concrete, a truck travels from the production site
to a customer to unload. Once the delivery is completed, the truck can either
return to a production station to reload and service another customer, or the
truck can return to the starting depot. Traveling between any location requires
t_{ij} time, $i, j \in \{0\} \cup C \cup P \cup \{n+1\}$, where 0 is the starting depot, and $n+1$ is
the ending depot (0 and $n+1$ can have the same physical location). Finally, note
that a single truck may also perform multiple deliveries for the same customer.

The routing problem can be formalized on a directed, weighted graph. Let
the sets P, C be defined as in Table 1. In addition, for each customer $i \in C$,
an ordered set consisting of deliveries, $C^i = \{1, \ldots, m(i), \ldots, n(i)\}$, is defined,

where $m(i) = \lceil \frac{q_i}{\max\limits_{k \in K}(l_k)} \rceil$ and $n(i) = \lceil \frac{q_i}{\min\limits_{k \in K}(l_k)} \rceil$ are respectively lower and up-per bounds on the number of deliveries required for customer i. As shorthand notation, c_j^i will be used to denote delivery j for customer i. A time window $[a_u, b_u]$ is associated with each delivery $u \in C^i$, $i \in C$, which is initialized to the time window of the corresponding customer $i \in C$, i.e. $[a_u, b_u] = [a_i, b_i]$ for all $i \in C, u \in C_i$. Finally, $D = \bigcup_{i \in C} C_i$ is the union of all deliveries.

Let $G(V, A)$ be the directed, weighted graph consisting of vertex set $V = \{0\} \cup D \cup \{n + 1\}$. The arc set A is defined as follows:

- the source, sink depots have outgoing resp. incoming edges to/from all other vertices.
- a delivery node c_h^i has a directed edge to a delivery node c_j^i if $h < j$, $i \in C$, $h, j \in C_i$.
- There is a directed edge from c_u^i to c_v^j, $i \neq j$, except if c_v^j needs to be scheduled earlier than c_u^i.

The arc costs are as follows:

- $c_{0,c_j^i} = \min_{p \in P} t_{0,p} + t_{p,i}$ for all $c_j^i \in D$
- $c_{c_u^i, c_v^j} = \min_{p \in P} t_{i,p} + t_{p,j}$ for all $c_u^i, c_v^j \in D, c_u^i \neq c_v^j$
- $c_{c_j^i, n+1} = t_{i,n+1}$
- $c_{0,n+1} = 0$

A solution to CDP consists of a selection of $|K|$ $s - t$ paths, which collectively satisfy the routing and scheduling constraints as outlined above. A summary of the notation used throughout this paper is provided in Table 1.

3   Related Research

This Section provides a brief overview on related works in the area of concrete production and distribution. An extensive overview and classification of these works has recently been published in Kinable et al. (2013).

A generalization of the CDP problem discussed in this paper is treated in Asbach et al. (2009). Next to the constraints discussed in Section 2, Asbach et al. (2009) consider vehicle synchronization at the loading depots, vehicles with specialized equipment, and additional time lags between deliveries for a single customer. In addition, Asbach et al. (2009) add constraints ensuring that some customers only receive concrete from a subset of production stations, and constraints limiting the time that concrete may reside in the vehicle's drum. Here we do not explicitly consider these constraints, as they can simply be incorporated in our model by modifying the arcs or costs in the underlying routing graph (Section 2). Asbach et al. (2009) present a MIP model to their problem, but solving it through Branch-and-Bound turned out intractable; instead, a heuristic procedure is used. The aforementioned MIP model served as the basis for the MIP approach in Kinable et al. (2013).

Schmid et al. (2009) and Schmid et al. (2010) consider a concrete delivery problem with soft time windows. In contrast to our work, their objective is to satisfy all customers, while minimizing total tardiness. Similar to Asbach et al. (2009), Schmid et al. (2009) and Schmid et al. (2010) also consider orders requiring vehicles with special equipment to be present during the entire delivery process, e.g. a pump or conveyor. Both works employ hybridized solution approaches combining MIP models and Variable Neighborhood Search heuristics.

Naso et al. (2007) propose a heuristic solution to a problem which involves the assignment of orders to concrete production centers, and the distribution of the concrete produced. Contrary to this work, deliveries are performed by a fleet of homogeneous vehicles. Furthermore, customers require an uninterrupted flow of concrete, meaning that the next truck must be available as soon as the previous truck finishes unloading. All orders must be fulfilled; unfulfilled orders are covered by external companies at a hire cost.

While in this work we consider both the routing and scheduling aspects of the concrete production and distribution, Yan and Lai (2007), Silva et al. (2005) focus primarily on the scheduling problems at the concrete production sites. In Yan and Lai (2007), only a single production center is considered, and in Silva et al. (2005) a single delivery takes a fixed amount of time which does not depend on the vehicle's location. Moreover, both works consider a homogeneous fleet of vehicles, as opposed to a heterogeneous fleet. This simplifies their scheduling problem considerably because the exact number of deliveries required to meet the customer's demand is known beforehand. This is not the case in our problem.

Similarly to this work, Hertz et al. (2012) present a decomposition approach for a concrete delivery problem. Their approach first assigns deliveries to vehicles, and then solves an independent routing problem for each vehicle. The latter is possible because the aforementioned work does not consider synchronization of the deliveries at the construction sites. Furthermore, no time windows on the deliveries are imposed.

Finally, in Durbin and Hoffman (2008), a decision support system is presented which assists in the dispatching of trucks, creating delivery schedules, and determining whether new orders should be accepted. The system is designed in such a way that it can coop with uncertainty and changes in the schedule caused by order cancellations, equipment failures, traffic congestions. The scheduling and routing problems handled by the system are represented by time-space networks, and are solved through MIP.

The Logic Based Benders' decomposition framework utilized in this work has recently been applied in a number of related related assignment, scheduling and routing problems. Applications include Round Robin Tournament Scheduling (Rasmussen and Trick, 2007), Tollbooth Allocation (Bai and Rubin, 2009), Parallel Machine Scheduling (Tran and Beck, 2012), Lock Scheduling (Verstichel et al., 2013) and Strip Packing (Côté et al., 2013).

4   A Logic-Based Benders' Decomposition

To solve the problem defined in the previous section, a logic-based Benders' decomposition is developed. The problem is decomposed in a master problem and a subproblem. The master problem, guided by the objective function, decides which customers are being serviced. To aid in this decision, a number of high-level problem characteristics are captured in the master problem, such as the availability of vehicles, their capacities, processing times and travel times between the customers and production centers. For a given subset of customers $\overline{C} \subseteq C$ selected by the master problem, the subproblem attempts to find a feasible delivery schedule in which the demand of all customers in $\overline{C}$ is satisfied, and all scheduling and routing constraints are met. Whenever such a schedule does not exist, a *feasibility cut* is generated and added to the master problem, effectively forcing the master problem to change the set $\overline{C}$. When, on the other hand, a feasible solution to the subproblem exist, a provable optimal solution to the CDP problem is obtained. An overview of the solution procedure is presented in Algorithm 1.

When compared to the MIP or CP approaches presented in Kinable et al. (2013), this decomposition approach decouples the allocation of concrete to customers from the actual routing and scheduling problem. As a consequence, many of the conditional constraints (big-M constraints) can be omitted or strengthened.

Algorithm 1. Combinatorial Benders Decomposition of CDP

Output: An optimal Concrete Delivery Schedule

1 repeat $\leftarrow$ true ;
2 **while** repeat **do**
3   Solve [MP];
4   get solution $(\overline{y_i})$, $i \in C$;
5   repeat $\leftarrow$ false ;
6   Solve [SP] for $\overline{y_i}$, $i \in C$;
7   **if** *[SP] is infeasible* **then**
8   repeat $\leftarrow$ true ;
9   add feasibility cut(s) to [MP] ;
10   **else**
11   get solution $(\overline{y}, x, C)$;

12 **return** *Optimal schedule* $(\overline{y}, x, C)$

4.1   Master Problem

The master problem is defined through the following MIP model.

$$MP:\ max \sum_{i \in C} q_i y_i \tag{1}$$

$$\sum_{k \in K} l_k z_{ki} \geq q_i y_i \qquad\qquad \forall i \in C \tag{2}$$

$$\sum_{i \in C} z_{ki} \leq \Delta_k \qquad \forall k \in K \qquad (3)$$

$$\sum_{i \in S} y_i \leq |S| - 1 \qquad \forall S \in \mathbb{S}, S \subseteq C \qquad (4)$$

$$y_i \in \{0, 1\} \qquad \forall i \in C \qquad (5)$$

$$0 \leq z_{ki} \leq \Delta_{ki} \qquad \forall i \in C \qquad (6)$$

Here, boolean variables y_i, $i \in C$, denote whether customer i is serviced and integer variables z_{ki} record the number of deliveries from vehicle $k \in K$ to customer $i \in C$. The auxiliary z_{ki} variables are used to produce stronger limits on the y_i variables; they are not used in the subproblem described in Section 4.2.

The first constraint, (2), links the variables y_i and z_{ki}, $i \in C, k \in K$: a customer is satisfied if sufficient concrete is delivered. Constraints (3), (6) restrict the number of deliveries made by a single vehicle through the bounds Δ_{ki}, Δ_k, for all $k \in K, i \in C$. Δ_{ki}, Δ_k, are resp. bounds on the maximum number of deliveries vehicle k can make for customer i, and bounds on the total number of deliveries a vehicle k can make. Finally, Constraints (4) are the feasibility cuts obtained through the subproblem, prohibiting certain combinations of customers.

Δ_{ki} is calculated via Algorithm 2, whereas Δ_k is calculated via the recursive Algorithm 3. The latter algorithm utilizes a sorted array of customers; a customer $i \in C$ precedes a customer $j \in C$ in the array if $b_i < b_j \vee (b_i = b_j \wedge a_i \leq a_j)$. Computing a bound on the maximum number of deliveries a vehicle can make, is achieved by calculating a route from the starting depot 0 to the ending depot $n + 1$ through a number of customers. At each customer, the vehicle makes as many deliveries as possible. The exact number of deliveries it can make for a given customer is limited by: (1) the demand of the customer, (2) the available time to make the deliveries. In turn, the available time to perform the deliveries is limited by the time window of the customer, the processing time of the vehicle, and the time required to reload the vehicle. Furthermore, whenever the vehicle completes its last delivery for a customer i at time t^i_{compl}, deliveries for the

---

Algorithm 2. Calculating an upper bound on the number of deliveries vehicle $k \in K$ can make for customer $i \in C$

Input: Vehicle $k \in K$, Customer $i \in C$

1 concreteDelivered $\leftarrow 0$;
2 timeConsumed $\leftarrow 0$;
3 $\Delta_{ki} \leftarrow 0$;
4 **while** $a_i +$ timeConsumed $+ p_k \leq b_i \wedge$ concreteDelivered $< q_i$ **do**
5 $\quad$ $\Delta_{ki} \leftarrow \Delta_{ki} + 1$;
6 $\quad$ concreteDelivered $\leftarrow$ concreteDelivered $+ l_k$;
7 $\quad$ timeConsumed $\leftarrow$ timeConsumed $+ p_k + c_{i,i}$;
8 **return** Δ_{ki}

Algorithm 3. Calculating an upper bound on the number of deliveries vehicle $k \in K$ can make

---

Input: Vehicle $k \in K$, Array of customers sortedCustomers[], sorted.
Output: Δ_k

1   **return** maxDeliveries(k, 0, 0, 0, 0) ;

2   **Function** *int* maxDeliveries($k \in K$, Δ_k, index, $i \in V$, complTime)
3   **if** index $= |C|$ **then**
4       $\lfloor$ **return** Δ_k;

5     $j \leftarrow$ sortedCustomers[index] ;
     /* Determine how many deliveries vehicle k can make for customer
      j */
6   concreteDelivered $\leftarrow 0$;
7   startTime $\leftarrow$ max(a_j,complTime $+c_{ij}$) ;   /* Start time 1st delivery for $j \in C$ */
8   timeConsumed $\leftarrow 0$;
9   deliveries $\leftarrow 0$;
10   **while** startTime + timeConsumed + $p_k \leq b_j \wedge$ concreteDelivered $< q_j$ **do**
11   deliveries $\leftarrow$ deliveries $+ 1$;
12   concreteDelivered $\leftarrow$ concreteDelivered $+ l_k$;
13   timeConsumed $\leftarrow$ timeConsumed $+ p_k + c_{jj}$;
14   complTimeNew $\leftarrow$ startTime + timeConsumed $- c_{jj}$;
15   **if** deliveries > 0 **then**
16   **return** max(maxDeliveries(k, Δ_k+deliveries,index $+1$,j,complTimeNew),
17   maxDeliveries(k, Δ_k, index $+1$, i, complTime));
18   **else**
19      $\lfloor$ **return** maxDeliveries(k, Δ_k, index $+1$, i, complTime);

---

next customer j cannot commence before $t^i_{compl} + c_{ij}$. The recursive algorithm outlined in procedure 3 iterates over all possible subsets of customers in an efficient manner. At each iteration, the algorithm tracks the total number of deliveries made, the last location $i \in V$ visited, an index to the customer it will visit next, and the time it departed from location $i \in V$.

4.2 Subproblem

Let $\overline{y_i}$, $i \in C$, be the optimal selection of customers obtained from problem MP, i.e. $\overline{C} = \{i \in C : \overline{y_i} = 1\}$. To assess the feasibility of this selection, a satisfiability subproblem (SP) is solved. Whenever no feasible solution to the subproblem exists, a cut is added to the Master problem:

$$\sum_{i \in \widehat{C}} y_i \leq |\widehat{C}| - 1$$

,where $\widehat{C} \subseteq \overline{C}$. The weakest cuts are obtained for $\widehat{C} = \overline{C}$. By reducing the size of $\widehat{C}$, stronger cuts may be obtained. The strongest cuts are based on Minimum Infeasible Subsets. In this context, a MIS is a subset of customers that cannot be accommodated in the same schedule; removing any of these customers from the set would however result in a subset of compatible customers. Note however that calculating a complete set of MIS is a difficult problem on its own.

The next paragraph outlines an exact procedure to establish the feasibility of a set $\overline{C}$. As this procedure is computationally expensive, we first try to solve the subproblem through the SD-heuristic proposed in Kinable et al. (2013). Furthermore, instead of solving the subproblem for the entire set $\overline{C}$ at once, we first solve the problem for a smaller set $\widehat{C} \subset \overline{C}$. If this smaller subproblem turns out to be feasible, we repeatedly add customers from $\overline{C}$ to $\widehat{C}$ and resolve the resulting subproblem.

A modified version of the CP model presented in Kinable et al. (2013) may be used to establish the feasibility of a set $\overline{C}$ of customers:

Algorithm 4. CP subproblem

Variable definitions:

1 $s = \{0, 0, 0, oblig.\}$
2 $t = \{0, \infty, 0, oblig.\}$
3 $d_j^i = \{r_i, d_i, 0, oblig.\}$ $\forall i \in \overline{C}, j \in \{1, \dots, m(i)\}$
4 $d_j^i = \{r_i, d_i, 0, opt.\}$ $\forall i \in \overline{C}, j \in \{m(i)+1, \dots, n(i)\}$
5 $d_{j,k}^i = \{r_i, d_i, p_k, opt.\}$ $\forall i \in \overline{C}, j \in \{1, \dots, n(i)\}, k \in K$

Constraints:

6 **forall** $i \in \overline{C}$
7 $\quad$ **forall** $j \in \{1, \dots, n(i)\}$
8 $\quad\quad \lfloor$ alternative$(d_j^i, \bigcup_{k \in K} d_{jk}^i)$
9 $\quad \sum_{k \in K} \sum_{j \in \{1, \dots, n(i)\}} l_k \cdot$ presenceOf$(d_{jk}^i) \geq q_i$
10 $\quad$ **forall** $j \in \{1, \dots, n(i)-1\}$
11 $\quad\quad$ endBeforeStart(d_j^i, d_{j+1}^i)
12 $\quad\quad \lfloor$ startBeforeEnd$(d_{j+1}^i, d_j^i, -\gamma)$
13 $\quad$ **forall** $j \in \{m(i), \dots, n(i)-1\}$
14 $\quad\quad$ $(\sum_{l \in \{1..j\}, k \in K} l_k \cdot$presenceOf$(d_{lk}^i) < q_i) \rightarrow$ presenceOf$(d_{j+1,k}^i)$
15 $\quad\quad \lfloor$ presenceOf$(d_{j+1}^i) \rightarrow$presenceOf(d_j^i)

16 **forall** $k \in K$
17 $\quad$ noOverlapSequence$(\bigcup_{i \in \overline{C}, j \in \{1, \dots, n(i)\}} d_{jk}^i \cup s \cup t)$
18 $\quad$ first$(s, \bigcup_{i \in \overline{C}, j \in \{1, \dots, n(i)\}} d_{jk}^i \cup t)$
19 $\quad \lfloor$ last$(t, \bigcup_{i \in \overline{C}, j \in \{1, \dots, n(i)\}} d_{jk}^i \cup s)$

The CP model utilizes a number of interval variables. For each interval variable, four parameters $\{a, b, d, o\}$ are specified, where a, b indicate resp. the earliest start time and latest completion time of the interval, and d is the minimum length of the interval. The last parameter o dictates whether an interval must (obligatory) or may (optional) be scheduled. The definitions of the constraints (Table 2) are taken from Kinable et al. (2013).

Variables d^i_j, $i \in \overline{C}, j \in \{1, \ldots, n(i)\}$, represent deliveries made for customer i. A delivery j for customer i is made if the corresponding interval variable d^i_j is present; otherwise it is absent. Variables $d^i_{j,k}$, $i \in \overline{C}, j \in \{1, \ldots, n(i)\}, k \in K$, link deliveries and the vehicles performing these deliveries. Clearly, each delivery $j \in \{1, \ldots, n(i)\}$ for a customer $i \in \overline{C}$ can only be made by a single vehicle (Line (8)), and the amount of concrete delivered for a customer should cover its demand (Line (9)). Deliveries for the same customer may not overlap (Line (11)) and must respect a maximum time lag γ (Line (12)). Similarly, deliveries made by a single vehicle cannot overlap in time and must comply with travel times (Line (17)). Trucks must start their trip at the starting depot, represented by variable s, and must return to some ending depot identified by variable t after the deliveries are completed (Lines (18), (19)). Finally, Line (15) ensures that deliveries are made in order, and Line (14) tightens the link between the d^i_j and d^i_{jk} variables.

Table 2. Description of CP constraints

Constraint	Description
presenceOf(α)	States that interval α must be present.
alternative(α, B)	If interval α is present, then exactly one of the intervals in set B is present. The start and end of interval α coincides with the start and end of the selected interval from set B.
endBeforeStart(α, β)	$endOf(\alpha) \leq startOf(\beta)$. Automatically satisfied if either of the intervals is absent.
startBeforeEnd(α, β, z)	$startOf(\alpha) + z \leq endOf(\beta)$. Automatically satisfied if either of the intervals is absent.
noOverlapSequence($B, dist$)	Sequences the intervals in set B. Ensures that the intervals in B do not overlap. Furthermore, the two-dimensional distance matrix $dist$ specifies a sequence dependent setup time for each pair of activities. Absent intervals are ignored.
first(α, B)	If interval α is present, it must be scheduled before any of the intervals in B.
last(α, B)	If interval α is present, it must scheduled after any of the intervals in B.

4.3   Generating an Initial Set of Cuts

Before invoking the Benders' procedure, first a number of initial cuts are computed and added to thet set $\mathbb{S}$ in the master problem. These custs are generated by enumerating all Minimum Infeasible Subsets consisting of up to three customers. In addition, the constructive heuristic presented in Kinable et al. (2013)

(Algorithm 2) may be used to compute an additional set of cuts. The constructive heuristic is initialized with an ordered set of customers. The heuristic schedules deliveries for these customers in an iterative fashion, where the time of a delivery and the vehicle performing the delivery are determined via a number of heuristic criteria. If at some point, the heuristic fails to schedule the next delivery for a customer due to the lack of an available vehicle, the heuristic would normally remove this customer from the schedule and continue with the next customer. Instead of simply removing the customer from the schedule, an additional check is performed. Given the subset of customers $\widehat{C}$ which have received concrete in the partially completed heuristic schedule, an exact algorithm, e.g. the CP model from Section 4.2, is used to determine whether there exists a feasible solution where each of the customers in $\widehat{C}$ is satisfied. If no such schedule exist, a cut is generated for the customers in $\widehat{C}$ and the heuristic removes the customer it could not satisfy from the schedule. If, on the other hand, the exact approach is capable of finding a feasible schedule satisfying all customers in $\widehat{C}$, the heuristic continues from the schedule generated by the exact method.

Naturally, the approach outlined in the previous paragraph may be repeated for several orderings of the customers.

5   Experimental Results

5.1   Data Sets

Experiments are conducted on the data set published by Kinable et al. (2013) (available online at (Kinable and Wauters, 2013)). A summary of the instances is provided in Table 3 of Kinable et al. (2013). The instances are subdivided into two sets, resp. A and B. The instances in set A range from $10 - 20$ *customers*, and $2 - 5$ *vehicles*. Larger instances, having up to 50 *customers* and 20 *vehicles*, can be found in data set B. Customer demands are within the range of $10 - 75$ for both sets. Each instance defines a number of different vehicle classes, where vehicles belonging to the same class have the same capacity and processing time. The location of the starting and ending depot, as well as the locations of the customers and up to 4 production depots are defined per instance. The travel time between two locations equals the euclidean distance, rounded upwards. The instances are constructed in such a way that for each customer, there is a production station within $1 - 25$ time units. The width of the delivery interval for a customer is set proportional to the demand of the customer.

5.2   Experiments

A number of experiments are conducted to assess the performance of the Benders' procedure outlined in this paper. The results of these experiments are reported in Tables 4, 5. In these tables, the first column provides the instance name, following a "$W\_x\_y\_z$" naming convention, where W identifies the data set, x is the number of vehicles, y is the number of customers and z is the number of concrete production stations. For each instance, we computed an initial

Table 3. Data sets Kinable et al. (2013)

	Set A	Set B
Instances	64	128
Customers	10-20	20-50
Demands	10-75	10-75
Time Windows	$q_i \times [1.1, 2.1]$	$q_i \times [1.1, 2.1]$
Time lags	5	5
Vehicles	2-5	6-20
Capacity	10-25	10-25
Vehicle classes	2-3	3
Processing time	$p_k = l_k$	$p_k = l_k$
Stations	1-4	1-4
Cust.-Station	1-30	1-30
Depot-Station	1-25	1-25

feasible solution using the CP procedure outlined in Kinable et al. (2013). These solutions, reported in the second column, are used to warm-start the Benders' procedure. The next 5 columns provide data on our Benders' procedure:

- obj: The objective of the best(feasible) solution obtained through the Benders' procedure.
- iCuts: The number of cuts added initially (Section 4.3)
- cuts: The number of cuts added during the Bender's procedure (Section 4.2)
- c-time: The time required to obtain the initial master problem, in seconds. This time is limited to 5 minutes, excluding the generation of the Minimum Infeasible subsets.
- s-time: The time required to solve the Benders' problem. For data set A, this time is limited to 10 minutes, and for data set B 15 minutes.

In Kinable et al. (2013) bounds on the optimal solutions are published. These bounds are computed through four different procedures, but for each instance only the strongest bound is reported in Kinable et al. (2013). The different procedures from Kinable et al. (2013) are:

- Optimal MIP solution (when available)
- Optimal CP solution (when available)
- LP relaxation, strengthened with cuts from all Minimum Infeasible Subsets (MIS) of size 2.
- Solution to the MIP problem consisting of the objective function (1) and all cuts generated from MIS of size k (Constraint (4)), where $k = 3$ for data set A. This bound has only been calculated for data set A in Kinable et al. (2013).

In columns 'bound*', and 'LP', resp. the bounds reported in Kinable et al. (2013) as well as the bounds obtained through the LP relaxation of the MIP model in Section 3.1.2 of Kinable et al. (2013) are shown. The gaps are computed with respect to the objective in column *obj*. Finally, the last two columns in the table provide the bounds obtained through the Benders' procedure.

When comparing the bounds attained through our Benders' procedure with the LP-bounds, we can observe that the LP based bounds are significantly weaker. The average gap between the LP bounds and the best primal solutions amounts to 9.26%, whereas the Benders' procedure produces an average gap of 1.81%. This gap is also smaller than the average gap (3.62%) obtained from the bounds reported in Kinable et al. (2013). In fact, none of the bounds computed through the Benders' procedure are weaker than the bounds reported in Kinable et al. (2013).

Table 4. Computational results Data Set A

Instance	iObj	Benders' decomposition					LP		Bound*		Benders'	
		obj	iCuts	cuts	c-time	s-time	bound	gap	bound	gap	bound	gap
A_2_5_1	85	85	28	0	1	0	85	0%	85	0%	85	0%
A_2_5_2	160	160	13	0	0	0	160	0%	160	0%	160	0%
A_2_5_3	105	105	26	0	0	0	105	0%	105	0%	105	0%
A_2_5_4	105	105	3	0	300	0	105	0%	105	0%	105	0%
A_2_10_1	50	50	142	0	3	0	50	0%	50	0%	50	0%
A_2_10_2	150	150	215	0	9	0	150	0%	150	0%	150	0%
A_2_10_3	220	220	154	0	3	0	230	4%	220	0%	220	0%
A_2_10_4	150	150	179	0	9	0	165	9%	150	0%	150	0%
A_2_15_1	215	215	567	0	10	0	225	4%	215	0%	215	0%
A_2_15_2	290	290	373	2	25	2	320	9%	320	9%	**290**	**0%**
A_2_15_3	205	205	502	0	77	0	215	5%	205	0%	205	0%
A_2_15_4	255	255	348	0	11	0	300	15%	255	0%	255	0%
A_2_20_1	255	255	549	0	38	0	260	2%	255	0%	255	0%
A_2_20_2	270	270	575	2	33	0	285	5%	270	0%	270	0%
A_2_20_3	260	260	651	0	15	0	280	7%	260	0%	260	0%
A_2_20_4	355	355	36	5	360	1	490	28%	380	7%	**355**	**0%**
A_3_5_1	205	205	0	0	3	0	205	0%	205	0%	205	0%
A_3_5_2	115	115	9	0	1	0	115	0%	115	0%	115	0%
A_3_5_3	125	125	0	0	0	0	125	0%	125	0%	125	0%
A_3_5_4	190	190	0	0	0	0	190	0%	190	0%	190	0%
A_3_10_1	205	205	100	0	5	0	205	0%	205	0%	205	0%
A_3_10_2	230	230	80	0	14	0	230	0%	230	0%	230	0%
A_3_10_3	305	305	114	0	23	0	305	0%	305	0%	305	0%
A_3_10_4	300	300	115	0	5	0	300	0%	300	0%	300	0%
A_3_15_1	330	330	485	0	35	0	330	0%	330	0%	330	0%
A_3_15_2	395	395	268	3	18	1	530	25%	425	7%	**395**	**0%**
A_3_15_3	290	290	378	23	49	7	430	33%	330	12%	**290**	**0%**
A_3_15_4	440	440	10	15	306	49	550	20%	475	7%	**440**	**0%**
A_3_20_1	340	340	667	1	350	157	410	17%	345	1%	**340**	**0%**
A_3_20_2	415	415	28	0	516	0	510	19%	415	0%	415	0%

Table 4. Computational results Data Set A

Instance	iObj	Benders' decomposition					LP		Bound*		Benders'	
		obj	iCuts	cuts	c-time	s-time	bound	gap	bound	gap	bound	gap
A_3_20_3	355	**360**	48	0	324	0	425	15%	360	0%	360	0%
A_3_20_4	480	480	31	0	319	0	590	19%	480	0%	480	0%
A_4_5_1	140	140	0	0	0	0	140	0%	140	0%	140	0%
A_4_5_2	150	150	0	0	0	0	150	0%	150	0%	150	0%
A_4_5_3	165	165	0	0	0	0	165	0%	165	0%	165	0%
A_4_5_4	230	230	0	0	0	0	230	0%	230	0%	230	0%
A_4_10_1	310	310	114	0	46	0	310	0%	310	0%	310	0%
A_4_10_2	370	370	48	0	4	0	390	5%	370	0%	370	0%
A_4_10_3	375	375	85	1	42	2	470	20%	445	16%	**375**	**0%**
A_4_10_4	285	285	23	0	1	0	285	0%	285	0%	285	0%
A_4_15_1	415	415	7	2	307	600	570	27%	545	24%	545	24%
A_4_15_2	475	475	4	18	312	600	650	27%	610	22%	**520**	**9%**
A_4_15_3	430	430	368	0	99	0	495	13%	450	4%	**430**	**0%**
A_4_15_4	490	490	5	0	304	600	515	5%	515	5%	515	5%
A_4_20_1	525	525	3	8	302	600	660	20%	585	10%	**575**	**9%**
A_4_20_2	425	425	497	4	38	1	490	13%	440	3%	**425**	**0%**
A_4_20_3	375	375	40	12	828	600	530	29%	425	12%	**405**	**7%**
A_4_20_4	465	465	18	17	1801	3	590	21%	500	7%	**465**	**0%**
A_5_5_1	200	200	0	0	0	0	200	0%	200	0%	200	0%
A_5_5_2	200	200	0	0	0	0	200	0%	200	0%	200	0%
A_5_5_3	220	220	0	0	0	0	220	0%	220	0%	220	0%
A_5_5_4	175	175	9	0	2	0	175	0%	175	0%	175	0%
A_5_10_1	350	350	0	0	0	0	350	0%	350	0%	350	0%
A_5_10_2	345	345	0	0	0	0	345	0%	345	0%	345	0%
A_5_10_3	285	285	25	0	5	0	300	5%	285	0%	285	0%
A_5_10_4	380	380	0	0	0	0	380	0%	380	0%	380	0%
A_5_15_1	455	455	2	4	308	600	590	23%	590	23%	**510**	**11%**
A_5_15_2	580	580	0	0	301	600	695	17%	695	17%	695	17%
A_5_15_3	350	350	2	3	576	600	435	20%	395	11%	**385**	**9%**
A_5_15_4	500	500	3	0	1218	600	600	17%	520	4%	520	4%
A_5_20_1	700	705	252	61	245	536	900	22%	760	7%	**705**	**0%**
A_5_20_2	555	555	15	4	388	600	810	31%	645	14%	**630**	**12%**
A_5_20_3	595	595	8	16	302	600	695	14%	645	8%	**615**	**3%**
A_5_20_4	520	520	13	2	355	600	705	26%	560	7%	**557.5**	**7%**
AVG								9.26%		3.72%		1.81%
Optimal								28		41		52

In summary, we reduced the average gap for the instances in data set A from 3.72% to 1.81%, 19 instances had their bounds improved, and 11 new optimal solutions were found. For almost 65% of the instances, the optimal solutions were already obtained at the first iteration of the master problem, i.e. no additional cuts had to be generated.

Table 5. Computational results Data Set B

Instance	iObj	Benders' decomposition					LP		Bound*		Benders'	
		obj	iCuts	cuts	c-time	s-time	bound	gap	bound	gap	bound	gap
B_6_20_1	725	725	0	102	301	356	805	10%	805	10%	**725**	**0%**
B_6_20_2	700	700	0	31	300	900	855	18%	855	18%	**830**	**16%**
B_6_20_3	675	675	0	0	300	900	760	11%	760	11%	760	11%
B_6_20_4	615	615	0	29	300	900	705	13%	705	13%	**680**	**10%**
B_6_30_1	860	860	0	1	300	900	1300	34%	1300	34%	**1290**	**33%**
B_6_30_2	850	850	0	34	300	900	1140	25%	1140	25%	**1105**	**23%**
B_6_30_3	725	725	0	120	301	900	1060	32%	1060	32%	**1035**	**30%**
B_6_30_4	625	625	2	1	302	900	1000	38%	1000	38%	**913.33**	**32%**
B_6_40_1	835	835	0	23	302	900	1545	46%	1545	46%	**1431.11**	**42%**
B_6_40_2	1045	1045	0	33	301	900	1635	36%	1635	36%	**1473.75**	**29%**
B_6_40_3	735	735	11	79	315	900	1570	53%	1570	53%	**1191.67**	**38%**
B_6_40_4	750	750	2	9	305	900	1450	48%	1450	48%	**1095**	**32%**
B_6_50_1	1010	1010	0	22	305	900	1890	47%	1890	47%	**1490**	**32%**
B_6_50_2	955	955	2	25	307	900	2250	58%	2250	58%	**1495**	**36%**
B_6_50_3	920	920	1	16	311	900	1740	47%	1740	47%	**1295**	**29%**
B_6_50_4	1000	1000	1	0	385	900	2080	52%	2080	52%	**1395**	**28%**
B_8_20_1	920	920	0	1	300	897	935	2%	935	2%	**920**	**0%**
B_8_20_2	850	850	0	0	300	900	865	2%	865	2%	865	2%
B_8_20_3	655	655	0	0	293	0	655	0%	655	0%	655	0%
B_8_20_4	820	820	0	0	69	0	820	0%	820	0%	820	0%
B_8_30_1	920	920	0	0	300	900	1085	15%	1085	15%	1085	15%
B_8_30_2	1005	1005	0	0	300	900	1115	10%	1115	10%	1115	10%
B_8_30_3	975	975	0	4	300	900	1155	16%	1155	16%	**1140**	**14%**
B_8_30_4	1110	1110	0	0	300	900	1320	16%	1320	16%	1320	16%
B_8_40_1	1180	1180	0	1	301	900	1665	29%	1665	29%	**1655**	**29%**
B_8_40_2	1190	1190	0	6	300	900	1415	16%	1415	16%	1415	16%
B_8_40_3	995	995	0	0	300	900	1495	33%	1495	33%	1495	33%
B_8_40_4	1105	1105	0	0	301	900	1730	36%	1730	36%	1730	36%
B_8_50_1	1130	1130	0	11	323	901	1980	43%	1980	43%	**1925**	**41%**
B_8_50_2	1150	1150	0	0	304	900	1935	41%	1935	41%	1935	41%
B_8_50_3	1210	1210	0	0	302	900	1960	38%	1960	38%	1960	38%
B_8_50_4	1105	1105	0	0	303	900	1835	40%	1835	40%	**1740**	**36%**
B_10_20_1	805	805	0	0	116	0	805	0%	805	0%	805	0%
B_10_20_2	825	825	0	0	237	0	825	0%	825	0%	825	0%
B_10_20_3	730	730	0	0	4	0	730	0%	730	0%	730	0%
B_10_20_4	765	765	0	0	1	0	765	0%	765	0%	765	0%
B_10_30_1	910	910	0	0	300	900	1215	25%	1215	25%	1215	25%
B_10_30_2	1170	1170	0	0	300	900	1355	14%	1355	14%	1355	14%
B_10_30_3	1135	1135	0	0	300	900	1210	6%	1210	6%	1210	6%
B_10_30_4	1165	1165	0	0	300	900	1235	6%	1235	6%	1235	6%
B_10_40_1	1210	1210	0	0	301	900	1475	18%	1475	18%	1475	18%
B_10_40_2	1485	1485	0	0	301	900	1580	6%	1580	6%	1580	6%
B_10_40_3	1375	1375	0	0	302	900	1605	14%	1605	14%	1605	14%
B_10_40_4	1365	1365	0	0	301	900	1455	6%	1455	6%	1455	6%
B_10_50_1	1425	1425	0	0	308	900	2265	37%	2265	37%	2265	37%
B_10_50_2	1010	1010	0	0	302	900	1900	47%	1900	47%	**1745**	**42%**
B_10_50_3	1260	1260	0	0	302	900	2005	37%	2005	37%	2005	37%
B_10_50_4	1455	1455	0	0	303	901	1925	24%	1925	24%	1925	24%
AVG								23.83%		23.83%		20.51%
Optimal								6		6		8

The instances in data set B are significantly harder to solver than the instances in data set A. Table 5 shows the results obtained for the instances up to 10 vehicles; no improvements could be made to the remaining instances. For the instances in Table 5, the average gap induced by the Benders' bounds is 10.08%, opposed to 11.32% from the bounds published in Kinable et al. (2013). Furthermore, optimality was attained for two additional instances; 21 instances had their bounds improved.

Future attempts to improve the Benders' decomposition approach should be targeted at improving the the runtime of the subproblem, as this procedure takes the largest amount of time. Especially for the larger instances, solving the subproblem is challenging.

6 Conclusion

In this work, we presented a Logic Based Benders' decomposition which decouples the CDP into a master problem and a subproblem. The master problem allocates concrete to customers, whereas the subproblem handles the routing and scheduling of the concrete delivery trucks. By decomposing the problem, part of the complexity is shifted to the subproblem. Furthermore, dedicated procedures may be used to solve these problems. Here, we solve the master problem through MIP, whereas the subproblem is solved through a dedicated scheduling heuristic and a CP model from (Kinable et al., 2013). Because the subproblem does not have to deal with the allocation of concrete as this is being handled by the master problem, we simplified and strengthened the latter CP model, thereby significantly improving its performance.

Computing bounds for CDP is a non-trivial task. Linear Programming-based bounds are generally very weak (see Section 5), as the problem has a large number of conditional constraints. Furthermore, exact approaches based on CP often provide little insight as to the quality of the solution. Our Benders' decomposition may however provide a viable alternative. Extensive computational tests show that the bounds computed through the Benders' decomposition are consistently stronger than the bounds in Kinable et al. (2013), which where obtained by aggregating the bounds of four different procedures.

By improving the bounds for a large number of instances, and simultaneously improving several primal solutions, we were able to solve a number of previously unsolved benchmark instances to optimality. To further enhance the performance of this Benders' procedure, one would have to find a way to speed up the subproblem. One possible direction would be to decouple the subproblem even further, for example by fixing more variables in the master problem. Alternatively, one could try to replace the current subproblem by a relaxation of the exact subproblem, which is easier to solve. Using this relaxation, it could still be possible to add a number of cuts to the master problem, thereby refining the master problem, with significantly less computational effort.

References

Asbach, L., Dorndorf, U., Pesch, E.: Analysis, modeling and solution of the concrete delivery problem. European Journal of Operational Research 193(3), 820–835 (2009)

Bai, L., Rubin, P.A.: Combinatorial benders cuts for the minimum tollbooth problem. Operations Research 57(6), 1510–1522 (2009)

Côté, J.-F., Dell'Amico, M., Iori, M.: Combinatorial benders' cuts for the strip packing problem. Interuniversity Research Centre on Enterprise Networks, Logistics and Transportation and Department of Computer Science and Operations Research, Université de Mon (CIRELT), Tech. Rep. CIRRELT-2013-27 (April 2013)

Durbin, M., Hoffman, K.: OR Practice - The Dance of the Thirty-Ton Trucks: Dispatching and Scheduling in a Dynamic Environment. Operations Research 56(1), 3–19 (2008)

Hertz, A., Uldry, M., Widmer, M.: Integer linear programming models for a cement delivery problem. European Journal of Operational Research 222(3), 623–631 (2012)

Kinable, J., Wauters, T.: CDPLib (2013), https://sites.google.com/site/cdplib/

Kinable, J., Wauters, T., Vanden Berghe, G.: The Concrete Delivery Problem. Computers & Operations Research (2014) (in press)

Naso, D., Surico, M., Turchiano, B., Kaymak, U.: Genetic algorithms for supply-chain scheduling: A case study in the distribution of ready-mixed concrete. European Journal of Operational Research 177(3), 2069–2099 (2007)

Rasmussen, R., Trick, M.: A benders approach for the constrained minimum break problem. European Journal of Operational Research 177(1), 198–213 (2007)

Schmid, V., Doerner, K.F., Hartl, R.F., Savelsbergh, M.W.P., Stoecher, W.: A hybrid solution approach for ready-mixed concrete delivery. Transportation Science 43(1), 70–85 (2009)

Schmid, V., Doerner, K.F., Hartl, R.F., Salazar-González, J.-J.: Hybridization of very large neighborhood search for ready-mixed concrete delivery problems. Computers and Operations Research 37(3), 559–574 (2010)

Silva, C., Faria, J.M., Abrantes, P., Sousa, J.M.C., Surico, M., Naso, D.: Concrete Delivery using a combination of GA and ACO. In: 44th IEEE Conference on Decision and Control, 2005 and 2005 European Control Conference, CDC-ECC 2005, pp. 7633–7638 (2005)

Tran, T.T., Beck, J.C.: Logic-based benders decomposition for alternative resource scheduling with sequence dependent setups. In: ECAI. Frontiers in Artificial Intelligence and Applications, vol. 242, pp. 774–779. IOS Press (2012)

Verstichel, J., Kinable, J., Vanden Berghe, G., De Causmaecker, P.: A combinatorial benders decomposition for the lock scheduling problem. KU Leuven, Tech. Rep., http://allserv.kahosl.be/ jannes/lockscheduling/ combinatorialBenders_19112013.pdf (September 2013)

Yan, S., Lai, W.: An optimal scheduling model for ready mixed concrete supply with overtime considerations. Automation in Construction 16(6), 734–744 (2007)

Evaluating CP Techniques to Plan Dynamic Resource Provisioning in Distributed Stream Processing

Andrea Reale, Paolo Bellavista, Antonio Corradi, and Michela Milano

Department of Computer Science and Engineering – DISI
Università di Bologna, Italy
{andrea.reale,paolo.bellavista,antonio.corradi,michela.milano}@unibo.it

Abstract. A growing number of applications require continuous processing of high-throughput data streams, e.g., financial analysis, network traffic monitoring, or big data analytics. Performing these analyses by using Distributed Stream Processing Systems (DSPSs) in large clusters is emerging as a promising solution to address the scalability challenges posed by these kind of scenarios. Yet, the high time-variability of stream characteristics makes it very inefficient to statically allocate the data-center resources needed to guarantee application Service Level Agreements (SLAs) and calls for original, dynamic, and adaptive resource allocation strategies. In this paper we analyze the problem of planning adaptive replication strategies for DSPS applications under the challenging assumption of minimal statistical knowledge of input characteristics. We investigate and evaluate how different CP techniques can be employed, and quantitatively show how different alternatives offer different trade-offs between problem solution time and stream processing runtime cost through experimental results over realistic testbeds.

1 Introduction

We are rapidly moving toward an always-connected world, where technology is an increasingly present mediator in the interactions between people and the environment [2]. Ever growing quantities of heterogeneous data are continuously generated and exchanged by moving or stationary sensors, smartphones, and wearable devices. This multitude of unbounded data flows must be handled effectively and efficiently.

Distributed Stream Processing Systems (DSPSs) [22] address the need of processing big data streams flexibly and in real-time by leveraging the parallel computational resources hosted inside data centers. A DSPS lets users define their own stream processing functionalities and encapsulate them in reusable components called *operators*. *Stream processing applications* (or, hereinafter, simply applications) are defined by arranging operators as vertices of *data-flow* graphs, directed and acyclic graphs that define the input-output relationships between different operators and between operators and external stream *data sources* and *data sinks*. Stream processing applications are deployed on a set of distributed

H. Simonis (Ed.): CPAIOR 2014, LNCS 8451, pp. 193–209, 2014.

resources, and their components are executed according to an event-based model that reacts to the arrival of new input data. One major problem in managing deployments of distributed stream processing applications lies in the proper management of the load fluctuations that arise due to sudden and possibly temporary variations in the rates of input data streams. If not handled properly load peaks can lead to increased processing latency due to data queuing, and to data loss due to queue overflows. To avoid these effects, it is necessary to allocate the correct amount of additional resources to overloaded applications either statically or dynamically when load variations are detected [1,5,12]. Another typical requirement is the fulfillment of *fault-tolerance* guarantees because applications usually run for (indefinitely) long time intervals and failures are unavoidable. A simple and commonly adopted solution is active replication of operator components [10], so that, if any replica fails, another can immediately take over and quickly mask the failure.

In [3,4] we have introduced LAAR, a Load Adaptive Active Replication technique that minimizes the cost of running replicated stream processing applications while guaranteeing that their deployment is never overloaded and, at the same time, that a user-specified fault-tolerance SLA is satisfied. It does so by dynamically deactivating and activating operator replicas, and by adapting the number of active ones to the changing application load according to a *replica activation strategy* precomputed before runtime. In this paper, we describe how we solve the challenging *replica activation problem*, i.e., the optimization problem whose solutions define how the LAAR runtime performs its dynamic activation of replicas depending on the currently observed configuration of input data rates. We propose a detailed study of this problem, and we present a quantitative comparison and evaluation of the effectiveness of different CP-based solution methods. The final goal is to highlight the trade-offs offered by different CP techniques considering the quality of their solutions and the associated cost.

2   Problem Definition

An application A consists of a set of components: a set I of data sources, a set P of operators, and a set O of data sinks, which collectively define the set $X = I \cup P \cup O = \{x_i\}$. The components in X are arranged in a directed acyclic application graph $W = (X, E)$. The set of edges E is described by the function:

$$pred : X \mapsto \mathcal{P}(X) \qquad (1)$$

which, for each component x_i, identifies the set of predecessors $\{x_j\}$ so that $x_j \in pred(x_i) \Leftrightarrow (x_j, x_i) \in E$.

We assume that the characteristics of the application inputs are known in terms of a probability mass function that describes the probability of a source to produce data at different rates. This information could be available thanks to previous knowledge of the application domain, or inferred through an initial profiling step [9]. We also assume that the continuous space of possible rates has been properly discretized through, e.g., binning techniques [8]. In particular, every data source $x_i \in I$ can produce output at one rate among a finite set of input rates R_i. The Cartesian product $C = R_1 \times \ldots \times R_t$, where t is the number of sources, is the set of all the possible *input configurations*, and $P_C : C \mapsto [0,1]$ is the probability mass function associated to the probability distribution of different input configurations in time. The output rate of data source $x_i \in I$ in a particular input configuration c is indicated as $\Delta(x_i, c)$.

Every operator receives one or more data streams from sources or from other operators and produces one data stream as output. For uniformity of notation, we use the symbol $\Delta(x_i, c)$ also to indicate the output rate of every operator $x_i \in P$.

Like what has been done previously in the literature (e.g., [10,11,21,20,23,25]), we summarize the characteristics of operators through a *selectivity* function δ and a *per-tuple CPU cost* function γ. For each couple (x_i, x_j) so that $x_i \in I \cup P$ and $x_j \in P$ and that $(x_i, x_j) \in E$, $\delta(x_i, x_j)$ defines the contribution of the stream generated by x_i to the output of operator x_j, so that, in absence of failures:

$$\Delta(x_j, c) = \sum_{x_i \in pred(x_j)} \delta(x_i, x_j) \Delta(x_i, c) \tag{2}$$

In a similar way, $\gamma(x_i, x_j)$ represents the per-tuple CPU cost for operator x_j to process tuples from x_i, so that the number of CPU cycles used by x_j per second can be expressed as:

$$\sum_{x_i \in pred(x_j)} \gamma(x_i, x_j) \Delta(x_i, c) \tag{3}$$

Each operator in P is actively replicated and all the replicated components are deployed on a set of distributed hosts $H = \{h_i\}$. We indicate the replicated set of operators as $\widetilde{P} = \{\tilde{x}_i^m\}$, where the symbol $\tilde{x}_i^m$ indicates the m-th replica of operator x_i. The assignment of replicas to hosts is represented by the function:

$$\vartheta : \widetilde{P} \mapsto H \tag{4}$$

For convenience, we also define $\vartheta^{-1} : H \mapsto \mathcal{P}(\widetilde{P})$ such that $\vartheta^{-1}(h) = \{\tilde{x}_i^j \in \widetilde{P} : \vartheta\left(\tilde{x}_i^j\right) = h\}$. We assume that ϑ is given, for example, because computed beforehand by an operator placement algorithm (e.g., [11,23]). In this paper, we only consider the case of twofold replication, i.e., two replicas per operator, since this scenario is the most commonly considered in real world stream applications. The model, however, can be very easily extended to k-fold replication.

A *replica activation strategy* is a function:

$$s : \widetilde{P} \times C \mapsto \{0, 1\} \tag{5}$$

that associates every operator replica – input configuration pair to one of the two possible active/inactive states. The goal of the optimization problem discussed in this paper is to find a replica activation strategy that suitably satisfies the application fault-tolerance quality requirements.

2.1   The Internal Completeness (IC) Metric

By activating/deactivating operator replicas according to the current input configuration, LAAR dynamically modifies the resilience of applications to failures. In order to measure the effect of LAAR on fault-tolerance guarantees, we define the *internal completeness* (IC) metric. Intuitively, given a failure model that describes how hosts and operators are expected to fail and a *replica activation strategy* s, the internal completeness measures, with respect to a time period T^1, the fraction of total tuples expected to be processed in case of failures compared to the number of tuples that would be processed in absence of failures.

In a no-failure scenario (best-case), the total number of tuples statistically expected to be processed by the application operators during T is:

$$BIC = T \cdot \sum_{\substack{c \in C, \\ x_i \in P, \\ x_j \in pred(x_i)}} P_C(c) \cdot \Delta(x_j, c) \tag{6}$$

Best-case internal completeness. (BIC) is the summation of the contributions of all the application operators in different input configurations, weighted by the probability of each configuration to occur.

Failure internal completeness. (FIC) measures the expected number of tuples processed with failure model ϕ and replica activation strategy s. It is defined as:

$$FIC(s) = T \cdot \sum_{\substack{c \in C, \\ x_i \in P, \\ x_j \in pred(x_i)}} P_C(c) \cdot \phi(x_i, c, s) \cdot \widehat{\Delta}(x_j, c, s) \tag{7}$$

$$\widehat{\Delta}(x_i, c, s) = \begin{cases} \Delta(x_i, c) & \text{if } x_i \in I \\ \phi(x_i, c, s) \cdot \sum_{x_j \in pred(x_i)} \delta(x_j, x_i) \widehat{\Delta}(x_j, c, s) & \text{if } x_i \in P \end{cases} \tag{8}$$

The function $\phi(x_i, c, s)$ depends on the chosen failure model and describes the probability that at least one replica of operator x_i is alive and active when the input configuration is c and the replica activation strategy is s. $\widehat{\Delta}(x_i, c, s)$, instead, represents the expected output of operator x_i under failure model ϕ, when the input configuration is c and the replica activation strategy is s; note that the definition of $\widehat{\Delta}$ is recursive, as the number of tuples produced by a operator depends not only on its possible failure status (described by ϕ) but also on the number of tuples produced by its predecessor (8). Let us rapidly note that the possible failures of data sources, which are components external to the application, are assumed to be handled externally.

Internal completeness (IC) is defined as the ratio between FIC and BIC:

$$IC(s) = \frac{FIC(s)}{BIC} \tag{9}$$

---

[1] We choose T long enough for the statistical characteristics of the sources to apply.

2.2   The Replica Activation Problem

In this section, we define the optimization problem that, solved off-line and before deployment time, outputs a replica activation strategy that fits the application fault tolerance requirements, and that is used at runtime by LAAR to activate operator replicas.

We call this problem *replica activation problem*, and we define it as follows:

$$\underset{s}{\text{minimize}} \quad cost\,(s) \tag{10a}$$

subject to:

$$IC(s) \geq G \tag{10b}$$

$$\sum_{\substack{\tilde{x}_i^m \in \vartheta^{-1}(h),\\ x_j \in pred(x_i)}} \gamma\,(x_j, x_i)\,\Delta(x_j, c)s(\tilde{x}_i^m, c) \leq K_h \qquad \substack{\forall h \in H,\\ \forall c \in C} \tag{10c}$$

$$s\left(\tilde{x}_i^0, c\right) + s\left(\tilde{x}_i^1, c\right) \geq 1 \qquad \substack{\forall x_i \in P,\\ \forall c \in C} \tag{10d}$$

The *cost* function in the minimization term represents the cost, in terms of resources, for a service provider to run the application using replica activation strategy s and the replicated assignment defined by ϑ. In this work, we assume the bandwidth available for cluster-local communication to be an abundant resource (a common assumption in data center contexts), and we model our cost function as the total number of CPU cycles used in a period T. It is defined as follows:

$$cost\,(s) = T \sum_{\substack{c \in C,\\ \tilde{x}_i^m \in \tilde{P},\\ x_j \in pred(x_i)}} P_C\,(c)\,\gamma\,(x_j, x_i)\,\Delta(x_j, c)s(\tilde{x}_i^m, c) \tag{11}$$

and is the summation over the CPU consumption of all the operator replicas.

Equation (10b) constrains IC to satisfy a requested fault-tolerance value G, while (10c) states that each host in the deployment should never be overloaded; K_h is a constant expressing the number of CPU cycles per second available at host h. The last constraint, expressed in (10d), requires that there is at least one active replica of every operator in every input configuration, and it ensures that the measured IC value is one in absence of failures.

2.3   Failure Model

We consider a simplified failure model ϕ, based on the following assumptions:

1. In any failure scenario, one replica of every operator fails.
2. Unless both the replicas are active at some point in time, the non-failed replica is assumed to be the one that was inactive according to the replica activation strategy.
3. Once failed, replicas never recover.

or, more formally:

$$\phi(x_i, c, s) = \begin{cases} 0 & \text{if } s(\tilde{x}_i^0, c) + s(\tilde{x}_i^1, c) < 2, \ \tilde{x}_{i,l} \in \tilde{P} \\ 1 & \text{otherwise} \end{cases} \quad (12)$$

This model will in general overestimate possible failure conditions because, in the actual stream processing deployment, it is highly unlikely that replicas of every operator fail at the same time, and because, in our runtime LAAR implementation, when an operator failure is detected, any corresponding and possibly deactivated replica returns to its active state, while an automatic recovery procedure promptly replaces the failed component [4]. For these reasons, we refer to ϕ as *pessimistic failure model*. While, overstating, on the one hand, the effects and consequences of failures compared to actual runtime conditions, on the other hand, this choice of ϕ provides two fundamental benefits: (i) the IC value computed using this model is a large lower bound to the real IC that will be observed on actual application deployments because any real failure condition is highly likely to be much less severe than those predicted by the model; (ii) its mathematical formulation simplifies the computation of IC values for different possible replica activation strategies and hence the optimization complexity.

Note that the solution space of this problem is still very large, as for every application there are $3^{|P| \cdot |C|}$ possible replica activation strategies. Note also that, in cost function (11), the IC constraint (10b) and the hosts CPU constraints (10c) depend on $\widehat{\Delta}(x_i, c, s)$ (8), which is a recursively defined exponential term. Hence, to find algorithms that can find optimal or good enough solutions to this problem is a major technical challenge.

3   Solving the Problem with CP

In this section, we analyze different CP-based approaches to solve the replica activation problem presented in Section 2.

As a first straightforward solution, we implemented the optimization model (10) "as-is" on the commercial IBM ILOG CP Optimizer solver [13], and used it to get a better understanding of the problem structure.

Looking at the problem from a user perspective, cost (11) and IC (9) are the most important parameters because, together, they determine the cost-quality trade-off for running stream processing applications with LAAR. Intuitively, since the basic mechanism to mask the effects of failures is to activate more replicas, in general, requiring higher IC (and hence better fault-tolerance) will correspond to higher runtime costs.

Fig. 1a gives some insight about the shape of the problem solution space when considering together cost and IC: it shows the space of possible feasible solutions of a problem instance consisting of 24 operator replicas distributed on 6 computing hosts and IC constraint — G in (10b) — set to 0.1. The continuous black line is a *loess* regression [6] of the solution points and confirms that, as a general trend, the cost of solutions is proportional to their IC value. However, the graph also shows

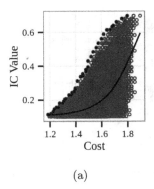

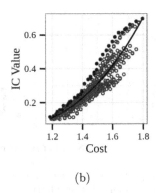

(a)   (b)

Fig. 1. Cost–IC relationship in the solution space of a problem instance consisting of 12 operators (2 replicas each) deployed on 6 hosts. Full circles represent the Pareto frontier of the problem space, while the continuous line is a regression of the solution points. Without (a) and with partial filtering of sub-optimal solutions (b).

that there is a very large number of sub-optimal solutions (empty circles) and that higher costs do not necessarily imply higher IC. Recall that the IC value does not only depend on the number of active replicas, but also on the particular choice of operators to activate and on the topology of the application data flow graph. As a consequence, a wrong choice of active operator replicas can easily lead to a useless waste of resources.

However, an important fraction of sub-optimal solutions not belonging to the Pareto frontier of the solution space can be discarded quickly with simple considerations. For example, think about a pipeline of operators where a first operator (O0) feeds a second one (O1). Given the pessimistic failure model in (12), having, in any input configuration, two active replicas of O1 and, at the same time, only one active replica for O0 does not contribute to the overall IC value because, in case of failures, O1 would not receive any sample to process from O0; that would, however, increase the solution cost. This is not only valid for pipelines but can be generalized for any graph shape: in particular, any feasible replica activation strategy s_x that, in some input configuration c, has two active replicas for some operator x_i whose predecessors all have only one active replica is sub-optimal with respect to a corresponding feasible replica activation strategy s_y that differs from s_x only for the fact that x_i has just one active replica. This relation can be used to add a new constraint to (10) that approximates the set of Pareto-optimal points by performing a *partial sub-optimal solution filtering* (PSF), which removes obviously sub-optimal solutions. We formulate this constraint as follows:

$$\exists\, x_i \in P, c \in C \text{ s.t. } \forall x_j \in pred\,(x) \sum_{l=0,1} s\left(\tilde{x}_j^m, c\right) = 1 \implies \sum_{l=0,1} s\left(\tilde{x}_i^m, c\right) = 1$$

$$(13)$$

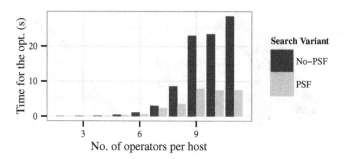

Fig. 2. Comparison of average time to find an optimal solution with and without the partial sub-optimal solution filtering

Fig. 1b shows the solution space of the same problem instance in Fig. 1a after the filtering based on (13). This important reduction in size also has a significant impact on the time needed to solve the problem. Fig. 2 summarizes the average search time needed to find the optimum solutions for a batch of small problem instances in which graphs of 2 to 11 operators are deployed on 4 hosts with two replicas per operator. As the graph complexity increases, the benefit of the additional constraint in (13) becomes more and more evident.

Given these characteristics of the solution space, we have developed three search strategies. The first (called *Basic*) is the straightforward implementation of the model in (10) with the additional constraint in (13) on the ILOG CP Optimizer solver. Since the realization of this first strategy is straightforward, we do not detail it further in this paper. On the contrary, in the following two subsections, we introduce the second and third search strategies, respectively a *Large Neighborhood Search (LNS)-based strategy* and a *decomposition-based* one.

3.1   LNS-Based Strategy

The basic idea behind LNS strategies [19] is to start from an initial solution and then proceed through incremental improvement steps that focus on *large neighborhoods* of the current best solution. We developed a strategy to solve our replica activation problem that is based on these concepts. The algorithm starts from a solution found either by using the *Basic* solver presented in the previous section, or by leveraging a simple *greedy* initial solution. The algorithm that finds this last type of initial solution is very simple: it starts with a replica activation strategy where two replicas of every operator are always active; iteratively, it deactivates the most resource hungry operator until all the non-overloading condition constraints (10c) are met. The advantage of using this greedy algorithm to find the initial solution four our LNS approach is its ability to terminate extremely quickly. These greedy solutions are not necessarily feasible because they can violate the constraint (10b), but our practical experience has shown that those infeasibilities can be often rapidly corrected through a few LNS moves.

Given the initial problem solution, the LNS-based strategy proceeds through a series of iterative improvement steps. At every round, a new optimization problem is built by *relaxing* the current best solution, i.e., by fixing the values of a subset of the search variables to those of the best solution so far and by focusing the exploration on the subspace of the remaining *relaxed* variables. We choose the variables to relax at each iteration according to one of two alternative methods. In the first (*simple random*), they are chosen completely random; in the second (*weighted random*), every search variable is assigned a weight that depends on its corresponding input configuration, so that variables associated to more resource-hungry input configurations (typically corresponding to load peaks) have higher chances to be chosen for relaxation. The weighted random strategy aims at relaxing these variables first because they usually require the highest number of operator replicas deactivated in order to satisfy (10c) and consequently have a stronger influence on the satisfiability of the IC requirement (10b). Every iterative improvement step explores the corresponding relaxed subspace either until it finds a solution that improves upon the previous one or until a local time limit expires. Every round dynamically adapts the number of relaxed variables by reducing it when the last step has produced improvements or by increasing it in case of deteriorations of the solution quality. The algorithm terminates either when there is no improvement for a configurable number of consecutive rounds or when a global time limit expires.

Note that, differently from the Basic search strategy, the LNS-based one does not recognize when an optimum has been reached, and, when greedy starting points are used, it cannot conclude whether a problem has any solutions if none are found.

3.2   Decomposition-Based Strategy

In this section we propose a solution approach that decomposes the problem in a number of orthogonal subproblems along its $|C|$ different input configurations. The goal of this decomposition-based approach is to provide scalability especially for instances with a large number of input configurations. This type of optimization is very important in many common real-world problems, where stream processing applications process data from tens of sources, each producing data at different possible data rates.

Let us consider once again the formulation of the replica activation problem (10). Separating the CPU constraints in (10c) and the minimum replicas constraints in (10d) is trivial, because each of them involves only terms relative to a single input configuration c. The search variables s can be equally easily separated by considering $|C|$ different replica activation strategies s_c such that:

$$s(\tilde{x}_i^m, c) = s_c(\tilde{x}_i^m), s_c : \widetilde{P} \mapsto \{0, 1\} \tag{14}$$

The IC constraint (10b) can be, instead, rewritten as follows:

$$\frac{FIC(s)}{T} \geq \underbrace{\frac{BIC}{T}(G)}_{G'}$$

$$\Leftrightarrow \underbrace{\sum_{c \in C} \sum_{\substack{x_i \in P, \\ x_j \in pred(x_i)}} P_C(c) \cdot \phi(x_i, c, s) \cdot \widehat{\Delta}(x_j, c, s)}_{\mu_c(s_c)} \geq G'$$

$$\Leftrightarrow \sum_{c \in C} \mu_c(s_c) \geq G' \tag{15}$$

Similarly, considering (11), the minimization term (10a) can be written as:

$$\min \sum_{c \in C} \underbrace{\sum_{\substack{\tilde{x}_i^m \in \tilde{P}, \\ x_j \in pred(x_i)}} P_C(c)\, \gamma(x_j, x_i)\, \Delta(x_j, c)s(\tilde{x}_i^m, c)}_{\lambda_c(s_c)}$$

$$\Leftrightarrow \min \sum_{c \in C} \lambda_c(s_c) \tag{16}$$

Note that, while the CPU and minimum replicas constraints can be evaluated and satisfied considering each input configuration c separately, the IC constraint and the cost minimization expression cannot; nonetheless, they both can be expressed as a sum of $|C|$ non negative terms, and each of this terms can be evaluated separately for different values of c.

Our decomposition approach consists in defining $|C|$ subproblems $prob_c$, one per input configuration; the solution of each problem is a partial replica activation strategy s_c that satisfies at least the corresponding CPU and minimum replication constraints (10c) and (10d). The subproblems' optimization goal and possible additional constraints, instead, depend on the particular phase the decomposition algorithm is in. Algorithm 1 sketches, in pseudo-code, the main steps of the decomposition-based solver.

The algorithm starts by maximizing the $\mu_c(s_c)$ values of each subproblem (Phase 1, lines 1–9). Note that, after this phase is complete, if a solution is found for every subproblem, an upper bound on the possible IC for the original problem can be obtained using (15): through it, it is possible to test immediately whether the original problem admits solutions (line 7) and, in case it does, to output an initial and in general sub-optimal solution. During Phase 2 (lines 10–22), this initial solution is improved by working separately and iteratively on each subproblem. At every iteration, the problem whose contribution to the overall IC is minimum with respect to its contribution to the cost (line 12) is chosen as a candidate for improvement, and the algorithm tries to decrease its cost while ensuring that the obtained $\mu_c(s_c)$ value still allows to satisfies the overall IC requirement (line 13). This iteration is repeated until no improvement

---

Algorithm 1. Decomposition-based Search Strategy

---

input : $\{prob_c\}$: the $|C|$ decomposed subproblems.
output: A replica activation strategy s, or None if no solution found

1 **Phase 1:**   /* μ_c maximization */
2   **foreach** $prob_c$ **do**
3      $s_c^{max} \leftarrow$ maximize μ_c in $prob_c$
4   **if** s_c^{max} *is* **None then return** None $\mu_c^{max} \leftarrow$ maximum μ_c for $prob_c$
5      $\lambda_c^{max} \leftarrow$ cost value corresponding to s_c^{max}
6   **end**
7   **if** $\sum_{c \in C} \mu_c^{max} < G'$ **then**   /* Feasibility test */
8   **return** None
9   **end**
10 **Phase 2:**   /* optimization */
11   **foreach** $prob_c$ **do** $\mu_c^{cur} \leftarrow \mu_c^{max}$; $\lambda_c^{cur} \leftarrow \lambda_c^{max}$ **while** *exists* $prob_c$ *that can be improved* **do**
12     $c' \leftarrow \max_c (\lambda_c^{cur}/\mu_c^{cur})$ /* Choose prob. to improve */
13     $\mu_{c'}^{limit} \leftarrow G' - \sum_{\substack{c \in C \\ c \neq c'}} \mu_c^{cur}$
14   Post $\mu_{c'} \geq \mu_{c'}^{limit}$ as constraint on $prob_{c'}$
15   Post $\lambda_{c'} < \lambda_{c'}^{cur}$ as constraint on $prob_{c'}$
16     $s_{c'}^{cur} \leftarrow$ findFirst$(prob_{c'})$ /* Solve $prob_c$ */
17   **if** $s_{c'}$ *is* **None then**
18       $prob_{c'}$ cannot be improved further
19   **else**
20   Update $\mu_{c'}^{cur}$ and $\lambda_{c'}^{cur}$ according to $s_{c'}^{cur}$
21   **end**
22   **end**
23 **Phase 3:**   /* End */
24   $s \leftarrow$ Combine all the s_c^{cur}
25   **return** s

---

can be obtained from any subproblem. In Phase 3, finally, the partial replica activation strategies are combined, and the result returned as output.

Like the LNS-based strategy, this algorithm can decide whether the problem is feasible, but cannot recognize an optimal solution. In the cases where the operator graph is particularly complex, it might be necessary to set a time limit for Phase 1 to avoid blocking the solver for too long; in such cases, the solution obtained after Phase 1 is no longer an upper bound on the obtainable IC, and so the algorithm cannot decide anymore about the feasibility of the entire problem. Let us note, finally, that the various subproblems optimizations (either the initial IC maximization or the subsequent cost minimizations) can be performed with any optimization technique.

4 Experimental Evaluation

The primary goals of our evaluation study are i) to compare the quality of the best solutions that the three strategies can find within a reasonable time limit and ii) to evaluate the scalability of the search strategies (in particular of the decomposition-based one) as the problem size grows.

For the first part of this evaluation, we consider a batch of 20 different stream processing applications with data flow graphs of 96 operators each. Every application has three data sources, each producing output at two possible data rates (for a total of 8 input configurations), and is associated with a replicated deployment (two replicas per operator, 192 replicas in total) on 24 computing hosts. The IC constraint in the related Replica Activation Problem is set to 0.5. We choose these applications as we believe their complexity to be well representative of real world stream processing deployments.

We compare the optimization algorithms in the following variants:

1. Basic solver with partial sub-optimal-filtering (*BASIC*).
2. LNS-based strategy using BASIC for the initial solution and simple weighted random relaxation method (*L_SRW*).
3. LNS-based strategy using a greedy initial solution and weighted random relaxation method (*L_GRW*).
4. LNS-based strategy using a greedy initial solution and simple random relaxation method (*L_GRS*).
5. Decomposition strategy using BASIC for Phase 1 (*DEC_S*).
6. Decomposition strategy using LNS_GRW for Phase 1 (*DEC_L*).

All the experiments are executed on a machine with an AMD Phenom II X6 1055T @2.8 GHz processor and 8 GB of main memory. The ILOG CP Optimizer is configured to use only one worker (single threaded solution), and its search time limit is set to 300 seconds wall time. Due to the complexity of the problems, for no instance it was possible to demonstrate the optimality of the solutions found; however, feasible solutions were found for all instances except four.

In this experimental campaign, we were primarily interested in two aspects, i.e., the ability to find good solution in a relatively large time frame, and the complementary capacity of quickly finding a first feasible solution. Both aspects are critical in our use-case scenario: the first is more significant during the deployment of new stream processing applications, when a larger time budget is usually available; the second is more relevant when replica activation strategies must be quickly adjusted at runtime due to dynamic variations of input characteristics. For reasons of space, in the following, we only report the results about the first aspect. Experimental data about the second are available in our on-line appendix [17], together with downloadable descriptions of the problem instances used in this evaluation.

The bar plot in Fig. 3 compares the various search algorithms to BASIC, which we choose as the base line solution method; the plot analyzes the best solution cost (BCOST) and its associated search time (BTIME) and is obtained by normalizing, separately for each problem instance, the values of BTIME and BCOST with respect to the results obtained in BASIC. The figure shows the average of these values along with the associated standard error.

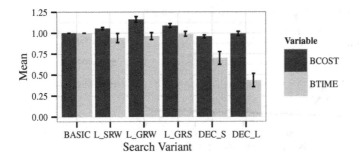

Fig. 3. Mean time to find the best solution within the time limit and associated solution cost. All the results are normalized w.r.t. the BASIC variant.

In general, the two decomposition-based variants find solutions that are at least as good or slightly better than those found by BASIC. In more detail, in the only six instances where the decomposition variants finds solutions worst than BASIC, that solution is at most 23% more expensive; at the same time, they can save considerable amounts of time (43% on average). The results show also that, on the one hand, DEC_L can find good solutions much faster than DEC_S (4% to 70% faster), probably due to the initial speed-up given by the LNS greedy strategy used to find the starting solutions for Phase 1; on the other hand, the solutions found by DEC_L tend to be a little more expensive than those found by DEC_S (from 1% to 24%): that is explained by considering that the use of the BASIC solver in Phase 1 usually gives tighter IC upper bounds, which, in turn, permit to use looser constraints on the μ_c^{limit} values in Phase 2. The solutions found by the LNS-based variants, finally, are in all but one case worst than those found by BASIC, with cost inflations up to 57%. Among the LNS variants, L_SRW is the one providing the best results thanks to its better (although slower) initial solutions, with higher costs (between 1% and 13%), but solution times that are 75% smaller to 26% bigger than BASIC.

Finally, we evaluate the scalability of the decomposition-based strategy when the number of input configurations grows. In order to measure it, we started from an application graph with 32 operators and one data source, with a replicated deployment (64 operators) on 8 hosts, and we randomly generated 40 different applications, for each, customizing the number of possible data source rates. The result is a set of 40 different replica activation problem instances sharing the same processing graph and deployment, but with a progressively growing number of input configurations (from 2 to 80 by steps of 2). We solved these instances through the BASIC and the DEC_S search variants. Fig. 4 shows the time taken by the two strategies to find their best solution as the number of input configurations grows. The results for the BASIC variants grow very quickly, and, for instances with more than 18 configurations, we could not find any solution within the time limit. On the contrary, by using DEC_S, the solution time grows much more slowly, and we easily solved all the problem instances.

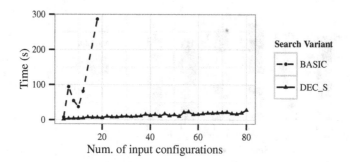

Fig. 4. Scalability of the BASIC and DEC_S solution strategies

5   Related Work

The problem of managing load variations in DSPSs has been widely investigated in the related literature, and different solution techniques have been proposed.

Tatbul et al. [20], for example, tries to avoid resources overload in spite of changing load condition by introducing controlled data drops in the data-flow graph. In a pre-deployment phase, they solve a set of LP problems to build load-shedding plans that decide where and how many tuples to drop to maximize application throughput. Like our *replica activation strategies*, load-shedding plans contain decisions for each input configuration. However, these plans do not take into account the interplay of fault-tolerance and variable input load: with load shedding, data is dropped even when no failure occurs, while LAAR guarantees that no data is lost in that case and it bounds the maximum amount of loss in case of failures. For these reasons, the problem model in [20] and the one presented in this paper are very different and difficult to compare.

Another common approach is to move operators between hosts to re-balance the system and accommodate new load conditions: in [12,24,25] this is done through continuous greedy improvement steps. Likewise, in [1], the authors develop a resource allocation algorithm that uses a dynamic flow-control algorithm based on a linear quadratic regulator [7] to maximize the application throughput. All these approaches assume that the available resources are enough to handle any input configuration, or that the data sources rate can be paced at will until there are enough resources to handle the load; in LAAR additional resources are dynamically provided by temporarily replication adjustments.

Using CP to manage replicas in distributed systems has been previously done by Michel et al. [16], who propose a CP model that solves the problem of deploying replicas on distributed nodes to minimize the communication cost in Eventually Serializable Data Service (ESDS) systems. Our replica activation problem is different because we do not deal with the assignment of replicas to computing resources, but we decide their dynamic activation strategy.

In [15], the authors solve a combined assignment and scheduling problem for conditional task graphs (CTG). Similarly to this work, the CP model includes stochastic elements, but they are used to describe the probability that branches in the task graphs are actually used at runtime.

Our problem formulation closely resembles stochastic optimization problems with value at risk (VaR) guarantees [18], and our decomposition strategy is based on the notion of separability of optimization problems commonly used in OR contexts [14].

6 Conclusions and Future Work

In this paper we have introduced an optimization problem whose solution is at the foundations of LAAR, a technique for dynamic and load-adaptive active replication in DSPSs. After having investigated the characteristics of the problem and of its solution space, we have presented three possible solution strategies: a naïve and straightforward model solver, a LNS-based search strategy, and a Decomposition-based strategy. Our experimental evaluation shows that when sufficient time budged is available, the decomposition approach can find good solutions and scale particularly well for instances where possibly many data sources produce input at many possible rates. On the contrary, the LNS-approaches represent an appropriate solution when finding quickly a good-enough feasible solution is the main concern [17]. Finally, the naïve solver provides the most consistent behavior across all the possible scenarios when used in combination with an ad-hoc sub-optimal solution filtering constraint.

As future work, we will continue our investigation on the problem trying to correlate specific problem characteristics (e.g., shape of the graph, properties of the deployment) to the behavior of different search strategies. In addition, we will continue experimenting with the current approaches on a broader set of problem instances in order to expand and further validate our findings. Let us finally note that, although we have introduced the replica activation problem and LAAR in the context of stream processing, the presented principles are applicable to the much larger domain of distributed data flow systems that can tolerate weaker fault-tolerance levels through dynamic active replication.

Acknowledgements. We would like to thank the Smarter Cities Technology Centre, IBM Dublin Research Laboratory and, in particular, Spyros Kotoulas for the precious contribution in the development of the original ideas behind LAAR.

References

1. Amini, L., Jain, N., Sehgal, A., Silber, J., Verscheure, O.: Adaptive control of extreme-scale stream processing systems. In: Proc. of the 26th IEEE ICDS Conference, pp. 71–78. IEEE (2006)
2. Atzori, L., Iera, A., Morabito, G.: The internet of things: A survey. Computer Networks 54(15), 2787–2805 (2010)

3. Bellavista, P., Corradi, A., Kotoulas, S., Reale, A.: Dynamic datacenter resource provisioning for high-performance distributed stream processing with adaptive fault-tolerance. In: Proc. of the 2013 ACM/IFIP/USENIX International Middleware Conference. Posters and Demos Track (2013)

4. Bellavista, P., Corradi, A., Kotoulas, S., Reale, A.: Adaptive fault-tolerance for dynamic resource provisioning in distributed stream processing systems. In: Proc. of the of 17th International EDBT Conference. ACM (2014)

5. Boutsis, I., Kalogeraki, V.: Radar: adaptive rate allocation in distributed stream processing systems under bursty workloads. In: Proc. of the 31st SRDS Symposium, pp. 285–290. IEEE (2012)

6. Cleveland, W.S., Devlin, S.J.: Locally weighted regression: an approach to regression analysis by local fitting. J. Amer. Statist. Assoc. 83(403), 596–610 (1988)

7. Cobb, D.: Descriptor variable systems and optimal state regulation. IEEE Transactions on Automatic Control 28(5), 601–611 (1983)

8. Dougherty, J., Kohavi, R., Sahami, M.: Supervised and unsupervised discretization of continuous features. In: Proc. of the 12th ICML Conference, pp. 194–202. Morgan Kaufmann (1995)

9. Gedik, B., Andrade, H., Wu, K.-L.: A code generation approach to optimizing high-performance distributed data stream processing. In: Proc. of the 18th CIKM Conference, pp. 847–856. ACM (2009)

10. Hwang, J.-H., Balazinska, M., Rasin, A., Çetintemel, U., Stonebraker, M., Zdonik, S.: High-availability algorithms for distributed stream processing. In: Proc. of the 21st ICDE Conference, pp. 779–790. IEEE (2005)

11. Khandekar, R., Hildrum, K., Parekh, S., Rajan, D., Wolf, J., Wu, K.-L., Andrade, H., Gedik, B.: Cola: Optimizing stream processing applications via graph partitioning. In: Bacon, J.M., Cooper, B.F. (eds.) Middleware 2009. LNCS, vol. 5896, pp. 308–327. Springer, Heidelberg (2009)

12. Kumar, V., Cooper, B., Schwan, K.: Distributed stream management using utility-driven self-adaptive middleware. In: Proc. of the 2nd ICAC Conference, pp. 3–14. IEEE (2005)

13. Laborie, P.: Ibm ilog cp optimizer for detailed scheduling illustrated on three problems. In: van Hoeve, W.-J., Hooker, J.N. (eds.) CPAIOR 2009. LNCS, vol. 5547, pp. 148–162. Springer, Heidelberg (2009)

14. Li, D., Sun, X.: Separable integer programming. In: Nonlinear Integer Programming, ch. 7, pp. 209–239. Springer (2006)

15. Lombardi, M., Milano, M.: Allocation and scheduling of conditional task graphs. Artificial Intelligence 174(78), 500–529 (2010)

16. Michel, L., Shvartsman, A.A., Sonderegger, E., Van Hentenryck, P.: Optimal deployment of eventually-serializable data services. In: Perron, L., Trick, M.A. (eds.) CPAIOR 2008. LNCS, vol. 5015, pp. 188–202. Springer, Heidelberg (2008)

17. Reale, A., Bellavista, P., Corradi, A., Milano, M.: Evaluationg cp techniques to plan dynamic resource provisioning in distributed stream processing: On-line appendix, http://middleware.unibo.it/people/ar/laar-rap/ (web page, last visited in February 2014)

18. Rockafellar, R.T., Uryasev, S.: Optimization of conditional value-at-risk. Journal of Risk 2, 21–42 (2000)

19. Shaw, P.: Using constraint programming and local search methods to solve vehicle routing problems. In: Maher, M.J., Puget, J.-F. (eds.) CP 1998. LNCS, vol. 1520, pp. 417–431. Springer, Heidelberg (1998)

20. Tatbul, N., Çetintemel, U., Zdonik, S.: Staying fit: efficient load shedding techniques for distributed stream processing. In: Proc. of the 33rd VLDB Conference. The VLDB Endowment (2007)
21. Tatbul, N., Çetintemel, U., Zdonik, S., Cherniacak, M., Stonebraker, M.: Load shedding in a data stream manager. In: Proc. of the 29th VLDB Conference, pp. 309–320. The VLDB Endowment (2003)
22. Turaga, D., Andrade, H., Gedik, B., Venkatramani, C., Verscheure, O., Harris, J., Cox, J., Szewczyk, W., Jones, P.: Design principles for developing stream processing applications. Soft. Pract. Exper. 40(12), 1073–1104 (2010)
23. Xing, Y., Hwang, J.-H., Çetintemel, U., Zdonik, S.: Providing resiliency to load variations in distributed stream processing. In: Proc. of the 32nd VLDB Conference. The VLDB Endowment (2006)
24. Xing, Y., Zdonik, S., Hwang, J.H.: Dynamic load distribution in the borealis stream processor. In: Proc. of the 21st ICDE Conference, pp. 791–802. IEEE (2005)
25. Zhou, Y., Ooi, B.C., Tan, K.-L., Wu, J.: Efficient dynamic operator placement in a locally distributed continuous query system. In: Meersman, R., Tari, Z. (eds.) OTM 2006. LNCS, vol. 4275, pp. 54–71. Springer, Heidelberg (2006)

Disregarding Duration Uncertainty in Partial Order Schedules? Yes, We Can!

Alessio Bonfietti, Michele Lombardi, and Michela Milano

DISI, University of Bologna
{alessio.bonfietti,michele.lombardi2,michela.milano}@unibo.it

Abstract. In the context of Scheduling under uncertainty, Partial Order Schedules (POS) provide a convenient way to build flexible solutions. A POS is obtained from a Project Graph by adding precedence constraints so that no resource conflict can arise, for any possible assignment of the activity durations. In this paper, we use a simulation approach to evaluate the expected makespan of a number of POSs, obtained by solving scheduling benchmarks via multiple approaches. Our evaluation leads us to the discovery of a striking correlation between the expected makespan and the makespan obtained by simply fixing all durations to their average. The strength of the correlation is such that it is possible to disregard completely the uncertainty during the schedule construction and yet obtain a very accurate estimation of the expected makespan. We provide a thorough empirical and theoretical analysis of this result, showing the existence of solid ground for finding a similarly strong relation on a broad class of scheduling problems of practical importance.

1 Introduction

Combinatorial Optimization approaches have a tremendous potential to improve decision making activities. Most optimization techniques, however, require complete problem knowledge to be available, and with good reason. Indeed, the dynamic and uncertain nature of real world problems is a true curse to deal with. With the exception of specific settings, the scalability of stochastic optimization approaches is orders of magnitude worse than their deterministic counterparts.

Scheduling problems make no exception to this rule and their stochastic variants tend to be even harder to solve than the (already hard) classical formulations. In the literature, the case of uncertain activity durations [2] has received the greatest attention. This is because it provides a convenient framework to model a number of unexpected events (delays, failure of unary resources, unknown input. . .) and because there is hardly any practical scheduling problem where the durations are really deterministic. Many have tackled this class of problems by shifting the activity start times [18], or by doing that *and* re-adjusting the schedule when a variation occurs [17].

A somehow orthogonal family of approaches has focused instead in providing flexible solutions as Partial Order Schedules (POSs). A POS is an augmentation of the original Project Graph, where a number of precedence constraints has

H. Simonis (Ed.): CPAIOR 2014, LNCS 8451, pp. 210–225, 2014.

been added to prevent the occurrence of resource conflicts, whatever the activity durations are. A POS can be obtained through a variety of methods (the reader may refer for details to [8, 5, 6, 11–14]). A POS can be designed to optimize some probabilistic performance metric, such as the expected makespan (a frequent pick) or the n-th quantile (see [3]). Unfortunately, even simply computing such metrics is by itself a difficult problem. Hence, a more convenient approach is to optimize a deterministic scenario (e.g. worst case durations) and rely on the flexibility of the POS for a good performance once the uncertain comes into play.

In this paper, we tackle the rather unusual problem of characterizing the runtime behavior of a POS. Other authors have evaluated the quality of a POS via intuition-based metrics [13, 14, 1], or even via simulation [15]. However, those efforts have been focused in comparing solution approaches. Conversely, *we focus on inferring general properties of the expected makespan of a POS*. The raw data for our investigation is obtained by simulating a large number of POSs, obtained by solving classical scheduling benchmarks via multiple methods. Our main, rather disconcerting, outcome is that in a large variety of situations there is a very strong correlation between the expected makespan and the makespan obtained when all the activities take their expected duration. In particular, their difference appears very resilient to scheduling decisions. *In such a situation it is possible to disregard the uncertainty altogether, solve a deterministic problem, and end up with a close-to-optimal solution for a (much harder!) stochastic problem.* We provide a reasonable explanation for the observed behavior, supported by a mathematical model that we check against the empirical data. We deduce that the behavior we observed, although not universally guaranteed, can be reasonably expected on a broad class of practical problems.

2   Experimental Setup

Origin of our POSs: The POSs for our analysis represent solutions of classical scheduling benchmarks. They have been obtained via two methods: 1) the constructive approach from [12] and 2) the application of a chaining step inspired by [13] to solutions obtained via ILOG CP Optimizer. For each considered instance, we kept all the solutions found by both solvers in the optimization process. In the following, whenever non-explicitly mentioned, the presented data refers to the chaining approach. As target benchmarks we have employed all the RCPSP instances in the PSPLIB [10] from the j30, j60, and j90 sets (named after the number of activities), plus most of the job shop scheduling instances by Taillard [16] (up to 30 jobs and 20 machines). The PSPLIB instances cover a wide range of resource usage patterns, while the job shop instances provide data for a radically different scheduling problem.

The Simulation Approach: In the simplest terms, our simulator is just a composition of Monte-Carlo sampling and critical path computation. Formally, we model the duration of each activity a_i as an *independent* random variable δ_i ranging in the interval $[d_i, D_i]$. The value D_i is the maximum duration and it is always equal to the (fixed) duration value from the benchmark instances. The minimum duration d_i is computed as $\alpha \cdot D_i$, where $\alpha \in \,]0,1[$. The value of α and the probability distribution for the δ_i variables are simulation parameters. The makespan is formally defined as a function $\tau(\delta)$ of the duration variables. We use the special notation T to refer to the value $\tau(D)$, i.e. the makespan for the worst case durations.

The makespan function has unknown probability distribution, and therefore unknown expected value $\mathbb{E}[\tau]$. Our simulator builds an approximation for $\mathbb{E}[\tau]$ by sampling the duration variables to obtain a set of scenarios Δ. Each scenario $\delta^{(k)} \in \Delta$ is an assignment of values to the δ_i variables. Then for each $\delta^{(k)}$ we compute the makespan by running a simple critical path algorithm: since the POS are resource-feasible, there is no need to take into account the capacity limits. By doing so, we obtain the sample average:

$$\overline{\tau} = \frac{1}{|\Delta|} \sum_{\delta^{(k)} \in \Delta} \tau(\delta^{(k)}) \tag{1}$$

which is an approximation for the expected value $\mathbb{E}[\tau]$. We simulated each of our POSs 10,000 times (i.e. $|\Delta| = 10,000$). We have considered two types of distribution for the δ_i variables, namely 1) a uniform distribution and 2) a discrete distribution where the duration can be either αD_i with probability p or D_i with probability $1 - p$. We have run tests with $p = 0.5$ and $p = 0.9$ and $\alpha = 0.1, 0.25, 0.5$ and 0.75. In the following, whenever non explicitly mentioned, the presented data refers to $\alpha = 0.5$ and uniformly distributed durations.

3   Our Main Result

As mentioned in Section 1, it is common to build a POS to minimize the worst case makespan, so as to provide worst case guarantees. Even in this setting, however, having a good expected performance is a highly desirable property. We therefore decided to start our investigation by comparing the expected makespan of a POS (or rather the approximation $\overline{\tau}$), with its worst case value T.

This comparison can be done effectively by plotting the two quantities one against each other (say, T on the x- and $\overline{\tau}$ on the y-axis). It is clear that $\overline{\tau} \leq T$. That said, given how complex the structure of a POS can be and the amount of variability involved, one could expected the plot to contain a cloud of points distributed almost uniformly below the main bisector. A significant discrepancy between different benchmarks and simulation settings could also be expected.

Surprisingly, we observe instead that in *all the considered cases there exists a strong, seemingly linear, correlation between the average makespan and the worst case one.* Figure 1 shows the described plots for 4 different configurations, chosen as examples: (a) the POSs for the j30 instances from the PSPLIB, solved with the chaining approach and uniform distributions; (b) the j30 set with the constructive approach and uniform distributions; (c) the j60 set with the constructive approach, discrete distribution and $p = 0.9$. Finally, (d) the POSs from the Taillard instances with the chaining approach and uniform distributions.

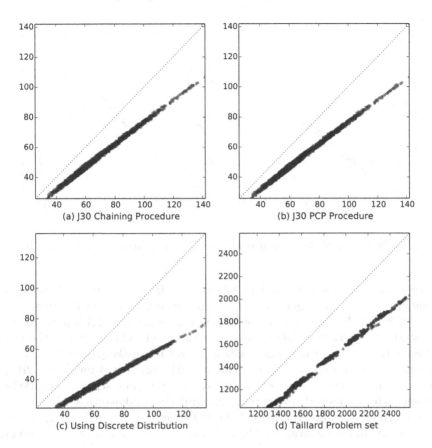

Fig. 1. Expected makespan (y-axis) over worst case makespan (x-axes) for various experimental settings

The Real Correlation: A necessary first step to understand the observed correlation is to identify exactly its terms. In our experimentation, we are assuming all activities to have uniformly distributed durations in the range $[d_i, D_i]$, so that their expected value is $\mathbb{E}[\delta_i] = 0.5 \cdot (d_i + D_i)$. Moreover, since we are also assuming that $d_i = \alpha D_i$, we have that $\mathbb{E}[\delta_i] = \frac{1}{2}(1 + \alpha)D_i$, i.e. the expected durations are proportional to the worst case ones. This means that by fixing all the activities to $\mathbb{E}[\delta_i]$ rather than to D_i, the critical path stays unchanged and the makespan becomes:

$$\tau(\mathbb{E}[\delta]) = \frac{1}{2}\sum_{i \in \pi}(\alpha D_i + D_i) = \frac{1+\alpha}{2}\sum_{i \in \pi} D_i = \frac{1+\alpha}{2}T \qquad (2)$$

Hence, the observed correlation between $\overline{\tau}$ and T could be just a side effect of a actual correlation between $\overline{\tau}$ and $\tau(\mathbb{E}[\delta])$, because T and $\tau(\mathbb{E}[\delta])$ are *proportional*. To test if this is the case, we have designed a second set of simulations where,

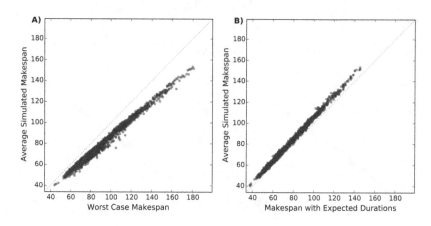

Fig. 2. Correlation of the simulated makespan with α varying on a per-task basis

for each task, we choose at random between $\alpha = 0.1$ or $\alpha = 0.75$. By doing so, Equation (2) does no longer hold. As a consequence, we expect the correlation between $\bar{\tau}$ and T to become more blurry. Figure 2A and 2B report, for the j60 benchmarks, scatter plots having $\bar{\tau}$ on the y-axis. On the x-axis the plots respectively have the worst case makespan and the makespan with expected durations. The data points are much more neatly distributed for Figure 2B. In summary, **there is evidence for the existence of a very strong correlation between the expected makespan $\mathbb{E}[\tau]$ and the makespan $\tau(\mathbb{E}[\delta])$.**

In the (very uncommon) case that the POS consists of a single path, this is a trivial result. In such a situation the makespan function τ can be expressed as:

$$\tau(\delta) = l(\pi) = \sum_{a_i \in \pi} \delta_i \tag{3}$$

where π is the path (a sequence of activity indices $\pi(0), \pi(1), \dots \pi(k-1)$) and $l(\pi)$ is the path length. For the linear properties of the expected value, it follows that $\mathbb{E}[\tau] = \sum_{i \in \pi} \mathbb{E}[\delta_i]$, i.e. the expected makespan is exactly $\tau(\mathbb{E}[\delta])$. On the other hand, a *general* POS has a much more complex structure, with multiple paths competing to be critical in a stochastic setting. In such case, the presence of the observed correlation is therefore far from an obvious conclusion.

Makespan Deviation: What is even more interesting, however, is that $\bar{\tau}$ and $\tau(\mathbb{E}[\delta])$ remain remarkably close one to each other, rather than diverging as their values grow. It is therefore worth investigating in deeper detail the behavior of the difference $\bar{\tau} - \tau(\mathbb{E}[\delta])$, as an approximation for $\mathbb{E}[\tau] - \tau(\mathbb{E}[\delta])$. We refer to this quantity as Δ and we call it *makespan deviation.*

Figure 3A shows the deviation for j60, over the makespan with expected durations. The plot (and every plot from now on) is obtained using a unique α value for a whole benchmark, because this allows to conveniently compute $\tau(\mathbb{E}[\delta])$ as

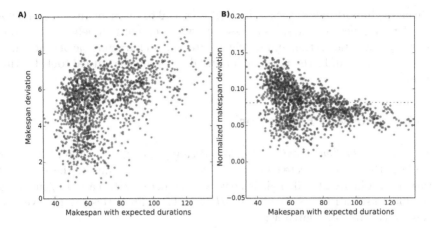

Fig. 3. Deviation and normalized deviation over $\tau(\mathbb{E}[\delta])$

$\frac{1+\alpha}{2}T$. The deviation appears to grow slowly as the makespan (with expected durations) increases.

It is possible to determine the rate of growth more precisely via a test that involves the normalized makespan deviation, i.e. the quantity $\Delta/\tau(\mathbb{E}[\delta])$. Specifically, let us assume that a linear dependence exists, then we should have:

$$\mathbb{E}[\tau] = a \cdot \tau(\mathbb{E}[\delta]) + b \qquad (4)$$

Now, note that $\tau(\mathbb{E}[\delta]) = 0$ can hold iff all the activities have zero expected duration. This is only possible if they have zero maximum duration, too. Hence, if $\tau(\mathbb{E}[\delta]) = 0$, then $\mathbb{E}[\tau] = 0$. This implies that $b = 0$, i.e. there is no offset. Hence, we can deduce that $\mathbb{E}[\tau]$ and $\tau(\mathbb{E}[\delta])$ are linearly dependent iff:

$$\frac{\Delta}{\tau(\mathbb{E}[\delta])} = a - 1 \qquad (5)$$

i.e. if the normalized makespan deviation is constant. Figure 3B shows for j60 the normalized Δ over $\tau(\mathbb{E}[\delta])$. We have added to the plot a line that shows what the data trend would be in case of a constant normalized deviation. As one can see, the normalized Δ tends instead to get lower as $\tau(\mathbb{E}[\delta])$ grows.

Therefore, **the makespan deviation $\mathbb{E}[\tau] - \tau(\mathbb{E}[\delta])$ appears to grow sub-linearly with the makespan obtained when all activities take their expected duration, i.e. $\tau(\mathbb{E}[\delta])$.** From another perspective, this means that the deviation has sub-linear sensitivity to scheduling decisions. In such a situation, **it becomes possible to disregard the duration uncertainty and optimize under the assumption that each activity a_i has duration $\mathbb{E}[\delta_i]$, being reasonably sure that the resulting POS will be close to optimal also in the stochastic setting.** This makes for a tremendous reduction in the complexity of scheduling under duration uncertainty, with far reaching consequences both on the theoretical and the practical side.

Given the importance of such a result, it is crucial to reach an understanding of the underlying mechanism, so as to assess to which extent this behavior depends on our experimental setting and to which extent it is characteristic of scheduling problems in general. In the next section, we develop mathematical tools for this analysis.

4   Our Analysis Framework

Independent Path Equivalent Model: We already know from Section 3 that for a single path π the expected makespan is equal to the sum of the expected durations and hence $\Delta = 0$. This is however a rather specific case. In general, it is possible to see a POS as a collection Π of paths π. From this perspective, the makespan can be formulated as:

$$\tau(\delta) = \max_{\pi \in \Pi} l_\pi(\delta) \tag{6}$$

i.e. for a given instantiation of the activity durations, the makespan is given by the longest path. The worst case makespan T will correspond to the worst case length of one or more particular paths: we refer to them as *critical paths*.

In general, many paths will not be critical. Moreover, several paths will likely share some activities. Finally, the number of paths $|\Pi|$ may be exponential in the Project Graph size. For those reasons, computing $\mathbb{E}[\tau]$ for the model from Equation (6) is not viable in practice. Hence, to understand the impact of multiple paths on $\mathbb{E}[\tau]$ we resort to a simplified model. Specifically, we assume that a POS consists of n identically distributed, independent, critical paths. Formally:

$$\tau(\delta) = \max_{i=0..n-1} \lambda_i(\delta) \tag{7}$$

where each λ_i is a random variable representing the length of a path i and ranging in $[\alpha T, T]$. The notation is (on purpose) analogous to that of our simulation. We refer to Equation (7) as the *Independent Path Equivalent Model* (IPE Model).

Effect of Multiple Paths: Now, let F_λ denote the (identical) Cumulative Distribution Function (CDF) for all λ_i, i.e. $F_\lambda(x)$ is the probability that a λ_i variable is less than or equal to x. Then for the CDF of τ we have:

$$F_\tau(x) = \prod_{i=0}^{n-1} F_\lambda(x) = F_\lambda(x)^n \tag{8}$$

in other words, for τ to be lower than a value x, all the path lengths must be lower than x. Therefore, **increasing the number of critical paths n reduces the likelihood to have small makespan values and leads to a larger $\mathbb{E}[\tau]$**. Figure 4A shows this behavior, under the assumption that all λ_i are uniformly distributed with $T = 1$ and $\alpha = 0$ and hence have an average length of 0.5. The drawing reports the Probability Density Function (PDF), i.e.

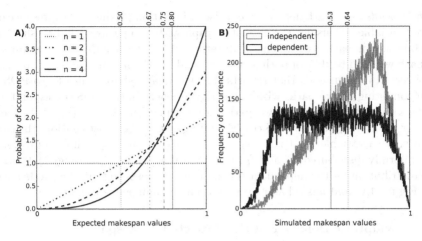

Fig. 4. Effect of multiple paths

the probability that $\tau = x$, with $n = 1, 2, 3, 4$. As n grows, the PDF is skewed to the right and the corresponding expected value (marked by a vertical red line) moves accordingly. Note that moving from a single path to multiple ones causes a relevant shift of $\mathbb{E}[\tau]$ from the average critical path length (i.e. 0.5). This means that the POS structure may in theory have a strong impact on the makespan deviation. However, this is not observed in Figure 1 and Figure 2, raising even more interest in the reasons for this behavior.

4.1   Worst Case Assumptions

The IPE Model differs from the exact formulation in Equation (6) by three main simplifications: 1) all paths are critical, 2) all paths are identically distributed, and 3) all paths are independent. Due to those simplifications, **the IPE Model actually corresponds to a worst case situation and is therefore well suited to obtain conservative estimates.**

Non-critical Paths: An actual POS may contain a number of non-critical paths. In the IPE Model, a non-critical path is modeled by introducing a random variable λ_i with a maximum value strictly lower than T. Any such variable can be replaced by one with the same (scaled) distribution and maximum T, for an increased expected makespan.

Non-identical Distributions: In general, each critical path λ_i will have a different probability distribution F_{λ_i}. However, it can be proved[1] that it is possible to obtain a worst case estimate by assuming all F_{λ_i} to be equal to the one having the lowest value for the integral $\int_0^\infty F_{\lambda_i}(x)dx$.

---

[1] The proof can be found in a technical report associated to this paper, see [4].

Path Dependences: Path dependences arise in Equation (6) due to the presence of shared variables. Consider the case of a graph with two paths π_i and π_j sharing a subpath. For each assignment of durations $\delta^{(k)}$, the length of the common subpath will add to that of both π_i and π_j, making them more similar. This can be observed in Figure 4B that reports the empirical distributions of two POSs, both consisting of two paths with two activities each. In the POS corresponding to the grey distribution, the two paths are independent, hence $l(\pi_i) = \delta_h + \delta_k$ and $l(\pi_j) = \delta_l + \delta_m$. In the POS corresponding to the black distribution, the two paths share a variable, i.e. $l(\pi_i) = \delta_h + \delta_k$ and $l(\pi_j) = \delta_h + \delta_m$. All the durations are uniformly distributed, δ_h and δ_l range in $[0, 0.8]$, while δ_j, δ_m range in $[0, 0.2]$. The distribution for the dependent setting is much closer to that of a single path (see Figure 4A) and has a lower average then the independent case.

4.2   Asymptotic Behavior of the Expected Makespan

We are now in a position to use the IPE Model to obtain a conservative estimate of $\mathbb{E}[\tau]$. In particular, we will show that **the makespan deviation Δ for the IPE Model is bounded above by a quantity that depends 1) on the variability of the activities on the paths and 2) on the number of paths**. For proving this, we will consider each critical path to be a sequence of m activities. This can be captured by assuming that $\lambda = \sum_{j=0}^{m-1} \rho_j$, where each ρ_j is an independent random variable with expected value μ_{ρ_j} and relative variability β_{ρ_j}. From this definition, it follows:

$$\sum_{j=0}^{m-1} \mu_{\rho_j} = \mathbb{E}[\lambda] \quad \text{and} \quad \sum_{j=0}^{m-1} \beta_{\rho_j} = (1 - \alpha)T \tag{9}$$

Note that with $m = 1$ we obtain the original IPE Model. Now, let Z be a random variable equal to $\tau - \mu_\lambda$, where $\mu_\lambda = \mathbb{E}[\lambda]$. The Z variable is designed so that in the model $\mathbb{E}[Z]$ corresponds to the makespan deviation Δ. Now, from Jensen's inequality [9], we know that:

$$e^{\mathbb{E}[tZ]} \leq \mathbb{E}[e^{tZ}] \tag{10}$$

because the exponential function e^{tx} is convex in x. The term t is a real valued parameter that will allow some simplifications later on. From the definition of Z and since the exponential is order preserving, we obtain:

$$\mathbb{E}[e^{tZ}] = \mathbb{E}\left[e^{t\max_i(\lambda - \mu_\lambda)}\right] = \mathbb{E}\left[\max_{i=0..n-1} e^{t(\lambda - \mu_\lambda)}\right] \tag{11}$$

By simple arithmetic properties and by linearity of the expectation $\mathbb{E}[]$:

$$\mathbb{E}\left[\max_{i=0..n-1} e^{t(\lambda - \mu_\lambda)}\right] \leq \sum_{i=0}^{n-1} \mathbb{E}\left[e^{t(\lambda - \mu_\lambda)}\right] = n\,\mathbb{E}\left[e^{t(\lambda - \mu_\lambda)}\right] \tag{12}$$

Note that using a sum to over-approximate a maximum is likely to lead to a loose bound. This is however mitigated by the use of a fast-growing function like the exponential. Now, since $\lambda_i = \sum_j \rho_j$, we can write:

$$n\,\mathbb{E}\left[e^{t(\lambda-\mu_\lambda)}\right] = n\,\mathbb{E}\left[e^{t\sum_j(\rho_j-\mu_{\rho_j})}\right] = n\,\mathbb{E}\left[\prod_{j=0}^{m-1} e^{t(\rho_j-\mu_{\rho_j})}\right] = n\prod_{j=0}^{m-1}\mathbb{E}\left[e^{t(\rho_j-\mu_{\rho_j})}\right]$$

because the ρ_j variables are independent. Now, each term $\rho_j - \mu_{\rho_j}$ has zero mean and spans by definition in a finite range with size β_{ρ_j}. Therefore we can apply Hoeffding's lemma [7] to obtain:

$$n\prod_{j=0}^{m-1}\mathbb{E}\left[e^{t(\rho_j-\mu_\rho)}\right] \leq n\prod_{j=0}^{m-1} e^{\frac{1}{8}t^2\beta_{\rho_j}^2} = ne^{\frac{1}{8}t^2\sum_j\beta_{\rho_j}^2} \tag{13}$$

By merging Equation (10) and Equation (13) we get:

$$\mathbb{E}[tZ] = t\,\mathbb{E}[Z] \leq \log\left(ne^{\frac{1}{8}t^2\sum_j\beta_{\rho_j}^2}\right) = \log n + \frac{1}{8}t^2\sum_j\beta_{\rho_j}^2 \tag{14}$$

which holds for every $t \in \mathbb{R}$. By choosing $t = \frac{\sqrt{8\log n}}{\sqrt{\sum_j\beta_{\rho_j}^2}}$, we finally obtain:

$$\mathbb{E}[Z] = \mathbb{E}[\tau] - \mu_\lambda = \Delta \leq \frac{1}{\sqrt{2}}\sqrt{\sum_{j=0}^{m-1}\beta_{\rho_j}^2}\sqrt{\log n} \tag{15}$$

where the two main terms in the product at the right-hand side depend respectively on the variability of the paths and on their number, thus proving our result. Note that Equation (15) identifies the terms that have an impact on Δ, and it also bounds the *degree* of such impact. Since the IPE Model represents a worst case, the provided bound is applicable to general POSs, too.

5   Empirical Analysis of the Asymptotic Behavior

We now start to employ the mathematical framework from Section 4 for an empirical evaluation, to get a better grasp of the behavior of the makespan deviation. Our main tool will be the bound from Equation (15). As a preliminary, when moving to real POSs some of the parameters of the IPE Model cannot be exactly measured and must be approximated. As a guideline, we use the parameters of the critical path in the POS as representative for those of the paths in the IPE model. Hence:

- the value μ_λ corresponds to $\tau(\mathbb{E}[\delta])$, i.e. to $\frac{1+\alpha}{2}T$.
- each β_{ρ_i} is equal to $(1-\alpha)D_i$.

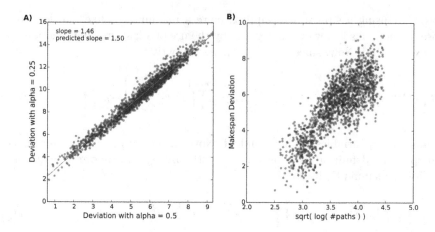

Fig. 5. A) Normalized Deviations with $\alpha = 0.5$ and $\alpha = 0.25$. B) Dependence of the deviation on the number of paths.

- for m, we take the number of activities on the critical path.
- for n, we take the total number of paths in the POS.

We expect the largest approximation error to come from considering all paths (regardless the degree of their dependence) and from assuming that all of them are critical. Now, we proceed by investigating how the makespan deviation Δ depends on the two main terms from Equation (15).

Dependence on the Variability: For our experimentation we have that:

$$\sqrt{\sum_{j=0}^{m-1} \beta_{\rho_j}^2} = \sqrt{\sum_{j=0}^{m-1} ((1-\alpha)D_i)^2} = (1-\alpha)\sqrt{\sum_{j=0}^{m-1} D_i^2} \qquad (16)$$

Hence, the variability term in Equation (15) grows linearly with $(1 - \alpha)$. Therefore, in order to check if the bound reflects correctly the dependence of Δ on the variability, we can repeat our simulation with different α values and see if we observe a linear change of the makespan deviation. We have experimented with $\alpha = 0.1, 0.25, 0.5$, and 0.75.

Figure 5A shows a scatter plot for j60 where the makespan deviation for $\alpha = 0.25$ and $\alpha = 0.5$ are compared. The presence of a linear correlation is apparent. Equation (16) allows one to predict the slope of the line for two values α'' and α', which should be $\frac{1-\alpha''}{1-\alpha'}$. For $\alpha'' = 0.25$ and $\alpha' = 0.5$, we get 1.5. By fitting a trend line it is possible to measure the actual slope, which in this case is 1.46, remarkably close to the predicted one. **This all points to the fact that the asymptotic dependence identified by our bound is in fact tight.**

Dependence on the Number of Paths: The path related term in the bound predicts that the makespan deviation will grow (in the worst case) with $\sqrt{\log n}$.

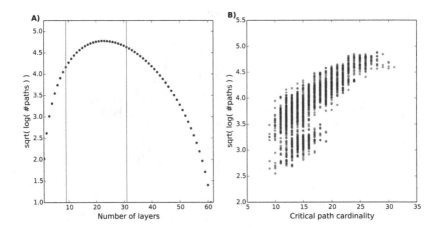

Fig. 6. The path-dependent term in the bound vs. the path cardinality

In this section, we investigate how accurate this estimate is in practice. In a first attempt to assess the correlation between Δ and the number of paths, we simply plot the deviation values against $\sqrt{\log n}$, using (we recall) the total number of paths in the graph as a proxy for the number of critical paths. If the bound reflects a tight asymptotic dependence, we should observe a linear correlation.

Figure 5B shows such a plot for j60 and the correlation does not appear to be linear. This may hint to an overestimation in our bound, but it could also be due to our approximations, or to the presence of correlations between the variability- and the path-related term in Equation (15). Now, we have recalled that the number of paths in a graph may be exponential in the graph size. The number of *independent* paths, however, is polynomially bounded by $|A|$, i.e. the size of the graph. In fact, every activity can be part of at most one independent path. **For the number of paths to be exponential, the activities must have multiple predecessors and successors.**

Layered Graph Approximation: Next, we will investigate the relation between the number of paths and their structure. This can be conveniently done on a layered graph, i.e. a graph where the nodes are arranged in m layers, such that layer $k - 1$ is totally connected with layer k. The number of paths in a layered graph is estimated by the quantity $(q/m)^m$, where $q = |A|$ is the graph size. There are two consequences: 1) to increase the number of paths exponentially, we must increase their cardinality; 2) since m appears also at the denominator, at some point the number of paths will start to *decrease* with growing m values.

This can be observed in Figure 6A, reporting the value of $\sqrt{\log (q/m)^m}$ for $q = 62$ and m ranging in $\{1 \ldots 62\}$. Figure 6B reports instead the value of $\sqrt{\log(n)}$ over the cardinality of the critical path for the j60 benchmark, which also features 62 activities per graph (60 + 2 fake nodes). The two red bars in Figure 6A mark the minimum and the maximum for the critical path cardinality

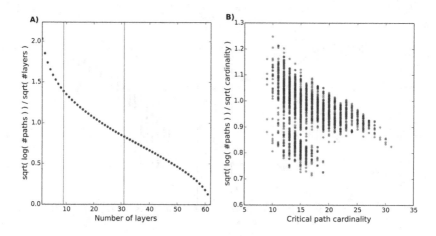

Fig. 7. The makespan deviation decreases with the cardinality of the critical path

in j60. The shape (and the range!) of the empirical plot matches closely enough that of the highlighted segment of the theoretical curve. At a careful examination, this makes a lot of sense. When resolving resource conflicts, a scheduling solver operates by arranging the activities into a sequence of parallel groups. **Tighter resource restrictions lead to longer sequences and fewer parallel activities in each group. In a POS, each of those groups translates to a layer, making a layered graph a fairly good approximate model.**

Path Cardinality and Variability: Now, we will show that the variability of a path π gets smaller if π contains a large number of activities with uncertain duration. If most of the activities in the graph have uncertain duration, this means that the variability of a path π will decrease with its cardinality. For our proof, we will assume for sake of simplicity that all ρ_i variables in the IPE Model are identical. Hence, in our experimental setting, their variability will be equal to $\frac{1}{m}(1-\alpha)T$. If we also assume to have a layered graph, Equation (15) becomes:

$$\Delta \le \frac{1}{\sqrt{2}}\sqrt{m\beta_{\rho_j}^2}\sqrt{\log\left(\frac{q}{m}\right)^m} \tag{17}$$

$$\text{i.e.} \quad \Delta \le \frac{1}{\sqrt{2}}\frac{1}{\sqrt{m}}(1-\alpha)T\sqrt{m\log\left(\frac{q}{m}\right)} \tag{18}$$

In Equation (18), the role of path cardinalities in the deviation bound becomes explicit. In particular, paths with a lot of variables are less likely to have very long extreme lengths, reducing the deviation by a $\sqrt{m}$ factor. The degree of the dependence is such to completely counter the exponential growth of the path number, so that the product $\frac{1}{\sqrt{m}}\sqrt{m\log(q/m)}$ behaves like $\sqrt{\log(q/m)}$.

Hopefully Equation (18) provides a reasonable approximation even when its simplifying assumptions are not strictly true. For testing this, in Figure 7A and

7B we report, next to each other, the value of $\sqrt{\log(q/m)}$ for m ranging in $\{1 \ldots 62\}$ (which is strictly decreasing) and the value of $\frac{1}{\sqrt{m}}\sqrt{\log(n)}$ over the critical path length for j60. As in Figure 6A we use red lines to mark on the theoretical plot the minimum/maximum of the critical path cardinality for j60. Again, the figures are similar enough in terms of both shape and range. Hence, **if most of the activities have uncertain duration, an increase of the path cardinality tends to have a beneficial effect on the deviation Δ.**

Back to our Main Result: In Equation (18) the deviation bound is proportional to T. However, under the simplifying assumption that all activities are identical, increasing T requires to increase m as well, causing a reduction of the term $\frac{1}{\sqrt{m}}\sqrt{m\log(q/m)}$. It follows that the deviation should grow sub-linearly with the makespan. We conjecture that this same mechanism is at the base of the sub-linear dependence we have observed in Figure 3A and checked by analyzing the normalized deviation $\Delta/\tau[\mathbb{E}[\delta]]$.

Now, we already know that the ratio $\frac{1}{\sqrt{m}}\sqrt{m\log(q/m)}$ does decrease with m for our benchmarks. Hence, if we observe a linear correlation between m and $\tau(\mathbb{E}[\delta])$, we should expect our bound to grow sub-linearly. In Figure 8A we report, for the j60 benchmark, the length of the critical path over the makespan with average duration, i.e. $\tau(\mathbb{E}[\delta])$. The plot confirms the existence of a linear correlation. At this point, repeating our experiment from Figure 3B with the (normalized) bound values instead of the deviation would be enough to confirm our conjecture. However, we have an even more interesting result. Figure 8B reports a scatter plot having on the x-axis the normalized bound and on the y-axis the normalized deviation. The plot shows the existence of a clear linear

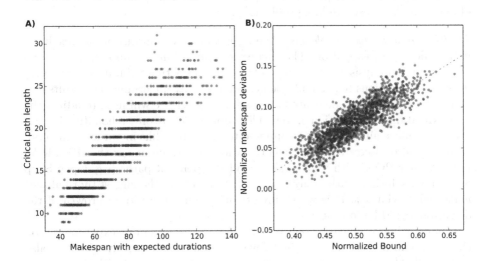

Fig. 8. A) Linear correlation between the makespan and the critical path cardinality. B) Normalized bound from Equation (15) over the normalize deviation $\Delta/\tau(\mathbb{E}[\delta])$

correlation between the two. This has two important consequences: **1) the hypothesis that the sub-linear growth is due to a correlation between the number of paths and their variability is consistent with the observed data.** Moreover, **2) the bound based on the IPE Model provides a tight prediction of the rate of growth of the normalized deviation.**

6   Concluding Remarks

Our results: In this paper we provide a detailed empirical and theoretical analysis of general run time properties of POSs. We observe, in a variety of settings, the existence of a very strong correlation between the expected makespan and the makespan obtained by assuming that all activities have their expected duration. Specifically, their difference (makespan deviation) appears to exhibit sub-linear variations against makespan changes. As a likely cause for this behavior, we suggest a strong link existing between the number of paths in a POS (that tends to increase the deviation) and the number of activities with uncertain durations in the paths (that makes the deviation smaller). We provide support for our hypothesis by means of a mathematical framework (the IPE Model) and of an extensive empirical evaluation. In the process, we end up identifying a number of important mechanisms that determine the behavior of a POS.

It is important to observe that the strong resilience of the deviation to makespan variations represents a resilience to *scheduling decisions*. As an immediately applicable consequence, this makes it possible to disregard the duration uncertainty and build a POS so as to optimize the makespan of the scenario where all the activities take their expected value. The solution of such *deterministic* problem will have a good chance to be close to optimal for its *stochastic* counterpart as well. These result can be profitably applied to dramatically lower the complexity of many real world scheduling problems.

Limitations and Open Problems: The applicability conditions of our result require a deeper investigation. The single most important assumption for the validity of our analysis is the independence of the activity durations. Strongly dependent durations (e.g. simultaneous failures) may undermine our results and will certainly deserve a dedicated analysis. Moreover, we will need to adjust the approximations from Section 5, to tackle problems where only a handful of activities have uncertain duration (as opposed to all of them like in our experiments).

We are also interested in the identification of extreme cases: the IPE Model predicts that POSs with a lot of short and independent paths (e.g. Open Shop Scheduling solutions with many resources) should have the largest (normalized) makespan deviations. It is worth investigating what the rate of growth of the deviation would be in that case.

Finally, while we have a good understanding of why in our experimentation the makespan deviation *varies* little, we still don't know exactly why it *is* also very small. This open question, which is also tightly connected with the problem of predicting the value of the deviation, rather than its rate of growth. We plan to address both topics in future research.

References

1. Aloulou, M.A., Portmann, M.C.: An efficient proactive reactive scheduling approach to hedge against shop floor disturbances. In: Proc. of MISTA, pp. 223–246 (2005)
2. Beck, J.C., Davenport, A.J.: A survey of techniques for scheduling with uncertainty (2002),
   http://www.eil.utoronto.ca/profiles/chris/gz/uncertainty-survey.ps
3. Beck, J.C., Wilson, N.: Proactive Algorithms for Job Shop Scheduling with Probabilistic Durations. Journal on Artificial Intelligence Research (JAIR) 28, 183–232 (2007)
4. Bonfietti, A., Lombardi, M., Milano, M.: Disregarding Duration Uncertainty in Partial Order Schedules? Yes, we can! Technical Report LIA-001-14, LIA Series no. 101. University of Bologna, Italy (February 2014),
   http://www.lia.deis.unibo.it/Research/TechReport/LIA-001-14.pdf
5. Cesta, A., Oddi, A., Smith, S.F.: Scheduling Multi-Capacitated Resources under Complex Temporal Constraints. In: Maher, M.J., Puget, J.-F. (eds.) CP 1998. LNCS, vol. 1520, pp. 465–465. Springer, Heidelberg (1998)
6. Godard, D., Laborie, P., Nuijten, W.: Randomized Large Neighborhood Search for Cumulative Scheduling. In: Proc. of ICAPS, pp. 81–89 (2005)
7. Hoeffding, W.: Probability inequalities for sums of bounded random variables. Journal of the American statistical association 58(301), 13–30 (1963)
8. Igelmund, G., Radermacher, F.J.: Preselective strategies for the optimization of stochastic project networks under resource constraints. Networks 13(1), 1–28 (1983)
9. Jensen, J.L.W.V.: Sur les fonctions convexes et les inégalités entre les valeurs moyennes. Acta Mathematica 30(1), 175–193 (1906)
10. Kolisch, R.: PSPLIB - A project scheduling problem library. European Journal of Operational Research 96(1), 205–216 (1997)
11. Laborie, P.: Complete MCS-Based Search: Application to Resource Constrained Project Scheduling. In: Proc. of IJCAI, pp. 181–186. Professional Book Center (2005)
12. Lombardi, M., Milano, M., Benini, L.: Robust Scheduling of Task Graphs under Execution Time Uncertainty. IEEE Transactions on Computers 62(1), 98–111 (2013)
13. Policella, N., Cesta, A., Oddi, A., Smith, S.F.: From precedence constraint posting to partial order schedules: A CSP approach to Robust Scheduling. AI Communications 20(3), 163–180 (2007)
14. Policella, N., Smith, S.F., Cesta, A., Oddi, A.: Generating Robust Schedules through Temporal Flexibility. In: Proc. of ICAPS, pp. 209–218 (2004)
15. Rasconi, R., Policella, N., Cesta, A.: SEaM: Analyzing schedule executability through simulation. In: Ali, M., Dapoigny, R. (eds.) IEA/AIE 2006. LNCS (LNAI), vol. 4031, pp. 410–420. Springer, Heidelberg (2006)
16. Taillard, E.: Benchmarks for basic scheduling problems. European Journal of Operational Research 64(2), 278–285 (1993)
17. Van de Vonder, S.: A classification of predictive-reactive project scheduling procedures. Journal of Scheduling 10(3), 195–207 (2007)
18. Van de Vonder, S.: Proactive heuristic procedures for robust project scheduling: An experimental analysis. European Journal of Operational Research 189(3), 723–733 (2008)

An Exact Branch and Bound Algorithm with Symmetry Breaking for the Maximum Balanced Induced Biclique Problem

Ciaran McCreesh and Patrick Prosser

University of Glasgow, Glasgow, Scotland
c.mccreesh.1@research.gla.ac.uk,
patrick.prosser@glasgow.ac.uk

Abstract. We show how techniques from state-of-the-art branch and bound algorithms for the maximum clique problem can be adapted to solve the maximum balanced induced biclique problem. We introduce a simple and effective symmetry breaking technique. Finally, we discuss one particular class of graphs where the algorithm's bound is ineffective, and show how to detect this situation and fall back to a simpler but faster algorithm. Computational results on a series of standard benchmark problems are included.

1 Introduction

Let $G = (V, E)$ be a graph (by which we always mean finite, undirected and with no loops) with vertex set V and edge set E. A *biclique*, or complete bipartite subgraph, is a pair of (possibly empty) disjoint subsets of vertices $\{A, B\}$ such that $\{a, b\} \in E$ for every $a \in A$ and $b \in B$. A biclique is *balanced* if $|A| = |B|$, and *induced* if no two vertices in A are adjacent and no two vertices in B are adjacent. The maximum balanced induced biclique problem is to find a balanced induced biclique of maximum size in an arbitrary graph. We illustrate an example in Fig. 1.

Finding such a maximum is NP-hard [1, Problem GT24], both in bipartite and arbitrary graphs. A naïve exponential algorithm could simply enumerate every possible solution to find a maximum. Here we develop a branch and bound algorithm

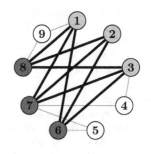

Fig. 1. A graph, with its unique maximum balanced induced biclique of size six, $\{\{1, 2, 3\}, \{6, 7, 8\}\}$, shown in light and dark blue

with symmetry breaking that substantially reduces the search space. We believe that this is the first attempt at tackling this problem. We are not yet aware of any practical applications, but the problem is interesting from an algorithmic perspective.

H. Simonis (Ed.): CPAIOR 2014, LNCS 8451, pp. 226–234, 2014.
© Springer International Publishing Switzerland 2014

If $G = (V, E)$ is a graph, we write $V(G)$ for the vertex set V. The *neighbour-hood* of a vertex v in a graph G is the set of vertices adjacent to v; we denote this $N_G(v)$. The *degree* of a vertex is the cardinality of its neighbourhood.

A graph $G' = (V', E')$ is a *subgraph* of $G = (V, E)$ if $V' \subseteq V$ and $E' \subseteq E$; the subgraph *induced by* V' is the subgraph with vertex set V' and all possible edges. A set of vertices, no two of which are adjacent, is called an *independent set*. A set of vertices, all of which are adjacent, is called a *clique*; the size of a maximum clique is denoted ω. A *clique cover* is a partition of the vertices in a graph into sets, each of which is a clique. We introduce the symbol $\ddot{\omega}$ for the size (i.e. $|A| + |B|$) of a maximum balanced induced biclique, which is always even (this simplifies comparisons with unbalanced biclique variants). A graph is *bipartite* if its vertices may be partitioned into two disjoint independent sets.

2   A Branch and Bound Algorithm

A very simple branch and bound algorithm for the maximum induced biclique problem is given in Algorithm 1. The algorithm works by recursively building up two sets A and B such that $\{A, B\}$ is a biclique. At each stage, P_a contains those vertices which may be added to A whilst keeping a feasible solution (i.e. each $v \in P_a$ is individually adjacent to every $b \in P_b$ and nonadjacent to every $a \in A$), and similarly P_b contains vertices which may be added to B. Initially, A and B are both empty, and P_a and P_b both contain every vertex in the graph (line 4).

At each recursive call to **expand**, a vertex v is chosen from P_a (line 8) and moved to be in A instead (lines 10 and 11). The algorithm then considers the implications of $v \in A$ (lines 12 to 17). A new P'_a is constructed on line 12 by filtering from P_a those vertices adjacent to v (since A must be an independent set), and a new P'_b is constructed on line 13 by filtering from P_b those vertices *not* adjacent to v (everything in B must be adjacent to everything in A).

Now if P'_b is not empty, we may grow B further. Thus we repeat the process with a recursive call on line 17, swapping the roles of A and B—we are adding vertices to the two sides of the growing biclique in alternating order.

Having considered the possibility of $v \in A$, we then consider $v \notin A$ (line 18). The algorithm loops back to line 8, selecting a new v from P_a, until P_a is empty. Finally, we backtrack by returning from the recursive call.

We keep track of the largest feasible solution $\{A_{max}, B_{max}\}$ that we have found so far; this is called the *incumbent*. Initially it is empty (line 3). Whenever we find a potential solution, we compare it to the incumbent (line 14), and if our new solution is larger then the incumbent is unseated (line 15). Note that at this point, the balance condition must be checked explicitly, since either $|A| = |B|$, or $|A| = |B| + 1$ could be true.

Knowing the size of the incumbent allows us to avoid exploring some of the search space—this is the bound part of branch and bound. The condition on line 9 checks how much further we can grow A and B: if there are not enough vertices available to potentially unseat the incumbent, search at the current position can be abandoned. (This is not a very good bound, and is only for illustrative purposes. We discuss a more sophisticated bound below.)

Algorithm 1. A simple, alternating branch and bound algorithm for the maximum balanced induced biclique problem.

```
1  simpleBiclique :: (Graph G) → (Set of Integer, Set of Integer)
2  begin
3  │  (A_max, B_max) ← (∅, ∅)          // Initially our best solution is empty
4  │  expand(G, ∅, ∅, V(G), V(G), A_max, B_max)
5  └  return (A_max, B_max)


6  expand :: (Graph G, Set A, Set B, Set P_a, Set P_b, Set A_max, Set B_max)
7  begin
8  │  for v ∈ P_a do
9  │  │  if |P_a| + |A| > |A_max| and |P_b| + |B| > |B_max| then
10 │  │  │  A ← A ∪ {v}                              // Consider v ∈ A
11 │  │  │  P_a ← P_a \ {v}
12 │  │  │  P'_a ← P_a ∩ N̄_G(v)        // Remove vertices adjacent to v
13 │  │  │  P'_b ← P_b ∩ N_G(v)      // Remove vertices not adjacent to v
14 │  │  │  if |A| = |B| and |A| > |A_max| then
15 │  │  │  └  (A_max, B_max) ← (A, B)    // We've found a better solution
16 │  │  │  if P'_b ≠ ∅ then
17 │  │  │  └  expand(G, B, A, P'_b, P'_a, B_max, A_max)      // Swap and recurse
18 │  │  └  A ← A \ {v}                         // Now consider v ∉ A
```

Improving the Algorithm. We now adapt Algorithm 1 to incorporate symmetry breaking, an improved bound based upon clique covers, and an initial sort order. The end result is Algorithm 2. We have explicitly designed the algorithm to permit a bitset encoding for the data structures. For the maximum clique problem, this technique has allowed an increase in performance of between two and twenty times, without altering the steps taken by the algorithm. We refer to work by San Segundo et al. [2,3] for implementation details.

Symmetry Breaking. The search space for Algorithm 1 is larger than it should be: it explores legal *ordered* pairs (A, B) of vertex sets rather than unordered pairs $\{A, B\}$. Having explored every possible solution with $v \in A$, the search then considers $v \notin A$. But there is nothing to stop it from then considering a new $v' \in A$, and later placing $v \in B$. This is wasted effort, since if such a solution existed we would already have considered an equivalent with A and B reversed.

We may break this symmetry as follows: if, at the top of search, we have considered every possibility with $v \in A$ then we may eliminate v from P_b to avoid considering $v \in B$. The modified **expand** function in Algorithm 2 includes this rule: lines 38 to 39 remove symmetric solutions.

This technique may be seen as a special case of the standard lex symmetry breaking technique used in constraint programming [4,5]. A constraint programmer would view A and B as binary strings, and impose the constraint $B \leq A$

Algorithm 2. An improved alternating branch and bound algorithm for the maximum balanced induced biclique problem.

```
1  improvedBiclique :: (Graph G) → (Set of Integer, Set of Integer)
2  begin
3      (A_max, B_max) ← (∅, ∅)          // Initially our best solution is empty
4      permute G so that the vertices are in non-increasing degree order
5      expand(G, ∅, ∅, V(G), V(G), A_max, B_max)
6      return (A_max, B_max) (unpermuted)

7  cliqueSort :: (Graph G, Set P) → (Array of Integer, Array of Integer)
8  begin
9      bounds ← an Array of Integer
10     order ← an Array of Integer
11     P' ← P                            // vertices yet to be allocated
12     k ← 1                             // current clique number
13     while P' ≠ ∅ do
14         Q ← P'                        // vertices to consider for the current clique
15         while Q ≠ ∅ do
16             v ← the first element of Q    // get next vertex to allocate
17             P' ← P' \ {v}
18             Q ← Q ∩ N(G, v)           // remove non-adjacent vertices
19             append k to bounds
20             append v to order
21         k ← k + 1                     // start a new clique
22     return (bounds, order)

23 expand :: (Graph G, Set A, Set B, Set P_a, Set P_b, Set A_max, Set B_max)
24 begin
25     (bounds, order) ← cliqueSort(G, P_a)
26     for i ← |P_a| downto 1 do
27         if bounds[i] + |A| > |A_max| and |P_b| + |B| > |B_max| then
28             v ← order[i]
29             A ← A ∪ {v}               // Consider v ∈ A
30             P_a ← P_a \ {v}
31             P'_a ← P_a ∩ N_G(v)̄       // Remove vertices adjacent to v
32             P'_b ← P_b ∩ N_G(v)       // Remove vertices not adjacent to v
33             if |A| = |B| and |A| > |A_max| then
34                 (A_max, B_max) ← (A, B)   // We've found a better solution
35             if P'_b ≠ ∅ then
36                 expand(G, B, A, P'_b, P'_a, B_max, A_max)   // Swap and recurse
37             A ← A \ {v}               // Now consider v ∉ A
38             if B = ∅ then
39                 P_b ← P_b \ {v}       // Avoid symmetric solutions
```

(or the other way around—after all, the order of A and B is arbitrary). We are doing the same thing, by saying that if the first n bits of A are 0 then the first n bits of B must also be 0. Unlike adding a lex constraint, this approach does not interfere with the search order and does not introduce the risk of disrupting ordering heuristics [6]. Additionally, this constraint always removes symmetric solutions from the search tree as early as possible [7].

Bounding. We know that A and B must be independent sets. Finding a maximum independent set is a well studied NP-hard problem (although the literature usually discusses finding a maximum clique, which is a maximum independent set in the complement graph), and the main inspiration for our algorithm comes from a series of maximum clique algorithms due to Tomita [8,9,10]. These are branch and bound algorithms which use graph colouring (i.e. a clique cover in the complement graph) both as a bound and an ordering heuristic.

If we can cover a graph G using k cliques, we know that G cannot contain an independent set of size greater than k (each element in an independent set must be in a different clique). Finding an optimal clique cover is NP-hard, but a greedy clique cover may be found in polynomial time. This gives us a bound on P_a which can be much better than simply considering $|P_a|$: we construct a greedy clique cover of the subgraph induced by P_a, and consider its size instead.

Constructing a clique cover gives us more information than just a bound on the size of an independent set in all of P_a. This is the main benefit of Tomita's approach: a constructive greedy clique cover gives us an ordering heuristic and a way of reducing the number of clique covers which must be computed.

Tomita has considered ways of producing and using greedy colourings; we refer to a computational study by Prosser [11] for a detailed comparison. Our greedy clique cover bound and ordering routine is presented in Algorithm 2. The approach we have taken is a variation by San Segundo [2,3] which allows a bitset encoding to be used.

The `cliqueSort` function in Algorithm 2 produces two arrays. The *bounds* array contains bounds on the size of a maximum independent set: the subgraph induced by vertices 1 to n of *order* cannot have a maximum independent set of size greater than *bounds*[n]. The *order* array contains the vertices of P in some order, and is to be traversed from right to left, repeatedly removing the rightmost value for the choice branching vertex v.

These arrays are constructed in the `cliqueSort` function as follows: the variable P' tracks which vertices have yet to be allocated to a clique, and initially (line 11) it contains every vertex in the parameter P. While there are unallocated vertices (line 13), we greedily construct a new clique. The variable Q (line 14) tracks which vertices may legally be added to this growing clique. On line 16 we select a vertex v from Q, add it to the clique, and on line 18 we remove from Q any vertices which are not adjacent to v (so every vertex remaining in Q is adjacent to every vertex in the growing clique). We continue adding vertices to the growing clique until Q is empty (line 15), indicating we can go no further. We then start a new clique (line 21, looping back to line 13) if some vertices remain unallocated.

To integrate this bound, we make the following changes: we begin by using `cliqueSort` to obtain the *bounds* and *order* variables (line 25). We explicitly iterate over *order* from right to left (lines 26 and 28), rather than drawing v from P_a arbitrarily. And we make use of the bound on P_a, rather than using $|P_a|$ (line 27).

Search Order. We use a static ordering for constructing clique covers, so the initial order of vertices must also be considered—experiments show that, as for the maximum clique problem, a static non-increasing degree order fixed at the top of search is a good choice. We achieve this ordering by permuting the graph (again, to allow the possibility of a bitset encoding).

Detecting when the Bound is Useless. Our bound considers how far A can grow, based upon what is in P_a, and how far B can grow based upon what is in P_b. If both P_a and P_b are independent sets, this does not help, and constructing the clique cover ordering is a substantial overhead. This situation occurs in particular if the input is a bipartite graph, or close to one. We can at least detect when P_a is an independent set: this happens precisely if $bounds[i] = i$ (assuming *bounds* is 1-indexed), since if the graph contains at least two non-adjacent vertices then at least one such pair will be placed in the same clique [12, Proposition 2].

Ideally we would be able to switch to a better bound in the case that both P_a and P_b are (potentially overlapping) independent sets. However the authors have been unable to find a better bound which is sufficiently cheap to compute to provide a benefit—approaches which reduce the search space but increase runtime include the use of degrees, indirect colouring, or the fact that finding an (unbalanced) induced biclique in a bipartite graph can be done in polynomial time via a matching algorithm. However, we may still decay to a version of the algorithm which includes symmetry breaking and uses cardinality bounds as in Algorithm 1. We do not demonstrate this technique in Algorithm 2, but it is simple to incorporate.

3 Computational Experiments

We now present experimental results on a range of standard benchmark problems. The algorithm was implemented using C++, with a bitset encoding. The experiments were run on a machine with four AMD Opteron 6366 HE processors, and single-threaded runtimes are given. The implementation does include detection for independent sets, and falls back to a simple algorithm when this happens. Timing results include pre-processing and the initial sorting step, but do not include the time taken to read a graph in from a file. For the maximum clique problem, a sequential implementation previously described by the authors [13] was used.

In Table 1 we present results from four datasets. First is all the graphs from the Second DIMACS Implementation Challenge[1]. Many of these graphs are dense,

---

[1] http://dimacs.rutgers.edu/Challenges/

and designed to be computationally challenging for maximum clique algorithms. The second dataset is the smallest family of graphs for BHOSLIB[2]. These graphs contain a hidden clique of known size; again, these are challenging for maximum clique algorithms. Thirdly, we look at some large sparse graphs from BioGRID [14]. Finally, we include some large sparse graphs from a collection by Mark Newman[3]. For each instance we show results for both maximum clique and maximum balanced induced biclique: we show the size of the result, the time taken, and the number of search nodes (recursive calls made). Longer-running problems were aborted after one day; such results are shown in parentheses.

Sometimes $\ddot{\omega} = \omega$, sometimes it is larger, and sometimes it is smaller. Often finding $\ddot{\omega}$ was easier than finding ω (and there are no problems where the biclique search was aborted after a day but where the clique succeeeded), but not always.

Further experiments show that the symmetry breaking technique is successful in reducing both runtimes and the size of the search space. In many instances the gain approaches 50% (this is expected: halving the number of solutions will not halve the size of the search space). In other cases the interaction of the bound and symmetry breaking reduces the benefit (sometimes to zero, when the bound can already eliminate symmetric solutions), but it is never a penalty.

Detecting when the bound is useless and decaying to a simpler algorithm provides a measurable benefit for several of the "p_hat" family of graphs and for "san1000", but does not generally make a substantial difference. On the other hand, for random bipartite graphs, this technique avoids a factor of five slowdown from the overhead of calculating a useless bound.

4   Conclusion and Future Work

We have shown that max clique techniques generalise to other graph-related problems, although not always in the most obvious way—despite the name, finding a biclique involves finding independent sets, not cliques. Unlike the maximum clique problem, symmetry is an issue, but we provided a very simple and effective way of avoiding this problem. We do not have a good bound for the case where both sides are already independent sets, although we can detect this and fall back to a faster algorithm; this limitation is this work's main weakness.

More detailed computational experiments would be beneficial, particularly with random and (once the weakness is addressed) random bipartite graphs. We intend to look in more detail at "where the hard problems are" for this problem [15]: there is a conflict between wanting to create two independent sets, and requiring those independent sets be interconnected, which means it is not obvious how the density of a random graph would affect the difficulty.

Finally, this approach can likely be extended to exploit multi-core parallelism— the sequential algorithms upon which this work is based have been threaded successfully [13,16].

---

[2] http://www.nlsde.buaa.edu.cn/~kexu/benchmarks/graph-benchmarks.htm
[3] http://www-personal.umich.edu/~mejn/netdata/

Table 1. Results for the balanced biclique problem in DIMACS, BHOSLIB and large sparse graphs from BioGRID and Mark Newman. For each we show the size of a maximum clique, the time taken to obtain this result, and the number of search nodes (recursive calls made). We then give the same information for maximum balanced induced bicliques. Results in parentheses were aborted after one day.

Problem	Size	Time	Nodes	Size	Time	Nodes
C125.9	34	91ms	50240	8	1ms	920
C250.9	44	3043s	1.1×10^9	8	12ms	12448
C500.9	(53)	1 day	2.0×10^{10}	10	174ms	1.1×10^5
C1000.9	(58)	1 day	1.3×10^{10}	10	13.9s	7.4×10^6
C2000.5	(16)	1 day	1.4×10^{10}	(16)	1 day	2.9×10^{10}
C2000.9	(62)	1 day	5.5×10^9	12	1478s	3.2×10^8
C4000.5	(17)	1 day	7.7×10^9	(18)	1 day	1.4×10^{10}
DSJC500.5	13	1.8s	1.2×10^6	14	63.4s	6.8×10^7
DSJC1000.5	15	222s	7.7×10^7	16	12996s	8.9×10^9
MANN_a9	16	0ms	71	6	0ms	32
MANN_a27	126	533ms	38019	6	4ms	1407
MANN_a45	345	383s	2.9×10^6	6	56ms	9852
MANN_a81	(1100)	1 day	8.7×10^7	8	974ms	53902
brock200_1	21	868ms	5.2×10^5	10	45ms	57931
brock200_2	12	5ms	3826	12	111ms	1.7×10^5
brock200_3	15	23ms	14565	12	92ms	1.2×10^5
brock200_4	17	85ms	58730	12	76ms	1.0×10^5
brock400_1	27	508s	2.0×10^8	12	2.3s	1.8×10^6
brock400_2	29	362s	1.5×10^8	12	2.0s	1.8×10^6
brock400_3	31	287s	1.2×10^8	12	2.1s	1.8×10^6
brock400_4	33	140s	5.4×10^7	12	2.0s	1.8×10^6
brock800_1	23	7725s	2.2×10^9	14	1424s	9.5×10^8
brock800_2	23	7711s	2.2×10^9	14	1367s	9.1×10^8
brock800_3	25	7138s	2.1×10^9	14	1448s	9.6×10^8
brock800_4	26	2705s	6.4×10^8	14	1401s	9.4×10^8
c-fat200-1	12	0ms	24	2	0ms	214
c-fat200-2	24	0ms	24	2	0ms	353
c-fat200-5	58	1ms	139	2	1ms	927
c-fat500-1	14	3ms	14	2	3ms	523
c-fat500-2	26	3ms	26	2	4ms	619
c-fat500-5	64	4ms	64	2	7ms	1398
c-fat500-10	126	4ms	126	2	41ms	4219
gen200_p0.9_44	44	4.8s	1.8×10^6	10	3ms	2628
gen200_p0.9_55	55	461ms	1.7×10^5	8	4ms	4201
gen400_p0.9_55	(50)	1 day	2.2×10^{10}	16	21ms	85562
gen400_p0.9_65	(49)	1 day	2.3×10^{10}	14	26ms	11709
gen400_p0.9_75	(53)	1 day	2.0×10^{10}	12	35ms	16388
hamming6-2	32	0ms	32	4	0ms	4
hamming6-4	4	0ms	82	14	1ms	1896
hamming8-2	128	2ms	128	4	1ms	4
hamming8-4	16	80ms	36452	32	2ms	303
hamming10-2	512	56ms	512	4	21ms	4
hamming10-4	(38)	1 day	1.0×10^{10}	40	390s	4.5×10^7
johnson8-2-4	14	0ms	24	6	0ms	460
johnson8-4-4	14	0ms	126	10	0ms	211
johnson16-2-4	(16)	97ms	2.6×10^5	14	402ms	2.2×10^6
johnson32-2-4	(30)	1 day	1.4×10^{11}	(30)	1 day	4.2×10^{11}
keller4	11	17ms	13725	18	69ms	82646
keller5	(27)	1 day	1.8×10^{10}	32	7294s	3.6×10^9
keller6	(53)	1 day	2.6×10^9	(62)	1 day	6.0×10^9
p-hat300-1	8	4ms	1480	12	195ms	2.8×10^5
p-hat300-2	25	18ms	4256	12	268ms	2.8×10^5
p-hat300-3	36	2.0s	6.2×10^5	12	265ms	2.3×10^5
p-hat500-1	9	18ms	9777	12	3.3s	3.9×10^6
p-hat500-2	36	461ms	1.1×10^5	14	6.4s	5.9×10^6
p-hat500-3	50	201s	3.9×10^7	12	9.0s	6.4×10^6
p-hat700-1	11	65ms	26649	12	36.8s	4.3×10^7
p-hat700-2	44	5.0s	7.5×10^5	14	56.3s	3.5×10^7
p-hat700-3	62	2665s	2.8×10^8	14	67.9s	3.1×10^7
p-hat1000-1	10	454ms	1.8×10^5	14	295s	2.5×10^8
p-hat1000-2	46	251s	3.4×10^7	16	546s	3.6×10^8
p-hat1000-3	(63)	1 day	8.9×10^9	16	1300s	5.6×10^8
p-hat1500-1	12	6.9s	1.2×10^6	16	11859s	5.2×10^9
p-hat1500-2	65	43166s	2.0×10^9	16	23677s	6.8×10^9
p-hat1500-3	(79)	1 day	3.2×10^9	16	25745s	5.5×10^9
san200_0.7_1	30	31ms	13399	14	6ms	4330
san200_0.7_2	18	3ms	464	24	3ms	1939
san200_0.9_1	70	206ms	87329	24	3ms	1850
san200_0.9_2	60	769ms	2.3×10^5	8	4ms	3540
san200_0.9_3	44	19.3s	6.8×10^6	10	3ms	2085
san400_0.5_1	13	15ms	2453	62	9ms	1315
san400_0.7_1	40	459ms	1.2×10^5	20	54ms	16229
san400_0.7_2	30	4.0s	8.9×10^5	28	33ms	7973
san400_0.7_3	22	2.3s	5.2×10^5	38	38ms	10361
san400_0.9_1	100	52.3s	4.5×10^6	10	38ms	19054
san1000	15	3.5s	1.5×10^5	134	167ms	10778
sanr200_0.7	18	235ms	1.5×10^5	10	96ms	1.3×10^5
sanr200_0.9	42	45.2s	1.5×10^7	8	4ms	4095
sanr400_0.5	13	543ms	3.2×10^5	14	14.1s	1.4×10^7
sanr400_0.7	21	159s	6.4×10^7	14	4.3s	3.4×10^6
frb30-15-1	30	1165s	2.9×10^8	30	58ms	15361
frb30-15-2	30	2187s	5.6×10^8	30	63ms	17091
frb30-15-3	30	655s	1.7×10^8	30	59ms	16120
frb30-15-4	30	3575s	9.9×10^8	30	61ms	16694
frb30-15-5	30	1056s	2.8×10^8	30	55ms	14850
fission-yeast	12	50ms	208	12	110ms	33253
fruitfly	7	518ms	47	16	584ms	11538
human	13	897ms	13	18	1.0s	10300
mouse	7	21ms	7	10	22ms	1267
plant	9	29ms	9	10	31ms	1578
worm	7	122ms	7	12	130ms	3778
yeast	33	375ms	68	14	13.4s	2.5×10^6
adjnoun	5	0ms	17	6	0ms	207
astro	57	2.7s	57	6	3.5s	28143
celegans	8	1ms	32	8	2ms	1853
condmat	30	17.2s	30	6	20.2s	63980
dolphins	5	0ms	10	4	0ms	66
football	4	0ms	9	4	0ms	422
internet	17	5.1s	50	10	5.5s	24477
karate	5	0ms	5	4	0ms	31
lesmis	10	0ms	10	8	0ms	77
netscience	20	24ms	20	20	26ms	1184
polblogs	20	23ms	60	12	83ms	36693
polbooks	6	0ms	11	4	0ms	168
power	6	235ms	6	4	252ms	4623

References

1. Garey, M.R., Johnson, D.S.: Computers and Intractability: A Guide to the Theory of NP-Completeness. W. H. Freeman & Co., New York (1979)
2. San Segundo, P., Rodríguez-Losada, D., Jiménez, A.: An exact bit-parallel algorithm for the maximum clique problem. Comput. Oper. Res. 38(2), 571–581 (2011)
3. San Segundo, P., Matia, F., Rodríguez-Losada, D., Hernando, M.: An improved bit parallel exact maximum clique algorithm. Optimization Letters (2011)
4. Crawford, J., Ginsberg, M., Luks, E., Roy, A.: Symmetry-breaking predicates for search problems. In: KR 1996: Principles of Knowledge Representation and Reasoning, pp. 148–159. Morgan Kaufmann (1996)
5. Gent, I.P., Petrie, K.E., François Puget, J.: Symmetry in constraint programming. In: Handbook of Constraint Programming, pp. 329–376. Elsevier (2006)
6. Gent, I.P., Harvey, W., Kelsey, T.: Groups and constraints: Symmetry breaking during search. In: Van Hentenryck, P. (ed.) CP 2002. LNCS, vol. 2470, pp. 415–430. Springer, Heidelberg (2002)
7. Backofen, R., Will, S.: Excluding symmetries in constraint-based search. Constraints 7(3-4), 333–349 (2002)
8. Tomita, E., Seki, T.: An efficient branch-and-bound algorithm for finding a maximum clique. In: Calude, C.S., Dinneen, M.J., Vajnovszki, V. (eds.) DMTCS 2003. LNCS, vol. 2731, pp. 278–289. Springer, Heidelberg (2003), http://dx.doi.org/10.1007/3-540-45066-1_22
9. Tomita, E., Kameda, T.: An efficient branch-and-bound algorithm for finding a maximum clique with computational experiments. Journal of Global Optimization 37(1), 95–111 (2007)
10. Tomita, E., Sutani, Y., Higashi, T., Takahashi, S., Wakatsuki, M.: A simple and faster branch-and-bound algorithm for finding a maximum clique. In: Rahman, M. S., Fujita, S. (eds.) WALCOM 2010. LNCS, vol. 5942, pp. 191–203. Springer, Heidelberg (2010)
11. Prosser, P.: Exact Algorithms for Maximum Clique: A Computational Study. Algorithms 5(4), 545–587 (2012)
12. Batsyn, M., Goldengorin, B., Maslov, E., Pardalos, P.M.: Improvements to MCS algorithm for the maximum clique problem. Journal of Combinatorial Optimization, 1–20 (2013)
13. McCreesh, C., Prosser, P.: Multi-threading a state-of-the-art maximum clique algorithm. Algorithms 6(4), 618–635 (2013)
14. Stark, C., Breitkreutz, B.J., Reguly, T., Boucher, L., Breitkreutz, A., Tyers, M.: Biogrid: A general repository for interaction datasets. Nucleic Acids Research 34(suppl. 1), D535–D539 (2006)
15. Cheeseman, P., Kanefsky, B., Taylor, W.M.: Where the really hard problems are. In: Proceedings of the 12th International Joint Conference on Artificial Intelligence, IJCAI 1991, vol. 1, pp. 331–337. Morgan Kaufmann Publishers Inc., San Francisco (1991)
16. Depolli, M., Konc, J., Rozman, K., Trobec, R., Janežič, D.: Exact parallel maximum clique algorithm for general and protein graphs. Journal of Chemical Information and Modeling 53(9), 2217–2228 (2013)

Domain k-Wise Consistency
Made as Simple as Generalized Arc Consistency

Jean-Baptiste Mairy[1], Yves Deville[1], and Christophe Lecoutre[2]

[1] ICTEAM, Université catholique de Louvain, Belgium
{jean-baptiste.mairy,yves.deville}@uclouvain.be
[2] CRIL-CNRS UMR 8188, Université d'Artois, F-62307 Lens, France
lecoutre@cril.fr

Abstract. In Constraint Programming (CP), Generalized Arc Consistency (GAC) is the central property used for making inferences when solving Constraint Satisfaction Problems (CSPs). Developing simple and practical filtering algorithms based on consistencies stronger than GAC is a challenge for the CP community. In this paper, we propose to combine k-Wise Consistency (kWC) with GAC, where kWC states that every tuple in a constraint can be extended to every set of $k - 1$ additional constraints. Our contribution is as follows. First, we derive a domain-filtering consistency, called Domain k-Wise Consistency (DkWC), from the combination of kWC and GAC. Roughly speaking, this property corresponds to the pruning of values of GAC, when enforced on a CSP previously made kWC. Second, we propose a procedure to enforce DkWC, relying on an encoding of kWC to generate a modified CSP called k-interleaved CSP. Formally, we prove that enforcing GAC on the k-interleaved CSP corresponds to enforcing DkWC on the initial CSP. Consequently, we show that the strong DkWC can be enforced very easily in constraint solvers since the k-interleaved CSP is rather immediate to generate and only existing GAC propagators are required: in a nutshell, DkWC is made as simple and practical as GAC. Our experimental results show the benefits of our approach on a variety of benchmarks.

1 Introduction

Constraint Propagation is a key concept to Constraint Programming (CP). Interleaved with (backtrack) search decisions such as classical variable assignments, it typically discards many useless substantial parts of the search space of Constraint Satisfaction Problems (CSPs) by filtering out inconsistent values and/or tuples. Different levels of filtering exist, and usually they can be characterized by properties, called consistencies, of constraints or constraint networks. The central consistency in CP is Generalized Arc Consistency (GAC) [13], also called Domain Consistency (DC): it is the highest filtering level of variable domains when constraints are considered one at a time. Consistencies weaker than GAC are cheaper to enforce but they lose ground progressively, at least for binary and table constraints[1], as they reduce (far) less the search space of CSPs. On the other hand, consistencies stronger than GAC are more and more studied, and often tested on difficult problem instances, where the cost of enforcing them can be counterbalanced by their large inference capabilities. However, such strong consistencies

---

[1] For example, this is the case of the partial form of GAC maintained in Forward Checking.

H. Simonis (Ed.): CPAIOR 2014, LNCS 8451, pp. 235–250, 2014.
© Springer International Publishing Switzerland 2014

need to reason with several constraints simultaneously, which makes the development of filtering algorithms complex, especially for integration into existing CP solvers.

Most of the consistencies can also be classified in two categories: domain-based (or domain-filtering) consistencies and constraint-based ones. Domain-based consistencies identify inconsistent values that can be removed from the domains of variables whereas constraint-based ones identify inconsistent tuples in the constraints, for which the removal is not always a possible option in constraint solvers. Different examples of such consistencies can be found in [1,4,3,8,10,17]. Interestingly enough, combining a constraint-based consistency with a domain-based one such as GAC allows the pruning achieved in term of tuples to further prune variable domains. This is what we propose in this paper by combining k-Wise Consistency (kWC) with GAC. The constraint-based kWC states that every tuple in the scope of a constraint can be extended to every set of $k - 1$ other constraints. Note that kWC and GAC have already been theoretically combined in [7,6]. They have also been practically combined under weaker forms in [3,17] and a practical sophisticated algorithm for the full combination has been proposed in [8].

Our contribution in this paper is two-fold. First, we derive a domain-filtering consistency, called Domain k-Wise Consistency (DkWC), from the combination of kWC and GAC. Roughly speaking, this property corresponds to kWC and GAC combined, but where only the outcome in term of pruned values is considered. Second, we propose a simple and practical filtering procedure to enforce DkWC on a given initial CSP containing table constraints, relying on an encoding of kWC to generate a modified CSP, called k-interleaved CSP. This encoding allows invalidating k-wise inconsistent tuples by means of dual variables, without effectively removing them from constraint scopes, as pure k-wise consistency would do. Formally, we prove that enforcing GAC on the k-interleaved CSP corresponds to enforcing DkWC on the initial CSP. Consequently, we show that the strong DkWC can be enforced very easily in constraint solvers since the k-interleaved CSP is rather immediate to generate (and only once) and only existing GAC propagators are required: in a nutshell, DkWC is made as simple and practical as GAC. We also define two weaker variants of our filtering procedure, that can be used when the problems are too large for the full filtering. Our experimental results show the benefits of our approach on a variety of benchmarks including table constraints.

2   Background

A Constraint Satisfaction Problem (CSP) $P = (X, D, C)$ is composed of an ordered set of n *variables* $X = \{x_1, \ldots, x_n\}$, a set of *domains* $D = \{D(x_1), \ldots, D(x_n)\}$ where $D(x_i)$ is the finite set of possible *values* for variable x_i, and a set of e *constraints* $C = \{c_1, \ldots, c_e\}$, where each constraint c_j restricts the possible combinations of values, called *allowed tuples*, on a subset of variables of X ; this subset is called the *scope* of c_j and denoted by $scp(c_j)$. Because variable domains may evolve (be reduced), $D(x)$ is referred to as the *current domain* of x, which is a subset of the *initial domain* of x denoted by $D^{init}(x)$. If $Y \subseteq X$ then $D[Y]$ is the restriction of D to variables in Y. For any value refutation $x \neq a$, $P|_{x \neq a}$ denotes the CSP P where the value a is removed from $D(x)$, and for any set of value refutations Δ, $P|_{\Delta}$ is defined similarly. The *arity* of a constraint c is $|scp(c)|$, i.e., the number of variables involved in c. In this paper, we shall refer to (positive) table constraints where a *table constraint* c is a constraint given in extension by its explicit list $rel(c)$ of allowed tuples. A table constraint c holds iff $(x_1, \ldots, x_r) \in rel(c)$, where $scp(c) = \{x_1, \ldots, x_r\}$. The size of a constraint c corresponds to its number of allowed tuples and will be denoted by $|rel(c)|$. If τ is a r-tuple $(a_1, \ldots, a_r)$ then $\tau \odot b$ denotes the $r + 1$ tuple $(a_1, \ldots, a_r, b)$.

We assume an implicit total ordering on X, and given $Y = \{y_1, \ldots, y_k\} \subseteq X$, the set of tuples in $D(y_1) \times \ldots \times D(y_k)$ will be denoted by $D(Y)$, and the set of tuples in $D^{init}(y_1) \times \ldots \times D^{init}(y_k)$ will be denoted by $D^{init}(Y)$. A constraint c is satisfied by a tuple $\tau \in D^{init}(scp(c))$ iff τ is *allowed* by c ; the test evaluating to true iff τ is allowed by c is denoted by $c(\tau)$, or equivalently by $\tau \in c$. We shall use the term literal to refer to a variable value pair ; a literal of a constraint c is a pair (x, a) where $x \in scp(c)$ and $a \in D(x)$. An assignment of a set of variables $Y = \{y_1, \ldots, y_k\}$ is a set of literals $\{(y_1, v_1), \ldots, (y_k, v_k)\}$ with $(v_1, \ldots, v_k) \in D^{init}(Y)$; it is a *valid* assignment if $(v_1, \ldots, v_k) \in D(Y)$. Note that any tuple τ in $D^{init}(Y)$ can be seen as an assignment of Y. Actually, for simplicity, we shall use both concepts (assignments and tuples) interchangeably, with the same notations $\tau, \tau', \ldots$ For any tuple or assignment τ, the ith value in τ, associated with the variable y_i, will be denoted by $\tau[y_i]$. A solution of P is a valid assignment of X that satisfies all constraints of P. Two CSPs P and P' are equivalent iff they have the same set of solutions.

Now, let us turn to consistencies. On the one hand, Generalized Arc Consistency (GAC), also called Domain Consistency (DC) in the literature, is a well-known domain-filtering consistency. To define it, we need first to introduce the notion of support as follows: a *support* on a constraint c is a tuple $\tau \in D(scp(c))$ such that $c(\tau)$, and a support (on c) for a literal (x, a) of c is a support τ on c such that $\tau[x] = a$. Note that supports are *valid* tuples, meaning that involved values are necessarily present in the current domains.

Definition 1. (GAC) *A constraint c is* generalized arc consistent *(GAC) iff there exists at least one support for each literal of c. A CSP P is GAC iff every constraint of P is GAC.*

Enforcing GAC is the task of removing from domains all values that have no support on a constraint. Many algorithms have been devised for establishing GAC according to the nature of the constraints.

On the other hand, k-Wise Consistency (kWC) [7,1] can be classified as a constraint-based consistency because it allows us to identify inconsistent tuples (initially accepted by constraints) instead of inconsistent values. It is based on the idea of extending (valid) assignments.

Definition 2. (Extension) *Let Y and Z be two sets of variables. An assignment τ' of $Y \cup Z$ is an* extension *on $Y \cup Z$ of an assignment τ of Y iff $\tau'[y] = \tau[y], \forall y \in Y$.*

Of course, a valid extension is simply an extension that corresponds to a valid assignment. We can now define k-wise consistency, which basically guarantees that every support on a constraint can be extended to any set of $k - 1$ additional constraints. This kind of property allows us to reason about connections between constraints through shared variables.

Definition 3. *(kWC) A CSP* $P = (X, D, C)$ *is* k-wise consistent *(kWC) iff* $\forall c_1 \in C$, $\forall \tau \in c_1 : \tau \in D(scp(c_1))$, $\forall c_2, \ldots, c_k \in C, \exists \tau'$ *valid extension of* τ *on* $\bigcup_{i=1}^{k} scp(c_k)$ *satisfying* $c_2, \ldots, c_k$.

Note that k-wise consistency is called pairwise consistency for k=2 and three-wise consistency for k=3. It is immediate that k-wise consistency implies (k-1)-wise consistency. Enforcing kWC on a CSP involves removing from the constraints (i.e., considering as no more allowed) the tuples that cannot be extended. It thus modifies constraints, not domains. As a result, kWC is incomparable with GAC : a CSP can be kWC but not GAC and reciprocally [6]. However, combining both consistencies allows us to make more pruning of the domains than domain consistency alone. In this paper, we consider such a combination.

Definition 4. *(GAC+kWC) A CSP P is GAC+kWC iff P is both GAC and kWC.*

At this stage, although already suggested earlier, we can observe that GAC, kWC and GAC+kWC are well-behaved consistencies. We recall that a consistency ψ is well-behaved [10] when for any CSP P, the ψ-closure of P exists, where the ψ-closure of P is the greatest CSP, denoted by $\psi(P)$, which is both ψ-consistent and equivalent to P. The underlying partial order on CSPs is: $P' = (X, D', C') \preceq P = (X, D, C)$ iff $\forall x \in X, D'(x) \subseteq D(x)$ and there exists a bijection μ from C to C' such that $\forall c \in C, \mu(c) \subseteq c$. Enforcing ψ on P means computing $\psi(P)$, and an algorithm that enforces ψ is called a ψ-algorithm.

Interestingly, from GAC+kWC, we can derive a domain-filtering consistency, called *domain k-wise consistency*, or DkWC in short. When a CSP P is domain k-wise consistent, it means that all variable domains of P cannot be reduced when enforcing GAC+kWC.

Definition 5. *(DkWC) A CSP* $P = (X, D, C)$ *is* domain k-wise consistent *(DkWC) iff* $GAC+kWC(P)$ *is a CSP* $Q = (X, D^Q, C^Q)$ *such that* $D = D^Q$.

GAC+kWC is both domain-filtering and constraint-filtering, which may render uneasy its implementation in constraint solvers, whereas DkWC is only domain-filtering. In this paper, we propose to enforce DkWC indirectly by considering a reformulation of the CSP to be solved.

3   Enforcing kWC Using k-dual CSPs

This section presents a filtering process for achieving kWC. The filtering procedure is a generalization to the k-wise case of the filtering process presented in [6], and different from the one presented in [7]. It will be useful for our DkWC algorithm, presented in the next section. Due to explicit access to the list of allowed tuples, table constraints are particularly adapted for strong constraint-based consistencies. The filtering procedures proposed in this paper are thus designed for such table constraints. From now on, all constraints will be assumed to be table constraints.

As kWC is a constraint-based consistency, the idea is to define and use a special dual form of the given CSP in order to obtain kWC by simply enforcing GAC on the dual representation. Because this dual form depends on the value of k, we call it *k-dual* CSP. This is a generalization of the dual used in [6] that is equivalent to the *order k constraint graph* defined in [7]. Specifically, the k-dual of a CSP contains one *dual variable* x_i' per constraint c_i in the original CSP and one *k-dual* constraint c_j' per group of k original distinct constraints. Each variable x_i' has a domain which is the set of indexes of the tuples in the original constraint c_i, and each constraint c_j' is a table constraint representing the join of k original constraints. Note that the tuples in those new tables are represented with the indexes of the tuples in the original constraints, which allows the new constraints to have arity k only.

Definition 6. (k-dual CSP) *Let* $P = (X, D, C)$ *be a CSP. The* k-dual *of* P *is the CSP* $P^{kd} = (X^{kd}, D^{kd}, C^{kd})$ *where:*
- *for each constraint* $c_i \in C$, X^{kd} *contains a variable* x_i' *with its domain defined as* $D^{kd}(x_i') = \{1, 2, \ldots, |rel(c_i)|\}$,
- *for each subset* S *of* k *constraints of* C, C^{kd} *contains a constraint* c' *such that* $scp(c') = \{x_i' \mid c_i \in S\}$ *and* c' *is a* k-ary *table constraint containing the join of all constraints in* S *(represented with the indexes of the original tuples).*

If P^{kd} is the k-dual of P, then variables and constraints of P are said to be *original* whereas variables and constraints of P^{kd} are said to be *dual*. An example of a k-dual CSP for $k = 3$ can be found in Example 1.

Example 1. Let $P = (X, D, C)$ be a CSP such that $X = \{u, v, w, x, y, z\}$, $D = \{1, 2, 3, 4\}^6$ and $C = \{c_1, c_2, c_3\}$ where:
- $scp(c_1) = \{u, v, w\}$ and $rel(c_1) = \{(1, 2, 3), (1, 2, 4)\}$,
- $scp(c_2) = \{u, x, y\}$ and $rel(c_2) = \{(1, 3, 4), (2, 3, 4)\}$,
- $scp(c_3) = \{v, x, z\}$ and $rel(c_3) = \{(2, 3, 1), (3, 3, 2)\}$.

The 3-dual of P is a CSP $P^{kd} = (X^{kd}, D^{kd}, C^{kd})$ such that $X^{kd} = \{x_1', x_2', x_3'\}$, $D^{kd} = \{1, 2\}^3$, and $C^{kd} = \{c'\}$ with $scp(c') = \{x_1', x_2', x_3'\}$ and $rel(c') = \{(1, 1, 1), (2, 1, 1)\}$. It represents the full join of the original constraints on $\{u, v, w, x, y, z\}$, which is composed of the tuples $(1, 2, 3, 3, 4, 1)$ and $(1, 2, 4, 3, 4, 1)$. For example, the first tuple is obtained by joining the first tuple of c_1, the first tuple of c_2 and the first one of c_3.

Property 1. A CSP is kWC iff its k-dual CSP is GAC. [6,7].

Property 1 is introduced in [6] for $k = 2$ and the general result is established in [7] for a similar k-dual CSP. A filtering procedure to enforce GAC+2WC (i.e., both pairwise consistency and generalized arc consistency) on a CSP P consists in (1) enforcing GAC on the 2-dual of P, then (2) restraining the constraints of P in order to only contain tuples corresponding to valid dual values, and finally (3) establishing GAC on P [7,3]. The generalization of this procedure to the k-wise case uses the k-dual instead of the 2-dual.

4   Enforcing DkWC Using k-interleaved CSPs

In this section, we propose to reformulate the CSP P to be solved in order to be able to enforce DkWC in a single step, just by applying classical GAC. Basically, to enforce DkWC with GAC propagators only, we first add to P all variables and all constraints from the k-dual CSP of P. Then, we link dual variables and original constraints because, otherwise, the removal of a value from a dual variable would not leverage its corresponding original constraint (and reciprocally). In the definition of GAC+kWC (on which DkWC is based), only valid tuples can serve as supports either for values (generalized arc consistency part) or for other tuples (k-wise consistency part). The link we make guarantees that original tuples corresponding to invalid dual values are invalidated, and reciprocally, ensuring that original constraints and dual variables keep the same pace during filtering. The link we propose involves transforming each original constraint c_i into a new *hybrid* constraint $\phi(c_i)$ involving the original variables in the scope of c_i as well as the dual variable x'_i that is associated with c_i. For each tuple τ in $rel(c_i)$, we generate a tuple in $rel(\phi(c_i))$ by simply appending to τ its position in $rel(c_i)$.

Definition 7. (Hybrid Constraints) *Let* $P = (X, D, C)$ *be a CSP. The set of hybrid constraints* $\phi(C)$ *of P is the set* $\{\phi(c_i) \mid c_i \in C\}$ *where:*
- $scp(\phi(c_i)) = scp(c_i) \cup \{x'_i\}$
- $rel(\phi(c_i)) = \{\tau_j \odot j \mid \tau_j \text{ is the } j^{th} \text{ tuple of } rel(c_i)\}$

with x'_i denoting the dual variable associated with c_i.

In this way, the removal of a value j from $D(x'_i)$ will be reflected in $\phi(c_i)$, as the tuple $\tau_j \odot j$ will be invalidated. Also, when the tuple $\tau_j \odot j$ becomes invalid due to a value removed from the domain of an original variable, j will be removed from $D(x'_i)$. We can now introduce *k-interleaved CSPs* .

Definition 8. (k-Interleaved CSP) *Let* $P = (X, D, C)$ *be a CSP. The* k-interleaved *of P is the CSP* $P^{ki} = (X^{ki}, D^{ki}, C^{ki}) = (X \cup X^{kd}, D \cup D^{kd}, \phi(C) \cup C^{kd})$ *where* (X^{kd}, D^{kd}, C^{kd}) *is the k-dual of P and $\phi(C)$ the hybrid constraints of P.*

The following property shows an interesting connection: enforcing GAC on the k-interleaved CSP of a CSP P is equivalent to enforcing GAC+kWC on P, when the focus is only on the domains of the variables of P.

Property 2. Let $P = (X, D, C)$ be a CSP and $P^{ki} = (X^{ki}, D^{ki}, C^{ki})$ be the k-interleaved CSP of P. If $Q = (X, D^Q, C^Q)$ is the GAC+kWC-closure of P and $R = (X^{ki}, D^R, C^{ki})$ is the GAC-closure of P^{ki}, then we have $D^Q = D^R[X]$ (i.e., $D^Q(x) = D^R(x), \forall x \in X$).

The intuition of the proof is as follows. On the one hand, each literal (x, a) of D^Q is supported on each constraint c^Q involving x by a valid tuple in Q. This tuple is k-wise consistent in Q. By Property 1, the dual variables in X^{ki} precisely encode this k-wise consistency. On the other hand, each literal (y, b) of $D^R[X]$ is supported on each constraint c^{ki} involving y by a valid tuple in R. As all supports on constraints of C^{ki} include a valid dual variable, we have that $D^Q = D^R[X]$.

Then, we can deduce the following corollary.

Corollary 1. *If the k-interleaved CSP of a CSP P is GAC then P is DkWC.*

It is important to note that "k-interleavedness" is preserved after refuting any value. This is stated by the following property (whose proof is omitted).

Property 3. Let $P = (X, D, C)$ be a CSP and P^{ki} be the k-interleaved CSP of P. $\forall x \in X, \forall a \in D(x), P^{ki}|_{x \neq a}$ is the k-interleaved CSP of $P|_{x \neq a}$.

From Properties 2 and 3, we can derive the following important corollary.

Corollary 2. *Let $P = (X, D, C)$ be a CSP and $P^{ki} = (X^{ki}, D^{ki}, C^{ki})$ be the k-interleaved CSP of P. Let Δ be a set of value refutations on variables of X. If $Q = (X, D^Q, C^Q)$ is the GAC+kWC-closure of $P|_\Delta$ and $R = (X^{ki}, D^R, C^{ki})$ is the GAC-closure of $P^{ki}|_\Delta$, then we have $D^Q = D^R[X]$.*

Corollary 2 is central to our approach. It allows us to achieve DkWC indirectly using GAC, and at any stage of a backtrack search. So, it is important to note that the generation of the k-interleaved CSP is only performed once since it can be used during the whole search.

The complexity of enforcing DkWC is the complexity of enforcing GAC on the k-interleaved CSP. As the k-interleaved CSP only contains table constraints, the complexity analysis will use the optimal time complexity for a table constraint given in [14]. Let P be a CSP with n variables, a maximum domain size d, e constraints, a maximum number t of tuples allowed by a constraint, and a maximum constraint arity r. The k-interleaved CSP of P is a CSP with $n' = n + e$ variables, a maximum domain size $d' = max(d, t)$, $e' = e + \binom{e}{k}$ constraints[2], an upper bound $t' = t^k$ of the maximum number of allowed tuples by a constraint, and a maximum constraint arity $r' = max(r + 1, k)$ Enforcing GAC on the k-interleaved CSP with optimal table constraint propagators has a complexity of $O(e' \cdot (r' \cdot t' + r' \cdot d')) = O(((\binom{e}{k} + e) \cdot (r' \cdot t' + r' \cdot d'))$.

Necessity of Hybrid Constraints. It is important to note that the filtering procedure for DkWC presented in this paper is stronger than the propagation that would be obtained by simply replacing the original constraints by their joins. The reason is that, in the second setting, the invalidation of a tuple in a join is not reflected in the other joins, whereas with the k-interleaved CSP, the supports for the tuples on a join must themselves be supported. This is illustrated by Example 2.

Example 2. Let $P = (X, D, C)$ be a CSP such that $X = \{x, y, u, v\}$, $D = \{0, 1\}^4$ and $C = \{c_1, c_2, c_3\}$ with $scp(c_1) = \{x, y, u, v\}$, $scp(c_2) = \{x, y\}$ and $scp(c_3) = \{u, v\}$, and $rel(c_1)$, $rel(c_2)$ and $rel(c_3)$ defined as in Figure 2(a). Let us compare domain pairwise consistency (D2WC) with the joins of any two pairs of constraints: in this CSP, the two possible joins and the 2-interleaved CSP are depicted in Figure 1. On the one hand, enforcing GAC on the two join constraints J_{12} and J_{13} has no effect (observe that values 0 and 1 are present in each column of both tables). On the other hand, enforcing GAC on the 2-interleaved CSP reduces $D(y)$ to $\{1\}$ and $D(v)$ to $\{0\}$. The reduction of $D(y)$ comes from the tuple $(0, 0)$ in $rel(c_2)$ which is the only support for $y = 0$ on c_2. This tuple is only supported in J'_{12} by the second tuple of c_1: $(0, 0, 0, 1)$. As $(0, 0, 0, 1)$ has no support on J'_{13}, we can safely remove 0 from $D(y)$.

5   Practical Use of k-interleaved CSPs

Enforcing GAC on the k-interleaved CSP may be expensive. One possible cause is the number of constraints from the k-dual CSP that are added to the k-interleaved CSP:

---

[2] $\binom{e}{k}$ is the binomial coefficient corresponding to the number of subsets of size k that can be formed using elements from a set of size e.

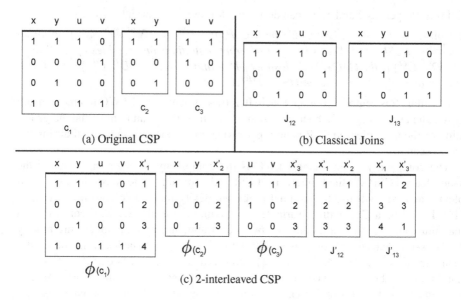

Fig. 1. Illustration of (a) the CSP, (b) the classical joins and (c) the 2-interleaved CSP from Example 2

$\binom{e}{k}$ for an original CSP with e constraints. Some of those constraints can safely be ignored. For instance, this is trivially the case for k-dual constraints that are based on original constraints sharing no variables. Of course, a trade-off can be made between propagation strength and time complexity by integrating only a subset of all possible $\binom{e}{k}$ constraints. In that case, we shall not achieve DkWC completely. Suppose that we limit the integration to the p most promising constraints from the k-dual CSP. The complexity, following our analysis performed above, becomes $O((e+p) \cdot (r' \cdot t' + r' \cdot d'))$: the term $\binom{e}{k}$ has been replaced by p. The most promising constraints can be selected, for example, according to the size of the joins. Indeed, small joins will induce more pruning whereas using large joins is another cause of inefficiency, as it is related to t'.

Following this discussion, we propose two weak variants of DkWC, and refer to them as *weak* DkWC: they only consider a subset of all possible k-dual constraints. The first one, called DkWC^{cy} only considers constraints from the k-dual CSP corresponding to cycles of original constraints (i.e., sequences of constraints at least sharing variables with previous and next constraints in a circular manner). There are typically far less cycles of k constraints than combinations of k constraints and besides they usually form smaller joins. The consistency level attained by DkWC^{cy} is weaker than DkWC but in practice, as we shall see, it shows good performances. Since all the original constraints are included in DkWC^{cy}, the consistency level attained is stronger than GAC. Unsurprisingly, for some problems, the size of the joins of some cycle constraints may be too large to be treated efficiently. For instance, in the modifiedRenault benchmark, some joins computed from cycles of length 3 exceed 10^6 tuples. This is why the second variant of DkWC, called DkWC^{cy-}, only considers constraints from the k-dual CSP corresponding to cycles of original constraints and admitting a join size smaller than a specified parameter (e.g., a percentage of the size of the largest table). In other words, a maximum size is imposed on the size of the joins. The consistency level attained by $DkWC^{cy-}$ is weaker than DkWC and $DkWC^{cy}$, but its practical interest will be shown on some problems. $DkWC^{cy-}$ attains a level of consistency stronger than GAC.

6 Related Work

GAC has already been combined with 2WC, 3WC and kWC [7,6,3,17,8,16]. A first approach consists in weakening the combination, to obtain a pure domain-based consistency. We obtain then max-restricted pairwise consistency (maxRPWC) [16,17,15].

Definition 9. (maxRPWC) *A CSP* $P = (X, D, C)$ *is max-restricted pairwise consistent (maxRPWC) iff* $\forall x \in X$, $\forall a \in D(x)$, $\forall c \in C \mid x \in scp(c)$, $\exists \tau \in D(scp(c))$ *such that* $\tau[x] = a$, $\tau \in c$ *and* $\forall c' \in C$ *there exists a valid extension of* τ *on* $scp(c) \cup scp(c')$ *satisfying* c'.

MaxRPWC is a domain-filtering consistency, close to the idea of DkWC and GAC+ 2WC but weaker than GAC+2WC [3]. In [11], the authors propose a specialized filtering procedure, called *eSTR*, for enforcing GAC+2WC (called full pairwise consistency in their work) on table constraints. Two techniques are combined: simple tabular reduction (STR) and tuple counting. This allows eSTR to keep and update a counter, for each tuple of each table, of the number of supports it has in the other tables. This counter can be used to detect and remove unsupported tuples. This approach is orthogonal to the one presented in this paper. Indeed, in [11], the authors lift up an existing GAC propagator, STR, to eSTR. Our approach is to propose a filtering procedure, relying on a modified CSP, only using existing pure GAC propagators and we are not restricted to GAC+2WC.

Other approaches compute the kWC-closure of a CSP in a first step and then apply GAC in a second step, as proposed in [7,6,3] for 2WC and [8] for kWC. The approach in [8] relies on a specialized propagator, inspecting each constraint with respect to each relevant group of k constraints. Inspecting a constraint means searching, for each tuple of the constraint, a support in each group. This search for support is performed using a backtracking search (Forward Checking), on the dual encoding of the CSP. This whole process is sped up by memorizing, for each constraint and each group, the last encountered support. A similar approach is developed in [20] for relational neighborhood inverse consistency. In [8], the authors also propose a slightly weaker consistency, considering only groups of constraints forming connected components in the minimal dual graph. Although attractive, these original forms of propagators can be hard to include in existing constraint solvers. For instance, in *Comet*, the context management system makes the start of an independent search inside a propagator impossible.

Other related approaches exist, although not trying to enforce directly GAC+kWC. In [12], the authors propose a consistent reformulation for the conjunction of two tables sharing more than one variable, keeping the space complexity low. In [2], the authors propose an algorithm to achieve GAC on global constraints. In this work, the global constraints are perceived as groups of constraints and the CSPs they define are solved on the fly to achieve GAC on them. GAC+kWC on a group of k constraints can be seen as solving the subproblem they define, but in our approach, the subproblems are not solved on the fly.

The easy integration of strong levels of consistency into existing solvers has been studied in [19]. The integration is performed within a generic scheme, incorporating the subset of the constraints involved in the local consistency into a global constraint.

7 Experimental Results

This section presents some experimental results concerning DkWC. For each test, we propose to maintain this property on k-interleaved CSPs at each node of the search trees developed by a backtrack search. However, as discussed in Section 5, including all k-dual constraints is unpractical for many problems, because of the number of additional constraints and/or because of their size. So, in our experiments, we have only used the weaker versions defined in Section 5, namely, DkWC^{cy} and DkWC^{cy-}, and we have focused our attention to weak D3WC and weak D4WC. Those values of k allow a significant search space reduction with respect to GAC while keeping the number and size of k-dual constraints tractable. Notice that labeling is only performed for original variables during search, and that all solutions are searched for. The GAC propagator used for the original constraints as well as the k-dual ones is the optimal state-of-the-art propagator from [14].

Our filtering procedure is compared with the GAC propagator from [14], the max-RPWC3 procedure from [3] and the state-of-the-art eSTRw propagator from [11]. The eSTRw propagator is weaker than eSTR but easier to incorporate into an existing solver, and is at least as good as eSTR on the benchmarks used in [11]. All the algorithms are (re-)implemented on top of Comet, but as mentioned in Section 6, it is unfortunately impossible to implement the filtering algorithm from [8] in Comet. Eight different benchmarks have been used. Two of them contain only binary table constraints, five of them contain binary and ternary table constraints and the last benchmark contains table constraints up to arity 10. The tests are executed on an Intel Xeon 2.53GHz using Comet 2.1.1. A timeout of 20 minutes on the total execution time is used for each instance. When comparing different techniques in terms of CPU time and search space sizes, we can only use the subset of instances for which none of the techniques timed out. In the results, we thus do not report measurements for some of the techniques on some benchmarks because including them would cause the common instance set to be empty or too small for a meaningful comparison. In the tables, a '-' thus represents a technique that timed out on the set of instances considered. The percentage of the instance set that is solved is however given for each technique on each benchmark.

The results are presented in Table 1. For each instance set and each technique, we report means of different quantities (times are in seconds): the execution time (T), the "posting" time (pT), the join selection time (jST) that corresponds to the amount of time used to select the joins for the k-interleaved CSPs, the join computation time (jT), the number of propagator calls (nP), the number of fails (nF) and the number of choice points (nC). Table 1 also reports the percentage to the best with respect to execution time (%b), the mean of the percentage to the best instance by instance (μ%b) and the percentage of instances from the sets that are solved (%sol). The total time (T) includes all precomputations the algorithms have to perform before search. This means that both times of join selection (jST) and join computation (jT) for our DkWC algorithm are included in T. The posting time (pT) is the time taken between the loading of the instance file and the start of search without the time for jT. It thus includes time for all precomputations except for join computation. The difference between %b and μ%b is the following. For %b, all execution times are averaged before computing it: there is thus one identified best algorithm. For μ%b, the percentages are first computed instance by instance, and then aggregated with a geometrical mean (as suggested in [5]): this measure takes into account the fact that different instances may have different best algorithms.

Binary Random Instances. This instance set contains 50 instances involving binary table constraints. These instances have 50 variables, a uniform domain size of 10 and 166 constraints whose proportion of allowed tuples is 0.5. They have been generated using the model RD [21], in or close to the phase transition. The search strategy used to solve them (for all techniques) is a lexicographic variable and value ordering. On this benchmark, D3WC^{cy} includes on average 48.4 3-dual constraints, and their tables contain on average 111.5 tuples. We can see that the pruning obtained by D3WC^{cy} on this benchmark allows it to reduce drastically the search space. Moreover, since the mean number of added constraints from the 3-dual CSP and their size is small, D3WC^{cy} has the lowest overall computation time. Propagators maxRPWC3 and eSTRw also reduce the search space with respect to GAC (partly due to the presence of constraints with identical scopes), but this reduction comes at the price of a greater total computation time.

Ternary Random Instances. This instance set contains 50 instances involving ternary table constraints. These instances have 50 variables, a uniform domain size of 5 and 75 constraints whose proportion of allowed tuples is 0.66. They have been generated using the model RD [21], in or close to the phase transition. The search strategy used to solve them is a lexicographic variable and value ordering. On this benchmark, D3WC^{cy} includes, on average 112.3 3-dual constraints, and their tables contain on average 529.5 tuples. As for the binary case, the search space reduction obtained by D3WC^{cy} is important. On this benchmark, the size and number of added constraints is small enough to allow D3WC^{cy} to be the fastest technique. Note that the search space reduction obtained by maxRPWC3 doesn't repay its cost, contrary to eSTRw.

AIM Instances. This instance set contains 24 instances from the AIM series used in the CSP solver competition [18] (100 variables, a majority of ternary constraints and binary ones). The search strategy used is a lexicographic variable and value ordering. D3WC^{cy} includes 3000 constraints from the 3-dual, on average, and the added constraints contain on average 30.3 tuples. On this benchmark, the filtering obtained by maxRPWC3, eSTRW and D3WC^{cy} allows each of them to be significantly faster than GAC. Although D3WC^{cy} achieves the best search space reduction, eSTRw remains the fastest technique. The greater computation time for D3WC^{cy} is due to the number of 3-dual constraints included in the 3-interleaved CSPs: 3000 on average while the number of original constraints lies between 150 and 570. Even if the 3-dual constraints have small tables, they still have to be propagated during search. Interestingly, D3WC^{cy} solves significantly more instances than the other techniques.

Table 1. Results of the experiments on the different benchmarks. T is the mean time in seconds, pT is the mean posting time in seconds, jST is the mean join selection time, jT is the mean join time, nP is the number of calls to the propagators, nF is the number of fails during the search, nC is the number of choice points during the search, %b is the percentage to the best, μ%b is the mean percentage to the best and %sol is the percentage of instances solved.

propagator	T	pT	jST	jT	nP	nF	nC	%b	μ%b	%sol
Binary Random										
GAC	9.9	0.0	0.0	0.0	3 M	7.7 k	1 213.2	337	218	**100**
maxRPWC3	72	0.7	0.0	0.0	148 k	2.2 k	340.1	2448	1833	98
eSTRw	11.4	0.1	0.0	0.0	293 k	2.2 k	340.1	389	283	**100**
D3WC^{cy}	**2.9**	0.1	0.0	0.1	963 k	0.5 k	68.4	100	113	**100**
Ternary Random										
GAC	23.1	0.0	0.0	0.0	4 M	42.4 k	11.8 k	183	223	**100**
maxRPWC3	124	0.2	0.0	0.0	237 k	8.2 k	2.2 k	982	1455	90
eSTRw	16.6	0.0	0.0	0.0	409 k	7.7 k	2.1 k	131	189	**100**
D3WC^{cy}	**12.6**	0.5	0.1	0.4	2 M	0.6 k	0.1 k	100	143	**100**
AIM										
GAC	82	0.1	0.0	0.0	35 M	941.6 k	522 k	6745	460	46
maxRPWC3	14.7	1.0	0.0	0.0	46 k	1.2 k	0.7 k	1204	1481	46
eSTRw	**1.2**	0.3	0.0	0.0	35 k	0.7 k	0.4 k	100	208	50
D3WC^{cy}	3.4	2.4	1.2	0.5	139 k	0.1 k	0.1 k	279	497	**88**
Pret										
GAC	160	0.0	0.0	0.0	58 M	7 M	5 M	121	121	**50**
maxRPWC3	977	0.0	0.0	0.0	26 M	7 M	5 M	741	741	**50**
eSTRw	504	0.0	0.0	0.0	30 M	7 M	5 M	382	382	**50**
D3WC^{cy}	**132**	0.0	0.0	0.0	57 M	4 M	3 M	100	100	**50**
Langford-2										
GAC	**0.5**	0.0	0.0	0.0	171 k	1.4 k	1 k	100	100	**58**
maxRPWC3	44.6	2.2	0.0	0.0	43 k	1.4 k	1 k	9637	3863	46
eSTRw	1.5	0.1	0.0	0.0	65 k	1.4 k	1 k	326	233	54
D3WC^{cy}	10.5	0.9	0.1	3.2	2 M	0.7 k	0.7 k	2270	1782	50
Dubois										
GAC	793	0.0	0.0	0.0	158 M	42 M	37 M	394	390	15
maxRPWC3	-	-	-	-	-	-	-	-	-	8
eSTRw	598	0.0	0.0	0.0	37 M	5 M	3 M	297	294	15
D4WC^{cy}	**201**	0.1	0.0	0.0	68 M	2 M	1 M	100	100	**30**
TSP-20										
GAC	**52**	0.5	0.0	0.0	14 M	17 k	7 k	100	100	**93**
maxRPWC3	-	-	-	-	-	-	-	-	-	33
eSTRw	233	1.5	0.1	0.0	5 M	17 k	7 k	447	438	80
D3WC^{cy}	-	-	-	-	-	-	-	-	-	40
D3WC^{cy-}	94	4.3	0.6	0.1	40 M	17 k	7 k	180	270	**93**
Modified Renault										
GAC	-	-	-	-	-	-	-	-	-	6
eSTRw	-	-	-	-	-	-	-	-	-	0
maxRPWC3	743	0.0	0.0	0.0	23.7	0.0	0.0	148	417	26
D3WC^{cy}	-	-	-	-	-	-	-	-	-	0
D3WC^{cy-}	**502**	497	3.9	5.1	33 k	0.0	0.0	100	111	**34**

Pret Instances. This instance set also comes from the CSP solver competition [18] and counts 8 instances (only ternary table constraints). The search strategy used is a lexicographic variable and value ordering. D3WC^{cy} includes on average 13 constraints that have a mean size of 8 tuples. On this benchmark, neither maxRPWC3 nor eSTRw is able to reduce the search space with respect to GAC. Their additional computations make them slower than GAC. The small number of small constraints from the 3-dual CSP included by D3WC^{cy} allows it to significantly reduce the search space and to be the fastest on this series. The mean percentage to the best ($\mu\%b$) of D3WC^{cy} means that it is the best technique on average but also on each instance. Note that the join selection, join computation and posting times are negligible on this problem.

Langford Problem. Langford number problem is Problem 24 of CSPLIB[3], here modeled with binary table constraints only. We used the set Langford-2 containing 24 instances that can be found in [9]. The search strategy used is *dom/deg* combined with a lexicographic value ordering. On this set, D3WC^{cy} includes, on average, 328 3-dual constraints whose tables contain 274.7 tuples on average. On this benchmark, GAC is the fastest technique on average (actually, it is the fastest technique on each instance). Neither maxRPWC3 nor eSTRw are able to reduce the search space with respect to GAC. However, they have a lower propagator call count. This is due to their ability to reach the fixed point faster. The number of constraints added from the 3-dual by D3WC^{cy} is large comparatively to the number of original constraints (the non-timeout instances are the smallest ones). The small search space reduction obtained by D3WC^{cy} does not compensate the cost to propagate all the added constraints. The number of propagator calls is significantly larger for D3WC^{cy}. We can also see that, on this benchmark, the time required to compute the joins is larger than the time required by GAC to solve the instances.

Dubois Instances. Those 13 instances also comes from the CSP solver competition [18] (ternary table constraints). These instances do not contain any cycle of original constraints of length 3. We thus present the results of D4WC^{cy}. On this series, D4WC^{cy} adds on average 156 4-dual constraints and their tables contain, on average, 15.2 tuples. Clearly, D4WC^{cy} is the fastest approach here and solves more instances than the other techniques. D4WC^{cy} is also the fastest on each instance, as shown by the mean percentage to the best ($\mu\%b$). The search space reduction obtained by eSTRw is less than that obtained by D4WC^{cy} but it allows it to be faster than GAC.

Travelling Salesman Problem. We used the set of 15 Travelling Salesman satisfaction instances *tsp-20* from [9] (table constraints of arity 2 and 3). The search strategy used here is *dom/deg* combined with a lexicographic value ordering. On this instance set, there is, on average, 1000 cycles of length 3 in the 3-dual CSP and they contain up to 2000 tuples. In that context, D3WC^{cy} only solves 40% of the instances. We thus present the results for D3WC^{cy-} where the limit on the size of the joins is set to one percent of the maximal original constraint size (200). D3WC^{cy-} includes 59.8 constraints from the 3-dual CSP on average, and their tables contain, on average, 26 tuples. As we can see, on those instances, neither eSTRw nor D3WC^{cy-} is able to reduce the search space. The extra computations of eSTRw and the extra propagation effort of D3WC^{cy-} make them slower than GAC (which is also the fastest approach on each

---

[3] www.csplib.org

Table 2. Summary of the results of the experimental section. T is the total solving time in seconds and %sol is the percentage of the instances solved.

Benchmark	GAC		maxRPWC3		eSTRw		wDkWC	
	T	%sol	T	%sol	T	%sol	T	%sol
Binary Random	9.9	**100**	72	98	11.4	**100**	**2.9**	**100**
Ternary Random	23.1	**100**	124	90	16.6	**100**	**12.6**	**100**
AIM	82	46	14.7	46	**1.2**	50	3.4	**88**
Pret	160	**50**	977	**50**	504	**50**	132	**50**
Langford2	**0.5**	58	44.6	46	1.5	54	10.5	50
Dubois	793	15	-	8	598	15	**201**	**30**
TSP-20	52	**93**	-	33	233	80	94	**93**
ModRenault	-	6	743	0	-	26	**502**	**34**

instance). However, $D3WC^{cy-}$ is faster than eSTRw and it is the only one able to solve the same number of instances as GAC.

Modified Renault Problem. The modified Renault problem instances originate from a real Renault Megane configuration problem, modified to generate 50 instances[9] (large tables and arities up to 10). The search strategy used is *dom/deg* variable ordering combined with a lexicographic value ordering. Since the tables of the original problem can count up to 50K tuples, $D3WC^{cy}$ is unpractical because of the size of the joins. We thus present the results for $D3WC^{cy-}$ where the limit on the size of the joins has been set to one percent of the largest original constraint size (500), as in the TSP benchmark. On those instances, $D3WC^{cy-}$ includes 481.6 3-dual constraints on average, and their tables contain on average 253.9 tuples. As we can see, both maxRPWC3 and $D3WC^{cy-}$ detect the inconsistencies of all instances without performing any search. However, despite the fact that $D3WC^{cy-}$ has a larger propagator call count, it is faster than maxRPWC3. $D3WC^{cy-}$ is also able to solve more instances than maxRPWC3.

Summary of the Experimental Results. A summary of the experimental results can be found in Table 2. This table contains the total execution time (T) and the percentage of instances solved (% sol) for each technique. The column wDkWC represents our weak DkWC approach: it is $D3WC^{cy}$ for binary random, ternary random, AIM, Pret and Langford-2 instances, $D4WC^{cy}$ for Dubois instances and $D3WC^{cy-}$ for TSP-20 and modified Renault instances. Weak DkWC is faster than maxRPWC3 and eSTRw, except for two benchmarks. It is also faster than GAC on all but two benchmarks, where GAC is faster than all strong consistencies. Weak DkWC is also the strong consistency leading to the largest reductions of search space. Except on Langford-2, weak DkWC solves the largest number of instances within the time limit.

For all these benchmarks, we insist that (full) DkWC is unpractical because of the number of possible joins and/or their size. This is the reason why we have introduced weak DkWC. On the majority of benchmarks, we used $DkWC^{cy}$ but on two benchmarks, even $D3WC^{cy}$ suffers from the number of 3-dual constraints and their sizes. Consequently, we also used $DkWC^{cy-}$, for which the best limit on the joins size has been empirically found to be equal to 1 percent of the maximum original constraint size. This parameter value allows $D3WC^{cy-}$ to include a significant number of small (highly filtering) 3-dual constraints without including too many of them. All these results show, on a large variety of benchmarks with constraints of various arities, that the weak DkWC filtering procedures defined in this paper are competitive.

8   Conclusion

In this paper, we have derived a domain-filtering consistency, DkWC, from the combination of kWC and GAC. We have shown how to establish and maintain this strong consistency by simply establishing and maintaining GAC on so-called k-interleaved CSPs. Such reformulated CSPs, which integrate dual variables, hybrid constraints and k-dual constraints, are simple to generate, and need to be generated only once before search. To manage the complexity of join operations, we have proposed a few solutions such as the ones relying on the presence of cycles or on the use of a limit on the maximal size of joins. The experimental results that we have obtained show, on a large variety of problems, that our weak DkWC filtering procedures are competitive.

Acknowledgments. The first author is supported as a Research Assistant by the Belgian FNRS. This research is also partially supported by the FRFC project 2.4504.10 of the Belgian FNRS, and by the UCLouvain Action de Recherche Concertée ICTM22C1. The third author benefits from the financial support of both CNRS and OSEO within the ISI project 'Pajero'.

References

1. Bessiere, C.: Constraint propagation. In: Rossi, F., van Beek, P., Walsh, T. (eds.) Handbook of Constraint Programming. Elsevier, New York (2006)
2. Bessière, C., Régin, J.-C.: Enforcing arc consistency on global constraints by solving sub-problems on the fly. In: Jaffar, J. (ed.) CP 1999. LNCS, vol. 1713, pp. 103–117. Springer, Heidelberg (1999)
3. Bessiere, C., Stergiou, K., Walsh, T.: Domain filtering consistencies for non-binary constraints. Artificial Intelligence 72(6-7), 800–822 (2008)
4. Debruyne, R., Bessière, C.: Domain filtering consistencies. Journal of Artificial Intelligence Research 14, 205–230 (2001)
5. Fleming, P., Wallace, J.: How not to lie with statistics: the correct way to summarize benchmark results. Communications of the ACM 29(3), 218–221 (1986)
6. Janssen, P., Jégou, P., Nouguier, B., Vilarem, M.-C.: A filtering process for general constraint-satisfaction problems: achieving pairwise-consistency using an associated binary representation. In: Proceedings of IEEE Workshop on Tools for Artificial Intelligence, pp. 420–427 (1989)
7. Jégou, P.: Contribution à l'étude des Problèmes de Satisfaction de Contraintes: Algorithmes de propagation et de résolution. Propagation de contraintes dans les réseaux dynamique. PhD thesis, Université de Montpellier II (1991)
8. Karakashian, S., Woodward, R., Reeson, C., Choueiry, B., Bessiere, C.: A first practical algorithm for high levels of relational consistency. In: Proceedings of AAAI 2010, pp. 101–107 (2010)

9. Lecoutre, C.: Instances of the Constraint Solver Competition,
   `http://www.cril.fr/~lecoutre/`
10. Lecoutre, C.: Constraint Networks: Techniques and Algorithms. ISTE/Wiley (2009)
11. Lecoutre, C., Paparrizou, A., Stergiou, K.: Extending STR to a higher-order consistency. In: Proceedings of AAAI 2013, pp. 576–582 (2013)
12. Lhomme, O.: Practical reformulations with table constraints. In: Proceedings of ECAI 2012, pp. 911–912 (2012)
13. Mackworth, A.K.: Consistency in networks of relations. Artificial Intelligence 8(1), 99–118 (1977)
14. Mairy, J.-B., Van Hentenryck, P., Deville, Y.: An optimal filtering algorithm for table constraints. In: Milano, M. (ed.) CP 2012. LNCS, vol. 7514, pp. 496–511. Springer, Heidelberg (2012)
15. Paparrizou, A., Stergiou, K.: An efficient higher-order consistency algorithm for table constraints. In: Proceedings of AAAI 2012, pp. 335–541 (2012)
16. Stergiou, K.: Strong inverse consistencies for non-binary CSPs. In: Proceedings of ICTAI 2007, pp. 215–222 (2007)
17. Stergiou, K.: Strong domain filtering consistencies for non-binary constraint satisfaction problems. International Journal on Artificial Intelligence Tools 17(5), 781–802 (2008)
18. van Dongen, M., Lecoutre, C., Roussel, O.: CSP solver competition (2008),
   `http://www.cril.univ-artois.fr/CPAI08/`
19. Vion, J., Petit, T., Jussien, N.: Integrating strong local consistencies into constraint solvers. In: Larrosa, J., O'Sullivan, B. (eds.) CSCLP 2009. LNCS (LNAI), vol. 6384, pp. 90–104. Springer, Heidelberg (2011)
20. Woodward, R., Karakashian, S., Choueiry, B., Bessiere, C.: Solving difficult CSPs with relational neighborhood inverse consistency. In: Proceedings of AAAI 2011, pp. 112–119 (2011)
21. Xu, K., Boussemart, F., Hemery, F., Lecoutre, C.: Random constraint satisfaction: easy generation of hard (satisfiable) instances. Artificial Intelligence 171(8-9), 514–534 (2007)

Representative Encodings to Translate Finite CSPs into SAT

Pedro Barahona[1], Steffen Hölldobler[2], and Van-Hau Nguyen[2]

[1] Departamento de Informática, Faculdade de Ciências e Tecnologia
Universidade Nova de Lisboa, 2829-516 Caparica, Portugal
pb@fct.unl.pt
[2]International Center for Computational Logic
Technische Universität Dresden, 01062 Dresden, Germany
sh,hau@iccl.tu-dresden.de

Abstract. Solving Constraint Satisfaction Problems (CSPs) by Boolean Satisfiability (SAT) requires suitable encodings for translating CSPs to equivalent SAT instances that should not only be effectively generated, but should also be efficiently processed by SAT solvers. In this paper we investigate hierarchical and hybrid encodings, focussing on two specific encodings: the *representative-sparse encoding*, already proposed albeit not thoroughly tested, and a new *representative-order encoding*. Compared to the *sparse* and *order* encodings, two widely used and efficient encodings, these representative encodings require a smaller number of variables and longer clauses to express complex CSP constraints. The paper shows that, concerning unit propagation, these encodings are incomparable. Nevertheless, the preliminary experimental results show that the new representative encodings are clearly competitive with the original sparse and order encodings, but further studies are needed to better understand the trading between fewer variables and longer clauses of the representative encodings.

1 Introduction

Propositional satisfiability solving (SAT) is having an increasing impact on applications in various areas, ranging from artificial intelligence to hardware design and verification, and its success inspired a wide range of real-world and challenging applications.

Many such applications can be expressed as constraint satisfaction problems (CSPs) [24], although hardly any problems are originally given by SAT formulas. Nevertheless, to benefit from powerful SAT solvers, many SAT-encodings of CSPs have been studied recently, and finding suitable SAT encodings for CSPs is of greatest interest [32,1,21,1,15,9,23,31,28,12].

In fact, choosing an appropriate encoding is regarded as important as choosing an efficient algorithm and is considered to be one of the most exciting challenges in SAT solving [15]. Nevertheless, mapping a CSP into a SAT instance largely remains more of an art than a science [32,11,22,21,23], and some guidance is

H. Simonis (Ed.): CPAIOR 2014, LNCS 8451, pp. 251–267, 2014.
© Springer International Publishing Switzerland 2014

required for choosing appropriate encoding schemes for specific SAT-encoded instances.

Although other alternatives have been proposed, e.g. [32,9,29,31], in practice only two types of encodings are being widely used to encode an integer variable V into SAT: the *sparse encoding* (as denoted in [14], and adopted among others by the *direct encoding* [16,32] or the *support encoding* [11]) and the *order encoding* [7,2].

The *sparse encoding* is a common and effective encoding of CSPs to SAT instances. For example, Cadoni and Schaerf introduced a compiler (NP-SPEC), that translates a problem specification to SAT [6], Berre and Lynce built a SAT-based solver which competed in many cases against CSP solvers [4], and Jeavons and Petke [14] recently showed that modern SAT solvers using the *sparse encoding* can efficiently deal with certain families of CSPs which are even challenging for conventional constraint-programming solvers.

By facilitating the propagation of bounds, many combinatorial problems can be solved more efficiently by the *order encoding*, adopted in Sugar [28], BEE [18] and in lazy-clause generation [20].

In both encodings the number of SAT variables required to represent a CSP variable is linear on its domain size, which is quite penalizing for large domains, significantly consuming the runtime of SAT solvers. In contrast, the number of SAT variables required by the *log encoding* is only logarithmic on the domain size; unfortunately, it requires much longer clauses to encode most CSP constraints, and unit propagation is much less powerful than that obtained with the former encodings. Unlike the *sparse encoding* or the *order encoding*, we are unaware of any SAT-based solvers based on the *log encoding*.

Hierarchical and hybrid SAT encodings proposed by Velev [31] use one simple SAT encoding, (a variant of) the direct encoding or (a variant of) the log encoding, to recursively partition the domain of a CSP variable into subdomains. The twelve simple encodings combined with a variety of structures led to a numerous way of translations of a domain to SAT. However, it is impractical to determine how these variants behave for a given problem. Nguyen et al. [19] studied in some detail two special cases of hierarchical and hybrid SAT encodings, namely the *log-direct* and the *log-order* encodings, which are significantly more efficient than the *direct* and the *order* encodings.

In this paper we investigate the *representative-sparse* and the *representative-order* encodings, and show that despite the fact that the propagation achieved with these encodings is not comparable, the experimental results obtained with these representative encodings are quite promising, highlighting their potential for handling hard and practical problems.

Our work generalizes the work presented in [19], in which Nguyen et al. studied two specific representative encodings, using a log encoding of the first level. Here we study a more general encoding with several values at the first level, and include not only a theoretical comparison regarding the propagation strength of these encodings (missing in [19]) but also a more substantial empirical evidence obtained from three distinct state-of-the-art SAT solvers.

The rest of the paper is organized as follows. In Section 2, we briefly overview the *sparse* and the *order* encodings in the context of CSP problems. Section 3 introduces the *representative* encodings and presents their main features, namely their complexity and their propagation strength. Section 4, presents the experimental results obtained in a number of typical benchmarks, and show the significant speed-ups obtained with the representative encodings, when compared with their flat counterparts. The final section summarizes the main results and proposes further work.

2 Background

This section briefly overviews the two most widely and successfully used SAT encodings of CSP variables and constraints, viz. the *sparse encoding* and the *order encoding*, together with the basic concepts and notation of CSPs and SAT.

2.1 Constraint Satisfaction Problems (CSPs)

A *constraint satisfaction problem (CSP)* is a triple $(\mathcal{V}, \mathcal{D}, \mathcal{C})$, where

- $\mathcal{V} = \{V_1, V_2, ..., V_k\}$ is a finite set of variables,
- $\mathcal{D} = \{D(V_1), D(V_2), ..., D(V_k)\}$ is a finite set of domains,
- $\mathcal{C} = \{C_1, C_2, ..., C_m\}$ is a finite set of constraints.

A tuple $\langle v_1, v_2, ..., v_k \rangle \in \langle D(V_1), D(V_2), ..., D(V_k) \rangle$ *satisfies* a constraint $C_i \in \mathcal{C}$ on the set of variables $\{V_{i1}, V_{i2}, ..., V_{il}\} \in \mathcal{V}$ if the projection of the tuple into these variables is a member of the constraint, i.e. $\langle v_{i1}, v_{i2}, ..., v_{il} \rangle \in C_i$. A tuple $\langle v_1, v_2, ..., v_k \rangle$ is a *solution* of a CSP iff it satisfies all constraints of $\mathcal{C}$. The CSP problem is to determine whether there exists one such tuple. In this paper, without loss of generality we assume that all variables have domain $\{1, ..., n\}$.

2.2 Boolean Satisfiability Problem (SAT)

A *formula in conjunctive normal form (CNF)* is a finite conjunction of clauses $C_1 \wedge C_2 \wedge ... \wedge C_m$, defined on a finite set $\{x_1, x_2, ..., x_n\}$ of propositional variables $(m, n \in \mathbb{N})$. A *clause* C is a finite disjunction of literals $l_1 \vee l_2 \vee ... \vee l_k$, where each literal l_i is either a Boolean variable x_l or its negation $(\neg x_l)$.

To each variable the truth values *false* (or 0) or *true* (or 1) can be assigned. A clause is *satisfied* by a truth assignment of the variables, if at least one of its literals is assigned value *true*. A CNF-formula $\mathcal{F}$ is *satisfiable* if there is a truth assignment that satisfies all its clauses, *unsatisfiable* otherwise. A *SAT problem* consisting of a CNF-formula $\mathcal{F}$ is the question whether a CNF-formula $\mathcal{F}$ is *satisfiable*.

2.3 Translating a Finite CSP to an Equivalent SAT Instance

The Sparse Encoding To encode a CSP variable V with domain $\{1, ..., n\}$ the *sparse encoding* uses n propositional variables d_i^V and the assignment $V = i$ is modelled by assigning d_i^V to *true* and all the other propositional variables to *false*.

Hence, the *sparse* encoding requires that exactly one d_i^V variable is assigned to *true*. Such constraint is achieved by means of a single *at-least-one* (ALO) clause:

$$d_1^V \lor d_2^V \lor ... \lor d_n^V$$

and a set of *at-most-one*(AMO) clauses, which may be naturally modelled by the pairwise encoding:

$$\bigwedge_{i=1}^{n-1} \bigwedge_{j=i+1}^{n} \neg(d_i^V \land d_j^V) \equiv \bigwedge_{i=1}^{n-1} \bigwedge_{j=i+1}^{n} (\neg d_i^V \lor \neg d_j^V)$$

where $\equiv$ denotes semantic equivalence. The pairwise encoding requires $\frac{n(n-1)}{2}$ binary clauses, but there are several efficient ways which only need $O(n)$ clauses (see [8,13]), like the sequential counter encoding [26] used in this paper.

The Conflict Clause The *direct encoding* [16] maps the CSP constraints onto a set of *conflict clauses*, modeling the disallowed variable assignments.

Adopting notions and notations from [23], let $K_{V,i}$ be the set of pairs (W, j) for which the assignments $(V = i, W = j)$ violate a CSP constraint. Then these constraints can be expressed by the following formula:

$$\bigwedge_{(w,j) \in K_{V,i}} \neg(d_i^V \land d_j^W) \equiv \bigwedge_{(w,j) \in K_{V,i}} (\neg d_i^V \lor \neg d_j^W)$$

The Support Clause. In contrast, the *support encoding* [11] maps the CSP constraints onto *support clauses* that specify the allowed variable assignments. Let $S_{V,j,W}$ be the values in the domain of V that support $W = j$ in some constraint. Then, the support clauses that model the constraint are expressed as:

$$d_j^W \to (\bigvee_{i \in S_{V,j,W}} d_i^V) \equiv \neg d_j^W \lor (\bigvee_{i \in S_{V,j,W}} d_i^V).$$

The Order Encoding. The *order encoding* [7,2] (a reformulation of the *sequential encoding* [13]) represents a CSP variable V with domain $\{1, ..., n\}$ by a vector of $n - 1$ Boolean variables $[o_1^V, ..., o_{n-1}^V]$. To specify the assignment $V = i$ the first $i - 1$ variables are assigned to *true* (or 1) and the remaining to *false* (or 0). For example, the assignments $V = 1$, $V = 3$, and $V = 5$ for a variable V with domain $\{1, 2, 3, 4, 5\}$, are represented as $[0, 0, 0, 0]$, $[1, 1, 0, 0]$, and $[1, 1, 1, 1]$, respectively.

This encoding may be specified by a set of binary clauses

$$\bigwedge_{i=1}^{n-2} \neg(\neg o_i^V \wedge o_{i+1}^V) \equiv \bigwedge_{i=1}^{n-2} (o_i^V \vee \neg o_{i+1}^V)$$

which guarantee the desired properties [25]:

- if $o_i^V = 1$, then $o_j^V = 1$ for all $1 \leq j \leq i \leq n-1$,
- if $o_i^V = 0$, then $o_j^V = 0$ for all $1 \leq i \leq j \leq n-1$.

A CSP assignment $V = i$ is modelled by imposing $o_{i-1}^V = 1$ and $o_i^V = 0$, whereas its negation $V \neq i$ is represented by $o_{i-1}^V = o_i^V$ [18] (to cope with $V = 1$, we assume an extra bounding variable $o_0^V = 1$).

According to [2] the main advantage of this encoding is in the representation of interval domains and the propagation of their bounds. Indeed, the value of V may be restricted to a range $i...j$, by setting $o_{i-1}^V = 1$ and $o_j^V = 0$.

As with the *sparse encoding*, CSP constraints can be represented in the *order encoding* either by conflict or support clauses, with the obvious adaptations.

3 The Representative Encodings

3.1 The Representative-Sparse Encoding

The *representative-sparse encoding* is a hierarchical hybrid encoding, where Boolean *representative* variables $g_1, g_2, ..., g_m (1 \leq m \leq \frac{n}{2})$ at level 1 divide the domain into m subdomains represented at level 2 with Boolean *sparse* variables. These variables of both levels require ALO and AMO constraints. An assignment in this encoding is as follows:[1]

$$v = v_i \Leftrightarrow \begin{cases} g_{\lfloor i/r \rfloor} \quad \wedge \ x_r & \text{if } i \bmod r = 0; \\ g_{\lfloor i/r \rfloor + 1} \wedge \ x_{i \bmod r} & \text{otherwise.} \end{cases} \tag{1}$$

Formula 1 translates a finite CSP domain $v = \{v_1, v_2, ..., v_n\}$ into SAT clauses by using Boolean representative variables, $\{g_1, ..., g_m\}$, and a set of Boolean sparse variables, $\{x_1, ..., x_r\}$, where $r = \lceil n/m \rceil$.

Fig 1 shows how the Boolean representative variables, $g_i, 1 \leq i \leq m$, assigns a value of V to the subdomains (note that when $n < m * r$ the prohibited values must be excluded).

Proposition 1. *When indexing the domain values of the CSP variables into SAT variables, the representative-sparse encoding is sound and complete.*

Proof: The *representative* variables $g_i, 1 \leq i \leq m \leq \frac{n}{2}$, at level one partition the domain of variable V into m subdomains $\{v_1, ..., v_{\lceil n/m \rceil}\}, ..., \{v_{(m-1)\lceil n/m \rceil + 1}, ..., v_n\}$. Moreover, the *at-least-one* and *at-most-one* clauses for

---

[1] $\lfloor x \rfloor$ ($\lceil x \rceil$) is the biggest (smallest) integer number not bigger (less) than x, and *mod* is the remainder operator.

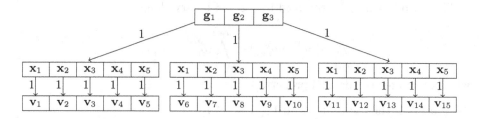

Fig. 1. An illustration of the representative-sparse encoding of the domain of a CSP variable using a group of *representative* variables at level one, $\{g_1, g_2, g_3\}$, and a set of Boolean *sparse variables* at level two, $\{x_1, x_2, x_3, x_4, x_5\}$, where exactly one variable is selected at each group

both sets of Boolean variables $\{g_1, \ldots, g_m\}$ and $\{x_1, \ldots, x_r\}$ lead to the selection of exactly one partition and exactly one value from the represented partition for the value of V. $\qquad\square$

Note that the sparse encoding requires n Boolean variables to encode a CSP variable V with domain $\{v_1, \ldots, v_n\}$, whereas the representative-sparse encoding requires only $m + \lceil n/m \rceil$ Boolean variables, $1 \leq m \leq \frac{n}{2}$.

The *sparse encoding* is adopted by both the *direct encoding* and the *support encoding*. Unit propagation on the *direct encoding* maintains a form of consistency called forward checking [32], while unit propagation on the support encoding preserves arc-consistency [11]. Propagation for *representative-sparse encoding* is not stronger nor weaker with respect to the *sparse encoding* as stated in the following proposition.

Proposition 2. *Unit propagation applied to the representative-sparse encoding (when $m \geq 2$) is not comparable to the sparse encoding.*

Proof: We prove for the case $m = 3$ (the others are similar). Let CSP variables V and W have domain $\{1, .., 15\}$, as in Fig 1, and be constrained by $V \neq 3 \vee W \neq 5$. In the *representative-sparse encoding* the constraint is represented by a clause $\neg g_1^V \vee \neg x_3^V \vee \neg g_1^W \vee \neg x_5^W$ whereas by a simpler clause $\neg d_3^V \vee \neg d_5^W$ in the *sparse encoding*. For a subsequent assignment $W = 5$ (obtained during search or propagation), Unit propagation (UP) in the *sparse encoding* results in the unit clause $\neg d_3^V$, that may be further propagated, whereas in the *representative-sparse encoding* it leads to the non-unit binary clause $\neg g_1^V \vee \neg x_3^V$. Hence, in this case, UP in the *sparse encoding* is stronger than in the *representative-sparse encoding*.

On the other hand, let $V = \{2, 7, 12, 13\}$ be a constraint to be encoded into SAT. The constraint is represented by clauses $(x_2^V \vee g_3^V) \wedge (x_2^V \vee x_3^V)$ in the *representative-sparse encoding*, and by clause $d_2^V \vee d_7^V \vee d_{12}^V \vee d_{13}^V$ in the *sparse encoding*. For a subsequent assignment $V < 11$ (i.e. $\neg g_3^V$), then UP in the *representative-sparse encoding* results to the unit clause x_2^V, whereas in the *sparse encoding* leads to $d_2^V \vee d_7^V$, that is not further propagated. Hence, in this case UP in the *representative-sparse encoding* is stronger than in the *sparse encoding*.   □

In general, we may tailor the representative-sparse encodings for specific problems. For example, when the elimination of all odd or even values from a domain are frequent, we could set $m = n/2$, with only two variables at the second level, one for the odd and the other for the even values.

If the *representative-sparse encoding* cannot be compared with the *sparse encoding*, it may be so with the *log encoding*.

Proposition 3. *Unit propagation with the representative-sparse encoding (with $m = 2$) is more effective than with the log encoding.*

Proof: Similar to the proof of Theorem 15 in [32], after noting that the *log encoding* and the *representative-sparse encoding* share the same one literal when representing a CSP value domain.   □

Finally, it is worth noting that one of the key strengths of the *representative-sparse encoding* is the ability to represent interval variables significantly better than the *sparse encoding* (in terms of the length clauses), namely when the interval does not cross the partitions. For example, to represent $V \geq 11$, the *sparse encoding* requires the "long "clause $d_{11}^V \vee d_{12}^V \vee d_{13}^V \vee d_{14}^V \vee d_{15}^V$, whereas the *representative-sparse encoding* simply requires g_3^V to be set to *true*.

3.2   The Representative-Order Encoding

The *representative-order encoding* is a new hierarchical hybrid encoding, where level one is composed of Boolean *representative* variables $g_1, g_2, ..., g_m$ $(1 \leq m \leq \frac{n}{2})$ dividing the domain into m partitions represented at level two with *order* encoded variables. The variables of level one require ALO and AMO constraints, while the variables of level two require the condition of the order encoding. An assignment in this encoding is as follows:

$$v = v_i \Leftrightarrow \begin{cases} g_{\lfloor i/(r+1) \rfloor} \wedge x_r & \text{if } i \bmod (r+1) = 0; \\ g_{\lfloor i/(r+1) \rfloor+1} \wedge x_1 & \text{if } i \bmod (r+1) = 1; \\ g_{\lfloor i/(r+1) \rfloor+1} \wedge x_{i \bmod (r+1)-1} \wedge x_{i \bmod (r+1)} & \text{otherwise.} \end{cases}$$

(2)

Formula 2 for translating a finite CSP domain $v = \{v_1, v_2, ..., v_n\}$ into SAT clauses by using Boolean representative variables, $\{g_1, ..., g_m\}$, and a set of Boolean order variables, $\{x_1, ..., x_r\}$, where $r = \lceil n/m \rceil - 1$.

Note that the order encoding requires $n - 1$ Boolean variables to encode a CSP variable V with domain $\{v_1, ..., v_n\}$, whereas the representative-order encoding requires only $m + \lceil n/m \rceil - 1$ Boolean variables, $1 \leq m \leq \frac{n}{2}$. Fig 2 shows how the representative variables, $g_i, 1 \leq i \leq m$, assigns a value V to the subdomains. As with the *representative-sparse encoding* when $n < m * r$ the prohibited values must be excluded. This assignment is correct as stated in the following proposition.

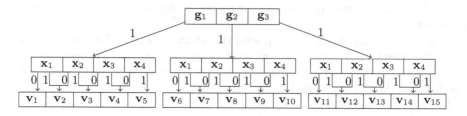

Fig. 2. An illustration of the representative-sparse encoding of the domain of a CSP variable using a group of *representative* variables at level one, $\{g_1, g_2, g_3\}$, where exactly one variable is selected, and a set of Boolean *order variables* at level two, $\{x_1, x_2, x_3, x_4\}$, where these variables are set in the constraint of the order encoding

Proposition 4. *When indexing the domain values of the CSP variables into SAT variables, the representative-order encoding is sound and complete.*

Proof: The representative variables at level one, $g_i, 1 \leq i \leq m \leq \frac{n}{2}$, divide the domain of variable V into m subdomains $\{v_1, \ldots, v_{\lceil n/m \rceil}\}, \ldots, \{v_{(m-1)\lceil n/m \rceil + 1}, \ldots, v_n\}$.

Given the *at-least-one* and *at-most-one* clauses for the representative Boolean variables $\{g_1, \ldots, g_m\}$ exactly one partition can be selected. Moreover, the constraints on the Boolean variables $\{x_1, \ldots, x_r\}$ imposed by the *order encoding*, lead to the selection for V of exactly one value in the selected partition. $\square$

The comparison of the strength of unit propagation with the *order* and *representative-order* encodings is not straightforward. In the case of convex domains (intervals) the *order* encoding is usually simpler, as it sets one single positive and one single negative literal for the interval limits, whereas the *representative-order* encoding requires the setting of the representative literals as well. In the above example of CSP variables V having domain $\{1, .., 15\}$, an interval $V \in \{3..10\}$ is imposed by setting $o_2^V \wedge \neg o_{10}^V$ in the *order* encoding whereas it requires setting $(g_1^V \wedge x_2^V) \vee g_2^V$ in the *representative-order* encoding.

On the other hand, for non-convex domains, the *representative-order* encoding may be more compact. For example domain $V \in \{[2..4]; [7..9]; [12..14]\}$ is represented as $x_1^V \wedge \neg x_4^V$ the *representative-order* encoding whereas the *order* encoding requires the setting of $(o_1^V \wedge \neg o_4^V) \vee (o_6^V \wedge \neg o_9^V) \vee (o_{10}^V \wedge \neg o_{14}^V)$, i.e., 8 ternary clauses. Admittedly, this is an extreme example. Nevertheless, if the *order* encoding is usually more compact, many cases exist where the *representative-order* is superior, especially with non-convex domains.

4 Experiments

All experiments reported in this section were performed on a 2.66 Ghz, Intel Core 2 Quad processor with 3.8 GB of memory, under Ubuntu 10.04. Runtimes reported in CPU-time are in seconds. The used solvers are *riss3G* [17] (SAT competition 2013 version), *lingeling* [5] (SAT competition 2013 version) and *clasp* [10]

($clasp2.1.3x86\_64linux$) with default configurations, conflict-driven clause learning SAT solvers, which were ranked first on application and craft benchmarks in different categories at recent SAT competitions. [2] We use the *sequential* counter encoding [26] for the at-most-one constraint in the *representative-sparse* encodings. The columns $S_1, S_2, S_{\sqrt{n}}$, and $S_{n/2}$ ($O_1, O_2, O_{\sqrt{n}}$, and $O_{n/2}$) refer to the *representative-sparse* encodings (the *representative-order* encodings) with corresponding partitions (1, 2, $\sqrt{n}$ and $n/2$).

4.1 The Pigeon-Hole Problem

The goal of the problem is to prove that p pigeons can not be fit in $h = p-1$ holes. This is a trivially unsatisfiable problem, composed exclusively of disequality constraints on the CSP variables, which SAT solvers have difficulty to handle (there is no all different global constraint in SAT). This is confirmed in Table 1, for different size of the partitions (2, $\sqrt{n}$ and $n/2$), which are very similar in the case of domains of this sizes of ($n \in 10 \ldots 15$). Nevertheless, the speedups show a large variation, depending on the solver, but the most relevant finding is that, despite their different performance, all solvers exhibit a speed-up of one to two orders of magnitude, when the problem is encoded with the representative encodings. The main reason for this speedups is the number of SAT variables required to encode the problems, that is much smaller with the representative encodings, as shown in Table 2, and compensate the fact that the clauses to encode disequality constraints are longer (but slightly less) in the representative encodings. Column *Inst* refers to p, the number of pigeons.

Table 1. The running time comparison of encodings performed by *riss3G*, *lingeling* and *clasp* on unsatisfiable Pigeon-Hole instances. Runtimes reported are in seconds.

Solvers	Inst	S_1	S_2	$S_{\sqrt{n}}$	$S_{n/2}$	O_1	O_2	$O_{\sqrt{n}}$	$O_{n/2}$
riss3G	11	64.23	1.59	0.73	0.82	0.28	0.27	0.75	0.93
	12	6773.73	13.26	2.82	2.97	0.61	1.08	2.34	1.04
	13	> 7200	22.14	43.74	263.73	4.23	1.71	9.51	27.02
	14	> 7200	240.24	944.55	773.06	22.69	6.28	31.37	36.72
	15	> 7200	3865.60	2085.45	> 7200	680.79	106.76	331.02	1820.30
lingeling	11	8.10	1.22	1.47	1.50	0.53	0.68	1.29	1.43
	12	863.76	7.69	5.43	2.61	1.45	3.91	4.38	6.80
	13	> 7200	33.96	12.40	37.55	10.02	15.68	5.57	27.38
	14	> 7200	514.35	101.09	273.52	24.31	36.15	79.14	53.97
	15	> 7200	4540.23	738.28	4293.51	136.83	137.08	98.79	2600.62
clasp	11	10.94	0.52	0.91	0.53	1.43	0.62	1.867	0.53
	12	110.11	2.84	4.92	3.56	8.26	2.73	14.15	4.37
	13	1127.97	12.93	14.58	12.17	45.45	9.95	11.26	12.22
	14	> 7200	75.10	94.71	77.50	602.51	32.34	100.32	73.67
	15	> 7200	386.52	333.03	386.26	4264.07	153.42	305.97	368.51

Finally, it is interesting to note that the *representative-order* encodings perform better than the *representative-sparse* encodings, even if difference constraints do not require handling of bounds of the domains, and this is possibly

---

[2] http://www.satcompetition.org

Table 2. The number of variables and clauses on unsatisfiable Pigeon-Hole instances

$Inst$	S_1	S_2	$S_{\sqrt{n}}$	$S_{n/2}$	O_1	O_2	$O_{\sqrt{n}}$	$O_{n/2}$
11	220| 869	121| 704	110| 715	88| 704	110| 638	66| 605	66| 638	66| 671
12	264|1086	156| 942	120| 894	120| 978	132| 834	84| 810	72| 810	84| 918
13	312|1391	169|1157	130|1105	117|1183	156|1066	91|1014	78|1014	91|1144
14	364|1715	210|1477	168|1435	154|1561	182|1337	112|1295	98|1309	112|1491
15	420|2085	225|1770	180|1725	150|1845	210|1650	120|1575	105|1590	120|1800

caused by the use of the *sequential counter* to code the AMO constraints in the *representative-sparse* encodings.

4.2　The Golomb Ruler Problem

A Golomb ruler (m,g) can be defined as a sequence $0 \le a_1 < a_2 < \cdots < a_m$ of m integers such that the $m(m-1)/2$ differences $a_j - a_i$ ($1 \le i < j \le m$) are all distinct. We executed the satisfiable version of the problem to check whether there is a ruler with $a_m \le g$.

Table 3. The running time comparison of encodings performed by *riss3G*, *lingeling* and *clasp* for finding one solutions on Golomb Ruler instances. Runtimes reported are in seconds.

Solvers	Inst	S_1	S_2	$S_{\sqrt{n}}$	$S_{n/2}$	O_1	O_2	$O_{\sqrt{n}}$	$O_{n/2}$
riss3G	G(9,44)	0.49	0.76	0.82	0.52	0.13	0.80	1.56	0.20
	G(10,55)	2.84	1.21	3.46	2.82	1.23	1.46	3.42	5.90
	G(11,72)	1.16	9.72	29.7	48.18	33.68	54.24	9.02	0.69
	G(12,85)	185.26	719.41	578.62	3121.12	235.89	35.06	705.06	5.85
lingeling	G(9,44)	0.51	1.22	0.94	4.62	0.34	2.16	1.30	2.09
	G(10,55)	9.27	2.26	13.33	5.53	1.09	6.78	13.42	16.72
	G(11,72)	112.13	35.40	20.47	76.47	125.24	51.07	73.83	152.17
	G(12,85)	566.23	753.03	458.22	84.60	685.95	356.89	2272.25	5200.11
clasp	G(9,44)	0.91	2.24	3.15	3.52	0.26	4.59	4.49	1.26
	G(10,55)	2.48	7.38	21.85	30.73	2.51	15.78	28.73	25.40
	G(11,72)	6.76	69.63	128.88	654.78	101.67	473.63	1155.79	854.12
	G(12,85)	409.53	1229.56	6992.35	> 7200	1217.28	5720.18	4703.78	4593.86

The problem includes disequality and inequality constraints on the CSP variables. Despite these latter constraints, the *representative-sparse* encodings seem to perform better than the *representative-order* encodings. The domains are now larger (from 44 to 85) and the execution times obtained in this problem suggest that in this problem the speed-ups obtained with the representative encodings are not very significant, with the notable exception of the *riss3G* solver.

Table 4. The number of variables and clauses of different encodings on Golomb Ruler instances

Inst	S_1		S_2		$S_{\sqrt{n}}$		$S_{n/2}$	
G(9,44)	3978	105694	2043	102733	945	101608	1143	110563
G(10,55)	6070	203175	3135	198735	1210	196535	1760	215235
G(11,72)	9526	411767	4840	404650	1672	401152	2596	439806
G(12,85)	13284	686628	6786	676830	2184	672150	3666	737670
Inst	O_1		O_2		$O_{\sqrt{n}}$		$O_{n/2}$	
G(9,44)	1898	135718	1044	133181	585	127610	1044	110410
G(10,55)	3035	168182	1595	260056	770	247681	1595	238041
G(11,72)	4763	350770	2453	536448	1067	512424	2453	439582
G(12,85)	6624	593115	3432	900206	1404	863516	3432	817664

Table 4 shows the number of variables and clauses required by different encodings. Clearly, the variability of the speedups suggest that the combined effect of less variables and longer clauses must be further investigated.

Table 5. The running time comparison of different encodings performed by $riss3G$ on Graph-Colouring instances. S/U indicates the satisfiability or unsatisfiability of instances. K is the number of colours used.

Inst	K	C	S_1	S_2	$S_{\sqrt{n}}$	$S_{n/2}$	O_1	O_2	$O_{\sqrt{n}}$	$O_{n/2}$
jean	9	U	11.03	0.48	0.19	0.18	0.13	0.18	0.16	0.19
	10	S	0.02	0.03	0.02	0.02	0.01	0.02	0.03	0.03
anna	10	U	204.87	2.02	0.76	2.62	0.53	0.85	0.49	2.51
	11	S	0.09	0.06	0.05	0.06	0.03	0.05	0.06	0.05
david	10	U	156.84	2.50	0.81	2.11	0.45	1.02	0.70	2.17
	11	S	0.06	0.05	0.04	0.04	0.02	0.04	0.06	0.04
DSJC125.9	10	U	940.19	6.10	6.03	5.11	1.92	5.23	5.36	5.90
	11	U	>7200	13.86	10.18	5.23	5.82	7.67	8.55	6.13
huck	10	U	213.12	2.39	1.09	2.90	0.42	0.71	1.07	2.22
	11	S	0.04	0.03	0.03	0.02	0.02	0.04	0.04	0.03
miles500	10	U	606.15	2.92	1.36	2.96	1.09	0.97	1.13	3.03
	11	U	5872.35	8.93	2.87	2.15	12.32	2.78	3.93	3.29
miles750	10	U	580.88	2.38	2.18	2.61	0.84	1.13	0.93	2.74
	11	U	>7200	15.50	10.92	2.93	4.51	5.60	5.93	2.98
miles1000	10	U	531.56	2.98	2.12	3.58	0.90	1.82	3.87	2.71
	11	U	>7200	8.54	2.31	2.45	2.59	3.60	4.00	7.75
miles1500	10	U	959.88	4.61	3.66	4.43	1.66	3.62	4.52	4.96
	11	U	>7200	11.75	6.58	4.12	3.49	6.67	7.21	4.65
queen12_12	10	U	270.75	2.51	1.71	1.96	1.02	1.64	1.34	3.30
	11	U	4926.12	7.97	4.12	4.49	2.93	4.44	11.73	4.72
queen13_13	10	U	258.80	3.35	2.00	2.86	0.93	1.82	1.90	2.90
	11	U	2359.51	11.20	7.10	7.54	7.12	5.61	7.05	6.69
queen14_14	10	U	276.38	4.38	1.66	2.72	1.06	2.22	2.32	2.61
	11	U	3057.18	19.35	2.46	2.76	3.64	5.16	4.34	18.61
queen15_15	10	U	472.29	3.36	3.24	3.14	1.22	2.47	2.59	3.95
	11	U	6478.20	23.94	7.95	2.96	6.74	6.04	9.75	6.55
queen16_16	10	U	170.77	3.74	2.89	3.40	1.68	3.02	3.44	4.12
	11	U	2829.32	16.24	9.52	14.33	10.25	6.56	4.32	9.35

4.3 The Graph Colouring Problem

The well known Graph Colouring problem consists of finding an assignment of k colours to the vertices of an undirected graph, such that no two adjacent vertices share the same colour. We experimented on several hard unsatisfiable instances from [30], and report the results obtained in Table 5 and Table 6 (results obtained with *clasp* were similar to those with *riss3G* and were omitted for lack of space).

Like with the pigeon hole problem, the *representative-order* encodings perform better than the *representative-sparse* encodings, and in the former the representative encodings have similar runtimes, with some marginal speedup. The effect of the representative encoding much stronger in the *sparse* case, where in both solvers, speed-ups of one or two orders of magnitude are achieved with the *representative-sparse* encodings, typically favoring a $\sqrt{n}$ number of partitions.

Table 6. The running time comparison of different encodings performed by *lingeling* on Graph-Colouring instances. S/U indicates the satisficability or unsatisfiability of instances. K is the number of colors used. Runtimes reported are in seconds.

Inst	K	S/U	S_1	S_2	$S_{\sqrt{n}}$	$S_{n/2}$	O_1	O_2	$O_{\sqrt{n}}$	$O_{n/2}$
jean	9	U	0.02	0.02	0.02	0.02	0.74	0.68	0.65	0.80
	10	S	20.86	0.81	1.53	0.84	0.01	0.01	0.02	0.02
anna	10	U	142.53	1.74	2.28	1.70	1.06	1.52	1.44	2.01
	11	S	0.04	0.07	0.07	0.07	0.04	0.05	0.04	0.05
david	10	U	130.68	2.24	2.03	3.17	1.03	1.52	2.02	2.03
	11	S	0.02	0.05	0.05	0.06	0.03	0.04	0.03	0.04
huck	10	U	119.64	1.77	1.78	1.86	0.98	1.36	1.90	1.44
	11	S	0.02	0.04	0.04	0.04	0.02	0.02	0.02	0.02
DSJC125.9	10	U	0.04	2.85	2.32	2.96	1.92	2.04	1.29	2.41
	11	U	0.04	13.17	10.49	6.96	13.65	4.67	4.70	5.73
miles500	10	U	1532.22	1.78	2.34	2.26	3.91	1.90	1.60	2.79
	11	U	>7200.0	25.77	9.62	8.36	5.48	3.71	5.37	3.68
miles750	10	U	2041.76	2.39	2.72	2.08	1.90	2.04	1.05	2.80
	11	U	6685.31	36.54	7.96	7.80	6.94	5.14	2.94	6.50
miles1000	10	U	2.76	2.49	2.63	2.07	2.32	2.23	1.17	2.11
	11	U	2331.27	47.70	10.36	6.35	8.08	4.96	3.56	6.56
miles1500	10	U	0.56	3.85	2.62	3.34	3.96	3.04	1.63	3.15
	11	U	4.18	43.68	8.17	7.62	15.23	9.22	3.90	6.48
queen12_12	10	U	721.29	2.70	3.04	2.28	1.22	2.08	1.06	2.09
	11	U	2185.73	20.20	5.42	6.62	2.68	5.60	3.01	3.89
queen13_13	10	U	488.37	2.35	1.28	2.08	1.73	2.07	1.15	2.64
	11	U	>7200.0	35.73	6.60	6.47	3.74	4.69	5.26	4.64
queen14_14	10	U	715.68	2.80	1.38	2.81	2.96	2.43	1.36	1.98
	11	U	>7200.0	29.44	6.50	16.59	6.79	5.56	6.16	5.74
queen15_15	10	U	1760.25	2.96	1.60	2.42	4.52	2.53	1.46	2.91
	11	U	>7200.0	27.35	7.59	5.76	15.30	5.10	3.75	4.93
queen16_16	10	U	1679.02	3.25	1.83	3.13	4.33	2.94	1.63	2.20
	11	U	>7200.0	34.58	5.68	6.49	11.94	6.12	3.97	5.10

4.4   The Open Shop Scheduling Problem

Given a set of jobs, each consisting of a set of tasks that must be processed once in any order on a set of machines, the goal of this combinatorial optimization problem is to find a schedule of all tasks so as to complete all jobs within a given makespan. All CSP constraints of this problem are inequalities. The benchmarks used here were taken from [27]. Column *Instance* identifies the instances of the problem, where a_b refers to the bth instance of the benchmark with a jobs and a machines. Again, results obtained with *clasp* were similar to those with *riss3G* and were omitted for lack of space.

Table 7. The running time comparison of encodings performed by *riss3G* on Open Shop Scheduling instances. M is the makespan used. S/U indicates the satisfiability or unsatisfiability of instances. Runtimes reported are in seconds.

$Inst$	M	C	S_1	S_2	$S_{\sqrt{n}}$	$S_{n/2}$	O_1	O_2	$O_{\sqrt{n}}$	$O_{n/2}$
4_1	192	U	2.47	0.38	0.13	1.01	0.04	0.06	0.13	0.82
	193	S	1.96	0.34	0.12	0.93	0.03	0.03	0.13	0.82
4_2	235	U	7.23	0.90	0.31	2.36	0.12	0.11	0.28	2.88
	236	S	5.45	0.66	0.29	2.61	0.03	0.05	0.26	1.40
4_3	270	U	7.92	1.03	0.37	3.97	0.10	0.08	0.25	3.48
	271	S	6.56	0.91	0.23	2.51	0.06	0.04	0.16	2.51
4_4	249	U	6.08	1.06	0.33	2.73	0.09	0.12	0.30	3.43
	250	S	8.25	0.95	0.28	2.38	0.10	0.12	0.32	3.16
4_5	294	U	11.63	1.80	0.42	4.85	0.14	0.12	0.46	4.52
	295	S	10.32	1.32	0.30	4.52	0.09	0.07	0.25	3.44
5_1	299	U	34.26	5.34	1.81	14.98	0.56	0.64	1.34	14.07
	300	S	28.44	3.96	1.20	16.76	0.26	0.36	1.16	17.72
5_2	261	U	22.97	3.52	1.56	13.38	0.41	0.80	1.32	11.89
	262	S	16.83	2.61	0.91	10.52	0.38	0.30	0.60	9 .13
5_3	322	U	57.62	8.71	2.08	27.43	0.74	1.02	1.71	27.35
	323	S	59.26	8.36	2.45	33.86	0.63	0.46	2.02	37.72
5_4	309	U	49.51	9.57	2.81	31.54	0.81	0.81	2.06	27.39
	310	S	55.83	7.74	2.59	30.57	0.64	0.78	1.53	28.57
5_5	325	U	89.77	14.22	4.74	48.54	1.32	1.62	3.27	45.41
	326	S	73.65	5.45	3.07	32.27	0.97	1.20	2.72	29.58
7_1	434	U	734 .63	76.14	53.58	438.75	1.52	2.03	19.52	406.17
	435	S	3091.31	428.58	75.24	2282.58	3.84	9.64	47.15	2454.82
7_2	442	U	613.37	28.63	9.23	124.79	1.41	2.24	14.24	471.24
	443	S	2763.21	96.94	63.52	2617.25	7.83	14.76	39.42	1990.25
7_3	467	U	234.10	18.38	14.62	251.62	1.69	6.07	37.15	229.25
	468	S	>7200	1666.27	267.08	3986.71	41.64	30.81	187.10	4756.72
7_4	462	U	434.92	23.73	13.41	195.62	1.38	6.05	27.93	339.06
	463	S	5729.13	567.61	92.15	3914.25	6.20	22.28	75.75	3905.14
7_5	415	U	272.00	20.22	30.14	226.33	1.24	7.12	7.72	477.94
	416	S	1938.91	489.42	84.89	1909.78	7.26	7.59	58.20	1057.52

As expected, in this problem where CSP inequality constraints dominate, the *order* encoding is much faster than the *sparse* encoding. The former is particularly adequate to propagate changes in the bounds, and no speedups were expected with representative encodings (and in fact a slow-down is observed).

What is more interesting is that the *representative-sparse* encodings perform very similarly to the *representative-order* encodings, in both solvers, possibly due to the fact that the partition of the large size domains (size 192 to 416) efficiently propagates the changes in the bounds, for which the sparse encoding is not suited. Overall, the representative encodings perform worse than the *order* encoding but significantly better than the *sparse* encoding, specially when the number of SAT variables in levels 1 and 2 are balanced ($S_{\sqrt{n}}$).

Table 8. The running time comparison of encodings performed by *lingeling* on Open Shop Scheduling instances. M is the makespan used. S/U indicates the satisfiability or unsatisfiability of instances. Runtimes reported are in seconds.

Inst	M	C	S_1	S_2	$S_{\sqrt{n}}$	$S_{n/2}$	O_1	O_2	$O_{\sqrt{n}}$	$O_{n/2}$
4_1	192	U	1.81	0.92	0.31	3.72	0.12	0.17	0.66	2.15
	193	S	8.31	1.10	0.32	3.78	0.12	0.16	0.69	2.23
4_2	235	U	22.14	2.40	1.11	4.60	0.28	0.28	0.98	4.88
	236	S	18.16	2.13	1.03	4.67	0.10	0.21	0.94	4.16
4_3	270	U	43.43	2.81	1.26	8.82	0.30	0.28	0.50	7.37
	271	S	47.99	2.80	1.19	6.91	0.24	0.26	0.49	7.12
4_4	249	U	23.03	2.67	1.12	6.71	0.25	0.29	0.93	6.88
	250	S	27.11	2.36	1.15	5.26	0.21	0.22	0.91	5.18
4_5	294	U	41.87	3.39	1.42	9.54	0.34	0.33	0.68	11.04
	295	S	42.16	2.77	1.29	7.03	0.28	0.31	0.51	7.11
5_1	299	U	192.11	47.74	3.87	37.16	1.07	1.23	3.46	34.22
	300	S	185.54	47.71	4.14	32.39	0.40	1.05	2.33	37.01
5_2	261	U	132.14	30.11	3.54	27.93	1.02	1.17	3.22	25.20
	262	S	106.02	2.53	2.67	19.09	0.61	0.70	2.79	20.82
5_3	322	U	242.74	66.64	4.41	40.80	1.29	2.46	3.52	48.04
	323	S	230.84	58.65	6.24	38.61	1.62	1.63	3.52	46.48
5_4	309	U	271.32	51.13	6.48	38.97	1.59	1.89	5.35	51.42
	310	S	114.86	47.44	5.92	36.84	0.94	1.24	5.06	51.57
5_5	325	U	351.43	76.02	8.64	53.36	2.91	0.30	4.56	62.49
	326	S	249.76	72.24	6.43	40.63	2.32	0.31	5.74	45.34
7_1	434	U	6883.68	602.19	55.61	661.07	1.95	1.62	51.16	628.88
	435	S	>7200	644.73	73.62	685.32	9.54	11.40	75.56	522.86
7_2	442	U	3675.26	389.77	50.41	358.64	4.03	2.14	46.13	468.23
	443	S	3454.09	414.08	76.16	594.33	8.86	9.70	50.07	822.75
7_3	467	U	6679.69	804.14	82.39	404.46	2.26	16.43	88.32	2108.54
	468	S	>7200	4026.52	162.19	2074.28	34.81	33.72	91.38	2236.85
7_4	462	U	>7200	1853.84	58.75	944.45	2.09	1.86	53.96	910.24
	463	S	6767.76	1668.02	62.32	913.24	23.04	20.57	73.70	520.93
7_5	415	U	4049.08	712.38	47.01	405.28	2.20	3.46	40.76	306.70
	416	S	2279.14	807.89	48.15	404.55	6.98	9.21	79.28	735.05

5   Conclusions and Future Works

This paper introduces two specific hierarchical hybrid encodings, the *representative-sparse* and *representative-order* encodings, for modeling CSPs as propositional SAT problems, which aim at taking advantage of the sparse and order encodings, but requiring significant smaller number of SAT variables. The new encodings, that use only two levels in the hierarchy (all experiments we did with three or more levels were clearly less efficient) can be parameterised by different sizes of the level one of the hierarchy, and in general these hierarchical encodings are incomparable with respect to their flat counterparts (the *sparse* and *-order* encodings which are special cases with one single partition). The experimental results in a set of representative benchmarks show that, regardless of the variability of run times in different SAT solvers, the representative encodings are quite competitive and usually outperform (sometimes very significantly) the *sparse* and *order* encodings, with the exception of CSP problems where inequality constraints dominate, in which case the *order* encoding is still the best option. More work remains to be done to further assess the merit of these new encodings and, in particular, we intend to further investigate how to tune the number of partitions for different CSP problems and SAT solvers. One appealing work into apply some redundancy for the second level of the representative encodings, as outlined in [3].

Acknowledgment. We thank Christoph Wernhard, Norbert Manthey, Peter Steinke, and Tobias Philipp for useful suggestions. We are grateful to Miroslav N. Velev for the previous works.

References

1. Ansótegui, C., del Val, A., Dotú, I., Fernández, C., Manyà, F.: Modeling Choices in Quasigroup Completion: SAT vs. CSP. In: McGuinness, D.L., Ferguson, G. (eds.) Proceedings of the 19th National Conference on Artificial Intelligence, Sixteenth Conference on Innovative Applications of Artificial Intelligence, pp. 137–142. AAAI Press / The MIT Press (2004)
2. Bailleux, O., Boufkhad, Y.: Efficient CNF Encoding of Boolean Cardinality Constraints. In: Rossi, F. (ed.) CP 2003. LNCS, vol. 2833, pp. 108–122. Springer, Heidelberg (2003)
3. Barahona, P., Hölldobler, S., Nguyen, V.H.: Efficient SAT-Encoding of Linear CSP Constraints. In: 13rd International Symposium on Artificial Intelligence and Mathematics - ISAIM, Fort Lauderdale, Florida, USA, January 6-8 (2014)
4. Berre, D.L., Lynce, I.: CSP2SAT4J: A Simple CSP to SAT Translator. In: van Dongen, M., Lecotre, C., Rossel, O. (eds.) Proceedings of the Second International CSP Solver Competition, pp. 43–54 (2006)
5. Biere, A.: Lingeling, Plingeling and Treengeling Entering the SAT Competition 2013. In: Adrian Balint, A.B., Heule, M., Järvisalo, M. (eds.) Proceedings of SAT Competition 2013, pp. 51–52 (2013)
6. Cadoli, M., Schaerf, A.: Compiling Problem Specifications into SAT. Artificial Intelligence 162(1-2), 89–120 (2005)

7. Crawford, J.M., Baker, A.B.: Experimental Results on the Application of Satisfiability Algorithms to Scheduling Problems. In: Hayes-Roth, B., Korf, R.E. (eds.) Proceedings of the 12th National Conference on Artificial Intelligence, vol. 2, pp. 1092–1097. AAAI Press / The MIT Press (1994)
8. Frisch, A.M., Giannoros, P.A.: SAT Encodings of the At-Most-k Constraint. Some Old, Some New, Some Fast, Some Slow. In: Proc. of the Tenth Int. Workshop of Constraint Modelling and Reformulation (2010)
9. Gavanelli, M.: The Log-Support Encoding of CSP into SAT. In: Bessière, C. (ed.) CP 2007. LNCS, vol. 4741, pp. 815–822. Springer, Heidelberg (2007)
10. Gebser, M., Kaufmann, B., Schaub, T.: The Conflict-Driven Answer Set Solver clasp: Progress Report. In: Erdem, E., Lin, F., Schaub, T. (eds.) LPNMR 2009. LNCS, vol. 5753, pp. 509–514. Springer, Heidelberg (2009)
11. Gent, I.P.: Arc Consistency in SAT. In: van Harmelen, F. (ed.) Proceedings of the 15th European Conference on Artificial Intelligence, ECAI 2002, pp. 121–125. IOS Press (2002)
12. Hölldobler, S., Manthey, N., Nguyen, V., Steinke, P.: Solving Hidokus Using SAT Solvers. In: Proc. INFOCOM-5, pp. 208–212 (2012) ISSN 2219-293X
13. Hölldobler, S., Nguyen, V.H.: On SAT-Encodings of the At-Most-One Constraint. In: Katsirelos, G., Quimper, C.G. (eds.) Proc. Twelfth International Workshop on Constraint Modelling and Reformulation, Uppsala, Sweden, September 16-20, pp. 1–17 (2013)
14. Jeavons, P., Petke, J.: Local Consistency and SAT-Solvers. J. Artif. Intell. Res (JAIR) 43, 329–351 (2012)
15. Kautz, H., Selman, B.: The State of SAT. Discrete Appl. Math. 155, 1514–1524 (2007)
16. de Kleer, J.: A Comparison of ATMS and CSP Techniques. In: IJCAI, pp. 290–296 (1989)
17. Manthey, N.: The SAT Solver RISS3G at SC 2013. Department of Computer Science Series of Publications B, vol. B-2013-1, pp. 72–73. University of Helsinki, Helsinki, Finland (2013)
18. Metodi, A., Codish, M.: Compiling Finite Domain Constraints to SAT with BEE. Theory and Practice of Logic Programming 12(4-5), 465–483 (2012)
19. Nguyen, V.H., Velev, M.N., Barahona, P.: Application of Hierarchical Hybrid Encodings to Efficient Translation of CSPs to SAT. In: Brodsky, A. (ed.) Proc. 2013 IEEE 25th International Conference on Tools with Artificial Intelligence (ICTAI 2013), Special Track on SAT and CSP. pp. 1028–1035. Conference Publishing Services (2013)
20. Ohrimenko, O., Stuckey, P., Codish, M.: Propagation via Lazy Clause Generation. Constraints 14(3), 357–391 (2009)
21. Prestwich, S.D.: Full Dynamic Substitutability by SAT Encoding. In: Wallace, M. (ed.) CP 2004. LNCS, vol. 3258, pp. 512–526. Springer, Heidelberg (2004)
22. Prestwich, S.D.: Local Search on SAT-encoded Colouring Problems. In: Giunchiglia, E., Tacchella, A. (eds.) SAT 2003. LNCS, vol. 2919, pp. 105–119. Springer, Heidelberg (2004)
23. Prestwich, S.D.: CNF Encodings, ch. 2, pp. 75–98. IOS Press (2009)
24. Rossi, F., Beek, P.V., Walsh, T.: Handbook of Constraint Programming (Foundations of Artificial Intelligence). Elsevier Science Inc., New York (2006)
25. Marques-Silva, J., Lynce, I.: Towards Robust CNF Encodings of Cardinality Constraints. In: Bessière, C. (ed.) CP 2007. LNCS, vol. 4741, pp. 483–497. Springer, Heidelberg (2007)

26. Sinz, C.: Towards an Optimal CNF Encoding of Boolean Cardinality Constraints. In: van Beek, P. (ed.) CP 2005. LNCS, vol. 3709, pp. 827–831. Springer, Heidelberg (2005)
27. Taillard, É.: http://mistic.heig-vd.ch/taillard/problemes.dir/ordonnancement.dir/ordonnancement.html
28. Tamura, N., Taga, A., Kitagawa, S., Banbara, M.: Compiling Finite Linear CSP into SAT. Constraints 14(2), 254–272 (2009)
29. Tanjo, T., Tamura, N., Banbara, M.: A Compact and Efficient SAT-Encoding of Finite Domain CSP. In: Sakallah, K.A., Simon, L. (eds.) SAT 2011. LNCS, vol. 6695, pp. 375–376. Springer, Heidelberg (2011)
30. Trick, M.: DIMACS Graph-Coloring Problems, http://mat.gsia.cmu.edu/COLOR/instances.html
31. Velev, M.N.: Exploiting Hierarchy and Structure to Efficiently Solve Graph Coloring as SAT. In: International Conference on Computer-Aided Design (ICCAD 2007), San Jose, CA, USA, pp. 135–142. IEEE (2007)
32. Walsh, T.: SAT v CSP. In: Dechter, R. (ed.) CP 2000. LNCS, vol. 1894, pp. 441–456. Springer, Heidelberg (2000)

SAT and Hybrid Models of the Car Sequencing Problem

Christian Artigues[1,2], Emmanuel Hebrard[1,2], Valentin Mayer-Eichberger[3,4],
Mohamed Siala[1,5], and Toby Walsh[3,4]

[1] CNRS, LAAS, 7 avenue du colonel Roche, F-31400 Toulouse, France
[2] Univ de Toulouse, LAAS, F-31400 Toulouse, France
[3] NICTA
[4] University of New South Wales
[5] Univ de Toulouse, INSA, LAAS, F-31400 Toulouse, France
{artigues,hebrard,siala}@laas.fr,
{valentin.mayer-eichberger,toby.walsh}@nicta.com.au

Abstract. We compare both pure SAT and hybrid CP/SAT models for solving
car sequencing problems, and close 13 out of the 23 large open instances in
CSPLib. Three features of these models are crucial to improving the state of
the art in this domain. For quickly finding solutions, advanced CP heuristics are
important and good propagation (either by a specialized propagator or by a sophisticated SAT encoding that simulates one) is necessary. For proving infeasibility, clause learning in the SAT solver is critical. Our models contain a number
of novelties. In our hybrid models, for example, we develop a linear time mechanism for explaining failure and pruning the AtMostSeqCard constraint. In our
SAT models, we give powerful encodings for the same constraint. Our research
demonstrates the strength and complementarity of SAT and hybrid methods for
solving difficult sequencing problems.

1 Introduction

In the car sequencing problem [26], we sequence a set of vehicles along an assembly
line. Each class of cars requires a set of options. However, the working station handling
a given option can only deal with a fraction of the cars passing on the line. Each option
j is thus associated with a fractional number u_j/q_j standing for its capacity (at most u_j
cars with option j occur in any sub-sequence of length q_j). Several global constraints
have been proposed in the Constraint Programming (CP) literature to model this family
of sequence constraints. At present, CP models with the AtMostSeqCard constraint
[23] or its combination with the *Global Sequencing Constraint (Gsc)* [20] have provided the best performance. However, pure CP approaches suffer when we consider
proving unsatisfiability. The goal of this paper is to show that by exploiting Boolean-
Satisfiability (SAT), we can improve upon the state of the art in this domain. We are
able to close 13 out the 23 large open instances in CSPLib.

We propose several approaches combining ideas from SAT and CP for solving the
car sequencing problem. First, we capture CP propagation in SAT by a careful formulation of the problem into conjunctive normal form (CNF). We propose a family of pure
SAT encodings for this problem and relate them to existing encoding techniques. They
are based on an extension of Sinz's encoding for the Cardinality constraint [24]

H. Simonis (Ed.): CPAIOR 2014, LNCS 8451, pp. 268–283, 2014.
© Springer International Publishing Switzerland 2014

and have similarities to the decomposition of the GEN-SEQUENCE constraint given in [2]. Second, we introduce a linear time procedure for computing compact explanations for the ATMOSTSEQCARD constraint. This algorithm can be used in a hybrid CP/SAT approach such as SAT Modulo Theory or lazy clause generating solver, where non-clausal constraints need a propagator and an explanation algorithm. In principle, the hybrid approach has access to all the advances from the SAT world, whilst benefiting from constraint propagation and dedicated branching heuristics from CP. However, our experiments reveal that in practice, SAT solvers maintain an edge when proving unsatisfiability. Due to the most up to date data structures and tuning of parameters for literal activity and clause deletion, encoding into SAT significantly outperforms the hybrid approach on hard unsatisfiable instances.

We made three observations based on these experiments: First, CP heuristics are good at quickly finding solutions. Whilst generic activity based heuristics are surprisingly robust, dedicated CP can be much faster. Second, propagation, either through finite domain propagators, or through unit propagation via a "strong" encoding, is extremely important for finding solutions reliably on harder instances. Strong propagation makes it less likely to enter an unsatisfiable subproblem during search. In conjunction with this, restarting ensures that these unlikely cases do not matter. Third, clause learning is critical for proving unsatisfiability. In this respect, the approaches that we introduce (especially the SAT encodings) greatly improve the state of the art for the car sequencing problem. Moreover, counter-intuitively, it does not seem that constraint propagation of the ATMOSTSEQCARD constraint nor the "strength" of the SAT encoding, has a significant impact on the ability of the solver to prove unsatisfiability.

The remainder of this paper is organized as follows. In Section 2, we give some background on CP, SAT and their hybridization. In Section 3, we recall the state of the art CP models for this problem and show the connection with SAT. In Section 4, we show that how to build explanations for the ATMOSTSEQCARD constraint based on its propagator. Then, we present advanced SAT encodings for this constraint in Section 5. Finally, in Section 6, we compare experimentally the approaches we introduce against pure CP and pseudo Boolean models.

2 Background

Constraint Programming. A constraint network is defined by a triplet $P = (\mathcal{X}, \mathcal{D}, \mathcal{C})$ where $\mathcal{X}$ is a set of variables, $\mathcal{D}$ is a mapping of variables to finite sets of values and $\mathcal{C}$ is a set of constraints that specify allowed combinations of values for subsets of variables. We assume that $\mathcal{D}(x) \subset \mathbb{Z}$ for all $x \in \mathcal{X}$. We denote $[x \leftarrow v]$ the assignment of the value v to the variable x, that is the restriction of its domain $\mathcal{D}(x)$ to $\{v\}$, similarly, we denote $[x \nleftarrow v]$ the pruning of the value v from $\mathcal{D}(x)$. A *partial instantiation S* is a set of assignments and/or pruning such that no variable is assigned more than one value and no value is pruned and assigned for the same variable. Let $\perp$ be a *failure* or a domain wipe-out, by convention equal to the set of all possible assignments and prunings. On finite domains, we can consider a *closure* of partial instantiations with respect to initial domains. That is, if the assignment $[x \leftarrow v]$ belongs to S, we also assume that $[x \nleftarrow v]$ for all $v \in \mathcal{D}(x) \setminus v$ belong to S. Similarly, if all but one of the values are pruned, the remaining value is added as an assignment. This is similar to *expanded* solutions in [16]. However, we shall restrict ourselves to Boolean domains in this paper. We therefore have $S \subseteq S'$ iff S' is a stronger (tighter) partial instantiation than S..Given an initial domain $\mathcal{D}$ and a partial instantiation S, we can derive a current domain taking into account the pruning and assignments of S in $\mathcal{D}$. There will not be ambiguities about the original domains, therefore we simply denote $S(x)$ the domain $\mathcal{D}(x)$ updated by the assignment or pruning associated to x in S.

A constraint C defines a relation $Rel(C)$, that is, a set of instantiations, over the variables in $Scope(C)$. Let S be a partial instantiation of $Scope(C)$. The constraint C is said *generalized arc consistent* (GAC) with respect to S iff, for every x in $Scope(C)$ and every $v \in S(x)$, there exists an instantiation $T \in Rel(C)$ such that $[x \leftarrow v] \notin T$ (T is a support) and $T \subseteq S$ (T is valid). We say that a constraint is *dis-entailed* with respect to a partial instantiation S iff there is no T in $Rel(C)$ such that $S \subseteq T$.

Throughout the paper we shall associate a *propagator* with a constraint C. This is a function mapping partial instantiations to partial instantiations or to the failure $\bot$. Given a partial instantiation S, we denote $C(S)$ for the partial instantiation (or failure) obtained by applying the propagator associated to C on S, with $S \subseteq C(S)$. We say that S implies the assignment or pruning p with respect to C iff $p \notin S$ & $p \in C(S)$. Finally, the *level* of an assignment or a pruning p is the order of appearance of the assignment (respectively pruning) in the search tree, and we denote it $lvl(p)$. For a comprehensive introduction to CP solving in general and its techniques we refer to [21].

SAT-Solving. The Boolean Satisfiability problem (SAT) is a particular case of CSP where domains are Boolean and constraints are only clauses (disjunction of literals). A SAT solver computes a satisfying instantiation of a formula of propositional logic in conjunctive normal form (CNF) or proves that no such instantiation exists. The most widely used method to solve SAT problems is based on the DPLL algorithm ([8]), which is a depth first search with backtracking and unit propagation. Unit propagation (UP) prunes the assignment of the remaining literal in a clause when all other literals have become false. An important improvement to the DPLL algorithms is Conflict-Driven Clause Learning (CDCL). A CDCL solver records for each conflict an appropriate reason in form of a clause, and adds it to the clause database. This can potentially prune unseen parts of the search tree. Furthermore, SAT solvers are equipped with robust domain-independent branching and decision heuristics (for instance VSIDS [17]). For a comprehensive introduction to SAT solving in general and its techniques we refer to [4].

Modelling in CNF is a crucial step for the success of solving problems with SAT. A natural approach to find a good SAT model is to describe the problem with higher level constraints and then translate these constraints to CNF. In accordance with this methodology, the representation of integer domains and encodings of a variety of global constraints have been proposed and analyzed [2,13,28]. Similarly the notion of GAC has been adapted to SAT. UP is said to maintain GAC on the CNF encoding of a constraint if it forces all assignments to the variables representing values that must be set to avoid unsatisfiability of the constraint. The quality of an encoding into SAT is measured by both its size and its level of consistency by UP. Moreover, we must taken into account that SAT solvers cannot distinguish between the original variables and any auxiliary variables introduced to produce a more compact encoding. Thus, when aiming for a good CNF encoding, we must consider how such auxiliary variables might increase propagation.

2.1 Hybrid CP/SAT

The notion of nogood learning in constraint programming is not new, in fact it pre-dates [22] similar concepts in SAT. However, CDCL learns and uses nogoods in a particular way, and such methods have been reintroduced into CP. For instance Katsirelos's generalized nogoods [15] [16] enable nogood learning with arbitrary domains. Another complexity is that propagation is now not restricted to unit propagation. A given constraint can be associated with a specific propagator. However, to perform clause learning, it is necessary to *explain* either a failure or the pruning of a domain value. We say that a partial instantiation S is an *explanation* of the pruning $[x \nleftarrow v]$ with respect to a constraint C if it implies $[x \nleftarrow v]$ (that is, $[x \nleftarrow v] \in C(S) \setminus S$). Moreover, S is a valid explanation iff $lvl([x \nleftarrow v]) > max(\{lvl(p) \mid p \in S\})$.

In this paper we use a solver with an architecture similar to *Lazy Clause Generation* with backward explanations [12]. In addition to a propagator, an explanation algorithm is associated with each constraint. However, as opposed to explanation based constraint programming [6,7], the explanations are used exactly as in CDCL, i.e., literals are replaced by their explanation until the current nogood contains a Unique Implication Point of the current level's pruning. In this sense it is very close to the way some Pseudo-Boolean CDCL solvers, such as PBS [1], PBChaff [9] or SAT4JPseudo [3] integrate unit propagations on clauses, dedicated propagators and explanations (cutting planes) for linear equations. On Boolean domains, the hybrid SAT/CP approach we use works as follows:

Propagation: The propagation is performed by a standard CP engine, except that for each pruned value we record the constraint responsible for this pruning (a simple pointer to the constraint is stored). Both original and learned clauses are handled by a dedicated propagator simulating the behavior of a clause base (i.e., using watched literals).

Learning: When a failure is raised, the standard CDCL conflict analysis algorithm is used. The constraint C responsible for the failure is asked to provide an explanation for this failure. The literals of this explanation form the base nogood Ng. Subsequently, any assignment $[x \leftarrow v]$ such that $lvl([x \leftarrow v]) \geq lvl(d)$ where d is the last decision, is removed from Ng and replaced by its explanation by the constraint marked as responsible for it. This process continues until Ng has a Unique Implication Point.

Search: Since a CP library (Mistral[1]) was used to implement this approach, it is possible to use hand made CP heuristics as well as built-in strategies such as VSIDS. However, as in CDCL algorithms, branching decisions are not refuted in a "right" branch. Instead, we backjump to the second highest level of literals in the learned clauses, and unit propagation for this clause is triggered.

---

[1] https://github.com/ehebrard/Mistral-2.0

3   The *Car Sequencing* Problem

In the car sequencing problem, n vehicles have to be produced on an assembly line. There are c classes of vehicles and m types of options. Each class $k \in \{1, \ldots, c\}$ is associated with a demand D_k^{class}, that is, the number of occurrences of this class on the line, and a set of options $\mathcal{O}_k \subseteq \{1, \ldots, m\}$. Each option is handled by a working station able to process only a fraction of the vehicles passing on the line. The capacity of an option j is defined by two integers u_j and q_j, such that no subsequence of size q_j may contain more than u_j vehicles requiring option j. A solution of the problem is then a sequence of cars satisfying both demand and capacity constraints. For convenience, we shall also define, for each option j, the corresponding set of classes of vehicles requiring this option $\mathcal{C}_j = \{k \mid j \in \mathcal{O}_k\}$, and the option's demand $D_j = \sum_{k \in \mathcal{C}_j} D_k^{class}$.

3.1   CP Modelling

As in a standard CP Model, we use two sets of variables. The first set corresponds to n integers $\{x_1, \ldots, x_n\}$ taking values in $\{1, \ldots, c\}$ and standing for the class of vehicles in each slot of the assembly line. The second set of variables corresponds to nm Boolean variables $\{o_1^1, \ldots, o_n^m\}$, where o_i^j stands for whether the vehicle in the ith slot requires option j. For the constraints, we distinguish three sets:

1. *Demand constraints*: for each class $k \in \{1..c\}$, $|\{i \mid x_i = k\}| = D_k^{class}$. This constraint is usually enforced with a Global Cardinality Constraint (GCC) [19] [18].
2. *Capacity constraints*: for each option $j \in \{1..m\}$, we post the constraint ATMOSTSEQCARD$(u_j, q_j, D_j, [o_1^j..o_n^j])$ using the propagator introduced in [23].
3. *Channelling*: Finally, we channel integer and Boolean variables: $\forall j \in \{1, \ldots, m\}$, $\forall i \in \{1, \ldots, n\}$, $o_i^j = 1 \Leftrightarrow x_i \in \mathcal{C}_j$

3.2   Default Pseudo-boolean and SAT Models

The above CP Model can be easily translated into a pseudo Boolean model since the majority of the constraints are sum expressions. We use the same Boolean variables o_i^j standing for whether the vehicle in the ith slot requires option j. Moreover, the class variables are split into nc Boolean variables c_i^j standing for whether the ith vehicle is of class j. We have the same constraints as in the CP model, albeit expressed slightly differently. Moreover, we need to constrain variables standing for class assignment of the same slot to be mutually exclusive.

1. *Demand constraints*: $\forall j \in [1..c]$, $\sum_i c_i^j = D_j$
2. *Capacity constraints*: $\sum_{l=i}^{i+q_j-1} o_l^j \leq u_j, \forall i \in \{1, \ldots, n - q_j + 1\}$
3. *Channelling*:
   - $\forall i \in [1..n]$, $\forall l \in [1..c]$, we have:
     - $\forall j \in \mathcal{O}_l$, $\overline{c_i^l} \vee o_i^j$
     - $\forall j \notin \mathcal{O}_l$, $\overline{c_i^l} \vee \overline{o_i^j}$

- For better propagation, we add the following redundant clause:

$$\forall i \in [1..n], j \in [1..m], \overline{o_i^j} \vee \bigvee_{l \in C_j} c_i^l$$

4. *Domain constraints*: a vehicle belong to only one class: $\forall i \in [1..n], \sum_j c_i^j = 1$

A SAT Encoding for this problem could translate each sum constraint (in this case only CARDINALITY constraints) into a CNF formula. We will show in Section 5 how such a translation can be improved.

4 Explaining the ATMOSTSEQCARD Constraint

We present here an algorithm explaining the ATMOSTSEQCARD constraint. This algorithm is based on the propagator for this constraint, which we therefore now recall. Let $[x_1, x_2..x_n]$ be a sequence of Boolean variables, u, q and d be integer variables. The ATMOSTSEQCARD constraint is defined as follows:

Definition 1.

$$\text{ATMOSTSEQCARD}(u, q, d, [x_1, \ldots, x_n]) \Leftrightarrow \bigwedge_{i=0}^{n-q} (\sum_{l=1}^{q} x_{i+l} \leq u) \wedge (\sum_{i=1}^{n} x_i = d)$$

In [23], the authors proposed a $O(n)$ filtering algorithm achieving AC on this constraint. We outline the main idea of the propagator.

Let $\mathcal{X} = [x_1..x_n]$ be a sequence of variables, and S a partial instantiation over these variables. The procedure `leftmost` returns an instantiation $\overrightarrow{w}_S \supseteq S$ of maximum cardinality by greedily assigning the value 1 from left to right while respecting the ATMOST constraints. Let $\overrightarrow{w}_S^i$ denote the partial instantiation $\overrightarrow{w}_S$ at the beginning of iteration i, and let $\overrightarrow{w}_S^1 = S$. The value $\max_S(i)$ denotes the maximum minimum cardinality, with respect to the current domain $\overrightarrow{w}_S^i$, of the q subsequences involving x_i. It is computed alongside $\overrightarrow{w}_S$ and will be useful to explain the subsequent pruning/failure. It is formally defined as follows (where $\min(\overrightarrow{w}_S^i(x_k)) = 0$ if $k < 1$ or $k > n$):

$$\max_S(i) = \max_{j \in [1..q]} (\sum_{k=i-q+j}^{i+j-1} \min(\overrightarrow{w}_S^i(x_k)))$$

Definition 2. *The outcome of the procedure* `leftmost` *can be recursively defined using* $\max_S$*: at each step* i*,* `leftmost` *adds the assignment* $[x_i \leftarrow 1]$ *iff this assignment is consistent with* $\overrightarrow{w}_S^i$ *and* $\max_S(i) < u$*, it adds the assignment* $[x_i \leftarrow 0]$ *otherwise.*

Example 1. For instance, consider the execution of the procedure `leftmost` on the constraint ATMOSTSEQCARD$(2, 4, 6, [x_1, x_2, x_3, x_4, x_5, x_6, x_7, x_8, x_9, x_{10}])$. We suppose that we start from the partial instantiation $\{[x_2 \leftarrow 0], [x_6 \leftarrow 1], [x_8 \leftarrow 0]\}$. Initially, we have the following structures, for each i representing an iteration (and also the index of a variable):

	1	2	3	4	5	6	7	8	9	10
$\overrightarrow{w}_S^{1}(x_i)$	{0,1}	0	{0,1}	{0,1}	{0,1}	1	{0,1}	0	{0,1}	{0,1}
$\overrightarrow{w}_S^{2}(x_i)$	1	0	{0,1}	{0,1}	{0,1}	1	{0,1}	0	{0,1}	{0,1}
$\overrightarrow{w}_S^{3}(x_i)$	1	0	{0,1}	{0,1}	{0,1}	1	{0,1}	0	{0,1}	{0,1}
$\overrightarrow{w}_S^{4}(x_i)$	1	0	1	{0,1}	{0,1}	1	{0,1}	0	{0,1}	{0,1}
$\cdots$										
$\overrightarrow{w}_S^{11}(x_i)$	1	0	1	0	0	1	1	0	0	1
$\max_S(i)$	0	1	1	2	2	1	1	2	2	1

The partial solution $\overrightarrow{w}_S^{1}$ is equal to S. Then at each step i, leftmost adds the assignment $[x_i \leftarrow 1]$ or $[x_i \leftarrow 0]$ according to Definition 2. For instance, at the begining of step 4, the subsequences to consider are $[x_1, x_2, x_3, x_4]$, $[x_2, x_3, x_4, x_5]$, $[x_3, x_4, x_5, x_6]$ and $[x_4, x_5, x_6, x_7]$, of cardinality 2, 1, 2 and 1, respectively, with respect to the instantiation $\overrightarrow{w}_S^{4}(x_i)$. The value of $\max_S(4)$ is therefore 2.

To detect failure, we simply need to run this procedure and check that the final cardinality of $\overrightarrow{w}_S$ is greater than or equal to the demand d. We shall see that we can explain pruning by using essentially the same procedure.

In order to express declaratively the full propagator, we need the following further steps: The same procedure is applied on variables in reverse order $[x_n..x_1]$, yielding the instantiation $\overleftarrow{w}_S$. Observe that the returned instantiations $\overrightarrow{w}_S$ and $\overleftarrow{w}_S$ assign every variable in the sequence to either 0 or 1. We denote respectively $L_S(i)$ and $R_S(i)$ the sum of the values given by $\overrightarrow{w}_S$ (resp. $\overleftarrow{w}_S$) to the i first variables (resp. $n - i + 1$ last variables). That is:

$$L_S(i) = \sum_{k=1}^{i} \min(\overrightarrow{w}_S(x_k)) \quad , \quad R_S(i) = \sum_{k=i}^{n} \min(\overleftarrow{w}_S(x_k))$$

Now we can define the propagator associated to the constraint ATMOSTSEQCARD described in [23], and which is a conjunction of GAC on the ATMOST (i.e. $\sum_{l=1}^{q} x_{i+l} \leq u$) constraints on each subsequence, of CARDINALITY constraint $\sum_{i=1}^{n} x_i = d$, and of the following:

$$\text{ATMOSTSEQCARD}(S) = \begin{cases} S, & \text{if } L_S(n) > d \\ \bot, & \text{if } L_S(n) < d \\ S \ \cup \ \{[x_i \leftarrow 0] \mid S(x_i) = \{0,1\} \\ \qquad \& \ L_S(i) + R_S(i) \leq d\} \\ \quad \cup \ \{[x_i \leftarrow 1] \mid S(x_i) = \{0,1\} \\ \qquad \& \ L_S(i-1) + R_S(i+1) < d\} \ \text{otherwise} \end{cases} \tag{4.1}$$

If a failure/pruning is detected by the CARDINALITY or an ATMOST constraint, then it is easy to give an explanation. However, if a failure or a pruning is due to the propagator defined in equation 4.1, then we need to specify how to generate a relevant explanation. We start by giving an algorithm explaining a failure. We show after that how to use this algorithm to explain pruning.

4.1 Explaining Failure

Suppose that the propagator detects a failure at a given level l. The original instantiation S would be a possible naive explanation expressing this failure. We propose in the following a procedure generating more compact explanations.

In example 2, the instantiation $S = \{[x_1 \leftarrow 1], [x_3 \leftarrow 0], [x_6 \leftarrow 0]\}$ is subject to ATMOSTSEQCARD$(2, 5, 3, [x_1..x_6])$. S is unsatisfiable since $L_S(6) < d$. Consider now the sequence $S^* = \{[x_6 \leftarrow 0]\}$. The result of leftmost on S and S^* is identical. Therefore, S^* and S are both valid explanations for this failure, however S^* is shorter. The idea behind our algorithm for computing shorter explanations is to characterise which assignments will have no impact on the behavior of the propagator, and thus are not necessary in the explanation.

$$
\textit{Example 2.} \quad
\begin{array}{r|cccccc}
S & 1 & . & 0 & . & . & 0 \\
\vec{w}(S) & 1 & 1 & 0 & 0 & 0 & 0 \\
max(S) & 1 & 1 & 2 & 2 & 2 & 1 \\
L(S) & 1 & 2 & 2 & 2 & 2 & 2 \\
 & & & d = 3 \\
 & & & L(6) = 2 \\
 & & & \rightarrow \text{Failure}
\end{array}
\quad
\begin{array}{r|cccccc}
S^* & . & . & . & . & . & 0 \\
\vec{w}(S^*) & 1 & 1 & 0 & 0 & 0 & 0 \\
max(S^*) & 1 & 1 & 2 & 2 & 2 & 1 \\
L(S^*) & 1 & 2 & 2 & 2 & 2 & 2 \\
 & & & d^* = 3 \\
 & & & L^*(6) = 2 \\
 & & & \rightarrow \text{Failure}
\end{array}
$$

Let $I = [x_{k+1}..x_{k+q}]$ be a (sub)sequence of variables of size q and S be a partial instantiation. We denote $card(I, S)$ the minimum cardinality of I under the instantiation S, that is: $card(I, S) = \sum_{x_i \in I} \min(S(x_i))$.

Lemma 1. *If* $S^* = S \setminus (\{[x_i \leftarrow 0] \mid max_S(i) = u\} \cup \{[x_i \leftarrow 1] \mid max_S(i) \neq u\})$ *then* $\vec{w}_S = \vec{w}_{S^*}$.

Proof. Suppose that there exists an index $i \in [1..n]$ s.t. $\vec{w}_S(x_i) \neq \vec{w}_{S^*}(x_i)$ and let k be the smallest index verifying this property. Since the instantiation S^* is a subset of S (i.e., S^* is weaker than S) and since leftmost is a greedy procedure assigning the value 1 whenever possible from left to right, it follows that $\vec{w}_S(x_k) = 0$ and $\vec{w}_{S^*}(x_k) = 1$. Moreover, it follows that $max_S(k) = u$ and $max_{S^*}(k) < u$. In other words, there exists a subsequence I containing x_k s.t the cardinality of I in $\vec{w}_S^k$ (i.e. $card(I, \vec{w}_S^k)$) is equal to u, and the cardinality of I in $\vec{w}_{S^*}^k$ ($card(I, \vec{w}_{S^*}^k)$) is less than u. From this we deduce that there exists a variable $x_j \in I$ such that $\min(\vec{w}_S^k(x_j)) = 1$ and $\min(\vec{w}_{S^*}^k(x_j)) = 0$.

First, we cannot have $j < k$. Otherwise, both instantiations $\vec{w}_S^k(x_j)$ and $\vec{w}_{S^*}^k(x_j)$ contain an assignment for x_j, and therefore we have $\vec{w}_S^k(x_j) = \{1\}$ and $\vec{w}_{S^*}^k(x_j) = \{0\}$, which contradicts our hypothesis that k is the smallest index of a discrepancy.

Second, suppose now that $j > k$. Since we have $card(I, \vec{w}_S^k) = u$, we can deduce that $card(I, \vec{w}_S^j) = u$. Indeed, when going from iteration k to iteration j, leftmost only adds assignments, and therefore $card(I, \vec{w}_S^j) \geq card(I, \vec{w}_S^k)$. It follows that $max_S(j) = u$, and by construction of S^*, we cannot have $[x_j \leftarrow 1] \in S \setminus S^*$. However, it contradicts the fact that $\min(\vec{w}_S^k(x_j)) = 1$ and $\min(\vec{w}_{S^*}^k(x_j)) = 0$. $\qquad \square$

Theorem 1. *If S is a valid explanation for a failure and $S^* = S \setminus (\{[x_i \leftarrow 0] \mid max_S(i) = u\} \cup \{[x_i \leftarrow 1] \mid max_S(i) \neq u\})$, then S^* is also a valid explanation.*

Proof. By Lemma 1, we know that the instantiations $\vec{w}_S$ and $\vec{w}_{S^*}$, computed from, respectively the instantiations S and S^* are equal. In particular, we have $L_S(n) = L_{S^*}(n)$ and therefore $\text{ATMOSTSEQCARD}(S) = \bot$ iff $\text{ATMOSTSEQCARD}(S^*) = \bot$. □

Theorem 1 gives us a linear time procedure to explain failure. In fact, all the values $max_S(i)$ can be generated using one call of `leftmost`. Example 3 illustrates the explanation procedure.

```
                  S      │ 1 0 1 0 0 . . 0 0 0 1 1 0 0 0 0 1 0 0 0 0 1
               max_S(i)  │ 2 2 2 2 2 1 2 2 2 2 2 2 2 2 2 1 1 1 1 1 1 1
               w_S(x_i)  │ 1 0 1 0 0 1 0 0 0 0 1 1 0 0 0 0 1 0 0 0 0 1
Example 3.     L_S(i)    │ 1 1 2 2 2 3 3 3 3 3 4 5 5 5 5 5 6 6 6 6 6 7
               S*       │ 1 . 1 . . . . . . . 1 1 . . . 0 . 0 0 0 0 .
               max_S*(i) │ 2 2 2 2 2 1 2 2 2 2 2 2 2 2 2 1 1 1 1 1 1 1
               w_S*(x_i) │ 1 0 1 0 0 1 0 0 0 0 1 1 0 0 0 0 1 0 0 0 0 1
```

We illustrate here the explanation of a failure on $\text{ATMOSTSEQCARD}(2, 5, 8, [x_1..x_{22}])$. The propagator returns a failure since $L_S(22) = 7 < d = 8$. The default explanation corresponds to the set of all the assignments in this sequence, whereas our procedure shall generate a more compact explanation by considering only the assignments in S^*. Bold face values in the $max_S(i)$ line represent the variables that will not be included in S^*. As a result, we reduce the size of the explanation from 20 to 9.

Observe, however, that the generated explanation is not minimal. Take for instance the assignment $[x_1 \leftarrow 1]$. Despite it does not meet Theorem 1 conditions (i.e. $max_S(1) = u$), the set of assignments $S^* \setminus [x_1 \leftarrow 1]$ is a valid explanation since `leftmost` would return the same result between S^* and $S^* \setminus [x_1 \leftarrow 1]$.

4.2   Explaining Pruning

Suppose that a pruning $[x_i \nleftarrow v]$ was triggered by the propagator in equation 4.1 at a given level l on S (i.e. propagating $\text{ATMOSTSEQCARD}(S)$ implies $[x_i \nleftarrow v]$). Consider the partial instantiation $S_{[x_i \leftarrow v]}$ identical to S on all assignments at level l except for $[x_i \leftarrow v]$ instead of $[x_i \nleftarrow v]$. By construction $S_{[x_i \leftarrow v]}$ is unsatisfiable. Let S^* be the explanation expressing this failure using the previous mechanism. We have then $S^* \setminus [x_i \leftarrow v]$ as a valid explanation for the pruning $[x_i \nleftarrow v]$.

5   SAT-Encoding for the ATMOSTSEQCARD Constraint

In this section we present several SAT-encodings for the ATMOSTSEQCARD constraint and relate them to existing encoding techniques. First we describe a translation of Boolean cardinality constraints by a variant of the sequential counter encoding [24]. This encoding can be used to translate the decomposition of ATMOSTSEQCARD into CARDINALITY and ATMOST. Then we introduce an encoding taking advantage of the globality of ATMOSTSEQCARD by reusing the auxiliary variables for the cardinality constraint and integrating the sequence of ATMOST constraints. Finally, we combine the two encodings and prove that in this case UP maintains GAC on ATMOSTSEQCARD.

5.1   Sequential Counter

We describe first a translation of the cardinality expression $l \leq \sum_{i \in [1..n]} x_i \leq u$ to CNF by a sequential counter where $l, u \in \mathbb{N}$ and $x_i \in \{0, 1\}$. For technical reasons we use an additional variable x_0 s.t. $D(x_0) = \{0\}$.

- Variables:
  - $s_{i,j}$: $\forall i \in [0..n]$, $\forall j \in [0..u+1]$, $s_{i,j}$ is *true* iff $|x_k; s.t. D(x_k) = \{1\}| \geq j$
- Encoding: $\forall i \in [1..n]$
  - Clauses for restrictions on the same level: $\forall j \in [0..u+1]$
    1. $\neg s_{i-1,j} \vee s_{i,j}$
    2. $x_i \vee \neg s_{i,j} \vee s_{i-1,j}$
  - Clauses for increasing the counter, $\forall j \in [1..u+1]$
    3. $\neg s_{i,j} \vee s_{i-1,j-1}$
    4. $\neg x_i \vee \neg s_{i-1,j-1} \vee s_{i,j}$
  - Initial values for the bounds of the counter:
    5. $s_{0,0} \wedge \neg s_{0,1} \wedge s_{n,l} \wedge \neg s_{n,u+1}$

In the rest of the section we refer to the clauses by numbers 1 to 5. The intuition is that the variables $s_{i,j}$ represent the bounds for cumulative sums of the sequence $x_1 \ldots x_i$. The encoding is best explained by visualising $s_{i,j}$ as a two dimensional grid with positions (horizontal) and cumulative sums (vertical). The binary clauses 1 and 3 ensure that the counter (i.e. the variables representing the cumulative sums) is monotonically increasing. Clauses 2 and 4 control the interaction with the variables x_i. If x_i is true, then the counter has to increase at position i whereas if x_i is false an increase is prevented at position i. The conjunction 5 sets the initial values for the counter to start counting at 0 and ensures that the partial sum at position n is between to l and u.

Example 4. We illustrate the auxiliary variables: Given a sequence of 8 variables and $l = u = 2$. To the left the initial condition of the variables, followed assigning x_2 to true and then to the right x_7 to true.

3	0 0 0 0 0 0 0 0		3	0 0 0 0 0 0 0 0		3	0 0 0 0 0 0 0 0
2	0 0 1		2	0 0 1		2	0 0 0 0 0 0 1 1
1	0 . . . . . . 1 1		1	0 . 1 1 1 1 1 1		1	0 0 1 1 1 1 1 1
0	1 1 1 1 1 1 1 1		0	1 1 1 1 1 1 1 1		0	1 1 1 1 1 1 1 1
$s_{i,j}$	0 1 2 3 4 5 6 7 8		$s_{i,j}$	0 1 2 3 4 5 6 7 8		$s_{i,j}$	0 1 2 3 4 5 6 7 8
x_i	. . . .		x_i	. 1		x_i	0 1 0 0 0 0 1 0

By variants of the above set of clauses we introduce two encodings C_C and C_A, that suffice to translate ATMOSTSEQCARD:

- $C_C(d, [x_1, x_2, \ldots x_n])$ encodes $\sum_{i \in [1..n]} x_i = d$ using clauses 1 to 5 with $u = l = d$.
- $C_A(u, q, [x_1, x_2, \ldots x_n])$ encodes $\bigwedge_{i=0}^{n-q}(\sum_{l=1}^{q} x_{i+l} \leq u)$ by a set of separate translations on each $\sum_{l=1}^{q} x_{i+l} \leq u$ with $i = 1 \ldots n - q$ using clauses 1 to 5 with $l = 0$ and u the upper bound.

Since each of the above encodings is a superset of the encoding introduced in [24], C_S and C_A have the following property regarding propagation:

Proposition 1. *Unit Propagation enforces* GAC *on*

1. $\sum_{i \in [1..n]} x_i = d$ *by the encoding* $C_C(d, [x_1, x_2, \ldots x_n])$.
2. ATMOSTSEQ$(u, q, [x_1, \ldots, x_n])$ *by the encoding* $C_A(u, q, [x_1, x_2, \ldots x_n])$.

With these encodings at hand we can completely translate ATMOSTSEQCARD to CNF and fulfil the first two properties of a GAC propagator as characterised in the end of Section 4. However, we are missing the global propagation of Equation 4.1.

The sequential counter encoding in [24] uses only clauses 1 and 4. Indeed, they are sufficient to enforce GAC by unit propagation in case of ATMOST. However, their encoding does not necessarily force all auxiliary variables when all x_i are assigned and this effectively increases the number of models which can lead to unnecessary search. Thus, we prefer the more constrained version of the counter encoding.

The encoding in [2] of the more general AMONG constraint has similarities to a counter encoding. This encoding builds on the translation of the REGULAR constraint and introduces variables for states and transitions of a deterministic finite automaton. Regarding propagation, this encoding is equivalent to the sequential counter, but on the clausal level, it is not identical. Our encoding consists only of binary and ternary clauses whereas their encoding introduces longer clauses using two types of auxiliary variables. Another difference is that the state variables represent exact partial sums whereas the encoding presented here relate to the idea of an order encoding.

5.2   Extension to ATMOSTSEQCARD

We still need to capture the missing propagation of a GAC propagator of ATMOSTSEQCARD. To do so we introduce the following binary clauses. They are referred to by $C_S(u, q, [x_1 \ldots x_n])$ and reuse the auxiliary variables $s_{i,j}$ introduced by C_C. $\forall i \in [q..n], \forall j \in [u..d+1]$:

6. $\qquad \neg s_{i,j} \vee s_{i-q, j-u}$

We will show that the binary clauses capture the missing propagation for ATMOSTSEQCARD as in Equation 4.1. For this, we precisely show how the auxiliary variables $s_{i,j}$ relate to L_S and R_S.

Proposition 2. *Let* C_C *and* C_S *be the decomposition of* ATMOSTSEQCARD$(u, q, d, [x_1 \ldots x_n])$. *Given a partial assignment* S *on* $\{x_1, x_2 \ldots x_n\}$ *and assuming that* $L_S(n) \leq d$ *and* $R_S(0) \leq d$, *for all* $i \in \{0 \ldots n\}$ *UP forces*

1. $s_{i, L_S(i)+1}$ *to false and*
2. $s_{i-1, d-R_S(i)}$ *to true.*

Proof. We concentrate on 1) since 2) is analogous. The proof follows an inductive argument on i. For $i = 0$ it holds from unit $s_{0,1}$ in the clauses 5 of C_C. For the inductive step we have to show, assuming $s_{i,L_S(i)+1}$ is set to false, that $s_{i+1,L_S(i+1)+1}$ is enforced to false by UP. There are two cases to analyse: a) $L_S(i+1) = L_S(i) + 1$, and b) $L_S(i+1) = L_S(i)$. The first case follows from clauses 3 in C_C. The second case involves a complicated step, essentially showing that with the induction hypothesis, clauses 6 in C_S and GAC on C_C, UP enforces $s_{i+1,L_S(i)+1}$ to false.

For case b) there are two situations to consider using the definition of L_S: x_{i+1} is assigned to false in S or x_{i+1} is unassigned. The first situation is covered by clauses 2. In the second situation it holds that $\overrightarrow{w}(x_{i+1}) = 0$. This is caused by leftmost not evaluating x_{i+1} to true. Hence, there exists a window $k+1 \ldots k+q$ that includes position $i+1$ and the maximal number of variables in this windows assigned to true by $\overrightarrow{w}$ is equal to u. Let there be α true assignements in $\overrightarrow{w}$ before $i+1$ and β after. We have $L_S(k) = L_S(i) - \alpha$. However, since $k \leq i$, we know by induction that $\neg s_{k,L_S(k)+1}$, that is, $\neg s_{k,L_S(i)-\alpha+1}$ holds.

Now, clauses 6 of C_S, instantiated to $\neg s_{k+q,L_S(i)-\alpha+u+1} \vee s_{k,L_S(i)-\alpha+1}$ infers by UP $\neg s_{k+q,L_S(i)+\beta+1}$ (recall that $\alpha + \beta = u$).

Finally, observe that when leftmost computed $L_S(i)$, no assignment were made on the interval $i+2 \ldots k+q$. Hence we have $\sum_{j=i+2}^{k+q} \min(x_j) = \beta$. Standard cardinality reasoning (clauses 1 and 4) is thus sufficient to show that $\neg s_{k+q,L_S(i)+\beta+1}$ implies $\neg s_{i+1,L_S(i)+1}$. Since we are in the case where $L_S(i+1) = L_S(i)$ we have shown that UP infers $\neg s_{i+1,L_S(i+1)+1}$. This concludes the inductive proof and it demonstrates that UP maintains the values for $L_S(i)$ for all positions i. The case for R_S follows a dual argument. □

The key idea of the binary clauses 6 can also be found behind the decomposition of GEN-SEQUENCE into cumulative sums as in [5]. Furthermore, there is a strong similarity between the combination of C_C with C_S and the encoding of GEN-SEQUENCE in [2] and it is possible to show that also here in fact it detects dis-entailment on ATMOSTSEQCARD similarly to Theorem 3 of [2]. The following case exemplifies what kind of propagation is missing with the combination of C_C and C_S.

Example 5. Consider the encodings C_C and C_S on $u = 1, q = 2, d = 2, n = 5$ and let x_3 be true, then UP does not enforce x_2 nor x_4 to false. Setting them to true will lead to a conflict by UP through clauses 4 and 6 on positions 2, 3 and 4.	$\begin{array}{c\|cccccc} 3 & 0 & 0 & 0 & 0 & 0 & 0 \\ 2 & 0 & 0 & 0 & . & . & 1 \\ 1 & 0 & . & . & 1 & 1 & 1 \\ 0 & 1 & 1 & 1 & 1 & 1 & 1 \\ \hline s_{i,j} & 0 & 1 & 2 & 3 & 4 & 5 \\ \hline x_i & & . & . & 1 & . & . \end{array}$

We see that the encoding C_A would propagate in the case of the previous example. If we combine all three encodings we can provide a CNF encoding that maintains the desired property of GAC on ATMOSTSEQCARD:

Theorem 2. *UP on $C_C + C_A + C_S$ enforces GAC on the ATMOSTSEQCARD constraint.*

Proof. The proof follows from Proposition 1 and 2 showing that this encoding fulfils all sufficient properties of a GAC propagator as described in Section 4.

In particular, UP maintains GAC on $\bigwedge_{i=0}^{n-q}(\sum_{l=1}^{q} x_{i+l} \leq u)$ and $\sum_{i \in [1..n]} x_i = d$ by C_C and C_A. We elaborate on the interesting cases of Equation 4.1:

1. Let $L_S(i) + R_S(i) \leq d$. By Proposition 2, UP forces $\neg s_{i,L_S(i)+1}$ and $s_{i-1,d-R_S(i)}$ on S, and by assumption we know that $d - R_S(i)$ is greater or equal to $L_S(i)$, so UP

forces also $s_{i-1,L_S(i)}$ to true. By clauses 4 instantiated to $\neg x_i \vee \neg s_{i-1,L_S(i)} \vee s_{i,L(i)+1}$ UP forces x_i to false. Hence if $L_S(i) + R_S(i) \leq d$ holds then UP forces x_i to false.

2. Let $L_S(i-1) + R_S(i+1) < d$. By Proposition 2, UP forces $\neg s_{i-1,L_S(i-1)+1}$ and $s_{i,d-R_S(i+1)}$. Since $L_S(i-1) < d - R_S(i+1)$ UP enforces by clauses 1 and 3 that $s_{i,L_S(i-1)+1}$ is true. Now clauses 2 trigger by $x_i \vee s_{i-1,L_S(i)+1} \vee \neg s_{i,L_S(i)+1}$ and set x_i to true by UP.   □

6 Experimental Results

We tested the different approaches on the three data sets available at CSPLib [14]. All experiments ran on Intel Xeon CPUs 2.67GHz under Linux. For each instance, we performed 5 randomized runs with a 20 minutes time cutoff. The first set contains 5 unsatisfiable and 4 satisfiable instances of relatively small size (100 cars). The second set contains 70 instances generated with varying usage rate. All instances in this set are satisfiable and involve 200 cars. The third set, proposed by Gagné, features larger instances divided into three sets of ten each, involving respectively 200, 300 and 400 cars. Seven of these instances were solved using local search algorithms. To the best of our knowledge the remaining 23 instances have never been proved unsatisfiable. To facilitate the analysis, we grouped the instances into three categories:

In the first category (sat[easy]), we consider the 70 satisfiable instances of the second set as well as the 4 satisfiable instances of the first set. All these instances are extremely easy for all the methods we introduce in this paper;

In the second category (sat[hard]), we consider the 7 known satisfiable instances of the second set. These instances are challenging and were often out of reach of previous systematic approaches;

In the third category (unsat*), we consider the remaining 5 unsatisfiable instances of the first set as well as the 23 unknown instances form the third set. Those instances are challenging and indeed open for 23 of them.

We ran the following methods:

SAT Encoding. We use Minisat (version 2.2.0) with default parameter settings on three variants of the SAT encoding. Links between classes and options as well as the constraint for exactly one class of vehicle per position are translated as in the basic model. For each option we encode one ATMOSTSEQCARD. The following three models differ only in how this translation is performed (w.r.t Section 5):

1. *SAT (1)* encodes the basic model by using $C_C + C_A$ for each ATMOSTSEQCARD.
2. *SAT (2)* uses $C_C + C_S$ for each ATMOSTSEQCARD.
3. *SAT (3)* combines all of three encodings $C_C + C_A + C_S$.

Hybrid CP/SAT. We use Mistral as a hybrid CP/SAT solver (Section 2) using our explanation for the ATMOSTSEQCARD constraint. We tested four branching heuristics:

1. *hybrid (VSIDS)* uses VSIDS;
2. *hybrid (Slot)* uses the following heuristic (denoted *Slot*): we branch on *option* variables from the middle of the sequence and towards the extremities following the first unassigned *Slot*. The options are firstly evaluated by their dynamic usage rate[25] then lexicographically compared.

3. *hybrid (Slot/VSIDS)* first uses the heuristic *Slot*, then switches after 100 non-improving restarts to VSIDS.
4. *hybrid (VSIDS/Slot)* reverse of above.

Baseline methods. We also use three "control" approaches run in the same setting:

1. *CP*: A pure CP approach, implemented using Mistral without clause learning on the model described in Section 3 using the *Slot* branching.
2. *PBO-clauses*: A pseudo-Boolean method relying on SAT encoding. We used MiniSat+ [11] on the pseudo-Boolean encoding described in Section 3 except that the AtMostSeqCard constraint is decomposed into Cardinality and AtMost.
3. *PBO-cutting planes*: A pseudo-Boolean method with dedicated propagation and learning based on cutting planes [10]. We used SAT4J [3] on the same model, with the "CuttingPlanes" algorithm.

For each considered data set, we report the total number of successful runs (*#suc*).[2] Then, we report the number of fail nodes (*fails*) and the CPU time (*time*) in seconds both averaged over all successful runs. We emphasize the statistics of the best method (w.r.t. *#suc*, ties broken by CPU time) for each data set using bold face fonts.

Table 1. Evaluation of the models

Method	sat[easy] (74 × 5)			sat[hard] (7 × 5)			unsat* (28 × 5)		
	#suc	avg fails	time	#suc	avg fails	time	#suc	avg fails	time
SAT (1)	370	2073	1.71	28	337194	282.35	**85**	**249301**	**105.07**
SAT (2)	370	1114	0.87	31	60956	56.49	65	220658	197.03
SAT (3)	370	612	0.91	34	32711	36.52	77	190915	128.09
hybrid (VSIDS)	370	903	0.23	16	207211	286.32	35	177806	224.78
hybrid (VSIDS/Slot)	370	739	0.23	35	76256	64.52	37	204858	248.24
hybrid (Slot/VSIDS)	370	132	0.04	34	4568	2.50	37	234800	287.61
hybrid (Slot)	370	132	0.04	**35**	**6304**	**3.75**	23	174097	299.24
CP	**370**	**43.06**	**0.03**	35	57966	16.25	0	-	-
PBO-clauses	277	538743	236.94	0	-	-	43	175990	106.92
PBO-cutting planes	272	2149	52.62	0	-	-	1	5031	53.38

We first observe that most of the approaches we introduce in this paper significantly improve the state of the art, at least for systematic methods. For instance, in the experiments reported in [23] several instances of the set sat[hard] were not solved within a 20 minutes cutoff. Moreover we are not aware of other systematic approaches being able to solve these instances. More importantly, we are able to close 13 out of the 23 large open instances proposed by Gagné. The set of open instances is now reduced to pb_200_02/06/08, pb_300_02/06/09, and pb_400_01/02/07/08.

Second, on satisfiable instances, we observe that pure CP approaches are difficult to outperform. It must be noticed that the results reported for *CP* are significantly better

---

[2] They all correspond to solutions found for the two first categories, and unsatisfiability proofs for the last.

than those previously reported for similar approaches. For instance, the best methods introduced in [27] take several seconds on most instances of the first category and were not able to solve two of them within a one hour time cutoff. Moreover, in [23], the same solver on the same model had a similar behavior on the category sat[easy], but was only able to solve 2 instances of the category sat[hard] due to not using restarts.

However, the best method on sat instances is the hybrid solver using a CP heuristic. Moreover, we can see that even with a "blind" heuristic, MiniSat on the strongest encodings has extremely good results (all sat instances were solved with a larger cutoff).

This study shows that propagation is very important to find solutions quickly, by keeping the search "on track" and avoiding exploring large unsatisfiable subtrees. There is multiple evidence for these claims: First, the pseudo Boolean models (*PBO-clauses* and *PBO-cutting planes*) perform limited propagation and are consequently very poor even on sat[easy]. Second, the best SAT models for sat[easy] and sat[hard] are those providing the tightest propagation. Last, previous CP approaches that did not enforce GAC on the ATMOSTSEQCARD constraint are all dominated by *CP*.

For proving unsatisfiability, our results clearly show that clause learning is by far the most critical factor. Surprisingly, stronger propagation is not always beneficial when building a proof using clause learning, as shown by the results of the different encodings. One could even argue for a negative correlation, since the "lightest" encodings are able to build more proofs than stronger ones. Similarly, the pure pseudo Boolean model performs much better comparatively to the satisfiable case. The hybrid models are slightly worse than pseudo Boolean but far better than the pure CP approach that was not able to prove any case of unsatisfiability. To mitigate this observation, however, notice that other CP models with strong filtering, using the Global Sequencing Constraint [20], or a conjunction of this constraint and ATMOSTSEQCARD [23,27] were able to build proofs for some of the 5 unsatisfiable instances of the CSPLib. However, these models were not able to solve any of the 23 larger unsatisfiable instances.

7   Conclusion

We proposed and compared hybrid CP/SAT models for the car sequencing problem against several SAT-encodings. Both approaches exploit the ATMOSTSEQCARD constraint. In particular, we proposed a linear time procedure for explaining failure and pruning as well as advanced SAT-encodings for this constraint. Experimental results emphasize the importance of advanced propagation for searching feasible solutions and of clause learning for building unsatisfiability proofs. Our models advance the state of the art in this domain, and close 13 out of the 23 large open instances in CSPLib.

References

1. Aloul, F.A., Ramani, A., Markov, I.L., Sakallah, K.A.: Generic ILP versus specialized 0-1 ILP: An update. In: Proceedings of ICCAD, pp. 450–457 (2002)
2. Bacchus, F.: GAC Via Unit Propagation. In: Bessière, C. (ed.) CP 2007. LNCS, vol. 4741, pp. 133–147. Springer, Heidelberg (2007)
3. Berre, D.L., Parrain, A.: The Sat4j library, release 2.2. Journal on Satisfiability, Boolean Modeling and Computation 7, 59–64 (2010)

4. Biere., A., Heule., M., van Maaren., H., Walsh, T.: Handbook of Satisfiability. Frontiers in Artificial Intelligence and Applications, vol. 185. IOS Press (2009)
5. Brand, S., Narodytska, N., Quimper, C.-G., Stuckey, P.J., Walsh, T.: Encodings of the SEQUENCE Constraint. In: Bessière, C. (ed.) CP 2007. LNCS, vol. 4741, pp. 210–224. Springer, Heidelberg (2007)
6. Cambazard, H.: Résolution de problmes combinatoires par des approches fondées sur la notion dexplication. PhD thesis, Ecole des mines de Nantes (2006)
7. Cambazard, H., Jussien, N.: Identifying and exploiting problem structures using explanation-based constraint programming. Constraints 11(4), 295–313 (2006)
8. Davis, M., Putnam, H.: A Computing Procedure for Quantification Theory. Journal of the ACM 7(3), 201–215 (1960)
9. Dixon, H.: Automating Pseudo-Boolean Inference within a DPLL Framework. PhD thesis, University of Oregon (2004)
10. Dixon, H.E., Ginsberg, M.L.: Inference Methods for a Pseudo-Boolean Satisability Solver. In: Proceedings of AAAI, pp. 635–640 (2002)
11. Eén, N., Sörensson, N.: Translating Pseudo-Boolean Constraints into SAT. Journal on Satisfiability, Boolean Modeling and Computation 2, 1–26 (2006)
12. Feydy, T., Schutt, A., Stuckey, P.: Semantic Learning for Lazy Clause Generation. In: TRICS workshop, held alongside CP (2013)
13. Gent, I.P.: Arc Consistency in SAT. In: Proceedings of ECAI, pp. 121–125 (2002)
14. Gent, I.P., Walsh, T.: CSPlib: A benchmark library for constraints (1999)
15. G. Katsirelos.: Nogood Processing in CSPs. PhD thesis, University of Toronto (2008)
16. Katsirelos, G., Bacchus, F.: Generalized NoGoods in CSPs. In: Proceedings of AAAI, pp. 390–396 (2005)
17. Moskewicz, M.W., Madigan, C.F., Zhao, Y., Zhang, L., Malik, S.: Chaff: Engineering an Efficient SAT Solver. In: DAC, pp. 530–535 (2001)
18. Quimper, C.-G., Golynski, A., López-Ortiz, A., Beek, P.V.: An Efficient Bounds Consistency Algorithm for the Global Cardinality Constraint. Constraints 10(2), 115–135 (2005)
19. Régin, J.C.: Generalized Arc Consistency for Global Cardinality Constraint. In: Proceedings of AAAI, vol. 2, pp. 209–215 (1996)
20. Régin, J.-C., Puget, J.-F.: A Filtering Algorithm for Global Sequencing Constraints. In: Smolka, G. (ed.) CP 1997. LNCS, vol. 1330, pp. 32–46. Springer, Heidelberg (1997)
21. Rossi, F., Beek, P.V., Walsh, T.: Handbook of Constraint Programming. Elsevier (2006)
22. Schiex, T., Verfaillie, G.: Nogood Recording for Static and Dynamic CSP. In: Proceeding of ICTAI, pp. 48–55 (1993)
23. Siala, M., Hebrard, E., Huguet, M.-J.: An optimal arc consistency algorithm for a particular case of sequence constraint. Constraints 19(1), 30–56 (2014)
24. Sinz, C.: Towards an Optimal CNF Encoding of Boolean Cardinality Constraints. In: van Beek, P. (ed.) CP 2005. LNCS, vol. 3709, pp. 827–831. Springer, Heidelberg (2005)
25. Smith, B.M.: Succeed-first or Fail-first: A Case Study in Variable and Value Ordering (1996)
26. Solnon, C., Cung, V.D., Nguyen, A., Artigues, C.: The car sequencing problem: Overview of state-of-the-art methods and industrial case-study of the ROADEF 2005 challenge problem. European Journal of Operational Research 191, 912–927 (2008)
27. van Hoeve, W.J., Pesant, G., Rousseau, L.-M., Sabharwal, A.: New Filtering Algorithms for Combinations of Among Constraints. Constraints 14(2), 273–292 (2009)
28. Walsh, T.: SAT v CSP. In: Dechter, R. (ed.) CP 2000. LNCS, vol. 1894, pp. 441–456. Springer, Heidelberg (2000)

Continuously Degrading Resource and Interval Dependent Activity Durations in Nuclear Medicine Patient Scheduling

Cyrille Dejemeppe and Yves Deville

ICTEAM, Université Catholique de Louvain (UCLouvain), Belgium
{cyrille.dejemeppe,yves.deville}@uclouvain.be

Abstract. Nuclear Medicine (NM) is a medical imaging technique in which patients are administered radioactive tracers. As the tracers decay in the human body, they emit photons, which are then captured to generate an image used for diagnostic purposes. The schedule of daily patients in an NM center is a hard and interesting problem due to the continuous decay of these radioactive tracers. This scheduling problem allows us to define two new scheduling concepts: continuously degrading resources and interval dependent activity durations. In this paper, we model the NM scheduling problem as a COP; we propose a resolution strategy using Constraint Programming with LNS, and we introduce a new propagator to deal with continuously degrading resources and interval dependent activity durations.

1 Introduction

Nuclear Medicine is a clinical practice in which patients are administered nuclear tracers in order to provide diagnostic information for a wide range of diseases. A nuclear tracer is a set of radioactive compounds. As defined in [3], the activity of a radioactive tracer decreases with time according to the following law of decay:

$$Rad(t) = Rad_0 \times e^{\frac{-t \ln(2)}{t_{0.5}}} \tag{1}$$

where Rad_0 is the initial activity of the decaying substance and $t_{0.5}$ is its half-life. As it decays, the radioactive component in the tracer emits gamma rays or high-energy photons. The energy levels of these emissions are such that a significant amount of energy can exit the body without being scattered or attenuated. External gamma-ray sensors allow capturing these emissions, and computers are then able to recreate an image from them. As stated in [2], NM has several advantages over other medical imaging techniques. The precision and the quality of the images obtained with NM makes it a technique widely used for medical imaging.

Nuclear Medicine Problem (NMP) consists in the optimization of the schedule of patient workflows inside an NM center. Patients coming to an NM center are first administered a nuclear tracer. Then they have to wait in a waiting room in order to allow their body to incorporate the tracer. The duration of this waiting time has to last a minimum amount of time, otherwise the human body will not have fully incorporated the

H. Simonis (Ed.): CPAIOR 2014, LNCS 8451, pp. 284–292, 2014.

nuclear tracer. On the other hand, the waiting time cannot be too long, since then the radioactive tracer would have decayed for too long, leading to its radioactive activity's being too low to provide satisfactory images. Once a patient's body has fully incorporated the tracer, the patient goes into a scanning room in which the image is captured. There can be several different scanning rooms which differ in their scanner equipment. The amount of time needed by the scanner equipment to capture the image directly depends on the amount of time the patient has been waiting after having been administered the radioactive tracer.

Even if the workflows of the patients are similar, they can differ by the duration of each step inside the workflow, by the type and quantity of tracer administered, by the minimal and maximal durations of their waiting times, and by the scanning room and equipment needed to capture the image. As stated in [7], the acquisition time can be expressed as a linear function of the waiting time as follows:

$$\text{acquisition time} = \alpha + \delta \times \text{waiting time} \qquad (2)$$

where α and δ are positive constants depending on the quantity and type of tracer administered to the patient. Figure 1b shows an example of how the acquisition time depends on the waiting time.

Patients are administered their tracer dose with a special injector and most NM centers only have a few of these. The scanning rooms exist in different instances and sometimes they have different kinds of equipment. There are not very many of these scanning rooms, and some NM centers even have only one scanning room. The day's radioactive tracers are delivered to the NM center every morning. These tracers decay with time, i.e. their radioactive activity decreases. As the patients need to be injected with a certain amount of radioactivity, the quantity of tracer they are injected with increases with time to compensate for the decrease in the radioactivity of the tracer that has occurred since the beginning of the day. An example of the radioactive activity of the widely used Fludeoxyglucose (FDG) tracer over time can be seen in Figure 1a.

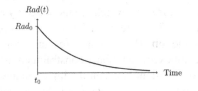

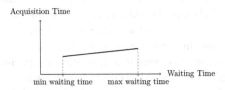

(a) Activity of the FDG tracer over time

(b) Time required for image acquisition in function of the length of the waiting time after injection

Fig. 1. Example of radioactivity decrease and its effects on acquisition time

NMP is an optimization problem with two distinct objectives. First, the purpose is to be able to treat patients in the shortest amount of time possible. This would possibly allow treating more patients per day. The second objective is to minimize the total consumption of radioactive tracer. Indeed, these substances are very expensive and being able to start the day with a smaller quantity would save money. These objectives are not aggregated, and NMP is thus a bi-objective optimization problem.

We model NMP as a scheduling problem. To model the decay of nuclear tracers, we have to introduce two new scheduling concepts: continuously degrading resources and interval dependent activity durations. The abstraction of a continuously degrading resource allows us to model a resource whose capacity decreases continuously over time. This allows us to represent the decay of the radioactive tracers, said capacity representing the decreasing radioactivity of the tracer. The abstraction of an interval dependent activity is used to model an activity whose duration depends on the interval between the activity itself and the activity preceding it. As we introduce these two scheduling abstractions, we define new constraints and their associated new propagators.

The two new abstractions we introduce could be used in several other scheduling applications. The continuously degrading resource abstraction could be used to model chemical compounds which react with each other and whose quantities thus continuously decrease over time. The interval dependent activity duration abstraction could be used to model other medical processes such as in the cure of an uncoupled shoulder: if the time between the radiography and the replacement of the shoulder is longer, the muscles will tighten more, so that the operation to replace the shoulder will take more time. This article also intends to illustrate how to solve cumulative scheduling problems with business constraints. As such, the modelling and search techniques described for NMP are as generic as possible.

2 The Model

In this section, we define a model for NMP defined in Section 1. NMP is a scheduling problem and can thus be defined by the four main components of a scheduling problem: activities, resources, constraints and objectives.

The activities of our model are the steps of the patient workflows. The workflow for each patient consists of a job j containing two activities: injection and image acquisition. The resources of NMP are the injectors and the scanning rooms, which are modelled by the cumulative resources, and the radioactive tracers, which are modelled by a new abstraction: a continuously degrading resource. There are precedence constraints imposing an order between the activities of the same job. There are also resource constraints to ensure that the resource capacities are not exceeded. Some constraints must also be imposed to ensure that the waiting time of a patient between the injection and the image acquisition remains bounded. Finally, some constraints are needed to model the durations of the activities: a fixed constraint for the injection, but one that depends on the length of the waiting time for the image acquisition. The objective of treating a patient in as little time as needed is modelled by a makespan minimization. On the other hand, the minimization of the total consumption of radioactive tracers is modelled by the sum of the consumptions of the tracer by the patients. We express our scheduling model more formally as a Constraint Optimization Problem (COP) $\langle X, D, C, O \rangle$ as follows.

X : **The Set of Variables.** The variables of our problem are the starting and ending times of each activity. We consider instances containing k jobs and inside each job there are n

activities. So for each activity $A_{i,j}$, the jth activity of the ith job, we define two decision variables: $start(A_{i,j})$ and $end(A_{i,j})$ with $1 \leq i \leq k$ and $1 \leq j \leq n$.

D: The Domains of the Variables   The start variables $start(A_{i,j})$ of the activities have the following domains:

$$D\left(start(A_{i,j})\right) = \begin{cases} [0; horizon - duration(A_{i,j})] \text{ if the duration is fixed} \\ [0; horizon] \text{ otherwise} \end{cases} \qquad (3)$$

with $1 \leq i \leq k$, $1 \leq j \leq n$, and $horizon$ being the horizon of the problem.
   Similarly, the end variables $end(A_{i,j})$ of the activities have the following domains:

$$D\left(end(A_{i,j})\right) = \begin{cases} [duration(A_{i,j}); horizon] \text{ if the duration is fixed} \\ [0; horizon] \text{ otherwise} \end{cases} \qquad (4)$$

with $1 \leq i \leq k$ and $1 \leq j \leq n$.

C: The Constraints.   The first constraints described here define an order between the activities belonging to the same job and impose a setup time between two successive activities when needed:

$$end(A_{i,j}) + setup(A_{i,j}, A_{i,j+1}) \leq start(A_{i,j+1}) \qquad (5)$$

with $1 \leq i \leq k$, $1 \leq j \leq n-1$ and $setup(A_{i,j}, A_{i,j+1})$ being the amount of time needed between the two activities. If the two successive activities are not subject to a setup time, then we have $setup(A_{i,j}, A_{i,j+1}) = 0$.
   Another constraint must impose a maximal delay between those pairs of activities which need it.

$$start(A_{i,j+1}) - end(A_{i,j}) \leq delay_{max}(A_{i,j}, A_{i,j+1}) \qquad (6)$$

with $1 \leq i \leq k$, $1 \leq j \leq n-1$, and $delay_{max}(A_{i,j}, A_{i,j+1})$ being the maximal amount of time allowed between the two activities. If the two successive activities are not subject to a maximum delay, then we have $delay_{max}(A_{i,j}, A_{i,j+1}) = horizon$.
   As explained above, some activities have fixed durations while others have interval dependent durations. In our model, some durations depend linearly on the length of the interval according to a monotonic increasing function $f(interval) = \alpha + \delta \times interval$ where α and β are positive constants. The following constraints explain these properties:

$$end(A_{i,j}) - start(A_{i,j}) = \begin{cases} duration(A_{i,j}) & \text{if fixed duration} \\ f(start(A_{i,j}) - end(A_{i,j-1})) & \text{otherwise} \end{cases} \qquad (7)$$

with $1 \leq i \leq k$, $1 \leq j \leq n-1$, and $A_{i,j}$ being an activity with interval dependent duration.
   The cumulative resource constraints can be expressed as follows:

$$\sum_A usage(A, R_{cum}, t) \leq capacity(R_{cum}) \qquad (8)$$

where $0 \leq t \leq horizon$, R_{cum} is a cumulative resource, $usage(A, R_{cum}, t)$ is the usage of R_{cum} by the activity A at time t, and $capacity(R_{cum})$ is the capacity of R_{cum}.

Finally, we have to express the constraint on the continuously degrading resources. We model these constraints not by diminishing the resource capacity over time, but by increasing the quantity required by the activities over time. We assume that the needed amount of a resource for an activity is consumed at its starting time in an atomic way. The quantity needed by an activity $A_{i,j}$ of the continuously degrading resource R_{dec} if starting at a time t is

$$q(A_{i,j}, R_{dec}, t) = \frac{C_i}{Rad(t)} = \frac{C_i}{Rad_0} \times e^{\frac{t\ln(2)}{t_{0.5}}} \tag{9}$$

where $Rad(t)$ is defined in Equation 1, Rad_0 and $t_{0.5}$ depend on R_{dec}, and C_i is a positive constant allowing to determine the amount of radioactivity needed by patient corresponding to job i. This formula expresses that the quantity of tracer a patient is injected is inversely proportional to the radioactivity of the tracer. We must ensure that the total consumption of each resource is less than or equal to the initial capacity of that resource. This constraint is expressed by

$$\sum_A q(A, R_{dec}, start(A)) \leq initialCapacity(R_{dec}) \tag{10}$$

where $initialCapacity(R_{dec})$ is the initial capacity of the resource R_{dec}.

O: **The Objectives.** The first objective is to minimize the makespan. This can be represented as follows:

$$\text{minimize } makespan = \max_A (end(A)). \tag{11}$$

The second objective is to minimize the total consumption of the continuously degrading resources. This can be expressed as follows:

$$\text{minimize } \sum_{R_{dec}} \sum_A w_{dec} \times q(A, R_{dec}, start(A)) \tag{12}$$

where the w_{dec} are positive weights.

3   Propagation

In this section, we describe how to perform propagation for the continuously degrading resource constraint and the interval dependent activity duration constraint.

3.1   Continuously Degrading Resource

The propagation procedure we describe achieves Bound Consistency (BC) for the constraint stated in Equation 10. For a given activity $A_{i,j}$ and a given continuously decreasing resource R_{dec}, the quantity of resource required by the activity is a monotonic function increasing with time t: $q(A_{i,j}, R_{dec}, t)$. This increasing quantity can be defined for NMP by Equation 1. The quantity of tracer a patient is injected is inversely proportional to the radioactivity of a tracer. We can thus express $q(A_{i,j}, R_{dec}, t)$ as follows:

$$q(A_{i,j}, R_{dec}, t) = \beta_{i,j} \times e^{\frac{t \ln(2)}{\gamma_{dec}}} \tag{13}$$

where $\beta_{i,j}$ is a constant depending on $A_{i,j}$ and γ_{dec} is a constant depending on R_{dec}. We are thus able to rewrite the constraint stated in Equation 10 as follows:

$$\sum \beta_{i,j} \times e^{\frac{start(A_{i,j}) \ln(2)}{\gamma_{dec}}} \leq initialCapacity(R_{dec}) \tag{14}$$

To express this constraint as a linear sum constraint, we will use the *view-based propagator derivation* technique proposed in [6]. Given a propagator p, a view is represented by two functions ϕ and ϕ^{-1} that are composed with p to obtain the desired propagator $\phi \circ p \circ \phi^{-1}$. The ϕ function transforms the input domain and ϕ^{-1} applies the inverse transformation to the propagator's output domain. To be able to use a linear bounded sum constraint propagator, we define a $\phi_{i,j}$ function for each variable $start(A_{i,j})$ involved in Equation 14 as follows:

$$\phi_{i,j}(v) = \beta_{i,j} \times e^{\frac{v \ln(2)}{\gamma_{dec}}} \tag{15}$$

The inverse functions $\phi_{i,j}^{-1}$ are defined as follows:

$$\phi_{i,j}^{-1}(v) = \frac{\gamma_{dec}}{\ln(2)} \times \ln\left(\frac{v}{\beta_{i,j}}\right) \tag{16}$$

As values returned by the ϕ function are real values and our start variables only accept integer values, we consider the domain of $start(A_{i,j})$ mapped by ϕ is a discrete domain in which each value correspond to a single value in the domain of $start(A_{i,j})$. The definitions of ϕ and ϕ^{-1} allow us to use a classical linear sum constraint propagator as proposed in [1].

3.2 Interval Dependent Activity Durations

The propagation procedure we describe achieves BC for the constraint stated in Equation 7. As stated earlier, the f function in Equation 7 is a linear monotonic increasing function. From this, we can rewrite the constraint depicted in Equation 7 as follows:

$$end(A_{i,j}) - start(A_{i,j}) = \varepsilon_{i,j} + \delta_{i,j} \times \left(start(A_{i,j}) - end(A_{i,j-1})\right) \tag{17}$$

where $\delta_{i,j}$ and $\varepsilon_{i,j}$ are positive constants. Similarly to what was described in Section 3.1, by using trivial views for variables, we are able to obtain a classical linear sum equality constraint. The propagation of the interval dependent activity duration can be achieved by using the propagator of a linear sum equality constraint and views.

4   Experimental Results

To give an overview of the complex nature of the NM problem, we propose to solve four different versions of the problem. Each version adds a new source of complexity to the previous version. These problems are solved using a Constraint Programming (CP) with Large Neighborhood Search (LNS) where the branching heuristic used is a binary first fail on the *start* variables. As defined in [8], a CP with LNS search favors the exploration of the search space at the loss of completeness. When a solution is found, several relaxations are applied to it and a new search begins from the partial solution obtained. For each version of our problem, a time limit of three minutes is imposed and the best values found for both objective functions (makespan and quantity of tracer consumed) are reported in Table 1. All experiments were conducted with the OscaR open-source solver [4]. The instances considered are lists from 10 to 50 patients obtained by a biased random generator we designed. The durations of patient activities and the resource capacities and decay parameters are generated using realistic values. However, typical NM centers with the considered configurations treat at most 25 patients per day and larger instances are considered to test the limits of the model.

The first problem, V_1, is a relaxation of the NM problem in which nor continuously decreasing resources nor the interval dependent activity durations (i.e. durations are fixed for all activities) are considered and the only objective is the minimization of the makespan. Hence, V_1 is a classical Cumulative Job-Shop Problem. In Table 1, we can observe that the quantity of tracer used (TQ) increases dramatically with the number of patients and the makespan (MS). This is due to the exponential nature of the quantity of tracer required with time by patients, as stated in Equation 9.

The second problem, V_2, adds the interval dependent activity duration constraint to the V_1 version of our problem. Hence, the durations of the acquisition activities of patients are not fixed any more and we add the constraint stated in Equation 7. Again, we only consider minimization of the makespan. When comparing the results from Table 1 for the problem V_1 with results for problem V_2, we can see the makespan and quantity of resource used are higher for V_2 than for V_1. This can be explained by two main reasons. First, the solutions for problem V_1 are not solutions for problem V_2. Indeed, V_2 adds a relation linking the waiting time of patients with the duration of imagery acquisition durations. This relation could not be respected in a solution for problem V_1. Second, as V_2 relaxes the duration of acquisition activities, the search space is larger for V_2 than for V_1. As both problems V_1 and V_2 have ran under the same conditions and with the same branching heuristics, it is normal that V_2 obtains solutions as good as those for V_1.

The third problem, V_3, adds the continuously decreasing resource constraint to the V_2 version of our problem. Furthermore, the search now focuses on minimizing the quantity of tracer used. As expected, when comparing results for problem versions V_2 and V_3 in Table 1, we observe that the quantity of tracer used is on average lower for V_3 than for V_2 as opposed to the makespan which is higher. To obtain solutions which are tradeoff between the two objective function, it is interesting to consider a bi-objective search version of our problem.

The fourth problem, V_4, considers the NM problem as a bi-objective problem minimizing the makespan and the quantity of tracer used. Hence, this version will find a set of non-dominated solutions instead of a single one. To solve this problem we use a variant of the constraint introduced in [5]. Results for the version V_4 of the problem are reported in Table 1. These results are the average best solutions obtained for both objectives. We can observe the reported average of the best solutions obtained are between the best and the worst values found for V_2 and V_3 for the makespan and the quantity of tracer used. In Figure 2, we report the Pareto front obtained by V_4 as well as the best solutions obtained by V_2 and $V3$ on instances with 20 and 40 patients. We observe that some solutions obtained by V_4 are dominated by the best solutions obtained by V_2 and V_3. Nevertheless some other solutions are not dominated by best solution for V_2 nor for V_3. As such, the problem version V_4 is well suited to obtain a set of tradeoffs between the two objectives considered.

Table 1. Average objective values for different versions of problem on different size of instances. MS is the makespan and TQ is the quantity of tracer used. For problems V_1, V_2 and $V3$, the values reported are the average values for the instance size considered. For problem V_4, values reported are the averages of the best values found for each objective for the instance size considered.

Problem Version	10 Patients		20 Patients		30 Patients		40 Patients		50 Patients	
	MS	TQ	MS	TQ	MS	TQ	MS	TQ	MS	TQ
Problem V_1	251	9.97	446	40.44	650	129.9	867	516	1,048	1,268
Problem V_2	253	11.49	486	51.16	737	242.6	994	1,065	1,211	2,770
Problem V_3	291	9.17	530	39.04	779	164.7	1,029	671	1,245	1,862
Problem V_4	266	9.42	495	38.32	751	182.2	1,011	757	1,234	1,885

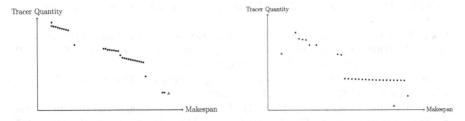

(a) Pareto front of an instance with 20 patients   (b) Pareto front of an instance with 40 patients

Fig. 2. Comparison of Pareto front solutions obtained by the problem version V_4 and the best solutions obtained by versions V_2 and V_3. The red squares are the best solutions obtained by version V_2, the blue triangles are the best solutions obtained by V_3 and the black circles are the points of the Pareto front obtained by the version V_4.

5   Conclusion

In this paper, we have described the NM problem and we modelled it as a scheduling problem. To deal with some characteristics of the problem, we introduced two new scheduling abstractions: continuously decreasing resources and interval dependent activity durations. These two scheduling abstractions were modelled with the help of views and sum constraints. Finally, we proposed an efficient method to solve the NM problem with CP and LNS. The resolution strategies and the problem tackled was declined in different versions. Each version allows to solve the problem according to the desired objective function or to perform bi-objective optimization to obtain a set of solutions which are tradeoffs between these objectives.

The proposed modelling and search techniques are generic and could be used for other cumulative scheduling problems with specific constraints. The only requirement is that these specific constraints combine existing constraints (i.e. with an existing propagator) on new variables which are defined as functions of variables of the initial problem (e.g. start and end activity variables). Thanks to the use of views, propagators of these constraints can be applied. Our approach allows a large range of cumulative scheduling problems with specific additional constraints.

Acknowledgments. The authors want to thank the anonymous reviewers for their helpful comments. This research is supported by the Mirror Project, the FRFC project 2.4504.10 of the Belgian FNRS, and the UCLouvain Action de Recherche Concerte ICTM22C1.

References

1. Apt, K.R.: Principles of Constraint Programming. Cambridge University Press, UK (2003)
2. Cherry, S., Sorenson, J., Phelps, M.: Physics in Nuclear Medicine. Elsevier Health Sciences (2012)
3. Fowler, J.S., Ido, T.: Initial and subsequent approach for the synthesis of 18fdg. Seminars in Nuclear Medicine 32(1), 6–12 (2002), Impact of FDG-PET Imaging on the Practice of Medicine
4. OscaR Team. OscaR: Scala in OR (2012), https://bitbucket.org/oscarlib/oscar
5. Schaus, P., Hartert, R.: Multi-objective large neighborhood search. In: Schulte, C. (ed.) CP 2013. LNCS, vol. 8124, pp. 611–627. Springer, Heidelberg (2013)
6. Schulte, C., Stuckey, P.J.: Efficient constraint propagation engines. Transactions on Programming Languages and Systems 31(1), 2:1–2:43 (2008)
7. Schder, H., Erdi, Y., Larson, S., Yeung, H.: Pet/ct: A new imaging technology in nuclear medicine. European Journal of Nuclear Medicine and Molecular Imaging 30(10), 1419–1437 (2003)
8. Shaw, P.: Using constraint programming and local search methods to solve vehicle routing problems. In: Maher, M.J., Puget, J.-F. (eds.) CP 1998. LNCS, vol. 1520, pp. 417–431. Springer, Heidelberg (1998)

Cost Impact Guided LNS

Michele Lombardi[1] and Pierre Schaus[2]

[1] DISI, University of Bologna
michele.lombardi2@unibo.it
[2] ICTEAM, Université Catholique de Louvain, Belgium
pierre.schaus@uclouvain.be

Abstract. In Large Neighborhood Search (LNS) [14], a problem is solved by repeatedly exploring (via tree search) a neighborhood of an incumbent solution. Whenever an improving solution is found, this replaces the current incumbent. LNS can improve dramatically the scalability of CP on large real world problems, provided a good neighborhood selection heuristic is available. Unfortunately, designing a neighborhood heuristic for LNS is still largely an art and on many problems beating a random selection requires a considerable amount of both cleverness and domain knowledge. Recently, some authors have advocated the idea to include in the neighborhood the variables that are most directly affecting the cost of the current solution. The proposed approaches, however, are either domain dependent or require non-trivial solver modifications. In this paper, we rely on constraint propagation and basic solver support to design a set of simple, cost based, domain independent neighborhood selection heuristics. Those techniques are applied on Steel Mill Slab problems illustrating the superiority of some of them over pure random relaxations.

1 Introduction

Large Neighborhood Search (LNS) is a powerful hybrid method that employs ideas from Local Search to dramatically improve the scalability of Constraint Programming (CP) on large scale optimization problems. Specifically, LNS is an iterative approach that starts from an incumbent solution and tries to improve it by using CP to explore a neighborhood. This neighborhood is usually defined by *freezing* a subset of variables, which are left assigned to the value they had in the incumbent solution. The remaining variables are instead *relaxed*, meaning that their domain is restored to its initial content. Typically, the neighborhood is explored under some search limit (e.g. maximum number of backtracks or time), to avoid spending too much time in a single iteration. If an improving solution is found, it becomes the new incumbent.

Formally, let $P = \langle z, X, D, C \rangle$ be a Constraint Optimization Problem, where X is the set of variables, D is the set of the variable domains (with D_i being the domain of x_i), and z is a cost variable. Without loss of generality we assume that z has initial domain $[-\infty, \infty]$ and must be minimized. The set C contains the problem constraints. Each constraint c_i is defined over a subset of variables $S(c_i)$, known as its *scope*. The scope can include both the X and the z variables.

H. Simonis (Ed.): CPAIOR 2014, LNCS 8451, pp. 293–300, 2014.

We view a (partial) assignment τ as a particular constraint that forces each variable x_i in its scope $S(\tau)$ to assume a specific value $v_i = \tau(x_i)$. An assignment τ is a solution if $S(\tau) = X$ and the problem $P_\tau = \langle z, X, D, C \cup \tau \rangle$ is consistent (none of the constraints detected an infeasibility). We use the special notation σ to refer to solutions.

In each LNS iteration we start from a solution σ, then we select a subset of variables to relax X_R and we build a partial assignment τ such that:

- the scope includes the variables to freeze, i.e. $S(\tau) = X \setminus X_R$
- $\tau(x_i) = \sigma(x_i)$ for each $x_i \in S(\tau)$

then we try to find a solution for P_τ with an improving cost. More precisely, let the notation $lb_\tau(x_i)$ refer to the lower bound of the domain of x_i after the propagation has reached a fix point on P_τ. Similarly, we use the notation $ub_\tau(x_i)$ for the upper bound. Then a solution σ' has better cost than σ iff $lb_{\sigma'}(z) < lb_\sigma(z)$.

The choice of the variables X_R to relax is crucial for the effectiveness of the approach. Currently, most of the best selection heuristics are domain specific and require a great deal of both expertise and knowledge to be formulated. While several researchers have addressed the topic of independent black-box search for CP [11,15,8,2], much less effort has been dedicated to make neighborhood selection in LNS problem independent. Some of the most relevant attempts in this direction are summarized in Section 2.

In this paper, we propose a novel cost driven, domain independent neighborhood selection method, based on the information collected by the progressive re-application of the current incumbent solution (a.k.a. a *dive*). The method is described in Section 3. In Section 4 we report results for the Steel Mill Slabs problem and in Section 5 we offer some concluding remarks.

2 Related Work

This section describes existing domain independent and cost based approaches for neighborhood selection in LNS. The discussion does not include *adaptive* approaches, where the goal is to automatically learn the best neighborhood (from a given pool) or the best parameters for a selection method. The interested reader is invited to check [5,3,12,6] for more details. As a remark, the integration of domain independent neighborhood selection and adaptive schemes offers a lot of opportunities and represents a very interesting topic for future research.

2.1 Propagation Guided LNS (PGLNS)

This approach is introduced in [10], where the authors define two neighborhood selection methods relying on information coming from constraint propagation. The first method defines the set of variable to *freeze* by incrementally building a partial assignment τ, starting from an empty scope. In particular:

1. First, a variable is selected at random from X and inserted in $S(\tau)$
2. The fix point for P_τ is reached
3. Then the next variable is selected at random among the 10 with the largest (non-zero) impact (defined as in [11]). If the list is empty the selection is done at random from $X \setminus S(\tau)$.
4. The process goes back to step 2, until the size of the search space of P_τ (actually, an approximation of that) is small enough.

The underlying idea is that of *freezing* related variables.

The second method, name *Reversed* PGLNS also follows an incremental scheme, but performs no propagation and relies on the availability of a closeness measure between pairs of variables. The method builds incrementally the set X_R of variables to *relax* by always choosing the next variable among the 10 with the largest (non-zero) closeness to the ones in X_R. The choice is made at random if the list is empty. As one can see, this approach is based on the idea of *relaxing* related variables. In their implementation, the authors interleave PGLNS and Reversed PGLNS and use the impacts from PGLNS to obtain the closeness scores. The Reversed PGLNS approach performed best in their experimentation.

2.2 Cost Based Neighborhoods for Scheduling Problems

As a major drawback, the PGLNS approach makes no effort to exploit the connection between variable assignment and the cost variable z. In [3], the authors propose a cost driven neighborhood selection method for scheduling problems. The main underlying idea is to include in the set X_R the variables that are most directly affecting the cost of the current solution. The authors successfully apply this idea to Job-Shop Scheduling by choosing the start variables to be relaxed among those having the smallest slack. Activities with a larger slack start to be considered only after a certain number of non-improving LNS iterations. Unfortunately, this approach cannot be considered really problem independent.

2.3 Generic Adaptive Heuristics for LNS

Several cost driven and domain independent neighborhood selection methods are proposed in [7], the most successful ones being based on the so-called *dynamic impact* of a variable. The dynamic impact tries to capture the effect that relaxing a variable would have on the problem cost. Specifically, let σ be the incumbent solution and let $\tau_{i,v}$ be an assignment that is identical to σ, except that $\tau_{i,v}(x_i) = v$. Then the impact of the pair (x_i, v) is defined as:

$$\mathcal{I}^d(x_i, v, \sigma) = lb_{\tau_{i,v}}(z) - lb_\sigma(z) \tag{1}$$

Note that $\tau_{i,v}$ is not guaranteed to be a solution, since it may make the problem infeasible. To avoid this problem, during the impact evaluation the authors disregard all constraints that are not needed for the cost computation.

The authors obtain their best results by selecting the variables to be relaxed with a probability proportional to their *mean dynamic impact*, defined as:

$$\overline{\mathcal{I}}^d(x_i, \sigma) = \frac{1}{|D_i|} \sum_{\substack{v \in D_i, \\ v \neq \sigma(x_i)}} \mathcal{I}^d(x_i, v, \sigma) \qquad (2)$$

which requires to compute the dynamic impact for each value in the original domain D_i. The author evaluated the neighborhood selection heuristic by performing a single LNS iteration starting from several reference solutions. The method based on mean impact was able to improve the solution more frequently than a random selection. This evaluation approach, although sound, may be biased by the choice of the reference solutions (e.g. improving a solution with loose constraints is very different improving one with tight constraints). The described approach has furthermore some drawbacks:

- Ignoring problem constraints (except for those needed for the cost computation) does not account for the indirect cost impact that a variable may have due to other constraints.
- Automatically detecting the constraints needed for the cost computation may not be doable. In such situation, the user would need to manually specify them, requiring a custom extension in the modeling interface.
- It is not necessarily true that measuring the cost impact of a variable under the assumption that it is the last to be assigned leads to a reliable evaluation.

3　Cost Impact Guided LNS

In this work we extend the idea introduced in [3] that an effective LNS neighborhood heuristics should be cost based. Our goal is to make this principle independent of the problem, by relying on the propagation over the cost variable, similarly to [7]. At the same time, however, we wish to avoid the drawbacks that we have identified in the previous section.

Our method relies on a cost impact metric based on the variation of the lower bounds of the cost variable[1]. Unlike [7], however, we collect those variations by incrementally re-applying the current solution in rearranged order, i.e. by performing a *dive*.

Specifically:

Definition 1. *Let π be a permutation of the variables in X and let k be the position of x_i in π. Then the cost impact of x_i w.r.t. a solution σ is the quantity:*

$$\mathcal{I}^z(x_i, \sigma, \pi) = lb_{\tau_{\pi,k}}(z) - lb_{\tau_{\pi,k-1}}(z) \qquad (3)$$

where:

$$S(\tau_{\pi,k}) = \{x_{\pi_j} \mid j = 0...k\} \qquad (4)$$

$$\tau_{\pi,k}(x_i) = \sigma(x_i) \quad \forall x_i \in S(\tau_{\pi,k}) \qquad (5)$$

i.e., $\tau_{\pi,k}$ forces the first $k+1$ variables in π to assume the value they have in σ.

---

[1] There is some similarity with the idea of pseudo-costs for MIP [1].

In other words, our impact measure is simply the variation of the cost lower bound recorded when adding the k-th assignment, during the re-application of the current solution σ in the order specified by π. It is possible to aggregate the cost impacts over a set of dives Π via their average, in this case we have:

$$\mathcal{I}^z(x_i, \sigma, \Pi) = \frac{1}{|\Pi|} \sum_{\pi \in \Pi} \mathcal{I}^z(x_i, \sigma, \pi) \tag{6}$$

A permutation-independent measure could be obtained by aggregating the cost impacts for every possible permutation. Since this would be prohibitive to obtain exactly, we propose to use the average impacts over a finite set of dives as an approximation. How often to perform the dives and how to choose the permutation π for each of them are some of the decisions that must be taken in order to design an actual neighborhood selection heuristic. Specifically, we have experimented with:

- For the diving frequency: 1) n dives per LNS iteration and 2) diving every n LNS iterations (which incurs less overhead).
- For the choice of the permutation: 1) uniformly randomized permutations and 2) decreasing-impact permutations (the variables are sorted in π by decreasing impact in an attempt to spread the cost variations).

Since the cost impacts depend on the incumbent solution, the accumulated impacts must be reinitialized whenever an improving solution is discovered. If no improving solution is found, each dive will add to the aggregated cost impacts, so that they will converge to the real average.

We experimented with different neighborhood selection strategies based on this information. Our most successful approach exploits the impacts for biasing the choice probability of the variables to be *relaxed*. The method is described in Algorithm 3 and consists in drawing a fixed number of variables from X, without replacement. The drawing probabilities are given by a score (see line 1), that in our case is a convex combination of the cost impact and a uniform quantity:

$$s_i = \alpha \cdot \mathcal{I}^z(x_i, \sigma, \Pi) + (1 - \alpha) \cdot \frac{1}{|X|} \sum_{x_j \in X} \mathcal{I}^z(x_j, \sigma, \Pi) \tag{7}$$

The presence of a uniform term ensures that even variables with zero impact have a chance to be relaxed. The strategy has a single additional parameter $\alpha \in [0, 1]$ (with $\alpha = 0$ corresponding to a pure random selection). In our experiments, we use $\alpha = 0.5$. We dive every 10 (failed) LNS attempts and every time an improving solution is found.

4   Experiments

The ability to diversify is one of the main reasons why on some benchmarks a pure random relaxation obtains very good results. For instance, on the Steel

---

Algorithm 1. Cost Impact Based Probability

---

1. assign a score s_i to each variable (see Equation 7)
2. let $r = \sum_{x_i} s_i$
3. **while** not enough variables selected for relaxation **do**
4.   pick a random value v in $[0, r]$
5.   **for all** not selected x_i **do**
6. $v = v - s_i$
7.   **if** $v \leq 0$ **then**
8. $r = r - s_i$
9.   select x_i for relaxation and continue at line 2

Mill Slab problem, a pure random relaxation was the best performer in [4,13]. Mairy et al. also concluded in [6] that their advanced reinforcement based learning strategy does not obtain better results than a random neighborhood on car sequencing problems. Given that our experimentation targets the Steel Mill Slab problem, it was natural to choose a pure random relaxation as a baseline for a comparison. Furthermore, we decided to include in our evaluation an implementation of the Reversed PGLNS from [10], because the strategy was demonstrated to be better than random relaxation on the car sequencing problem.

From the Steel Mill Slabs benchmarks most commonly employed in the literature[2], we selected the instances with 2,3,4 and 5 slab capacities (80 instances in total), because they were found to be the most difficult in [13]. We limited the number of LNS iterations per run to 1,000, so that the best solution was stable enough for each of the 3 considered relaxation strategies. The size of the neighborhood is 5 (i.e. we relax five variables) and each LNS iteration is stopped after 50 backtracks, as in [13]. All experiments were performed using the OscaR solver [9]. We report detailed results on Table 1.

As it can be seen in Table 1, the Cost Impact based relaxation dominates the Random and Reversed PGLNS on most of the instances. Surprisingly PGLNS seems inferior to Random on this problem[3]. For all of our benchmarks, we used the Student's t-test to check the statistical significance of the performance differences. On 75 over 80 instances the Cost Impact based relaxation obtains smaller average costs at a 5% significance level.

The results we obtained with other Cost Impact based relaxation strategies (not reported due to lack of space), confirmed that retaining some diversification ability is a key feature to obtain a good performance on the Steel Mill Slabs Problem. This is achieved by the proposed neighborhood selection method via the inclusion of a uniform term in the variable scores. This set of experiments confirmed also how beating a pure random relaxation strategy is far from trivial on this problem, which stresses the relevance of our results.

---

[2] The instances and best known results are available at
http://becool.info.ucl.ac.be/steelmillslab
[3] However, the description of the PGLNS approach from [10] lacks some details, hence implementation differences may exist.

Table 1. Results obtained on the instances from [13] with 2,3,4 and 5 capacities averages over 100 executions with different seeds

#capa	instance	0	1	2	3	4	5	6	7	8	9
	Random	52.98	74.1	177.62	100.02	36.4	86.32	96.87	77.47	531	114.34
2	PGLNS	55.24	75.76	178	99.64	38.75	97.06	151.72	77.82	531	110.1
	Cost Impact	**45.67**	**71.4**	**176.83**	**98.59**	**33.02**	**72.02**	**93.12**	**72.32**	**531**	**107.46**
	Random	18.05	74.11	30.38	63.6	23	67.58	67.59	78.98	118.84	235.38
3	PGLNS	26.01	79.74	36.45	65.64	18.98	75.35	69.67	86.4	133.56	236.59
	Cost Impact	**13.92**	**69.11**	**24.8**	**55.1**	**18.32**	**64.59**	**49.48**	**71.63**	**116.86**	**227.57**
	Random	38.32	40.94	**33.64**	32.31	16.16	22.1	21.78	26.59	16.03	27.25
4	PGLNS	39.36	40.12	41.33	32.2	16.36	22.25	24.05	29.25	18.14	27.5
	Cost Impact	**38.03**	**38.43**	42.59	**28.01**	**13.64**	**16.56**	**14.25**	**20.16**	**11.81**	**23.44**
	Random	5.93	32.07	15.68	10.19	21.19	18.83	9.09	17.34	14.63	27.43
5	PGLNS	7.25	32.51	16.32	13.23	21.25	21.34	12.57	18.58	13.21	28.4
	Cost Impact	**4.51**	**31.38**	**15.28**	**8.64**	**17.99**	**17.72**	**5.96**	**15.52**	**11.54**	**20.95**
#capa	instance	10	11	12	13	14	15	16	17	18	19
	Random	97.13	118.26	58.03	166.16	**159.63**	**296**	160.08	196.14	65.04	45.09
2	PGLNS	112.54	134.58	53.98	199.56	182.39	296.06	194.66	195.73	71.67	45.41
	Cost Impact	**92.69**	**123.66**	**47.2**	**157.81**	172.67	296.06	**159.72**	**195.46**	**60.64**	**45**
	Random	51.28	50.65	20.93	**84.39**	28.99	47.52	53.99	28.27	63.6	48.9
3	PGLNS	49.5	54.52	25.73	84.52	37.49	48.4	55.85	30.89	65.42	56.51
	Cost Impact	**49.01**	**40.26**	**15.53**	85.34	**23.64**	**47.05**	**47.74**	**24.58**	**54.31**	**38.93**
	Random	27.09	26.46	19.35	42.47	11.45	29.55	43.89	19.62	27.35	14.07
4	PGLNS	32.54	33.84	20.99	42.02	19.08	30.89	45.84	19.47	27.53	14.76
	Cost Impact	**25.1**	**22.81**	**9.72**	**36.43**	**11.06**	**25.88**	**36.06**	**9.02**	**20.81**	**13.9**
	Random	15.97	18.05	35.21	24.97	8.2	26.81	12.12	20.61	31.89	10.21
5	PGLNS	17.54	18.72	38.09	28.29	7.92	29.46	12.49	22.03	33.52	11.68
	Cost Impact	**14.02**	**14.61**	**32.54**	**23.16**	**5.68**	**22.07**	**9.21**	**19.77**	**29.94**	**8.51**

5   Conclusion

In this paper, we have introduced the Cost Impact, a measure of the propagation on the cost variable obtained when replaying the incumbent solution in a random or customized order. Obtaining Cost Impacts is easy and requires only basic support from the solver. In particular, our technique still allows to treat the problem constraints as black-boxes. A second contribution, we have described a simple and effective relaxation strategy based on Cost Impacts.

Our contribution can be seen as a mix of the ideas presented in [3], [10] and [7]. As in [7], we rely on the solver propagation to measure the impact on the cost. We also recognize that variables affecting the most the cost should be relaxed similarly to [3]. Finally as for PGLNS [10], our approach is problem independent and does not require to disable the propagation of any constraint when diving.

Our results have illustrated the superiority of the approach on the Steel Mill Slabs problem over a pure random relaxation and an implementation of Reversed PGLNS. Such outcome proves the potential of the proposed technique, providing motivation for future research.

Our method has been explained by representing solutions as assignments of decision variables, but it could easily be extended to more complex branching decisions (such as ordering activities in scheduling). As future work we plan to experiment the method on a broader set of problems, including scheduling variants.

References

1. Benichou, M., Gauthier, J.M., Girodet, P., Hentges, G., Ribiere, G., Vincent, O.: Experiments in mixed-integer linear programming. Mathematical Programming 1(1), 76–94 (1971)
2. Boussemart, F., Hemery, F., Lecoutre, C., Sais, L.: Boosting systematic search by weighting constraints. In: 16th European Conference on Artificial Intelligence (ECAI 2004), pp. 146–150 (2004)
3. Carchrae, T., Beck, J.C.: Principles for the design of large neighborhood search. Journal of Mathematical Modelling and Algorithms 8, 245–270 (2009)
4. Gargani, A., Refalo, P.: An efficient model and strategy for the steel mill slab design problem. In: Bessière, C. (ed.) CP 2007. LNCS, vol. 4741, pp. 77–89. Springer, Heidelberg (2007)
5. Laborie, P., Godard, D.: Self-adapting large neighborhood search: Application to single-mode scheduling problems. In: Proceedings MISTA-2007, Paris, pp. 276–284 (2007)
6. Mairy, J.-B.: Reinforced adaptive large neighborhood search. In: The Seventeenth International Conference on Principles and Practice of Constraint Programming (CP 2011), p. 55 (2011)
7. Mairy, J.-B., Schaus, P., Deville, Y.: Generic adaptive heuristics for large neighborhood search. In: Seventh International Workshop on Local Search Techniques in Constraint Satisfaction (LSCS 2010). A Satellite Workshop of CP (2010)
8. Michel, L., Van Hentenryck, P.: Activity-based search for black-box constraint programming solvers. In: Beldiceanu, N., Jussien, N., Pinson, É. (eds.) CPAIOR 2012. LNCS, vol. 7298, pp. 228–243. Springer, Heidelberg (2012)
9. OscaR Team. OscaR: Scala in OR (2012), https://bitbucket.org/oscarlib/oscar
10. Shaw, P., Furnon, V.: Propagation guided large neighborhood search. In: Wallace, M. (ed.) CP 2004. LNCS, vol. 3258, pp. 468–481. Springer, Heidelberg (2004)
11. Refalo, P.: Impact-based search strategies for constraint programming. In: Wallace, M. (ed.) CP 2004. LNCS, vol. 3258, pp. 557–571. Springer, Heidelberg (2004)
12. Ropke, S., Pisinger, D.: An adaptive large neighborhood search heuristic for the pickup and delivery problem with time windows. Transportation science 40(4), 455–472 (2006)
13. Schaus, P., van Hentenryck, P., Monette, J.-N., Coffrin, C., Michel, L., Deville, Y.: Solving steel mill slab problems with constraint-based techniques: Cp, lns, and cbls. Constraints 16(2), 125–147 (2011)
14. Shaw, P.: Using constraint programming and local search methods to solve vehicle routing problems. In: Maher, M.J., Puget, J.-F. (eds.) CP 1998. LNCS, vol. 1520, pp. 417–431. Springer, Heidelberg (1998)
15. Zanarini, A., Pesant, G.: Solution counting algorithms for constraint-centered search heuristics. In: Bessière, C. (ed.) CP 2007. LNCS, vol. 4741, pp. 743–757. Springer, Heidelberg (2007)

Proteus: A Hierarchical Portfolio
of Solvers and Transformations

Barry Hurley, Lars Kotthoff, Yuri Malitsky, and Barry O'Sullivan

Insight Centre for Data Analytics
Department of Computer Science, University College Cork, Ireland
{b.hurley,l.kotthoff,y.malitsky,b.osullivan}@4c.ucc.ie

Abstract. In recent years, portfolio approaches to solving SAT problems and CSPs have become increasingly common. There are also a number of different encodings for representing CSPs as SAT instances. In this paper, we leverage advances in both SAT and CSP solving to present a novel hierarchical portfolio-based approach to CSP solving, which we call Proteus, that does not rely purely on CSP solvers. Instead, it may decide that it is best to encode a CSP problem instance into SAT, selecting an appropriate encoding and a corresponding SAT solver. Our experimental evaluation used an instance of Proteus that involved four CSP solvers, three SAT encodings, and six SAT solvers, evaluated on the most challenging problem instances from the CSP solver competitions, involving global and intensional constraints. We show that significant performance improvements can be achieved by Proteus obtained by exploiting alternative view-points and solvers for combinatorial problem-solving.

1 Introduction

The pace of development in both CSP and SAT solver technology has been rapid. Combined with portfolio and algorithm selection technology impressive performance improvements over systems that have been developed only a few years previously have been demonstrated. Constraint satisfaction problems and satisfiability problems are both NP-complete and, therefore, there exist polynomial-time transformations between them. We can leverage this fact to convert CSPs into SAT problems and solve them using SAT solvers.

In this paper we exploit the fact that different SAT solvers have different performances on different encodings of the same CSP. In fact, the particular choice of encoding that will give good performance with a particular SAT solver is dependent on the problem instance to be solved. We show that, in addition to using dedicated CSP solvers, to achieve the best performance for solving a CSP the best course of action might be to translate it to SAT and solve it using a SAT solver. We name our approach Proteus, after the Greek god Proteus, the shape-shifting water deity that can foretell the future.

Our approach offers a novel perspective on using SAT solvers for constraint solving. The idea of solving CSPs as SAT instances is not new; the solvers Sugar, Azucar, and CSP2SAT4J are three examples of SAT-based CSP solving. Sugar [29]

H. Simonis (Ed.): CPAIOR 2014, LNCS 8451, pp. 301–317, 2014.

has been very competitive in recent CSP solver competitions. It converts the CSP to SAT using a specific encoding, known as the order encoding, which will be discussed in more detail later in this paper. Azucar [30] is a related SAT-based CSP solver that uses the compact order encoding. However, both Sugar and Azucar use a single predefined solver to solve the encoded CSP instances. Our work does not assume that conversion using a specific encoding to SAT is the best way of solving a problem, but considers multiple candidate encodings and solvers. CSP2SAT4J [21] uses the SAT4J library as its SAT back-end and a set of static rules to choose either the direct or the support encoding for each constraint. For intensional and extensional binary constraints that specify the supports, it uses the support encoding. For all other constraints, it uses the direct encoding. Our approach does not have predefined rules but instead chooses the encoding and solver based on features of the problem instance to solve.

Our approach employs algorithm selection techniques to dynamically choose whether to translate to SAT, and if so, which SAT encoding and solver to use, otherwise it selects which CSP solver to use. There has been a great deal of research in the area of algorithm selection and portfolios; we refer the reader to a recent survey of this work [20]. We note three contrasting example approaches to algorithm selection for the constraint satisfaction and satisfiability problems: CPHYDRA (CSP), SATZILLA (SAT), and ISAC (SAT). CPHYDRA [24] contains an algorithm portfolio of CSP solvers which partitions CPU-TIME between components of the portfolio in order to maximize the probability of solving a given problem instance within a fixed time limit. SATZILLA [34], at its core, uses cost-sensitive decision forests that vote on the SAT solver to use for an instance. In addition to that, it contains a number of practical optimizations, for example running a pre-solver to quickly solve the easy instances. ISAC [17] is a cluster-based approach that groups instances based on their features and then finds the best solver for each cluster. The Proteus approach is not a straightforward application of portfolio techniques. In particular, there is a series of decisions to make that affect not only the solvers that will be available, but also the information that can be used to make the decision. Because of this, the different choices of conversions, encodings and solvers cannot simply be seen as different algorithms or different configurations of the same algorithm.

The remainder of this paper is organised as follows. Section 2 motivates the need to choose the representation and solver in combination. In Section 3 we summarise the necessary background on CSP and SAT to make the paper self-contained and present an overview of the main SAT encodings of CSPs. The detailed evaluation of our portfolio is presented in Section 4. We create a portfolio-based approach to CSP solving that employs four CSP solvers, three SAT encodings, and six SAT solvers. Finally, we conclude in Section 5.

2   Multiple Encodings and Solvers

To motivate our work, we performed a detailed investigation for two solvers to assess the relationship between solver and problem encoding with features of

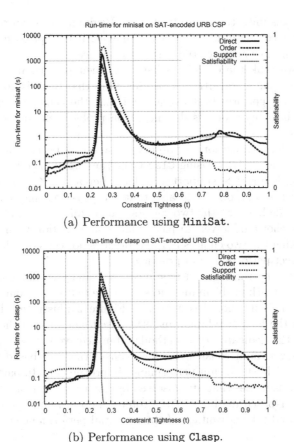

(a) Performance using MiniSat.

(b) Performance using Clasp.

Fig. 1. MiniSat and Clasp on random binary CSPs

the problem to be solved. For this experiment we considered uniform random binary (URB) CSPs with a fixed number of variables, domain size and number of constraints, and varied the constraint tightness. The constraint tightness t is a measure of the proportion of forbidden to allowed possible assignments to the variables in the scope of the constraint. We vary it from 0 to 1, where 0 means that all assignments are allowed and 1 that no assignments are part of a solution, in increments of 0.005. At each tightness the mean run-time of the solver on 100 random CSP instances is reported. Each instance contains 30 variables with domain size 20 and 300 constraints. This allowed us to study the performance of SAT encodings and solvers across the phase transition.

Figure 1 plots the run-time for MiniSat and Clasp on uniformly random binary CSPs that have been translated to SAT using three different encodings. Observe that in Figure 1(a) there is a distinct difference in the performance of MiniSat on each of the encodings, sometimes an order of magnitude. Before the phase transition, we see that the order encoding achieves the best performance and maintains this until the phase transition. Beginning at constraint tightness 0.41, the order encoding gradually starts achieving poorer performance and the support encoding now achieves the best performance.

Notably, if we rank the encodings based on their performance, the ranking changes after the phase transition. This illustrates that there is not just a single encoding that will perform best overall and that the choice of encoding matters, but also that this choice is dependent on problem characteristics such as constraint tightness.

Around the phase transition, we observe contrasting performance for Clasp, as illustrated in Figure 1(b). Using Clasp, the ranking of encodings around the phase transition is direct ≻ support ≻ order; whereas for MiniSat the ranking is order ≻ direct ≻ support. Note also that the peaks at the phase transition differ in magnitude between the two solvers. These differences underline the importance of the choice of solver, in particular in conjunction with the choice of encoding – making the two choices in isolation does not consider the interdependencies that affect performance in practice.

In addition to the random CSP instances, our analysis also comprises 1493 challenging benchmark problem instances from the CSP solver competitions that involve global and intensional constraints. Figure 2 illustrates the respective performance of the best CSP-based and SAT-based methods on these instances. Unsurprisingly the dedicated CSP methods often achieve the best performance. There are, however, numerous cases where considering SAT-based methods has the potential to yield significant performance improvements. In particular, there are a number of instances that are unsolved by any CSP solver but can be solved quickly using SAT-based methods. The Proteus approach aims to unify the best of both worlds and take advantage of the potential performance gains.

3   Background

3.1   The Constraint Satisfaction Problem

Constraint satisfaction problems (CSP) are a natural means of expressing and reasoning about combinatorial problems. They have a large number of practical applications such as scheduling, planning, vehicle routing, configuration, network design, routing and wavelength assignment [26]. An instance of a CSP is represented by a set of variables, each of which can be assigned a value from its domain. The assignments to the variables must be consistent with a set of constraints, where each constraint limits the values that can be assigned to variables.

Finding a solution to a CSP is typically done using systematic search based on backtracking. Because the general problem is NP-complete, systematic search algorithms have exponential worst-case run times, which has the effect of limiting the scalability of these methods. However, thanks to the development of effective heuristics and a wide variety of solvers with different strengths and weaknesses, many problems can be solved efficiently in practice.

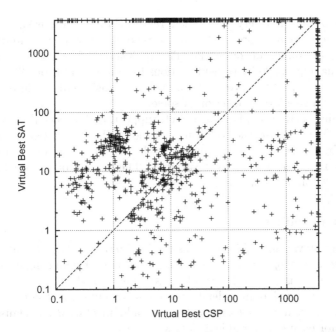

Fig. 2. Performance of the virtual best CSP portfolio and the virtual best SAT-based portfolio. Each point represents the time in seconds of the two approaches. A point below the dashed line indicates that the virtual best SAT portfolio was quicker, whereas a point above means the virtual best CSP portfolio was quicker. Clearly the two approaches are complementary: there are numerous instances for which a SAT-based approach does not perform well or fails to solve the instance but a CSP solver does extremely well, and vice-versa.

3.2 The Satisfiability Problem

The satisfiability problem (SAT) consists of a set of Boolean variables and a propositional formula over these variables. The task is to decide whether or not there exists a truth assignment to the variables such that the propositional formula evaluates to *true*, and, if this is the case, to find this assignment.

SAT instances are usually expressed in conjunctive normal form (CNF). The representation consists of a conjunction of *clauses*, where each clause is a disjunction of *literals*. A literal is either a variable or its negation. Each clause is a logical *or* of its literals and the formula is a logical *and* of each clause. The following SAT formula is in CNF:

$$(x_1 \lor x_2 \lor \neg x_4) \land (\neg x_2 \lor \neg x_3) \land (x_3 \lor x_4)$$

This instance consists of four SAT variables. One assignment to the variables which would satisfy the above formula would be to set $x_1 = true$, $x_2 = false$, $x_3 = true$ and $x_4 = true$.

SAT, like CSP, has a variety of practical real world applications such as hardware verification, security protocol analysis, theorem proving, scheduling, routing, planning, digital circuit design [5]. The application of SAT to many of these problems is made possible by transformations from representations like the constraint satisfaction problem. We will study three transformations into SAT that can benefit from this large collection of solvers.

The following sections explain the direct, support, and direct-order encodings that we use. We will use the following notation. The set of CSP variables is represented by the set $\mathcal{X}$. We use uppercase letters to denote CSP variables in $\mathcal{X}$; lowercase x_i and x_v refer to SAT variables. The domain of a CSP variable X is denoted $D(X)$ and has size d.

3.3   Direct Encoding

Translating a CSP variable X into SAT using the *direct encoding* [32], also known as the *sparse encoding*, creates a SAT variable for each value in its domain: $x_1, x_2, \ldots, x_d$. If x_v is *true* in the resulting SAT formula, then $X = v$ in the CSP solution. This means that in order to represent a solution to the CSP, exactly one of $x_1, x_2, \ldots, x_d$ must be assigned *true*. We add an *at-least-one* clause to the SAT formula for each CSP variable as follows:

$$\forall X \in \mathcal{X} : (x_1 \vee x_2 \vee \ldots \vee x_d).$$

Conversely, to ensure that only one of these can be set to *true*, we add *at-most-one* clauses. For each pair of distinct values in the domain of X, we add a binary clause to enforce that at most one of the two can be assigned *true*. The series of these binary clauses ensure that only one of the SAT variables representing the variable will be assigned *true*, i.e.

$$\forall v, w \in D(X) : (\neg x_v \vee \neg x_w).$$

Constraints between CSP variables are represented in the direct encoding by enumerating the conflicting tuples. For binary constraints for example, we add clauses as above to forbid both values being used at the same time for each disallowed assignment. For a binary constraint between a pair of variables X and Y, we add the conflict clause $(\neg x_v \vee \neg y_w)$ if the tuple $\langle X = v, Y = w \rangle$ is forbidden. For intensionally specified constraints, we enumerate all possible tuples and encode the disallowed assignments.

Example 1 (Direct Encoding). Consider a simple CSP with three variables $\mathcal{X} = \{X, Y, Z\}$, each with domain $\langle 1, 2, 3 \rangle$. We have an all-different constraint over the variables: alldifferent(X, Y, Z), which we represent by encoding the pairwise disequalities. Table 1 shows the complete direct-encoded CNF formula for this CSP. The first 12 clauses encode the domains of the variables, the remaining clauses encode the constraints between X, Y, and Z. There is an implicit conjunction between these clauses.

Table 1. An example of the direct encoding

Domain Clauses	$(x_1 \lor x_2 \lor x_3)$ $(\neg x_1 \lor \neg x_2)$ $(\neg x_1 \lor \neg x_3)$ $(\neg x_2 \lor \neg x_3)$
	$(y_1 \lor y_2 \lor y_3)$ $(\neg y_1 \lor \neg y_2)$ $(\neg y_1 \lor \neg y_3)$ $(\neg y_2 \lor \neg y_3)$
	$(z_1 \lor z_2 \lor z_3)$ $(\neg z_1 \lor \neg z_2)$ $(\neg z_1 \lor \neg z_3)$ $(\neg z_2 \lor \neg z_3)$
$X \neq Y$	$(\neg x_1 \lor \neg y_1)$ $(\neg x_2 \lor \neg y_2)$ $(\neg x_3 \lor \neg y_3)$
$X \neq Z$	$(\neg x_1 \lor \neg z_1)$ $(\neg x_2 \lor \neg z_2)$ $(\neg x_3 \lor \neg z_3)$
$Y \neq Z$	$(\neg y_1 \lor \neg z_1)$ $(\neg y_2 \lor \neg z_2)$ $(\neg y_3 \lor \neg z_3)$

3.4 Support Encoding

The *support encoding* [9, 18] uses the same mechanism as the direct encoding to encode CSP domains into SAT – each value in the domain of a CSP variable is encoded as a SAT variable which represents whether or not it takes that value. However, the support encoding differs on how the constraints between variables are encoded. Given a constraint between two variables X and Y, for each value v in the domain of X, let $S_{Y,X=v} \subset D(Y)$ be the subset of the values in the domain of Y which are consistent with assigning $X = v$. Either x_v is *false* or one of the consistent assignments from $y_1 \ldots y_d$ must be true. This is encoded in the support clause

$$\neg x_v \lor \left(\bigvee_{i \in S_{Y,X=v}} y_i \right).$$

Conversely, for each value w in the domain of Y, a support clause is added for the supported values in X which are consistent with assigning $Y = w$.

An interesting property of the support encoding is that if a constraint has no consistent values in the corresponding variable, a unit-clause will be added, thereby pruning the values from the domain of a variable which cannot exist in any solution. Also, a solution to a SAT formula without the *at-most-one* constraint in the support encoding represents an arc-consistent assignment to the CSP. Unit propagation on this SAT instance establishes arc-consistency in optimal worst-case time for establishing arc-consistency [9].

Example 2 (Support Encoding). Table 2 gives the complete support-encoded CNF formula for the simple CSP given in Example 1. The first 12 clauses encode the domains and the remaining ones the support clauses for the constraints. There is an implicit conjunction between clauses.

3.5 Order Encoding

Unlike the direct and support encoding, which model $X = v$ as a SAT variable for each value v in the domain of X, the order encoding (also known as the regular encoding [2]) creates SAT variables to represent $X \leq v$. If X is less than or equal to v (denoted $x_{\leq v}$), then X must also be less than or equal to $v + 1$

Table 2. An example of the support encoding

Domain Clauses	$(x_1 \lor x_2 \lor x_3)\ (\neg x_1 \lor \neg x_2)\ (\neg x_1 \lor \neg x_3)\ (\neg x_2 \lor \neg x_3)$ $(y_1 \lor y_2 \lor y_3)\ (\neg y_1 \lor \neg y_2)\ (\neg y_1 \lor \neg y_3)\ (\neg y_2 \lor \neg y_3)$ $(z_1 \lor z_2 \lor z_3)\ (\neg z_1 \lor \neg z_2)\ (\neg z_1 \lor \neg z_3)\ (\neg z_2 \lor \neg z_3)$
$X \neq Y$	$(\neg x_1 \lor y_2 \lor y_3)\ (\neg x_2 \lor y_1 \lor y_3)\ (\neg x_3 \lor y_1 \lor y_2)$ $(\neg y_1 \lor x_2 \lor x_3)\ (\neg y_2 \lor x_1 \lor x_3)\ (\neg y_3 \lor x_1 \lor x_2)$
$X \neq Z$	$(\neg x_1 \lor z_2 \lor z_3)\ (\neg x_2 \lor z_1 \lor z_3)\ (\neg x_3 \lor z_1 \lor z_2)$ $(\neg z_1 \lor x_2 \lor x_3)\ (\neg z_2 \lor x_1 \lor x_3)\ (\neg z_3 \lor x_1 \lor x_2)$
$Y \neq Z$	$(\neg y_1 \lor z_2 \lor z_3)\ (\neg y_2 \lor z_1 \lor z_3)\ (\neg y_3 \lor z_1 \lor z_2)$ $(\neg z_1 \lor y_2 \lor y_3)\ (\neg z_2 \lor y_1 \lor y_3)\ (\neg z_3 \lor y_1 \lor y_2)$

$(x_{\leq v+1})$. Therefore, we add clauses to enforce this consistency across the domain as follows:

$$\forall_v^{d-1} : (\neg x_{\leq v} \lor x_{\leq v+1}).$$

This linear number of clauses is all that is needed to encode the domain of a CSP variable into SAT in the order encoding. In contrast, the direct and support encodings require a quadratic number of clauses in the domain size.

The order encoding is naturally suited to modelling inequality constraints. To state $X \leq 3$, we would just post the unit clause $(x_{\leq 3})$. If we want to model the constraint $X = v$, we could rewrite it as $(X \leq v \land X \geq v)$. $X \geq v$ can then be rewritten as $\neg X \leq (v-1)$. To state that $X = v$ in the order encoding, we would encode $(x_{\leq v} \land \neg x_{\leq v-1})$. A conflicting tuple between two variables, for example $\langle X = v, Y = w \rangle$ can be written in propositional logic and simplified to a CNF clause using De Morgan's Law:

$$\neg((x_{\leq v} \land x_{\geq v}) \land (y_{\leq w} \land y_{\geq w}))$$
$$\neg((x_{\leq v} \land \neg x_{\leq v-1}) \land (y_{\leq w} \land \neg y_{\leq w-1}))$$
$$\neg(x_{\leq v} \land \neg x_{\leq v-1}) \lor \neg(y_{\leq w} \land \neg y_{\leq w-1})$$
$$(\neg x_{\leq v} \lor x_{\leq v-1} \lor \neg y_{\leq w} \lor y_{\leq w-1})$$

Example 3 (Order Encoding). Table 3 gives the complete order-encoded CNF formula for the simple CSP specified in Example 1. There is an implicit conjunction between clauses in the notation.

3.6 Combining the Direct and Order Encodings

The direct encoding and the order encoding can be combined to produce a potentially more compact encoding. A variable's domain is encoded in both representations and clauses are added to chain between them. This gives flexibility in the representation of each constraint. Here, we choose the encoding which gives the most compact formula. For example, for inequalities we use the order encoding since it is naturally suited, but for a (dis)equality we would use the direct encoding. This encoding is referred to as direct-order throughout the paper.

Table 3. An example of the order encoding

Domain Clauses	$(\neg x_{\leq 1} \vee x_{\leq 2})\ (\neg x_{\leq 2} \vee x_{\leq 3})\ (x_{\leq 3})$
	$(\neg y_{\leq 1} \vee y_{\leq 2})\ (\neg y_{\leq 2} \vee y_{\leq 3})\ (y_{\leq 3})$
	$(\neg z_{\leq 1} \vee z_{\leq 2})\ (\neg z_{\leq 2} \vee z_{\leq 3})\ (z_{\leq 3})$
$X \neq Y$	$(\neg x_{\leq 1} \vee \neg y_{\leq 1})$
	$(\neg x_{\leq 2} \vee x_{\leq 1} \vee \neg y_{\leq 2} \vee y_{\leq 1})$
	$(\neg x_{\leq 3} \vee x_{\leq 2} \vee \neg y_{\leq 3} \vee y_{\leq 2})$
$X \neq Z$	$(\neg x_{\leq 1} \vee \neg z_{\leq 1})$
	$(\neg x_{\leq 2} \vee x_{\leq 1} \vee \neg z_{\leq 2} \vee z_{\leq 1})$
	$(\neg x_{\leq 3} \vee x_{\leq 2} \vee \neg z_{\leq 3} \vee z_{\leq 2})$
$Y \neq Z$	$(\neg y_{\leq 1} \vee \neg z_{\leq 1})$
	$(\neg y_{\leq 2} \vee y_{\leq 1} \vee \neg z_{\leq 2} \vee z_{\leq 1})$
	$(\neg y_{\leq 3} \vee y_{\leq 2} \vee \neg z_{\leq 3} \vee z_{\leq 2})$

3.7 Algorithm Portfolios

The Algorithm Selection Problem [25] is to select the most appropriate algorithm for solving a particular problem. It is especially relevant in the context of algorithm portfolios [11,16], where a single solver is replaced with a set of solvers and a mechanism for selecting a subset to use on a particular problem.

Algorithm portfolios have been used with great success for solving both SAT and CSP instances in systems such as SATzilla [34], ISAC [17] or CPHYDRA [24]. Most approaches are similar in that they relate the characteristics of a problem to solve to the performance of the algorithms in the portfolio. The aim of an algorithm selection model is to provide a prediction as to which algorithm should be used to solve the problem. The model is usually induced using some form of machine learning.

There are three main approaches to using machine learning to build algorithm selection models. First, the problem of predicting the best algorithm can be treated as a classification problem where the label to predict is the algorithm. Second, the training data can be clustered and the algorithm with the best performance on a particular cluster assigned to it. The cluster membership of any new data decides the algorithm to use. Finally, regression models can be trained to predict the performance of each portfolio algorithm in isolation. The best algorithm for a problem is chosen based on the predicted performances.

Our approach makes a series of decisions – whether a problem should be solved as a CSP or a SAT problem, which encoding should be used for converting into SAT, and finally which solver should be assigned to tackle the problem. Approaches that make a series of decisions are usually referred to as hierarchical models. [33] and [12] use hierarchical models in the context of a SAT portfolio. They first predict whether the problem to be solved is expected to be satisfiable or not and then choose a solver depending on that decision. Our approach is closer to [10], which first predicts what level of consistency the `alldifferent` constraint should achieve before deciding on its implementation.

To the best of our knowledge, no portfolio approach that potentially transforms the representation of a problem in order to be able to solve it more efficiently exists at present.

4   Experimental Evaluation

4.1   Setup

The hierarchical model we present in this paper consists of a number of layers to determine how the instance should be solved. At the top level, we decide whether to solve the instance using as a CSP or using a SAT-based method. If we choose to leave the problem as a CSP, then one of the dedicated CSP solvers must be chosen. Otherwise, we must choose the SAT encoding to apply, followed by the choice of SAT solver to run on the SAT-encoded instance.

Each decision of the hierarchical approach aims to choose the direction which has the potential to achieve the best performance in that sub-tree. For example, for the decision to choose whether to solve the instance using a SAT-based method or not, we choose the SAT-based direction if there is a SAT solver and encoding that will perform faster than any CSP solver would. Whether this particular encoding-solver combination will be selected subsequently depends on the performance of the algorithm selection models used in that sub-tree of our decision mechanism. For regression models, the training data is the best performance of any solver under that branch of the tree. For classification models, it is the label of the sub-branch with the virtual best performance.

This hierarchical approach presents the opportunity to employ different decision mechanisms at each level. We consider 6 regression, 19 classification, and 3 clustering algorithms, which are listed below. For each of these algorithms, we evaluate the performance using 10-fold cross-validation. The dataset is split into 10 partitions with approximately the same size and the same distribution of the best solvers. One partition is used for testing and the remaining 9 partitions as the training data for the model. This process is repeated with a different partition considered for testing each time until every partition has been used for testing. We measure the performance in terms of PAR10 score. The PAR10 score for an instance is the time it takes the solver to solve the instance, unless the solver times out. In this case, the PAR10 score is ten times the timeout value. The sum over all instances is divided by the number of instances.

Instances. In our evaluation, we consider CSP problem instances from the CSP solver competitions [1]. Of these, we consider all instances defined using global and intensional constraints that are not trivially solved during 2 seconds of feature computation. We also exclude all instances which were not solved by any CSP or SAT solver within the time limit of 1 hour. Altogether, we obtain 1,493 non-trivial instances from problem classes such as Timetabling, Frequency Assignment, Job-Shop, Open-Shop, Quasi-group, Costas Array, Golomb Ruler, Latin Square, All Interval Series, Balanced Incomplete Block Design, and many others. This set includes both small and large arity constraints and all of the global constraints used during the CSP solver competitions: all-different, element, weighted sum, and cumulative.

For the SAT-based approaches, Numberjack [15] was used to translate a CSP instance specified in XCSP format [27] into SAT (CNF).

Features. A fundamental requirement of any machine learning algorithm is a set of representative features. We explore a number of different feature sets to train our models: *i*) features of the original CSP instance, *ii*) features of the direct-encoded SAT instance, *iii*) features of the support-encoded SAT instance, *iv*) features of the direct-order-encoded SAT instance and *v*) a combination of all four feature sets. These features are described in further detail below.

We computed the 36 features used in CPHYDRA for each CSP instance using `Mistral`; for reasons of space we will not enumerate them all here. The set includes static features like statistics about the types of constraints used, average and maximum domain size; and dynamic statistics recorded by running `Mistral` for 2 seconds: average and standard deviation of variable weights, number of nodes, number of propagations and a few others. Instances which are solved by `Mistral` during feature computation are filtered out from the dataset.

In addition to the CSP features, we computed the 54 SAT features used by SATZILLA [34] for each of the encoded instances and different encodings. The features encode a wide range of different information on the problem such as problem size, features of the graph-based representation, balance features, the proximity to a Horn formula, DPLL probing features and local search probing features.

Constraint Solvers. Our CSP models are able to choose from 4 complete CSP solvers:

- Abscon [22],
- Choco [31],
- Gecode [8], and
- Mistral [14].

Satisfiability Solvers. We considered the following 6 complete SAT solvers:

- clasp [7],
- cryptominisat [28],
- glucose [3],
- lingeling [4],
- riss [23], and
- MiniSat [6].

Learning Algorithms. We evaluate a number of regression, classification, and clustering algorithms using WEKA [13]. All algorithms, unless otherwise stated use the default parameters. The regression algorithms we used were Linear-Regression, PaceRegression, REPTree, M5Rules, M5P, and SMOreg. The classification algorithms were BayesNet, BFTree, ConjunctiveRule, DecisionTable, FT, HyperPipes, IBk (nearest neighbour) with 1, 3, 5 and 10 neighbours, J48, J48graft, JRip, LADTree, MultilayerPerceptron, OneR, PART, RandomForest, RandomForest with 99 random trees, RandomTree, REPTree, and SimpleLogistic. For clustering, we considered EM, FarthestFirst, and SimplekMeans. The FarthestFirst and SimplekMeans algorithms require the number of clusters to

Table 4. Performance of the learning algorithms for the hierarchical approach. The 'Category Bests' consists of the hierarchy of algorithms where at each node of the tree of decisions we take the algorithm that achieves the best PAR10 score for that particular decision.

Classifier	Mean PAR10	Number Solved
VBS	97	1493
Proteus	1774	1424
M5P with csp features	2874	1413
Category Bests	2996	1411
M5Rules with csp features	3225	1398
M5P with all features	3405	1397
LinearRegression with all features	3553	1391
LinearRegression with csp features	3588	1383
MultilayerPerceptron with csp features	3594	1382
lm with csp features	3654	1380
RandomForest99 with csp features	3664	1379
IBk10 with csp features	3720	1377
RandomForest99 with all features	3735	1383

be given as input. We evaluated with multiples of 1 through 5 of the number of solvers in the respective data set given as the number of clusters. The number of clusters is represented by $1n$, $2n$ and so on in the name of the algorithm, where n stands for the number of solvers.

We use the LLAMA toolkit [19] to train and test the algorithm selection models.

4.2   Portfolio and Solver Results

The performance of each of the 6 SAT solvers was evaluated on the three SAT encodings of 1,493 CSP competition benchmarks with a time-out of 1 hour and limited to 2GB of RAM. The 4 CSP solvers were evaluated on the original CSPs. Our results report the PAR10 score and number of instances solved for each of the algorithms we evaluate. The PAR10 is the sum of the runtimes over all instances, counting 10 times the timeout if that was reached. Data was collected on a cluster of Intel Xeon E5430 Processors (2.66Ghz) running CentOS 6.4. This data is available online.[1]

The performance of a number of hierarchical approaches is given in Table 4. The hierarchy of algorithms which produced the best overall results for our dataset involves M5P regression with CSP features at the root node to choose SAT or CSP, M5P regression with CSP features to select the CSP solver, LinearRegression with CSP features to select the SAT encoding, LinearRegression with CSP features to select the SAT solver for the direct encoded instance, LinearRegression with CSP features to select the SAT solver for the direct-order

---

[1] http://4c.ucc.ie/~bhurley/proteus/

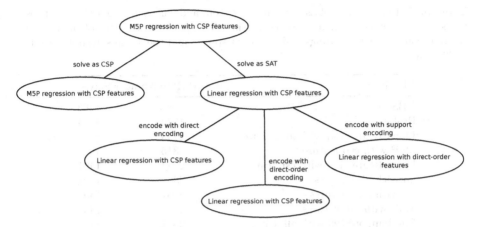

Fig. 3. Overview of the machine learning models used in the hierarchical approach

encoded instance, and LinearRegression with the direct-order features to select the SAT solver for the support encoded instance. The hierarchical tree of specific machine learning approaches we found to deliver the best overall performance on our data set is labelled Proteus and is depicted in Figure 3.

We would like to point out that in many solver competitions the difference between the top few solvers is fewer than 10 additional instances solved. In the 2012 SAT Challenge for example, the difference between the first and second place single solver was only 3 instances and the difference among the top 4 solvers was only 8 instances. The results we present in Table 4 are therefore very significant in terms of the gains we are able to achieve.

Our results demonstrate the power of Proteus. The performance it delivers is very close to the virtual best (VBS), that is the best performance possible if an oracle could identify the best choice of representation, encoding, and solver, on an instance by instance basis. The improvements we achieve over other approaches are similarly impressive. The results conclusively demonstrate that having the option to convert a CSP to SAT does not only have the potential to achieve significant performance improvements, but also does so in practice.

An interesting observation is that the CSP features are consistently used in each of the top performing approaches. One reason for this is that it is quicker to compute only the CSP features instead of the CSP features, then converting to SAT and computing the SAT features in addition. The additional overhead of computing SAT features is worthwhile in some cases though, for example for LinearRegression, which is at its best performance using all the different feature sets. Note that for the best tree of models (cf. Figure 3), it is better to use the features of the direct-order encoding for the decision of which solver to choose for a support-encoded SAT instance despite the additional overhead.

We also compare the hierarchical approach to that of a flattened setting with a single portfolio of all solvers and encoding solver combinations. The flattened portfolio includes all possible combinations of the 3 encodings and the 6 SAT

Table 5. Ranking of each classification, regression, and clustering algorithm to choose the solving mechanism in a flattened setting. The portfolio consists of all possible combination of the 3 encodings and the 6 SAT solvers and the 4 CSP solvers for a total of 22 solvers.

Classifier	Mean PAR10	Number Solved
VBS	97	1493
Proteus	1774	1424
LinearRegression with all features	2144	1416
M5P with csp features	2315	1401
LinearRegression with csp features	2334	1401
lm with all features	2362	1407
lm with csp features	2401	1398
M5P with all features	2425	1404
RandomForest99 with all features	2504	1401
SMOreg with all features	2749	1391
RandomForest with all features	2859	1386
IBk3 with csp features	2877	1378

solvers and the 4 CSP solvers for a total of 22 solvers. Table 5 shows these results. The regression algorithm LinearRegression with all features gives the best performance using this approach. However, it is significantly worse than the performance achieved by the hierarchical approach of Proteus.

4.3   Greater than the Sum of Its Parts

Given the performance of Proteus, the question remains as to whether a different portfolio approach that considers just CSP or just SAT solvers could do better. Table 6 summarizes the virtual best performance that such portfolios could achieve. We use all the CSP and SAT solvers for the respective portfolios to give us VB CSP and VB SAT, respectively. The former is the approach that always chooses the best CSP solver for the current instance, while the latter chooses the best SAT encoding/solver combination. VB Proteus is the portfolio that chooses the best overall approach/encoding. We show the actual performance of Proteus for comparison. Proteus is better than the virtual bests for all portfolios that consider only one encoding. This result makes a very strong point for the need to consider encoding and solver in combination.

Proteus outperforms four other VB portfolios. Specifically, the VB CPHYDRA is the best possible performance that could be obtained from that portfolio if a perfect choice of solver was made. Neither SATZILLA nor ISAC-based portfolios consider different SAT encodings. Therefore, the best possible performance either of them could achieve for a specific encoding is represented in the last three lines of Table 6.

These results do not only demonstrate the benefit of considering the different ways of solving CSPs, but also eliminate the need to compare with existing portfolio systems since we are computing the best possible performance that any of

Table 6. Virtual best performances ranked by PAR10 score

Method	Mean PAR10	Number Solved
VB Proteus	97	1493
Proteus	1774	1424
VB CSP	3577	1349
VB CPHydra	4581	1310
VB SAT	17373	775
VB DirectOrder Encoding	17637	764
VB Direct Encoding	21736	593
VB Support Encoding	21986	583

those systems could theoretically achieve. Proteus impressively demonstrates its strengths by significantly outperforming oracle approaches that use only a single encoding.

5 Conclusions

We have presented a portfolio approach that does not rely on a single problem representation or set of solvers, but leverages our ability to convert between problem representations to increase the space of possible solving approaches. To the best of our knowledge, this is the first time a portfolio approach like this has been proposed. We have shown that, to achieve the best performance on a constraint satisfaction problem, it may be beneficial to translate it to a satisfiability problem. For this translation, it is important to choose both the encoding and satisfiability solver in combination. In doing so, the contrasting performance among solvers on different representations of the same problem can be exploited. The overall performance can be improved significantly compared to restricting the portfolio to a single problem representation.

We demonstrated empirically the significant performance improvements Proteus can achieve on a large set of diverse benchmarks using a portfolio based on a range of different state-of-the-art solvers. We have investigated a range of different CSP to SAT encodings and evaluated the performance of a large number of machine learning approaches and algorithms. Finally, we have shown that the performance of Proteus is close to the very best that is theoretically possible for solving CSPs and significantly outperforms the theoretical best for portfolios that consider only a single problem encoding.

In this work, we make a general decision to encode the entire problem using a particular encoding. A natural extension would be to mix and vary the encoding depending on attributes of the problem. An additional avenue for future work would be to generalize the concepts in this paper to other problem domains where transformations, like CSP to SAT, exist.

Acknowledgements. This work is supported by Science Foundation Ireland (SFI) Grant 10/IN.1/I3032 and FP7 FET-Open Grant 284715. The Insight Centre for Data Analytics is supported by SFI Grant SFI/12/RC/2289.

References

1. CSP Solver Competition Benchmarks (2009),
   http://www.cril.univ-artois.fr/~lecoutre/benchmarks.html
2. Ansótegui, C., Manyà, F.: Mapping Problems with Finite-Domain Variables to Problems with Boolean Variables. In: H. Hoos, H., Mitchell, D.G. (eds.) SAT 2004. LNCS, vol. 3542, pp. 1–15. Springer, Heidelberg (2005)
3. Audemard, G., Simon, L.: Glucose 2.3 in the SAT 2013 Competition. In: Proceedings of SAT Competition 2013, p. 42 (2013)
4. Biere, A.: Lingeling, Plingeling and Treengeling Entering the SAT Competition 2013. In: Proceedings of SAT Competition 2013 (2013)
5. Biere, A., Heule, M.J.H., van Maaren, H., Walsh, T. (eds.): Handbook of Satisfiability. Frontiers in Artificial Intelligence and Applications, vol. 185. IOS Press (February 2009)
6. Een, N., Sörensson, N.: Minisat 2.2 (2013), http://minisat.se
7. Gebser, M., Kaufmann, B., Neumann, A., Schaub, T.: Clasp: A conflict-driven answer set solver. In: Baral, C., Brewka, G., Schlipf, J. (eds.) LPNMR 2007. LNCS (LNAI), vol. 4483, pp. 260–265. Springer, Heidelberg (2007)
8. Gecode Team: Gecode: Generic Constraint Development Environment (2006), http://www.gecode.org
9. Gent, I.P.: Arc Consistency in SAT. In: Proceedings of the 15th European Conference on Artificial Intelligence — ECAI 2002, pp. 121–125 (2002)
10. Gent, I.P., Kotthoff, L., Miguel, I., Nightingale, P.: Machine learning for constraint solver design – a case study for the alldifferent constraint. In: 3rd Workshop on Techniques for Implementing Constraint Programming Systems (TRICS), pp. 13–25 (2010)
11. Gomes, C.P., Selman, B.: Algorithm portfolios. Artificial Intelligence 126(1-2), 43–62 (2001)
12. Haim, S., Walsh, T.: Restart strategy selection using machine learning techniques. In: Kullmann, O. (ed.) SAT 2009. LNCS, vol. 5584, pp. 312–325. Springer, Heidelberg (2009)
13. Hall, M., Frank, E., Holmes, G., Pfahringer, B., Reutemann, P., Witten, I.H.: The WEKA data mining software: An update. SIGKDD Explor. Newsl. 11(1), 10–18 (2009)
14. Hebrard, E.: Mistral, A Constraint Satisfaction Library. In: Proceedings of the Third International CSP Solver Competition (2008)
15. Hebrard, E., O'Mahony, E., O'Sullivan, B.: Constraint Programming and Combinatorial Optimisation in Numberjack. In: Lodi, A., Milano, M., Toth, P. (eds.) CPAIOR 2010. LNCS, vol. 6140, pp. 181–185. Springer, Heidelberg (2010)
16. Huberman, B.A., Lukose, R.M., Hogg, T.: An economics approach to hard computational problems. Science 275(5296), 51–54 (1997)
17. Kadioglu, S., Malitsky, Y., Sellmann, M., Tierney, K.: ISAC – Instance-Specific Algorithm Configuration. In: Coelho, H., Studer, R., Wooldridge, M. (eds.) ECAI. Frontiers in Artificial Intelligence and Applications, vol. 215, pp. 751–756. IOS Press (2010)
18. Kasif, S.: On the Parallel Complexity of Discrete Relaxation in Constraint Satisfaction Networks. Artificial Intelligence 45(3), 275–286 (1990), http://dx.doi.org/10.1016/0004-3702(90)90009-0
19. Kotthoff, L.: LLAMA: leveraging learning to automatically manage algorithms. Tech. Rep. arXiv:1306.1031, arXiv (June 2013), http://arxiv.org/abs/1306.1031

20. Kotthoff, L.: Algorithm Selection for Combinatorial Search Problems: A Survey. AI Magazine (to appear, 2014)
21. Le Berre, D., Lynce, I.: CSP2SAT4J: A Simple CSP to SAT Translator. In: Proceedings of the Second International CSP Solver Competition (2008)
22. Lecoutre, C., Tabary, S.: Abscon 112, Toward more Robustness. In: Proceedings of the Third International CSP Solver Competition (2008)
23. Manthey, N.: The SAT Solver RISS3G at SC 2013. In: Proceedings of SAT Competition 2013, p. 72 (2013)
24. O'Mahony, E., Hebrard, E., Holland, A., Nugent, C., O'Sullivan, B.: Using Case-based Reasoning in an Algorithm Portfolio for Constraint Solving. In: Proceeding of the 19th Irish Conference on Artificial Intelligence and Cognitive Science (2008)
25. Rice, J.R.: The algorithm selection problem. Advances in Computers 15, 65–118 (1976)
26. Rossi, F., van Beek, P., Walsh, T.: Handbook of Constraint Programming. Foundations of Artificial Intelligence. Elsevier, New York (2006)
27. Roussel, O., Lecoutre, C.: XML Representation of Constraint Networks: Format XCSP 2.1. CoRR abs/0902.2362 (2009)
28. Soos, M.: Cryptominisat 2.9.0 (2011)
29. Tamura, N., Tanjo, T., Banbara, M.: System Description of a SAT-based CSP Solver Sugar. In: Proceedings of the Third International CSP Solver Competition, pp. 71–75 (2009)
30. Tanjo, T., Tamura, N., Banbara, M.: Azucar: A SAT-Based CSP Solver Using Compact Order Encoding — (Tool Presentation). In: Cimatti, A., Sebastiani, R. (eds.) SAT 2012. LNCS, vol. 7317, pp. 456–462. Springer, Heidelberg (2012)
31. choco team: choco: An Open Source Java Constraint Programming Library (2008)
32. Walsh, T.: SAT v CSP. In: Dechter, R. (ed.) CP 2000. LNCS, vol. 1894, pp. 441–456. Springer, Heidelberg (2000)
33. Xu, L., Hoos, H.H., Leyton-Brown, K.: Hierarchical hardness models for SAT. In: Bessière, C. (ed.) CP 2007. LNCS, vol. 4741, pp. 696–711. Springer, Heidelberg (2007)
34. Xu, L., Hutter, F., Hoos, H.H., Leyton-Brown, K.: SATzilla: Portfolio-based Algorithm Selection for SAT. Journal of Artificial Intelligence Research pp. 565–606 (2008)

Buffered Resource Constraint: Algorithms and Complexity

Christian Bessiere[1], Emmanuel Hebrard[2], Marc-André Ménard[3],
Claude-Guy Quimper[3], and Toby Walsh[4]

[1] CNRS, Université Montpellier, LIRMM
bessiere@lirmm.fr
[2] CNRS, Université de Toulouse, LAAS
hebrard@laas.fr
[3] Université Laval
marc-andre.menard.2@ulaval.ca, claude-guy.quimper@ift.ulaval.ca
[4] NICTA, University of New South Wales
toby.walsh@nicta.com.au

Abstract. The notion of buffered resource is useful in many problems. A buffer contains a finite set of items required by some activities, and changing the content of the buffer is costly. For instance, in instruction scheduling, the registers are a buffered resource and any switch of registers has a significant impact on the total runtime of the compiled code.
We first show that sequencing activities to minimize the number of switches in the buffer is NP-hard. We then introduce an algorithm which, given a set of already sequenced activities, computes a buffer assignment which minimizes the number of switches in linear time, i.e., $O(nd)$ where n is the length of the sequence and d the number of buffered items. Next, we introduce an algorithm to achieve bound consistency on the constraint SWITCH, that bounds the number of changes in the buffer, in $O(n^2d + n^{1.5}d^{1.5})$ time. Finally, we report the results of experimental evaluations that demonstrate the efficiency of this propagator.

1 Introduction

We consider a special type of resource, a *buffer*, corresponding to a set of items required by some tasks. In order to process a task, all items required by the task must be present in the buffer. However, the buffer has a limited capacity, and adding a new item is costly. Therefore, one may want to minimize the total number of changes, or *switches*. For instance, in instruction scheduling, the buffer can stand for memory caches, and minimizing the number of switches corresponds to minimizing page faults. Alternatively, the buffer may correspond to the reels of colored threads on an embroidery machine, and minimizing the number of reels changes over a sequence helps reducing the overall processing time. Yet another example arises in the design of validation plans for satellite payload [3]. Here, each test requires some components of the payload to be in a given configuration, and again the total number of configuration changes during the test campaign is a significant factor of its total duration.

H. Simonis (Ed.): CPAIOR 2014, LNCS 8451, pp. 318–333, 2014.
© Springer International Publishing Switzerland 2014

We show that achieving hybrid consistency on the BUFFEREDRESOURCE constraint, i.e., domain consistency on the integer variables and bound consistency on the set variables, is NP-hard. We therefore consider a decomposition involving an ALLDIFFERENT constraint to enforce that the sequence is a permutation of the original set of tasks, and the constraint SWITCH, that counts the number of *switches* along the sequence.

We introduce an algorithm for finding a support of SWITCH in linear time, that is, $O(nd)$ where n is the length of the sequence and d the number of items. Moreover, we show how bound consistency can be enforced in $O(n^2d + n^{1.5}d^{1.5})$ time using a flow representation.

Finally, we compare our filtering algorithm against a decomposition on two crafted optimization problems, albeit derived from industrial applications. In both cases the objective function is defined using one or several SWITCH constraints. In these experiments, the proposed propagation algorithm for SWITCH greatly outperforms the standard decomposition.

The paper is organized as follows: In Section 2 we recall some background about consistency on constraints involving both integer and set variables. In Section 3 we define the BUFFEREDRESOURCE and SWITCH constraints and discuss their complexities. Then, in Section 4, we introduce an algorithm for achieving bound consistency on the SWITCH constraint. Finally, in Section 5, we report experimental results.

2   Formal Background

A constraint satisfaction problem consists of a set of variables, each with a finite domain of values, and a set of constraints specifying allowed combinations of values for subsets of variables. We write $\mathrm{dom}(X)$ for the domain of a variable X. For totally ordered domains, we write $\min(X)$ and $\max(X)$ for the minimum and maximum values. A solution is an assignment of values to the variables satisfying the constraints. We consider both *integer* and *set* variables. A set variable S is represented by its lower bound $\mathrm{lb}(S)$ which contains the definite elements and an upper bound $\mathrm{ub}(S)$ which also contains the potential elements.

Constraint solvers typically explore partial assignments enforcing a local consistency property using either specialized or general purpose propagation algorithms. Given a constraint C, a *bound support* on C is a tuple that assigns to each integer variable a value between its minimum and maximum, and to each set variable a set between its lower and upper bounds which satisfies C. According to [1], a bound support in which each integer variable is assigned a value in its domain is called a *hybrid support*. If C involves only integer variables, a hybrid support is a *support*. A constraint C is *bound consistent* (*BC*) iff for each integer variable X_i, its minimum and maximum values belong to a bound support, and for each set variable S_j, the values in $\mathrm{ub}(S_j)$ belong to S_j in at least one bound support and the values in $\mathrm{lb}(S_j)$ belong to S_j in all bound supports. A constraint C is *hybrid consistent* (*HC*) iff for each integer variable X_i, every value in $\mathrm{dom}(X_i)$ belongs to a hybrid support, and for each set variable S_j, the values in $\mathrm{ub}(S_j)$ belong to S_j in at least one hybrid support, and the values in $\mathrm{lb}(S_j)$ are all those from $\mathrm{ub}(S_j)$ that belong to S_j in all hybrid supports. A constraint C involving only integer variables is *generalized arc consistent* (*GAC*) iff for each variable X_i, every value in $\mathrm{dom}(X_i)$ belongs to a support.

If all variables in C are integer variables, hybrid consistency reduces to generalized arc-consistency, and if all variables in C are set variables, hybrid consistency reduces to bound consistency.

To illustrate these concepts, consider the constraint $C(X_1, X_2, S)$ that holds iff the set variable S is assigned exactly the values used by the integer variables X_1 and X_2. Let $\mathrm{dom}(X_1) = \{1, 3\}$, $\mathrm{dom}(X_2) = \{2, 4\}$, $\mathrm{lb}(S) = \{2\}$ and $\mathrm{ub}(S) = \{1, 2, 3, 4\}$. BC does not remove any value since all domains are already bound consistent. On the other hand, HC removes 4 from $\mathrm{dom}(X_2)$ and from $\mathrm{ub}(S)$ as there does not exist any tuple satisfying C in which X_2 does not take value 2. Note that as BC deals with bounds, value 2 was considered as possible for X_1.

3   The BUFFEREDRESOURCE and SWITCH Constraints

We consider the problem of performing a set of tasks, each requiring a set of resources to be available on a buffer, whilst bounding the number of switches of resources on the buffer. We are given:

- A maximum buffer size $\overline{k}_i$ and a minimum buffer usage $\underline{k}_i$ at time i;
- A set of resources $R = \{r_1, \ldots, r_m\}$;
- A set of tasks $\mathcal{T} = \{T_i\}_{1 \le i \le n}$, where each task $T_i \in \mathcal{T}$ is associated with the set of resources it requires, that is, $\forall i \in [1, n]$, $T_i \subseteq R$.

Example 1. For instance, suppose that we want to embroid n garments. Each garment requires a set of colors. However, only $\overline{k}$ reels and therefore $\overline{k}$ different colors of thread can be loaded on the embroidery machine. Hence, whenever we embroid a garment requiring a color of thread that is not already mounted on the machine, we need to switch it with one of the currently mounted reels. Each such switch is time consuming. Therefore, the goal is to sequence the garments so that we minimize the number of reel changes. In other words, we want to compute a permutation of the tasks $p : [1, n] \mapsto [1, n]$ and an assignment $\sigma : [1, n] \mapsto 2^R$ of the buffer over time such that the items required by each task are buffered ($\forall 1 \le i \le n$, $T_i \subseteq \sigma(p_i)$), the size of the buffer is not exceeded ($\underline{k}_i \le |\sigma(i)| \le \overline{k}_i$) and the number of switches $\sum_{1 \le i < n} |\sigma(i+1) \setminus \sigma(i)|$ is minimized.

We introduce the BUFFEREDRESOURCE constraint to model this pattern.

Definition 1 (BUFFEREDRESOURCE). *Let $X_1, \ldots, X_n$ be integer variables, $S_1, \ldots, S_n$ be set variables, $\underline{k}_1, \ldots, \underline{k}_n$ and $\overline{k}_1, \ldots, \overline{k}_n$ be integers, and M an integer variable. The constraint* BUFFEREDRESOURCE($[X_1, \ldots, X_n], [S_1, \ldots, S_n], [\underline{k}_1, \ldots, \underline{k}_n], [\overline{k}_1, \ldots, \overline{k}_n], M$) *holds if and only if:*

1. $\forall i, j \in [1, n]$, $i \ne j \rightarrow X_i \ne X_j$ ($X_1, \ldots, X_n$ is a permutation)
2. $\forall i \in [1, n]$, $\underline{k}_i \le |S_i| \le \overline{k}_i$ (the buffer has a bounded capacity)
3. $\forall i$, $T_i \subseteq S_{X_i}$ (when a task is processed, all required resources are buffered)
4. $\sum_{1 \le i < n} |S_{i+1} \setminus S_i| \le M$ (the number of switches is less than or equal to M)

We shall see that this constraint is NP-hard (even if the buffer's size is fixed). Hence, throughout the rest of the paper we shall consider a decomposition:

$$\text{BUFFEREDRESOURCE}(X_1, \ldots, X_n, S_1, \ldots, S_n, [\underline{k}_1, \ldots, \underline{k}_n], [\overline{k}_1, \ldots, \overline{k}_n], M) \Leftrightarrow$$
$$\text{ALLDIFFERENT}(X_1, \ldots, X_n)$$
$$\wedge \qquad \forall i \in [1, n], \ \underline{k}_i \leq |S_i| \leq \overline{k}_i$$
$$\wedge \qquad \forall i, \ T_i \subseteq S_{X_i}$$
$$\wedge \qquad \text{SWITCH}([S_1, \ldots, S_n], [\underline{k}_1, \ldots, \underline{k}_n], [\overline{k}_1, \ldots, \overline{k}_n], M)$$

And in particular the constraint SWITCH, defined as follows:

Definition 2 (SWITCH). *Let $S_1, \ldots, S_n$ be set variables, $\underline{k}_i$ a lower bound on the cardinality of S_i, $\overline{k}_i$ an upper bound on the cardinality of S_i, and M an integer variable. The constraint $\text{SWITCH}([S_1, \ldots, S_n], [\underline{k}_1, \ldots, \underline{k}_n], [\overline{k}_1, \ldots, \overline{k}_n], M)$ holds if and only if: $\forall i \in [1, n], \ \underline{k}_i \leq |S_i| \leq \overline{k}_i \wedge \sum_{1 \leq i < n} |S_{i+1} \setminus S_i| \leq M$.*

Example 1 (Continued). Assume that we want to embroid 5 garments, each requiring one of 5 colors as shown in Fig. 1a, on a machine with 3 reels of thread. Let the domains shown Fig. 1b represent the possible permutations at some point during search. To this sequence of variables corresponds a sequence of set variables shown in Fig. 1c. Whereas the BUFFEREDRESOURCE constraint defines the possible combinations for all X's and S's, the constraint SWITCH involves only the set variables. We illustrate a support for SWITCH with $M = 2$ in Fig. 1d and a feasible solution for BUFFEREDRESOURCE corresponding to the permutation $2, 1, 3, 4, 5$ with 4 switches in Fig. 1e.

$T_1 = \{B,G,Y\}$	$X_1 = \{1,2\}$	$\{B,G\} \subseteq S_1 \subseteq \{B,G,R,Y\}$	$S_1 = \{B,G,R\}$	$S_1 = \{B,G,R\}$
$T_2 = \{B,G,R\}$	$X_2 = \{1,2\}$	$\{B,G\} \subseteq S_2 \subseteq \{B,G,R,Y\}$	$S_2 = \{B,G,R\}$	$S_2 = \{B,G,Y\}$
$T_3 = \{W,Y\}$	$X_3 = \{3,4,5\}$	$\{R\} \subseteq S_3 \subseteq \{B,R,W,Y\}$	$S_3 = \{B,W,R\}$	$S_3 = \{B,W,Y\}$
$T_4 = \{B,R,W\}$	$X_4 = \{3,5\}$	$\{W,Y\} \subseteq S_4 \subseteq \{R,W,Y\}$	$S_4 = \{Y,W,R\}$	$S_4 = \{B,W,R\}$
$T_5 = \{R,W,Y\}$	$X_5 = \{4,5\}$	$\{R,W\} \subseteq S_5 \subseteq \{B,R,W,Y\}$	$S_5 = \{Y,R,W\}$	$S_5 = \{Y,W,R\}$
(a) Garments	(b) Perm.	(c) Buffer bounds	(d) Support	(e) Solution

Fig. 1. Illustration of Example 1

Theorem 1. *Achieving HC on* BUFFEREDRESOURCE *is NP-hard.*

Proof. We reduce Hamiltonian path (on undirected graphs) to the problem of finding a satisfying solution to BUFFEREDRESOURCE. Let $G = (V, E)$ be an undirected graph with $|V| = n$ and $|E| = m$. We build an instance of BUFFEREDRESOURCE on the integer variables $[X_1, \ldots, X_n]$, the set variables $[S_1, \ldots, S_n]$ and switch variable M.

- For each $v_i \in V$, we have a task T_i requiring m items, that is, a set T_i of cardinality m that contains one value j for every edge $e_j \in E$ such that $v_i \in e_j$. In order to make sure that $|T_i| = m$, we use as fillers values appearing in no other task ($\{i * m + j\}_{1 \leq j \leq m - d(v_i)}$ where $d(v_i)$ is the degree of v_i).

- There is one integer variable X_i per vertex $v_i \in V$ with domain $\{1, \ldots, n\}$.
- There are as many set variables S_i as vertices in the graph G, with domains ranging from the empty set to the whole universe of values: $\{\} \subseteq S_i \subseteq \{1, \ldots, (n+1)m\}$ for each $v_i \in V$. The cardinality of each one of these set variables is $\underline{k}_i = \overline{k}_i = m$.
- The domain of M is set to the single value $(n-1)(m-1)$.

We first show that the existence of an Hamiltonian path entails the existence of a solution of BUFFEREDRESOURCE on the construction above. We set the value of X_i to the rank of the vertex v_i in the Hamiltonian path. Observe that given a permutation, there is a unique valuation of the set variables satisfying the constraint: $\forall i \in [1, n]$, $S_{X_i} = T_i$. Consider any two consecutive set variables S_j, S_{j+1}. Their domains correspond to two consecutive vertices in the Hamiltonian path, hence they share exactly one value: the common edge between the two nodes. The number of switches between these two set variables is thus equal to $m - 1$, hence the total number of switches is $(n-1)(m-1)$.

Next we show that the existence of a solution entails the existence of an Hamiltonian path in G. Consider any two consecutive set variables S_j, S_{j+1}, and assume that $X_i = j$ and $X_k = j + 1$. There are two cases, either there exists an edge $e_x = (v_i, v_k)$ in E and then $T_i \cap T_k = \{x\}$ or such edge does not exist, and therefore $T_i \cap T_k = \emptyset$. Since there are $(n-1)$ consecutive pairs of set variables, and since the only possible value for M is $(n-1)(m-1)$, the first case must hold for every consecutive pair. We can therefore conclude that there exists a path visiting every vertex of the graph exactly once.   □

Observe that the cardinality of each set variable S_i can be as low as 3 since Hamiltonian path is still NP-hard when the maximum degree of a vertex is 3. On the other hand, the proof above requires both the total number of resources and the bound on the number of switches to be large.

4   Filtering Algorithm for SWITCH

In this section, we show how bound consistency can be enforced on SWITCH in $O(n^2 d + n^{1.5} d^{1.5})$ time. First we introduce a greedy algorithm that finds an assignment minimizing the number of switches in $O(nd)$ time where d is the total number of resources. Let L be the number of switches of that assignment. Then we introduce a filtering procedure based on a network flow representation. The cost of the flow represents the number of switches and for each pair S_i, v such that $v \in ub(S_i)$ there is an edge in the network that can receive a unit of flow if and only if the set S_i may contain the value v in a support. Observe that forcing $v \in S_i$ may never entail more than two extra switches in the optimal assignment with L switches. Therefore, we need to prune the set variables only if the difference between L and the upper bound of M is at most 1.

The algorithm we propose therefore proceeds as follows. First, we compute a flow of minimum cost, which is a support, and provides a lower bound on M. Then, if $\max(M) - \min(M) \leq 1$, we consider the residual graph with respect

Table 1. Summary of the algorithm and its complexity

Step	Complexity
Finding an optimal assignment	$O(nd)$
Re-weighting the residual graph	$O((nd)^{1.5})$
Finding null cycles	$O(nd)$
Finding cycles of weight 1	$O(n^2d)$
Total	$O(n^2d + n^{1.5}d^{1.5})$

to this flow, and re-weight the edges so as to eliminate negative costs. Then we find all null cycles and, if $\max(M) - \min(M) = 1$, all cycles of weight 1. Table 1 summarizes the complexity of these four steps.

4.1   Finding a Support

We present an algorithm that greedily constructs a support for the SWITCH constraint. This support is optimal in the sense that it minimizes the number of switches. The algorithm FindSupport (Algorithm 1) successively assigns the variables S_1 to S_n to sets σ_1 to σ_n. At each step i, the algorithm computes a priority for each value. The lower the priority is for value v, the more likely the value v will belong to σ_i. While processing the variables S_i for $i = 1..n$, we maintain for each value v the index $\text{next}_\in(v) = \min(\{j \mid v \in \text{lb}(S_j), j \geq i\} \cup \{n+1\})$ that is the smallest variable index no smaller than i such that $v \in \text{lb}(S_j)$. We also maintain the index $\text{next}_\notin(v) = \min(\{j \mid v \notin \text{ub}(S_j), j \geq i\} \cup \{n+1\})$ that is the smallest variable index no smaller than i such that $v \notin \text{ub}(S_j)$. If $\text{next}_\in(v) < \text{next}_\notin(v)$, the value v will be required in the sequence $S_i \ldots S_n$ before it gets forbidden. We assign such a value a priority between 1 and n. If $\text{next}_\in(v) > \text{next}_\notin(v)$, the value v will be forbidden in the sequence $S_i, \ldots, S_n$ before it gets required. We assign such a value a priority between $n+2$ and $2n+1$. Finally, if $\text{next}_\in(v) = \text{next}_\notin(v)$, we assign to value v a priority of $n+1$. This later case only occurs if the value v is allowed to appear but not required to appear in every variable of the sequence $S_i, \ldots, S_n$. The insertion of a value that is not required or that does not belong to the previous set σ_{i-1} induces a unnecessary switch. Such a value is given a penalty of $2n+1$ on its priority. Once the priority is computed, we add the value $l_v = \text{prio}(v) \times (d+1) + v$ to a set L. From l_v, we can retrieve the value using the arithmetic operation $l_v \bmod (d+1)$. Moreover, the smaller the priority is, the smaller the value l_v is. The algorithm keeps a counter k of the number of values that can be added to σ_i without causing an unnecessary switch, i.e. a switch that is not due to the requirement $v \in \text{lb}(S_i)$. This counter is the cardinality that will be given to σ_i. If $k \notin [\underline{k}_i, \overline{k}_i]$, we update k to the closest value between $\underline{k}_i$ and $\overline{k}_i$. We call the algorithm Selection(L, k) to retrieve the kth smallest element in L. This algorithm has a running time complexity of $O(d)$ when implemented using a divide-and-conquer strategy and

---

Algorithm 1. FindSupport($[S_1, \ldots, S_n], [\underline{k}_1, \ldots, \underline{k}_n], [\overline{k}_1, \ldots, \overline{k}_n]$)

---

    for $v = 1..d$ **do** $\text{next}_\in(v) \leftarrow \text{next}_{\notin}(v) \leftarrow 1$;
    $\sigma_0 \leftarrow \emptyset$;
    for $i = 1..n$ **do**
        $L \leftarrow \emptyset,\ k \leftarrow 0$;
        for $v \in \text{ub}(S_i)$ **do**
            if $\text{next}_\in(v) < i$ **then** $\text{next}_\in(v) \leftarrow i$;

1   **while** $\text{next}_\in(v) \leq n \wedge v \notin \text{lb}(S_{\text{next}_\in(v)})$ **do** $\text{next}_\in(v) \leftarrow \text{next}_\in(v) + 1$;
            if $\text{next}_{\notin}(v) < i$ **then** $\text{next}_{\notin}(v) \leftarrow i$;

2   **while** $\text{next}_{\notin}(v) \leq n \wedge v \in \text{ub}(S_{\text{next}_{\notin}(v)})$ **do** $\text{next}_{\notin}(v) \leftarrow \text{next}_{\notin}(v) + 1$;

$$\text{prio} \leftarrow \begin{cases} \text{next}_\in(v) & \text{if } \text{next}_\in(v) < \text{next}_{\notin}(v) \\ n+1 & \text{if } \text{next}_\in(v) = \text{next}_{\notin}(v)\ ; \\ 2(n+1) - \text{next}_{\notin}(v) & \text{if } \text{next}_\in(v) > \text{next}_{\notin}(v) \end{cases}$$

3   **if** $v \notin \text{lb}(S_i) \wedge v \notin \sigma_{i-1}$ **then** prio $\leftarrow$ prio $+2n+1$ **else** $k \leftarrow k+1$
           $L \leftarrow L \cup \{\text{prio} \times (d+1) + v\}$;
        $k \leftarrow \min(\max(k, \underline{k}_i), \overline{k}_i)$;
        $l_{\max} \leftarrow \texttt{Selection}(L, k)$;
        $\sigma_i \leftarrow \{l \bmod (d+1) \mid l \in L \wedge l \leq l_{\max}\}$;
    return $[\sigma_1, \ldots, \sigma_n]$;

---

a randomized partition algorithm [4]. We finally retrieve the k smallest elements l_v from L and include their respective value v in the set σ_i. Some values might share a same priority, the algorithm breaks ties on the lexicographical order of the values in order not to obtain sets with cardinality greater than k. The vector $[\sigma_1, \ldots, \sigma_n]$ constitutes an optimal support to the constraint.

Theorem 2. *The algorithm* FindSupport *returns a support of* SWITCH *that minimizes the number of switches.*

Proof. Let $\text{next}^i_\in(v)$, $\text{next}^i_{\notin}(v)$, and $\text{prio}^i(v)$ be the values of $\text{next}_\in(v)$, $\text{next}_{\notin}(v)$ and $\text{prio}(v)$ at iteration i.

We consider an optimal solution σ that cannot be built by FindSupport. Let θ_i be the instantiation of S_i by FindSupport and let i be the first index such that $\theta_i \neq \sigma_i$. One of the three following propositions is true:

1. $\sigma_i \subset \theta_i$, hence $\exists v$ such that $|\{w \mid \text{prio}^i(w) < \text{prio}^i(v)\}| \leq |\theta_i|$, $v \notin \sigma_i$;
2. $\theta_i \subset \sigma_i$, hence $\exists v$ such that $|\{w \mid \text{prio}^i(w) < \text{prio}^i(v)\}| > |\theta_i|$, $v \in \sigma_i$;
3. $\sigma_i \not\subset \theta_i$ & $\theta_i \not\subset \sigma_i$, hence $\exists v, w$ s.t. $\text{prio}^i(v) < \text{prio}^i(w)$, $v \notin \sigma_i$ and $w \in \sigma_i$;

Case 1: Since $v \notin \sigma_i$, we have $v \notin \text{lb}(S_i)$ and $|\theta_i| > \underline{k}_i$. Therefore, $|\sigma_{i-1} \cup \text{lb}(S_i)| > \underline{k}_i$. It follows that $v \in \sigma_{i-1}$, hence adding v to σ_i does not add any v-switch (and might prevent one latter).

Case 2: Since $v \notin \theta_i$, we have $v \notin \mathrm{lb}(S_i)$ and since $v \in \sigma_i$ we have $|\theta_i| > \underline{k}_i$. Therefore, $|\sigma_{i-1} \cup \mathrm{lb}(S_i)| < \overline{k}_i$. It follows that $v \notin \sigma_{i-1}$, hence removing v from σ_i suppresses one v-switch (and might entail one latter).

Case 3: From now on, we assume $\mathrm{prio}^i(v) < \mathrm{prio}^i(w)$, $v \notin \sigma_i$ and $w \in \sigma_i$. We now show that for some $j > i$, we can swap all instances of w for v in the sets $\sigma_i, \ldots, \sigma_{j-1}$ whilst not increasing the number of switches. Let σ' be the solution obtained by this transformation on σ. For each such operation, we get strictly closer to a solution that can be obtained by FindSupport. We say that there is a v-switch at index i in solution σ iff $v \in \sigma_i \setminus \sigma_{i-1}$. Since the transformation only changes indices i to $j-1$ and values v and w, we only have to count v-switches and w-switches from index i to j.

Notice that $\mathrm{prio}^i(v) < \mathrm{prio}^i(w)$ & $w \in \sigma_{i-1}$ implies $v \in \sigma_{i-1}$. Indeed, in Line 3 of Algorithm 1, we make sure that values in $\sigma_{i-1} \cup \mathrm{lb}(S_i)$ have the best priority. Moreover, since w and v can be interchanged, none of them is in $\mathrm{lb}(S_i)$.

We first consider the case where only v is in the previous buffer: $w \notin \sigma_{i-1}$ and $v \in \sigma_{i-1}$. Let j be the minimum integer greater than i such that either $w \notin \sigma_j$ or $v \in \sigma_j$ or $w \in \mathrm{lb}(S_j)$ or $v \notin \mathrm{ub}(S_j)$ or $j = n+1$.

Now we count v- and w-switches on the solutions σ and σ'.

- On σ there is a w-switch at index i, and there may be a v-switch at index j.
- On σ' there is no v-switch, however there may be a w-switch at index j.

Therefore, in this case the transformation can only decrease the number of switches, or leave it unchanged.

Now we consider the case where both v and w are either in or out the previous buffer: $w \in \sigma_{i-1} \Leftrightarrow v \in \sigma_{i-1}$. Let j be the minimum integer greater than i such that either $w \notin \sigma_j$ or $v \in \sigma_j$ or $j = n+1$. We show that $v \in \mathrm{ub}(S_l)$ and $w \notin \mathrm{lb}(S_l)$ for all $i \leq l < j$, i.e.,

$$j \leq \mathrm{next}^i_{\in}(w) \ and \ j \leq \mathrm{next}^i_{\notin}(v) \tag{1}$$

Now, by hypothesis, $\mathrm{prio}^i(w) > \mathrm{prio}^i(v)$. There are two cases:

1. $\mathrm{next}^i_{\in}(v) < \mathrm{next}_{\notin}(i)v$: In this case, $\mathrm{next}^i_{\in}(v) < \mathrm{next}^i_{\in}(w)$. However, $v \in \sigma_{\mathrm{next}^i_{\in}(v)}$ entails $j \leq \mathrm{next}^i_{\in}(v)$, therefore proposition 1 is correct in this case.
2. $\mathrm{next}^i_{\in}(v) \geq \mathrm{next}_{\notin}(i)v$: In this case, $\mathrm{next}^i_{\in}(w) \geq \mathrm{next}^i_{\notin}(w)$ and $\mathrm{next}_{\notin}(i)v > \mathrm{next}^i_{\notin}(w)$. However, $w \notin \sigma_{\mathrm{next}^i_{\notin}(w)}$ we have $j \leq \mathrm{next}^i_{\notin}(w)$, therefore proposition 1 is correct in this case too.

Now we count v- and w-switches on the solutions σ and σ'.

- On σ there is a w-switch at index i iff $w \notin \sigma_{i-1}$. Moreover, if $v \in \sigma_j$ there is a v-switch at index j.
- On σ': there is a v-switch at index i iff $v \notin \sigma_{i-1}$. Moreover, if $w \in \sigma_j$ there is a w-switch at index j.

By definition of j, $w \in \sigma_j$ implies that $v \in \sigma_j$, hence there is a w-switch at index j in σ' only if there is a v-switch at index j in σ. Moreover, by hypothesis, $w \notin \sigma_{i-1} \Leftrightarrow v \notin \sigma_{i-1}$. It follows that the number of switches may not increase after the transformation. □

The increments on line 1 and 2 are executed at most n times for each value v. The Selection algorithm is called exactly n times and has a running time complexity of $O(d)$. Consequently, the algorithm FindSupport runs in time $O(nd)$.

4.2   Network Flow Model

Let $\sigma_1, \ldots, \sigma_n$ be an optimal solution with L switches, such as the one computed by FindSupport. For any set S_i, any value $v \in \mathrm{ub}(S_i)$ can be inserted into σ_i by adding at most 2 more switches (unless $|\mathrm{lb}(S_i)| = \overline{k}_i$). Indeed we can add a value v to S_i or replace any value $w \in S_i \setminus \mathrm{lb}(S_i)$ with v, entailing at most one switch with S_{i-1} and one switch with S_{i+1}. Hence no pruning is required unless $\max(M) - L < 2,$.

We therefore focus on the case where $\max(M) - L < 2$ and we describe an algorithm that finds the values that can be used without additional switches and the values that require exactly one additional switch.

We construct a network flow G_s where every flow of value d represents a solution to the SWITCH constraint. The network has the following nodes. For every $v \in \mathrm{ub}(S_i)$, we have the nodes a_i^v and b_i^v. For $0 \leq i \leq n$, we have a *collector* node C_i. The collector node C_0 is the source node and the collector node C_n is the sink node.

Each edge has a capacity of the form $[l, u]$ where l is the required amount and u is the allowed amount of flow that can circulate through the edge. The network G_s has the following edges:

1. An edge between the node a_i^v and b_i^v for every $v \in \mathrm{ub}(S_i)$
   (a) with capacity $[1, 1]$ if $v \in \mathrm{lb}(S_i)$;
   (b) with capacity $[0, 1]$ otherwise.
2. An edge between b_i^v and a_{i+1}^v of capacity $[0, 1]$ if these two nodes exist.
3. An edge between b_i^v and C_i of capacity $[0, 1]$ for $1 \leq i \leq n$ and $v \in \mathrm{ub}(S_i)$.
4. An edge between C_{i-1} and a_i^v of capacity $[0, 1]$ for $1 \leq i \leq n$ and $v \in \mathrm{ub}(S_i)$.
5. An edge between C_{i-1} and C_i of capacity $[d - \overline{k}_i, d - \underline{k}_i]$ for $1 \leq i \leq n$.

All edges have a cost of zero except the edges (C_{i-1}, a_i^v) for $2 \leq i \leq n$ and $v \in \mathrm{ub}(S_i)$ that have a cost of 1. The cost $c(f)$ of a flow f is the sum, over the edges, of the cost of the edge times the amount of flow on this edge, i.e. $c(f) = \sum_{(x,y) \in E} f(x, y) c(x, y)$. Figure 2 shows the network flow of Example 1.

Lemma 1. *Each flow of value d in G_s corresponds to an assignment of the* SWITCH *constraint where capacities are satisfied.*

Proof. We construct a solution $\sigma_1, \ldots, \sigma_n$ by setting $v \in \sigma_i$ if and only if the edge (a_i^v, b_i^v) accepts a unit of flow. Since the flow value is d and there are between $d - \overline{k}_i$ and $d - \underline{k}_i$ units of flow circulating through the edge (C_{i-1}, C_i), there are between $\underline{k}_i$ and $\overline{k}_i$ edges (a_i^v, b_i^v) accepting a unit of flow, thus $\underline{k}_i \leq |\sigma_i| \leq \overline{k}_i$. □

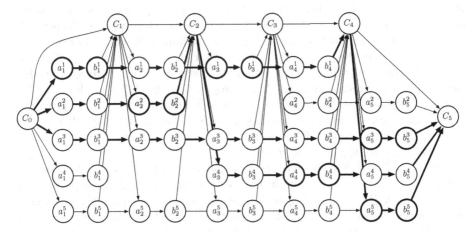

Fig. 2. The network flow associated to the SWITCH constraint of Example 1. Bold edges represent the flow. A pair of nodes (a_i^v, b_i^v) in bold indicates that $v \in \mathrm{lb}(S_i)$ and that the flow must go through the edge (a_i^v, b_i^v).

Lemma 2. *The cost of a flow in G_s gives the number of switches in the solution.*

Proof. The amount of flow going out of the collector C_{i-1} to the nodes a_i^v indicates the number of switches between S_{i-1} and S_i. Since these edges have a cost of 1, the cost of the flow equals the number of switches in the solution. Note that a flow can go through the nodes b_{i-1}^v, C_{i-1}, and then a_i^v which counts as a switch. Such cases do not occur in a minimum-cost flow as the flow could go through the edge (b_{i-1}^v, a_i^v) at a lesser cost. $\square$

4.3   Bound Consistency

The function `FindSupport` computes a minimum-cost flow with a flow value of d. Let $f(x, y)$ be the amount of flow circulating on the edge (x, y) that has capacity $[l, u]$. The residual graph G_r has the same nodes as the original network but has the following edges. If $f(x, y) < u$, there is an edge (x, y) with capacity $[l - f(x, y), u - f(x, y)]$ and cost $c(x, y)$. If $f(x, y) > l$ there is an edge (y, x) with capacity $[f(x, y) - u, f(x, y) - l]$ and cost $-c(x, y)$. Note that in the residual graph, all costs are either -1, 0, or 1.

Theorem 3. *Let f be a flow of minimum cost in G_s, and G_r the corresponding residual graph, SWITCH is BC if and only if:*

1. *For all $v \in \mathrm{ub}(S_i)$, $f(a_i^v, b_i^v) > 0$ or the edge (a_i^v, b_i^v) belongs to a cycle of cost lower than or equal to $\max(M) - c(f)$ in G_r.*
2. *For all $v \in \mathrm{lb}(S_i)$, $f(a_i^v, b_i^v) > 0$ and there is no cycle of cost lower than or equal to $\max(M) - c(f)$ in G_r that involves the edge (b_i^v, a_i^v).*

Proof. By Lemma 1, we know that a flow in G_s corresponds to an assignment of $S_1, \ldots, S_n$ that satisfies the capacities of the sets. Moreover, by Theorem 2, we know that the support corresponding to f minimizes the number of switches. Hence, f witnesses the existence of a support for all $v \in S_i$ such that $f(a_i^v, b_i^v) > 0$. Now, if $f(a_i^v, b_i^v) = 0$, it is known that there is a flow going through the edge (a_i^v, b_i^v) iff there exists a cycle in G_r passing by the edge (a_i^v, b_i^v). Moreover, by Lemma 2, the cost of the minimum cycle passing by the edge (a_i^v, b_i^v) gives the number of required additional switches if $v \in S_i$ is forced. It follows that in this case, $v \in S_i$ has a support iff there exists a cycle of cost $\max(M) - c(f)$ or less going through (a_i^v, b_i^v) in the graph G_r.

The flow f satisfies the bounds and capacities of the sets. Therefore, if $f(a_i^v, b_i^v) = 0$ then v cannot be in the lower bound of S_i. Now, suppose that $f(a_i^v, b_i^v) > 0$. By a reasoning similar as above, we have that there exists a bound support where $v \notin S_i$ iff there exists a cycle of cost $\max(M) - c(f)$ or less going through (b_i^v, a_i^v) in the graph G_r. Therefore, $v \in \mathrm{lb}(S_i)$ is entailed iff there is no such cycle. $\qquad\square$

Recall that we assume $\max(M) \geq c(f) + 2$ where $c(f)$ is the number of switches in an optimal solution. Therefore, we are interested to find all cycles of cost 0 and 1. We now describe how can this be done efficiently.

Since computing the minimum cycles is as hard as finding shortest paths, we first perform a preprocessing operation that will eliminate negative weights in the residual graph. This preprocessing is the same as the one used in Johnson's all-pair shortest path algorithm and was also used by Régin to filter COST-GCC [7].

We add a dummy node z to the residual graph that is connected to all other nodes with an edge of null cost. We compute the shortest path from the dummy node z to all other nodes. This can be done using Goldberg's algorithm [5] which computes the shortest path in a graph in $O(\sqrt{|V|}|E| \log W)$ time where $|V|$ is the number of vertices, $|E|$ is the number of edges, and W is the greatest absolute cost of an edge. In our case, we have $|V| \in O(nd)$, $|E| \in O(nd)$, and $W = 1$ which leads to a complexity of $O((nd)^{1.5})$. Let $\pi(x)$ be the shortest distance between the dummy node z and the node x. Let (x, y) be an edge in the residual graph with cost $c(x, y)$. We re-weight this edge with the cost function $c_\pi(a, b) = c(a, b) + \pi(a) - \pi(b)$. It is known that with the new cost function, there are no negative edges and that the cost of any cycle remains unchanged.

Finding null cycles in the new re-weighted residual graph becomes an easy problem. Since there are no negative edges, a cycle is null if and only if all its edges have a null cost. We can thus compute the strongly connected components in the graph induced by the edges of null costs. All values $v \in \mathrm{ub}(S_i)$ such that the edge (a_i^v, b_i^v) belongs to a null cycle has a support without extra switches.

To compute the supports with one additional switch, we modify once more the residual graph. We delete all edges with a cost greater than 1. Such edges necessarily lie on a path of weight greater than 1 and are not relevant. Only edges with weight 0 and 1 remain in the graph. We create two copies of the graph. We connect the two copies as follows: if there is an edge (x, y) of cost $c_\pi(x, y) = 1$, then we delete this edge in both copies and add an edge from x in the first graph to y in the second graph. The intuition is that to travel from a node in the first graph to a node in the second graph, one must pass through an edge of cost 1. Moreover, it is not possible to cross twice such an edge. So all paths in the resulting graph have cost at most 1.

To find a cycle of cost 1 passing by the edge (a_i^v, b_i^v) in the original graph, one needs to find a path from b_i^v in the first graph to a_i^v in the second graph. The problem is therefore transformed to a problem of reachability.

Computing whether there is a path from each of the $O(nd)$ nodes b_i^v to their associated nodes a_i^v can be done using $O(nd)$ depth first search (DFS) for a total computational time of $O(n^2 d^2)$. However, we use a key information to decrease this complexity. We know that to generate one more switch, the flow needs to pass by a collector. We can restrict our search to the cycles passing by a collector. For each of the n collectors C_i, we can compute with a DFS the nodes F_i^0 that can be reached from this collector with a forward path of cost 0 and the nodes F_i^1 that can be reached with a path of cost 1. While doing the DFS, we use two bitsets to represent F_i^0 and F_i^1 and flag the nodes in the appropriate bitsets depending whether the node belongs to the first copy of the graph or the second copy. We perform the same operation on the transposed graph to compute the nodes B_i^0 and B_i^1 that can be reached with a backward path from collector C_i with a cost of at most 0 and at most 1. We then compute the set of nodes P_{ub} that lie on a cycle of cost at most 1 that passes by a collector as follows.

$$P_{ub} = \bigcup_{i=1}^{n} \left(F_i^0 \cap B_i^1 \cup F_i^1 \cap B_i^0 \right) \tag{2}$$

For every $v \in ub(S_i)$ such that $v \notin \sigma_i$ and $a_i^v \in P_{ub}$, then it is possible to modify the solution σ to obtain a new solution σ' with $v \in \sigma_i'$. One simply needs to push one unit of flow on the cycle on which lies the node a_i^v. Since this node has (a_i^v, b_i^v) has unique outgoing edge in the residual graph, the new flow will pass by (a_i^v, b_i^v) and will have at most one more switch.

We use the same idea to test whether $v \notin S_i$ has a bound support whenever $f(a_i^v, b_i^v) > 0$. We compute the set P_{lb} that contains all indices i such that there does not exist a cycle of cost at most 1 that passes by a collector and by the edge (b_i^v, a_i^v) as follows.

$$P_{lb} = \{i \mid \not\exists j \text{ s.t. } (a_i^v \in B_j^0 \wedge b_i^v \in F_j^1) \vee (a_i^v \in B_j^1 \wedge b_i^v \in F_j^0)\} \tag{3}$$

For every $v \notin lb(S_i)$ such that $v \in \sigma_i$ and $i \in P_{lb}$, then it would not be possible to modify the solution σ to obtain a new solution σ' with $v \notin \sigma_i'$. Indeed there is no alternative for the unit of flow going through (a_i^v, b_i^v) without increasing the cost above $\max(M)$. We can therefore deduce that v should be added to $lb(S_i)$.

Each of the n DFS runs in time $O(nd)$. The union and intersection operations required for the computation of P_{ub} and P_{lb} can be done using bitwise conjunction and disjunctions in time $O(n^2 d)$. So the computation of the cycles are done in time $O(n^2 d)$.

5   Experimental Evaluation

We tested our propagator for SWITCH on two optimization problems based on industrial applications. However, they have been somewhat abstracted and simplified for the purpose of our experiments. Moreover, we randomly generated two sets of instances.[1] All experiments were run on Intel Core i5 2.30GHz machine with 6GB of RAM on Windows 7. For each problem, we have generated 50 instances of four classes, each defined by a tuple of parameters. We compare two Choco programs that differ in the representation of the objective function. In the first model it is decomposed into a sum of reified LESSTHAN ($<$) constraints. In the second model, the objective function is represented using a single SWITCH constraint (for the first problem), or several (for the second problem). All other constraints are the same in both models. Finally, we used two search heuristics (Impact-based Search [6] and the Domain over Weighted Degree Heuristic [2], denoted respectively Impact and Wdeg).

Embroidery Scheduling. This first problem is derived from a real life scheduling problem in the textile industry involving job-dependent setup times.

A set of n *garments* have to be embroidered using m *machines*. Each garment is characterized by a set $col_i \subseteq \{1, \ldots, k\}$ of *colors* required for embroidering a given pattern. Last, at most c_j reels of threads can be mounted on machine j (i.e., at most c_j colors of threads can be used without changing the reels).

The load on each machine must be balanced, so we assume that each machine will process n/m garments. A feasible solution for this problem is a mapping of the garments to the machines $f : \{1, \ldots, n\} \mapsto \{1, \ldots, m\}$. Moreover, each garment i must be assigned a position s_i on the machine it is assigned to, and for each machine j the set of colors of threads S_g^j available when processing the the gth garment must be sufficient for embroidering that garment (Eq. 5) while taking into account the maximum number of reels that can be mounted on each machine (Eq. 6). However, whenever the next garment to be embroidered requires a color of thread that is not loaded on the machine, one need to turn the machine off, change some reels and restart the machine. The number of reel changes must therefore be minimized (Eq. 4).

$$\text{minimize :} \qquad \sum_{1 \le j \le m} \sum_{1 \le g < n/m} |S_{g+1}^j \setminus S_g^j| \qquad (4)$$

$$\text{subject to :} \qquad \forall i \in \{1, \ldots, n\} \quad col_i \subseteq S_{s(i)}^{f(i)} \qquad (5)$$

$$\forall j \in \{1, \ldots, m\}, g \in \{1, \ldots, n/m\} \quad |S_g^j| \le c_j \qquad (6)$$

We have a set of n integer variables with domain $\{1, \ldots, m\}$ standing for the mapping of garments to position in the sequence (we consider here the whole sequence obtained by concatenating the sequences on each machine). Then we have n set variables, one for each garment and standing for the set of colors

---

[1] Available at http://homepages.laas.fr/ehebrard/switch/

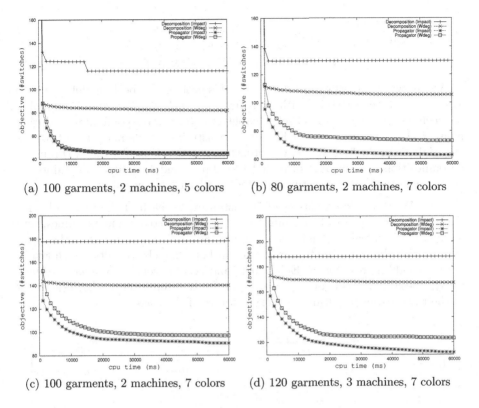

(a) 100 garments, 2 machines, 5 colors   (b) 80 garments, 2 machines, 7 colors

(c) 100 garments, 2 machines, 7 colors   (d) 120 garments, 3 machines, 7 colors

Fig. 3. Embroidery: #switches over time

of threads on the machine when embroidering the ith garment. Then we used a set of ELEMENT constraints to channel these two sets of variables. In the first model, the objective function is implemented through a decomposition using a SUM of reified LESSTHAN and MEMBER constraints. In the second model, the objective is stated as a sum of m SWITCH constraints (one per machine).

Test Sequencing. Next we consider the design of validation tests. A set of n *tests* have to be performed in order to verify a system involving a set F of k *features*. A test i is defined by a set $on_i \subseteq F$ of features that must be turned ON, and a set $off_i \subseteq F \setminus on_i$ of features that must be turned OFF while doing the test. A *configuration* is a complete characterization of the system, represented by the set $C_j \subseteq F$ of features that are ON (all other features are considered OFF).

The verification will go through m *phases* during which the system will be in a given configuration, and a subset of the tests will be performed. A feasible solution is a sequence of m configurations $[S_1, \ldots, S_m]$ and a mapping of the tests to a phase $f : \{1, \ldots, n\} \mapsto \{1, \ldots, m\}$ such that each test is compatible with the configuration in which it is done (Eq. 8). Moreover, there are restrictions on the number of features that can be ON simultaneously (Eq. 9). Finally, the total duration of the test campaign depends on the set of features that need to be turned ON during the transitions between two configurations (Eq. 7).

$$\text{minimize :} \qquad \sum_{1 \le j < m} |S_{j+1} \setminus S_j| \qquad (7)$$

$$\text{subject to :} \qquad \forall j \in \{1, \dots, m\} \; l \le |S_j| \le u \qquad (8)$$

$$\forall i \in \{1, \dots, n\} \; on_i \subseteq S_{f(i)} \; \& \; off_i \cap S_{f(i)} = \emptyset \qquad (9)$$

We use two straightforward models for this problem. In both models, we have a set of n integer variables with domain $\{1, \dots, m\}$ standing for the mapping between tests to phases and m set variables standing for the configuration during phase j. Then we use a set of ELEMENT constraints to channel these two sets of variables. In the first model, the objective function is implemented through a decomposition using a SUM of reified LESSTHAN and MEMBER constraints. In the second model, the objective is stated as a SWITCH constraint.

Results. We generated 4 classes of 50 instances, for each of the two problems, parameterized respectively by number of garments, colors and machines, and number of tests, features and configurations. For each class, we report the mean value of the objective function over time in both models. Results are shown in Figure 3 and Figure 4, for the Embroidery and Test Sequencing problems, respectively. These curves were obtained by averaging the step functions corresponding to the runs of one algorithm on every instance of the class.

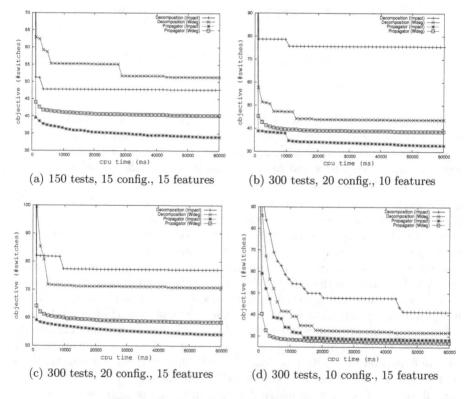

(a) 150 tests, 15 config., 15 features   (b) 300 tests, 20 config., 10 features

(c) 300 tests, 20 config., 15 features   (d) 300 tests, 10 config., 15 features

Fig. 4. Test Sequencing: #switches over time

We observe that the model using the propagator for SWITCH clearly outperforms the decomposition model. In particular, the decomposition does not seem able to significantly improve the objective after the initial drop. It is interesting to notice that whilst Wdeg outperforms Impact on the decomposition model, it is slightly worse in the model using the global constraint (except for the classes $\langle 100, 2, 5 \rangle$ in the Embroidery benchmark and $\langle 300, 10, 15 \rangle$ for the Test Sequencing benchmark). This is actually not surprising since Wdeg relies heavily on the shape of the constraint network and is severely hindered when using a global constraint. Despite this, the propagator yields better results, irrespective of the heuristic that is used. On the Embroidery problem, the improvement on the objective value obtained by using the propagator ranges from 32 to 48% (counted on the best heuristic in each case). On the Test Sequencing problem, we observe more modest but still sizeable improvements, ranging from 12 to 29%.

6   Conclusion

We have introduced the constraints BUFFEREDRESOURCE and SWITCH to reason about the number of item switches in a buffered resource. The former constraint is NP-hard, however it can be effectively decomposed using the latter in conjunction with an ALLDIFFERENT. We have introduced a linear algorithm to find a support to the SWITCH constraint, that is, to assign a sequence of set variables standing for the buffer so that the number of switches is minimized. Furthermore, using this algorithm and a flow-based model, we have shown that bound consistency can achieved in $O(n^2 d + n^{1.15} d^{1.5})$ time. Finally, our experimental results show that this propagator is a significant improvement with respect to expressing this relation with primitive constraints.

References

1. Bessiere, C., Hebrard, E., Hnich, B., Kiziltan, Z., Walsh, T.: The range and roots constraints: Specifying counting and occurrence problems. In: Proceedings of the Nineteenth International Joint Conference on Artificial Intelligence, pp. 60–65 (2005)
2. Boussemart, F., Hemery, F., Lecoutre, C., Sais, L.: Boosting Systematic Search by Weighting Constraints. In: Proceedings of the 16th European Conference on Artificial Intelligence - ECAI 2004, pp. 146–150 (2004)
3. Verfaillie, G., Maillet, C., Cabon, B.: Constraint programming for optimising satellite validation plans. In: 7th International Workshop on Planning and Scheduling for Space, IWPSS 2011 (2011)
4. Cormen, T.H., Leiserson, C.E., Rivest, R.L., Stein, C.: Introduction to Algorithms, 3rd edn. The MIT Press (2009)
5. Goldberg, A.V.: Scaling Algorithms for the Shortest Path Problem. SIAM Journal on Computing 24, 494–504 (1995)
6. Refalo, P.: Impact-Based Search Strategies for Constraint Programming. In: Wallace, M. (ed.) CP 2004. LNCS, vol. 3258, pp. 557–571. Springer, Heidelberg (2004)
7. Régin, J.-C.: Cost-based arc consistency for global cardinality constraints. Constraints 7(3-4), 387–405 (2002)

Combining Discrete Ellipsoid-Based Search and Branch-and-Cut for Binary Quadratic Programming Problems

Wen-Yang Ku and J. Christopher Beck

Department of Mechanical & Industrial Engineering
University of Toronto, Toronto, Ontario M5S 3G8, Canada
{wku,jcb}@mie.utoronto.ca

Abstract. We propose a hybrid algorithm that combines discrete
ellipsoid-based search (DEBS) and a branch-and-cut (B&C) MIP solver
to solve binary quadratic programming (BQP) problems, an important
class of optimization problems with a number of practical applications.
We perform experiments on benchmark instances for the BQP problem and compare the performance of two B&C based solvers, the DEBS
method that is commonly used in the communications community, and
the new hybrid algorithm. Our experimental results demonstrate that
the new hybrid algorithm outperforms both the well-known MIP solvers
and the DEBS approach. Further comparison against two state-of-the-art
special-purpose algorithms in the literature demonstrates that the hybrid approach is competitive: achieving the same or better performance
on six of seven benchmark sets against one algorithm and performing
competitively against the semi-definite programming (SDP) based algorithm for moderate size problems and some dense problems, while
under-performing on larger problems.

1 Introduction

Binary quadratic programming (BQP) problems arise in many combinatorial
optimization problems such as task allocation [1], quadratic assignment [2], and
max-cut problems [3]. A variety of exact methods exist for solving BQP problems
including the linearization method [4], discrete ellipsoid-based search (DEBS)
[5,6], and mixed integer programming (MIP) [7,8].

Semi-definite programming (SDP) based branch-and-bound approaches [9,3]
are often regarded as the state-of-the-art approach for solving BQPs. In this
approach, the semi-definite relaxation bound of the objective function is used
to prune nodes during the branch-and-bound process. Krislock et al. [3] showed
that their SDP algorithm dominates existing approaches for the BQP problems
in the Biq-Mac library [10].

Recently, Li et al. [11] proposed a specialized branch-and-bound approach and
demonstrated its strong performance on benchmark instances from a number of
sources. Li et al.'s techniques are derived from the geometric structure of the
BQP problem based on perturbation analysis. The results of the analysis are

H. Simonis (Ed.): CPAIOR 2014, LNCS 8451, pp. 334–350, 2014.
© Springer International Publishing Switzerland 2014

implemented in the form of problem-specific lower bounding techniques and inference rules to fix values. Variable and value ordering heuristics, as well as a primal heuristic to find high quality feasible solutions, are also proposed to accelerate the convergence of the search. Empirical results demonstrate that the algorithm is one of the state-of-the-arts for finding exact solutions to these benchmark BQP problems.

The common generic approach in Operations Research for exactly solving BQPs is the use of a commercial MIP solver such as CPLEX or Gurobi, solvers that have been extended over the past few years to be able to reason about quadratic constraints [12]. A modern MIP solver is able to outperform several other exact approaches [13] and MIP is commonly used as a standard comparison for evaluating heuristic methods [1].

In this paper we develop a new hybrid algorithm that combines discrete ellipsoid-based search (DEBS) and a branch-and-cut (B&C) MIP solver. DEBS is a specialized search used in the communications literature (e.g., see [14]) to solve integer least squares problems based on the clever enumeration of integer points within the hyper-ellipsoid defining the feasible space. As BQPs can be reformulated as integer least squares problems, we perform this transformation and, for the first time, evaluate DEBS on BQP. We then hybridize DEBS with a MIP solver (SCIP [15]), incorporating aspects of DEBS into presolving, into global-constraint-based inference, and as a primal heuristic.

Our experimental results demonstrate that DEBS performs much better than CPLEX on average, though approximately at the same level as default SCIP (without the hybrid extensions) on seven standard benchmark sets. Interestingly, DEBS is both significantly better and significantly worse than SCIP on different problem sets. The hybrid approach, implemented in SCIP, outperforms CPLEX, SCIP, and DEBS on all seven problem sets, though on some sets the improvement is marginal. Li et al.'s approach dominates the B&C solvers and DEBS overall while being comparable to our new hybrid approach: on one problem set Li et al.'s approach is superior, on another the hybrid performs approximately an order-of-magnitude better, and on the remaining five sets, the performance of the two algorithms is essentially equivalent. When compared to the SDP approach, the hybrid algorithm performs as efficiently for moderate size problems but lags behind the SDP approach for large problems. However, on some dense problems, the hybrid algorithm greatly outperforms the SDP approach. Overall, the SDP approach still appears to be the strongest for solving BQPs.

The contribution of this paper is two-fold. First, it is the first study that applies the DEBS approach from communications to BQPs. The somewhat different approach of DEBS compared to standard OR approaches provides new insight into this well-studied problem and may inspire future innovation. Second, we make use of this inspiration to propose a novel, competitive hybrid algorithm that combines DEBS and a B&C MIP solver. A particular advantage of this hybrid approach is that, unlike the SDP solver or Li et al.'s B&B, the hybrid algorithm can be applied to a broader class of problems, such as the unconstrained integer quadratic programming (IQP) problem and IQP problems with general integer bounds on the variables [16].

The rest of the paper is organized as follows. In Section 2 we define the BQP problem. Section 3 presents the necessary background, including a literature review for the DEBS method, B&C MIP solvers for the BQP problem, and previous results. Section 4 describes the hybrid algorithm. Sections 5 and 6 provide computational results and discussions. We conclude in Section 7.

2   Problem Definition

In this section we first define the BQP problem and its equivalent binary integer least squares (ILS) problem that can be solved with the DEBS method. Then we present the transformation between these two problems.

The BQP problem is defined as follows:

$$\min_{x \in \{0,1\}} \frac{1}{2} x^T H x + f^T x, \tag{1}$$

where $H \in \mathbb{R}^{n \times n}$ is a symmetric semi-definite positive matrix and $f \in \mathbb{R}^n$ is a vector.

Current MIP solvers can solve problems with quadratic objectives when the objective is semi-definite posiitve [12]. When H is not symmetric, the symmetric form can be obtained by setting $H = (H + H^T)/2$. Another common issue for real world applications is the non-convexity property of the matrix H. Convex-ifying H is possible in the BQP case since, for every variable, the relationship $x_i = x_i^2$ holds. Therefore, we can perturb the H matrix and make it semi-definite positive by using a vector u until $(H + \text{diag}(u))$ is semi-definite positive. The convex equivalent BQP problem can be obtained as follows:

$$\min_{x \in \{0,1\}} \frac{1}{2} x^T (H + \text{diag}(u)) x + \left(f - \frac{1}{2} u \right)^T x. \tag{2}$$

A computationally inexpensive way to find such a u vector consists of com-puting the eigenvectors of H [13].

To solve the BQP problem with the DEBS method, we transform the BQP problem into its equivalent binary ILS problem through the relationship $H = A^T A$ and $f = -y^T A$. Thus, the binary ILS problem can be defined as follows

$$\min_{x \in \{0,1\}} \| y - Ax \|_2^2. \tag{3}$$

3   Background

In this section we present the discrete ellipsoid-based search technique in detail and provide other the necessary background.

3.1   Discrete Ellipsoid-Based Search

Discrete Ellipsoid-Based Search (DEBS)[1] was originally proposed to solve the integer least squares (ILS) problems that arise in some communication applications [14]. For example, to achieve a very high precision in a global navigation satellite system, an ILS problem has to be solved to estimate the double-difference carrier phase ambiguities obtained from pairs of satellite and receivers [14]. Loosely, this problem is to find the integer number of wavelengths between the satellite and the receiver in order to precisely estimate their range. The DEBS method is the most common method for solving the ILS problem to optimality in the communication literature [5].

The DEBS method consists of two phases: reduction and search. The reduction is a preprocessing step that transforms the given ILS problem into one for which the search process is more efficient [5]. Search consists of the enumeration of the integer points in a bounded region [17].

Geometrically, the optimal solution is found by searching discretely inside the ellipsoid defined by the objective function of the binary ILS problem (3). Suppose the optimal solution z to the binary ILS problem (3) satisfies the following bound

$$\|y - Az\|_2^2 < \beta, \tag{4}$$

where β is a constant. This is a hyper-ellipsoid with center $A^{-1}y$. The DEBS method systematically searches for the optimal integer solution inside the ellipsoid.

Reduction. Reduction can be regarded as a preprocessing technique that transforms A to an upper triangular matrix R with certain properties such that the search is more efficient. It has been shown that the order of the diagonal entries of R can greatly affect the performance of the DEBS method [5]. It is desirable to have the diagonal entries satisfying the following relationship:

$$|r_{11}| \leq |r_{22}| \leq \ldots \leq |r_{kk}| \leq \ldots \leq |r_{n-1,n-1}| \leq |r_{nn}|.$$

In order to keep the solution of the binary ILS problem (3) unchanged, a transformation applied to A from the left-hand side must be an orthogonal matrix. To keep the integer nature of the unknown integer vector and the bounds on the variables unchanged, a transformation to A from the right-hand side has to be a permutation matrix. Therefore, the reduction can be described as the process of transforming A to R, by finding an orthogonal matrix $Q \in \mathbb{R}^{m \times m}$ and a permutation matrix $P \in \mathbb{Z}^{n \times n}$ such that

$$Q^T A P = \begin{bmatrix} R \\ 0 \end{bmatrix}, \quad Q = [Q_1, Q_2],$$

---

[1] There does not seem to be a standard name for this approach in the communications literature and so we have adopted this term.

where $Q \in \mathbb{R}^{m \times m}$ is orthogonal, $R \in \mathbb{R}^{n \times n}$ is upper triangular, and $P \in \mathbb{Z}^{n \times n}$ is a permutation matrix. With the QR factorization, we have

$$\|y - Ax\|_2^2 = \left\|Q_1^T y - RP^T x\right\|_2^2 + \left\|Q_2^T y\right\|_2^2,$$

where

$$\bar{y} = Q_1^T y, \quad z = P^T x, \quad \bar{l} = P^T l, \quad \bar{u} = P^T u.$$

Since the original lower and upper bounds on the variables are all 0 and 1 for the binary ILS problem, the new bounds after applying the permutation matrix P remain the same. Therefore, the original binary ILS problem (3) is transformed to the new reduced binary ILS problem

$$\min_{z \in \{0,1\}} \|\bar{y} - Rz\|_2^2, \tag{5}$$

where

$$z = P^T x.$$

After the optimal solution z^* to the new binary ILS problem (5) is found, the optimal solution, x^*, to the original binary ILS problem (3) can be recovered with the following relationship:

$$x^* = P^T z^*.$$

An effective reduction algorithm for ILS problems with bounds on the variables was proposed by Chang & Han [5], in which different reduction strategies are empirically compared. In our implementation of the DEBS method, we adopt Chang & Han's reduction algorithm for the binary ILS problem.

Search. After the reduction phase, we have the equivalent reduced problem (5). To illustrate the search strategy, we first explain the search with the unconstrained ILS problem, i.e., $z \in \mathbb{Z}^n$ instead of $z \in \{0,1\}$ then specialize the search for the binary ILS problem.

Suppose the optimal solution z satisfies the following bound

$$\|\bar{y} - Rz\|_2^2 < \beta,$$

or equivalently

$$\sum_{k=1}^{n} (\bar{y}_k - \sum_{j=k}^{n} r_{kj} z_j)^2 < \beta,$$

where β is a constant which can be obtained by substituting any feasible integer solution to equation (5). This is a hyper-ellipsoid with center $R^{-1} \bar{y}$. The goal is to search this ellipsoid to find the optimal integer solution.

Among several search strategies, the Schnorr & Euchner strategy is usually considered the most efficient [17]. Let $z_i^n = [z_i, z_{i+1}, \ldots, z_n]^T$. Define the so far unknown (apart from c_n) and usually non-integer variables

$$c_n = \bar{y}_n / r_{nn}, \quad c_k = c_k(z_{k+1}^n) = (\bar{y}_k - \sum_{j=k+1}^{n} r_{kj} z_j) / r_{kk}, \quad k = n-1, \ldots, 1.$$

Notice that c_k is fixed when z_{k+1}^n is fixed. The above equation can be rewritten as

$$\sum_{k=1}^{n} r_{kk}^2 (z_k - c_k)^2 < \beta.$$

This inequality is equivalent to the following n inequalities:

$$\text{level } n: \quad (z_n - c_n)^2 < \frac{1}{r_{nn}^2}\beta,$$

$$\text{level } n-1: \quad (z_{n-1} - c_{n-1})^2 < \frac{1}{r_{n-1,n-1}^2}[\beta - r_{nn}^2(z_n - c_n)^2],$$

$$\vdots$$

$$\text{level } k: \quad (z_k - c_k)^2 < \frac{1}{r_{kk}^2}[\beta - \sum_{i=k+1}^{n} r_{ii}^2(z_i - c_i)^2],$$

$$\vdots$$

$$\text{level } 1: \quad (z_1 - c_1)^2 < \frac{1}{r_{11}^2}[\beta - \sum_{i=2}^{n} r_{ii}^2(z_i - c_i)^2].$$

During the search process, z_k is determined at level k, where $z_n, z_{n-1}, \ldots, z_{k+1}$ have already been determined, but $z_{k-1}, z_{k-2}, \ldots, z_1$ are still unassigned (note that c_k depends on $z_{k+1}, z_{k+2}, \ldots, z_n$). Therefore, c_k is only changed when we move to level k from level $k+1$ but remains unchanged when we move to level k from level $k-1$. When we move from level $k+1$ to level k, we first compute c_k and then choose $z_k = \lfloor c_k \rceil$, the nearest integer to c_k. When we move from level $k-1$ to level k, we choose z_k to be the next nearest integer to c_k. At each level k, z_k takes on values in the order

$$\lfloor c_k \rceil, \lfloor c_k - 1 \rceil, \lfloor c_k + 1 \rceil, \lfloor c_k - 2 \rceil, \ldots, \quad \text{if } c_k \leq \lfloor c_k \rceil,$$

or

$$\lfloor c_k \rceil, \lfloor c_k + 1 \rceil, \lfloor c_k - 1 \rceil, \lfloor c_k + 2 \rceil, \ldots, \quad \text{if } c_k > \lfloor c_k \rceil.$$

In the Schnorr & Euchner strategy, the initial bound β can be set to ∞. When we first find the integer point, the k-th entry of the first integer point is $\lfloor c_k \rceil$ for $k = n, \ldots, 1$. We can use this integer point to update β, reducing the hyper-ellipsoid in what is, from an OR perspective, a bounding operation.

Alternatively, we can obtain an initial bound by rounding the solution of the continuous least squares (LS) problem. Let z_{LS} be the continuous LS solution, i.e., $z_{LS} = R^{-1}\bar{y}$. Then we round each entry of z_{LS} to its nearest integer to get the integer vector $\lfloor z_{LS} \rceil$. As the problem is unconstrained, $\lfloor z_{LS} \rceil$ is feasible and therefore the optimal solution must be as least as good. So we can set $\beta = \|\bar{y} - R\lfloor z_{LS} \rceil\|_2^2$.

The search algorithm for the binary ILS problem is modified to take into account the variable bounds, i.e., $z \in \{0, 1\}$. Chang & Han [5] proposed an efficient algorithm for ILS problems with bounds on the variables based on the same search framework, which is considered the state-of-the-art search algorithm. We use this algorithm in our implementation for the binary ILS problem.

3.2 B&C MIP Solvers

BQP problems can be modeled as a MIP, a well-known approach in OR that has been widely used for combinatorial optimization problems since the 1950s. It is also one of the most common approaches for solving discrete optimization problems. MIP can be further categorized into mixed integer linear programming (MILP) and mixed integer nonlinear programming (MINLP).

BQP is a special type of the more general integer quadratic programming problem (IQP), both of which belong to the sub-class of MINLP that is generally solved with the following four approaches based on MIP-related techniques: branch-and-bound and branch-and-cut (B&C) [18], outer approximation [19], extended cutting-planes [20], and generalized Benders decomposition [21]. Of the above methods, the outer-approximation and the B&C methods have received the most attention and both commercial and non-commercial solvers are available based on these two approaches. An empirical study of these two methods on binary quadratic programming problems concluded that the two approaches performs significantly differently on various classes of problems and it is not clear which is generally superior [7]. Though note that this conclusion was made in 1997.

The state-of-the-art MIP solvers are typically hybrid algorithms that combine many of the above elements. Many solvers such as SCIP [15] and FilMINT [22] incorporate the outer-approximation into the B&C framework and are efficient for solving convex MINLP problems. Commercial software such as CPLEX, GUROBI and BARON also provide functionality to solve convex mixed IQP problems and so can all be applied to solve the BQP problems. We refer interested readers to Bussieck & Vigerske [12] for a recent review of MINLP solvers.

3.3 Previous Results

Our preliminary results in solving general IQP problems with variable bounds demonstrated that the DEBS method performs especially well on problems where the residual $\|y - Az^*\|_2^2$ is small, where z^* is the optimal integer solution [16]. However, a MIP solver was more efficient than DEBS for problems with small variable domains (i.e., narrow variable bounds) and larger residual. These results provide motivation to combine the techniques and design a DEBS/B&C hybrid algorithm for the BQP problem: the BQP problem has the smallest possible variable domains and the residual is generally large. A detailed analysis of the residuals of the problem instances used here is given below in Section 6.

4 A DEBS/B&C Hybrid Algorithm

In this section we describe the three techniques that we integrated into the B&C-based MIP solver SCIP: preconditioning, axis-aligned circumscribed box constraints, and a primal heuristic using DEBS. First, preconditioning applies the same reduction techniques from the DEBS method to the BQP problem to transform it into a new BQP problem with the hope of reducing the size of the search tree explored by the B&C algorithm. Second, the axis-aligned circumscribed box represents globally valid bounds of the variables derived from the geometry of the objective function of the BQP problem. The box is computed and used to update the bounds whenever a new incumbent solution is found. Finally, we also apply DEBS as a primal heuristic during the search to help find high-quality solutions.

4.1   Preconditioning

We first describe the procedure of preconditioning for the BQP problem and then explain the logic behind it. Preconditioning applies the reduction techniques from the DEBS method to the BQP problem to transform the original BQP problem into a new BQP problem, defined as follows:

$$\min_{x \in \{0,1\}} \frac{1}{2} x^T \bar{H} x + \bar{f}^T x, \tag{6}$$

where $\bar{H} = R^T R$ and $\bar{f} = -\bar{y}^T R$. Note that R and $\bar{y}$ are from the reduced binary ILS problem (5).

The reason for performing preconditioning can be explained with the idea of *distance to integrality*. Let x^R denote the optimal solution to the continuous relaxation of the ILS problem and x^I the solution to the ILS problem. We define the distance to integrality as

$$d(x^R, x^I) = \|x^R - x^I\|_2 \, .$$

Thus, the distance to integrality is the Euclidean distance between the real and integer optimal solutions to the ILS problem. It has been shown that the above distance is a useful indicator of the size of the search tree explored in a B&C based MIP solvers [23]. The intuition is that if the root node x^R is closer to the leaf node x^I, fewer branching decisions are required and hence the search tree will be smaller. Previous results on communication applications showed that the experimental correlation between the number of nodes and the distance to integrality in logarithmic scales is around 0.7 [23]. Preconditioning aims at decreasing this distance.

In the following we describe the relationship between preconditioning and the distance between the real and integer optimal solutions to the ILS problem.

Using $A^\dagger$, the Moore-Penrose generalized inverse of A, we can write $x^R = A^\dagger y$ and $A^\dagger A = I$. Therefore

$$d(x^R, x^I) = \|x^R - x^I\|_2 = \|A^\dagger y - x^I\|_2 = \|A^\dagger (y - A x^I)\|_2,$$

leading to

$$d(x^R, x^I) \leq \|A^\dagger\|_2 \|y - A x^I\|_2. \tag{7}$$

Therefore, it is possible to decrease the distance to integrality by transforming the problem using a unimodular matrix U that minimizes $\|(AU)^{\dagger}\|_2$. We refer the process of using the unimodular transformation to reduce the conditioning factor to as *preconditioning*.

Preconditioning is effective for the general unconstrained IQP problems in communication applications [23]. In this paper we apply the preconditioning technique to the BQP problem. In our implementation, we perform the preconditioning in the presolving stage of the B&C algorithm to the BQP problem (3) and transform the BQP problem to its reduced formulation (5), then we perform the B&C search on the reduced problem.

4.2   Axis-Aligned Circumscribed Box Constraints

Based on the geometry of the objective function of the BQP problem, the axis-aligned circumscribed box can be computed and used to fix the values of the (binary) variables. Chang & Golub [6] proposed an efficient way to compute the smallest hyper-rectangle whose edges are parallel to the axes of the coordinate system and that includes the hyper-ellipsoid defined by Equation (4). Let $\mathcal{B}$ be the smallest hyper-rectangle including the reduced hyper-ellipsoid $\|\bar{y} - Rz\|_2^2 \leq \beta$, the lower bound l and the upper bound u can be computed as follows:

$$u_k = \left\lfloor \sqrt{\beta} \left\| R^{-T} e_k \right\|_2 + e_k^T R^{-1} \bar{y} \right\rfloor,$$

$$l_k = \left\lceil -\sqrt{\beta} \left\| R^{-T} e_k \right\|_2 + e_k^T R^{-1} \bar{y} \right\rceil.$$

In our implementation, the axis-aligned circumscribed box is computed every time a new incumbent is found. Since this box is globally feasible, the global bounds of the variables are updated if the box is tighter than the current global bounds of the variables. As our decision variables are binary, any bound tightening corresponds to fixing a value.

4.3   DEBS as a Primal Heuristic

Primal heuristics are important components in state-of-the-art MIP solvers [15] used to find feasible solutions. Good quality feasible solutions are beneficial in a number of ways [8]. First, the feasible solution proves that the problem is feasible, and the solution might already be good enough. Second, the feasible solution can be used to derive valid bound for pruning the search tree. Third, the feasible solution can be used for dual fixing or reduction to strengthen the problem formulation.

When applying the DEBS as a primal heuristic, we first employ a resource bound (e.g., a short time limit) for running the DEBS method at the root node after MIP presolving. If the DEBS method finds an optimal solution, SCIP returns it and the algorithm terminates. If the optimal solution is not found within the time limit, the best incumbent found by the DEBS method is given to SCIP as the starting integer solution. In addition, the DEBS method is executed during the search process. In our implementation, we follow SCIP's primal heuristics parameters: frequency and offset. The offset is the depth where the primal heuristic is executed first and frequency + offset defines the subsequent depths. We chose the values of 10 for frequency and 0 for offset meaning that the DEBS method is only used as a primal heuristic at the root and at nodes of depth 10. If the solution found by the DEBS is better than the current incumbent, the current incumbent is updated, and we re-compute the axis-aligned circumscribed box (Section 4.2) to further tighten the bounds of the variables.

During the B&C process, variables are fixed either by branching decisions or other techniques. When such fixing takes place, we reduce the size of the binary ILS problem. Let U be the set of indices of variables that are not yet fixed at a given node in the search tree and let A_i be the i-th column vector of A of the original binary ILS problem in Equation (3). The updated binary ILS can be defined as follows:

$$\min_{\tilde{x} \in \{0,1\}} \left\| \tilde{y} - \tilde{A}\tilde{x} \right\|_2^2,$$

where

$$\tilde{A} = [A_i], \quad i \in U,$$

$$\tilde{y} = y - \sum_{i \in U} A_i x_i, \quad i \in U,$$

$\tilde{A} \in \mathbb{R}^{n \times |U|}$ and $\tilde{y} \in \mathbb{R}^n$. The resulting problem is $(n - |U|)$ dimensional smaller than the original problem.

5   Computational Results

In this section we compare the performance of the B&C-based MIP solvers CPLEX and SCIP, the DEBS method, and the new hybrid algorithm implemented in SCIP. We also provide the best current results from Li et al. [11] and Krislock et al. [3]. For the problems we did not find results in the SDP literature (Carter and William type problems), we use the online SDP solver BiqCrunch [24] for comparison. We present results for CPLEX and SCIP in order to compare DEBS and our hybrid both to a state-of-the-art commercial MIP solver and to isolate the impact that our DEBS-based hybridizations have on default SCIP performance.

5.1   Experimental Setup

All experiments were performed on a Intel Core i7 3.40 GHz machine (in 64 bit mode) with 8GB memory running Red Hat Enterprise 6.2 with one thread. We perform experiments with CPLEX Optimization Studio v12.5 and SCIP v3.0.2. We use MATLAB 7.7.0 for generating the problem instances. The DEBS approach is written in C and the new hybrid algorithm is implemented in SCIP. The CPU time limit for each run on each problem instance is 3600 seconds.

We use a subset of the benchmark instances presented by Li et al. [11], excluding the max-cut problems as they require additional transformations and cannot be solved with MIP or DEBS directly. We perform experiments on medium size BQP problems on seven problem sets: Carter type problems [25], William type problems [26], bqp50 and bqp100 problems, and gkaia, gkaib, and gkaid problems [10]. We generate ten problem instances for each of the Carter and William type problems. In order to ensure convexity, we perform the same transformation as Li et al. We compute the smallest eigenvalue for the H matrix of each problem and let it be λ_{min}. Then we apply the perturbation vector $u = (-\lambda_{min} + 0.001)e$ to the BQP problem (2) when λ_{min} is negative. Note that the transformed problem has the same optimal solution as the original problem. For the DEBS method, we use Cholesky decomposition on matrix H in the BQP problem to obtain matrix A in the binary ILS problem, and we obtain y from the equation $f = -y^T A$, which gives us $y = -(fA^{-1})^T$.

5.2   Results

The results for all seven problem sets are presented in Table 1. For the bqp50, bqp100, gkaia, gkaib, and gkaid problems, we report the CPU time for each instance. For the Carter and Williams type problems, we report the arithmetic mean CPU time "arith", and the shifted geometric mean CPU time "geo" on the ten instances for each problem size.[2]

The comparison of the identical MIP models in CPLEX and SCIP shows, somewhat surprisingly, that CPLEX performs substantially better than SCIP on only two of the seven problem sets (Williams and Carter). SCIP is clearly superior for bqp50, bqp100, gkaia, and gkaib sets while solving one problem more for gkaid. We attribute this strong performance of SCIP to the quadratic constraint handler described in Berthold et al. [8].

DEBS is noticeably more efficient than B&C (both CPLEX and SCIP) on the William and Carter type problems and superior to CPLEX on the bqp50, gkaia, and gkaib problems. While CPLEX does not out-perform DEBS for any problem instances, SCIP has the edge over DEBS on bqp50, bqp100, and gkaia. For the final problem set, gkaid, SCIP is able to solve one more problem than DEBS. Comparing the DEBS method and CPLEX, the DEBS method is noticeably more efficient than CPLEX for five problem sets: William type, Carter type, bqp50, gkaia and gkaib.

With respect to the first contribution of this paper, therefore, we conclude that the DEBS approach from the communications literature is a competitive approach to solving BQPs that are of interest to the OR community. In fact, DEBS is superior to a state-of-the-art commercial solver, CPLEX.

Turning to the hybridization of B&C and DEBS, we see that the new hybrid approach out-performs CPLEX on all problem sets while achieving clearly

---

[2] The shifted geometric mean time is computed as follows: $\prod(t_i + s)^{1/n} - s$, where t_i is the actual CPU time, n is the number of instances, and s is chosen as 10. Using geometric mean can decrease the influence of the outliers of data [15].

Table 1. A comparison of four approaches plus Li et al.'s results and Krislock et al.'s SDP results for the seven problem sets. Bold numbers indicate the best approach for a given problem. The symbol '-' means that no problem instances were solved to optimality within 3600 seconds. For the Carter and William type problems, n is the size of the problem, and p and d are problem generation parameters. The superscripts indicate the number of instances for which no optimal solution was found.

Carter type problems

n	p	CPLEX		SCIP		DEBS		Hybrid		Li	SDP
		arith	geo	arith	geo	arith	geo	arith	geo	arith	arith
40	0.2	0.21	0.21	29.55	21.06	**0**	0	0.02	0.02	0.32	0.32
40	0.3	0.38	0.38	34.01	26.35	**0.01**	0.01	0.03	0.03	0.26	0.62
50	0.2	2.38	2.26	896.53	655.17	**0.15**	0.15	0.24	0.24	1.5	0.66
50	0.3	2.85	2.69	891.71	767.32	**0.80**	0.80	1.10	0.85	1.2	1.15
80	0.4	1170.01^2	349.34^2	-	-	1020.20^2	249.31^2	98.37	67.40	89.9	**21.39**
80	0.5	659.99	369.26	-	-	225.33	60.03	48.09	31.02	**2.5**	16.01
100	0.5	3222.90^8	2955.28^8	-	-	2766.31^7	2226.48^7	2249.53^5	1584.12^5	**10.8**	57.97

William type problems

n	d	CPLEX		SCIP		DEBS		Hybrid		Li	SDP
		arith	geo	arith	geo	arith	geo	arith	geo	arith	arith
40	0.5	0.25	0.25	8.29	7.59	**0.01**	0.01	**0.01**	0.01	0.44	0.33
40	0.7	0.22	0.21	14.16	12.16	**0.01**	0.01	**0.01**	0.01	0.47	0.34
50	0.2	1.75	1.70	12.00	10.51	**0.35**	0.32	0.44	0.40	6.6	0.52
50	0.4	1.74	1.68	148.64	115.64	**0.09**	0.09	0.14	0.14	24	0.75
50	1	2.78	2.61	1463.02^2	948.47^2	**0.13**	0.09	0.19	0.18	2.5	0.85
60	0.1	21.83	15.29	4.61	4.17	1.57	1.51	1.93	1.85	**0.72**	**0.72**
60	0.2	22.81	11.20	387.15	163.29	2.10	1.84	2.65	2.26	5.3	**1.33**
60	0.4	14.18	11.49	2568.63^4	1833.75^4	31.21	10.48	6.24	5.17	18.9	**2.32**
80	0.1	2014.32^5	1047.60^5	1688.28^3	850.38^3	157.52	80.99	178.82	89.52	7	**3.01**
80	0.2	1023.81^1	530.40^1	-	-	188.92	97.89	140.83	82.79	123.1	**2.19**

bqp50 problems

Instance	CPLEX	SCIP	DEBS	Hybrid	Li	SDP
bqp50-1	11.66	0.07	0.65	**0.06**	<0.2	0.25
bqp50-2	1.84	0.06	1.2	**0.03**	<0.2	0.24
bqp50-3	0.60	0.04	**0**	0.03	<0.2	0.17
bqp50-4	2.84	0.05	0.30	**0.04**	<0.2	0.22
bqp50-5	3.72	0.03	0.13	**0.02**	<0.2	0.21
bqp50-6	0.46	**0.01**	0.15	0.02	<0.2	0.21
bqp50-7	1.70	0.05	**0.02**	0.05	<0.2	0.26
bqp50-8	0.93	0.06	**0.03**	0.06	<0.2	0.19
bqp50-9	5.32	0.09	0.35	**0.07**	<0.2	0.25
bqp50-10	14.30	0.07	0.25	**0.06**	<0.2	0.24

bqp100 problems

Instance	CPLEX	SCIP	DEBS	Hybrid	Li	SDP
bqp100-1	-	55.50	-	46.13	1648	**1.40**
bqp100-2	-	6.74	-	7.12	63.4	**2.06**
bqp100-3	-	4.99	-	9.10	66.3	**1.6**
bqp100-4	-	8.83	-	8.96	165.7	**1.42**
bqp100-5	-	10.28	-	9.17	1230	**1.69**
bqp100-6	-	566.92	-	55.54	175.2	**11.12**
bqp100-7	-	42.52	-	13.37	427	**1.8**
bqp100-8	-	4.63	-	5.96	25.6	**1.63**
bqp100-9	-	3.50	1282.42	5.90	21.9	**1.46**
bqp100-10	-	4.43	-	4.45	57.2	**1.67**

gkaia problems

Instance	n	CPLEX	SCIP	DEBS	Hybrid	Li	SDP
gka1a	50	11.32	0.03	4.55	**0.02**	<1	0.25
gka2a	60	1.60	**0.02**	0.84	**0.02**	<1	0.42
gka3a	70	2817.3	0.93	1335.20	**0.79**	<1	0.93
gka4a	80	88.86	1.12	348.61	**0.86**	<1	1.07
gka5a	50	3.22	0.94	**0.11**	**0.11**	<1	0.29
gka6a	30	0.11	0.37	0	0.02	<1	0.06
gka7a	30	0.05	0.41	0	0.02	<1	0.06
gka8a	100	-	**0.08**	634.88	**0.08**	<1	2.11

gkaib problems

Instance	n	CPLEX	DEBS	SCIP	Hybrid	Li	SDP
gka1b	20	0.08	0	0.29	0.01	<1	0.1
gka2b	30	0.1	0	0.36	0.01	<1	2.18
gka3b	40	0.58	0	0.6	0.01	<1	9.44
gka4b	50	1.7	**0.01**	1.15	0.03	<1	18.42
gka5b	60	5.86	**0.04**	1.44	0.07	<1	39.2
gka6b	70	21.75	**0.15**	1.95	0.26	<1	86.56
gka7b	80	59.88	**0.34**	2.52	0.53	<1	212.73
gka8b	90	160.79	**0.98**	3.67	1.43	<1	789.1
gka9b	100	502.79	**2.66**	6.04	3.67	<1	1608.04
gka10b	125	-	20.7	9.51	12.06	<1	5104.21

gkaid problems

Instance	n	CPLEX	DEBS	SCIP	Hybrid	Li	SDP
gka1d	100	-	-	5.90	5.84	24	**1.96**
gka2d	100	-	-	-	-	5671	**1.63**
gka3d	100	-	-	-	-	1713	**1.85**
gka4d	100	-	-	-	-	3835	**3.25**
gka5d	100	-	-	-	-	5466	**55.72**
gka6d	100	-	-	-	-	1534	**2.14**
gka7d	100	-	-	-	-	4273	**46.04**
gka8d	100	-	-	-	-	683	**2.94**
gka9d	100	-	-	-	-	2481	**34.08**
gka10d	100	-	-	-	-	1878	**30.97**

better performance than SCIP on the Williams and Carter type problems and marginally better performance on the other five problem sets: there are a number of problem instances in the latter five sets where the hybrid makes meaningful improvements on the default SCIP performance and none where it is substantially inferior. Similarly, the hybrid improves on the pure DEBS results on all problem sets though the improvement is marginal for bqp50, gkaib and gkaid.

While our experimental environment is different from that of Li et al., the new hybrid algorithm appears competitive on the Carter type, William type, bqp50, gkaia, and gkaib problems. The hybrid algorithm performs better on the bqp100 problems and worse on the gkaid problems. For the bqp100 problems, the hybrid algorithm outperforms Li et al.'s algorithm by about an order of magnitude.

Compared to the SDP approach, the hybrid algorithm performs basically the same or slightly better for moderate size problems, but significantly less efficiently for larger problems. However for some dense problem instances (gkaib), the hybrid algorithm greatly outperforms the SDP approach. The hybrid algorithm was able to solve all of them quickly, while the SDP approach is in general two orders of magnitudes slower. Overall, the SDP approach still appears to be most efficient for the BQP problems.

6   Discussion

Our preliminary work showed that the DEBS method performs especially well on ILS problems with bounds on the variables when the residual, $\|y - Az^*\|_2^2$, is small (z^* is the optimal integer solution). However, a MIP solver performs much better than DEBS with small variable domains and large residual [16]. To explain the role of the residual, we define the $noise$[3] in the data with the following linear model: $y = Ax + v$, where $v \in \mathbb{R}^n$ denotes the noise vector. A large noise vector results a large residual. The residual directly determines the search space of the DEBS method, since the minimum ellipsoid is defined by the optimal integer solution z^* and its associated residual (Eqn. (4)). In the extreme case, when $v = 0$, there is no noise. When such a problem (3) is solved to optimality, $\|y - Az^*\|_2^2$ is equal to zero and so the ellipsoid reduces to a single point. As v increases, the DEBS method needs to search a larger space.

In communication applications, a noise vector v of $\sigma * \texttt{randn(m,1)}$, $\sigma \geq 5$ is generally considered as a very large noise in the data [23]. Note that $\texttt{randn(m,1)}$ is a MATLAB function which returns an m-dimensional vector containing pseudorandom values drawn from the standard normal distribution $N(0,1)$. We computed the noise vectors in the seven problem sets by substituting the optimal solution or the best solution found within the time limit into the linear model and found that $\sigma > 5$ for all seven problem sets. Therefore these problems can be regarded as problems with very large noise and the poor performance of the DEBS method on the larger problems is consistent with both our previous results and the need to search a larger ellipsoid.

---

[3] Noise has a recognized physical meaning in the communications literature [5].

However, it is surprising that the DEBS method performs so well compared to the MIP solvers, especially CPLEX, given the large noise and tightest possible variable bounds in the problem instances. This contradicts our previous results on general IQPs that CPLEX performs better than the DEBS method for problems with large noise and small variable domains [16]. One interesting direction for future work is to further investigate the reason that the DEBS method performs well on the BQP problem and extract more information to enhance the hybrid approach.

Our results show that combining the DEBS method with a B&C based MIP solver indeed results in better performance when large noise is present in the problem instances. The additional techniques from MIP such as relaxation, cutting planes, and inference appear to complement the presolving, axis-aligned box constraints, and primal heuristic adopted from the DEBS approach.

Analysis of CPLEX solving behaviour shows that the reason that CPLEX is unable to prove optimality for bigger problems (e.g., bqp100 and gkaid) is mainly weakness in the dual bound. Solutions with good quality are found early in the search but the dual bound only improves slowly. For the DEBS method, optimal or near optimal solutions are typically found, though not as quickly as with CPLEX. However, similarly to CPLEX, DEBS is not able to prove optimality since the search space is too large and it does not have an effective dual bounding mechanism. SCIP, however, performs well on the gka1d, bqp50, and bqp100 problems because it is able to make rapid improvements in the dual bound. The lack of such an ability, conversely, appears to explain the very poor performance of SCIP on the Carter and Williams problems. The performance of the hybrid suggests that the poor performance of SCIP on Carter and Williams may be due to not being able to find a high-quality feasible solution. One area for future work is to investigate the performance differences between problem sets, in particular, to understand the structure that SCIP appears to be taking advantage of in the problems in which it performs well. The comparison of the hybrid algorithm and the SDP approach shows that the hybrid algorithm seems to outperform the SDP approach on dense problems (e.g., gkaib). Further investigation is required to identify the reasons.

7   Conclusions

In this paper we conducted the first empirical study comparing the performance of a B&C based MIP solver and the DEBS method for the BQP problem. We also proposed a new hybrid algorithm that combines techniques from MIP solving and DEBS, implementing DEBS-based presolving, axis-aligned box constraints, and a primal heuristic in a B&C MIP solver. Though we only examine binary quadratic problems here, the resulting hybrid algorithm can be used to solve problems with general integer bounds on the variables.

We compared the performance of the B&C based MIP solvers CPLEX and SCIP, the DEBS method, and the new hybrid algorithm for the BQP problems on seven sets of benchmark instances. Results show that the DEBS performs much better than CPLEX and about even with SCIP, though different problem sets show markedly different relative performance. The DEBS algorithm from the communications literature, therefore, is at least competitive with state-of-the-art B&C MIP approaches to binary quadratic problems.

The hybridization of DEBS and B&C out-performs both the B&C and DEBS based approaches across all problem sets. The improvement is substantial for some problem sets, though only marginal in others. Compared to the best special-purpose algorithms, Li et al.'s branch-and-bound algorithm [11] and an SDP approach [3], the new hybrid algorithm is competitive though overall not as strong as the SDP algorithm. We, therefore, conclude that the hybridization of DEBS with B&C is a strong contender on BQP problems, though the SDP approach remains the state of the art.

In the future, we are interested in exploring other possibilities of using geometric information to enhance MIP solving. We expect to extend the hybrid algorithm to solve generic MIP problems (i.e., problems with linear constraints) while preserving the ability to solve the BQP problems efficiently. We are investigating applications in communications where such constrained quadratic problems may be a natural model.

References

1. Lewis, M., Alidaee, B., Kochenberger, G.: Using xqx to model and solve the uncapacitated task allocation problem. Operations research letters 33(2), 176–182 (2005)
2. Finke, G., Burkard, R., Rendl, F.: Quadratic assignment problems. Surveys in combinatorial optimization 61 (2011)
3. Krislock, N., Malick, J., Roupin, F.: Improved semidefinite bounding procedure for solving max-cut problems to optimality. Mathematical Programming, 1–26 (2012)
4. Watters, L.J.: Reduction of integer polynomial programming problems to zero-one linear programming problems. Operations Research 15(6), 1171–1174 (1967)
5. Chang, X.W., Han, Q.: Solving box-constrained integer least squares problems. IEEE Transactions on Wireless Communications 7(1), 277–287 (2008)
6. Chang, X.W., Golub, G.H.: Solving ellipsoid-constrained integer least squares problems. SIAM Journal on Matrix Analysis and Applications 31(3), 1071–1089 (2009)
7. Borchers, B., Mitchell, J.E.: A computational comparison of branch and bound and outer approximation algorithms for 0–1 mixed integer nonlinear programs. Computers & Operations Research 24(8), 699–701 (1997)
8. Berthold, T., Heinz, S., Vigerske, S.: Extending a CIP framework to solve MIQCPs. In: Mixed-Integer Nonlinear Programming. The IMA Volumes in Mathematics and its Applications, vol. 154, pp. 427–445. Springer (2012)
9. Rendl, F., Rinaldi, G., Wiegele, A.: Solving max-cut to optimality by intersecting semidefinite and polyhedral relaxations. Mathematical Programming 121(2), 307–335 (2010)
10. Wiegele, A.: Biq mac library–a collection of max-cut and quadratic 0–1 programming instances of medium size. Preprint (2007)
11. Li, D., Sun, X., Liu, C.: An exact solution method for unconstrained quadratic 0–1 programming: a geometric approach. Journal of Global Optimization 52(4), 797–829 (2012)

12. Bussieck, M.R., Vigerske, S.: MINLP solver software. Wiley Encyclopedia of Operations Research and Management Science. Wiley, Chichester (2010)
13. Billionnet, A., Elloumi, S.: Using a mixed integer quadratic programming solver for the unconstrained quadratic 0-1 problem. Mathematical Programming 109(1), 55–68 (2007)
14. Teunissen, P.J., Kleusberg, A., Teunissen, P.: GPS for Geodesy, vol. 2. Springer (1998)
15. Achterberg, T.: Constraint Integer Programming. PhD thesis, Technische Universität Berlin (2007)
16. Ku, W.Y., Beck, J.C.: Combining discrete ellipsoid-based search and branch-and-cut for integer least squares problems. Technical Report MIE-OR-TR2013-07, University of Toronto, Toronto (2013)
17. Schnorr, C.P., Euchner, M.: Lattice basis reduction: Improved practical algorithms and solving subset sum problems. Mathematical programming 66(1), 181–199 (1994)
18. Gupta, O.K., Ravindran, A.: Branch and bound experiments in convex nonlinear integer programming. Management Science 31(12), 1533–1546 (1985)
19. Duran, M.A., Grossmann, I.E.: An outer-approximation algorithm for a class of mixed-integer nonlinear programs. Mathematical programming 36(3), 307–339 (1986)
20. Westerlund, T., Pettersson, F.: An extended cutting plane method for solving convex minlp problems. Computers & Chemical Engineering 19, 131–136 (1995)
21. Geoffrion, A.M.: Generalized benders decomposition. Journal of Optimization Theory and Applications 10(4), 237–260 (1972)
22. Abhishek, K., Leyffer, S., Linderoth, J.: Filmint: An outer approximation-based solver for convex mixed-integer nonlinear programs. INFORMS Journal on computing 22(4), 555–567 (2010)
23. Ku, W.Y., Anjos, M.F., Chang, X.W.: Lattice preconditioning for the real relaxation branch-and-bound approach for integer least squares problems. Technical report, Group for Research in Decision Analysis, Montreal (2013)
24. Krislock, N., Malick, J., Roupin, F.: BiqCrunch online solver (2012), http://lipn.univ-paris13.fr/BiqCrunch/solver (retrieved: December 22, 2013)
25. Carter, M.W.: The indefinite zero-one quadratic problem. Discrete Applied Mathematics 7(1), 23–44 (1984)
26. Williams, A.: Quadratic 0-1 Programming Using the Roof Dual: With Computational Results. RUTCOR, Hill Center, Rutgers University (1985)

Parallel Combinatorial Optimization
with Decision Diagrams

David Bergman[1], Andre A. Cire[2], Ashish Sabharwal[3],
Horst Samulowitz[3], Vijay Saraswat[3], and Willem-Jan van Hoeve[2]

[1] School of Business, University of Connecticut, Stamford, CT 06901
david.bergman@business.uconn.edu
[2] Tepper School of Business, Carnegie Mellon University, Pittsburgh, PA 15213
{acire,vanhoeve}@andrew.cmu.edu
[3] IBM Watson Research Center, Yorktown Heights, NY 10598
{samulowitz,ashish.sabharwal,vsaraswa}@us.ibm.com

Abstract. We propose a new approach for parallelizing search for combinatorial optimization that is based on a recursive application of approximate Decision Diagrams. This generic scheme can, in principle, be applied to any combinatorial optimization problem for which a decision diagram representation is available. We consider the maximum independent set problem as a specific case study, and show how a recently proposed sequential branch-and-bound scheme based on approximate decision diagrams can be parallelized efficiently using the X10 parallel programming and execution framework. Experimental results using our parallel solver, DDX10, running on up to 256 compute cores spread across a cluster of machines indicate that parallel decision diagrams scale effectively and consistently. Moreover, on graphs of relatively high density, parallel decision diagrams often outperform state-of-the-art parallel integer programming when both use a single 32-core machine.

1 Introduction

In recent years, hardware design has increasingly focused on multi-core systems and parallelized computing. In order to take advantage of these systems, it is crucial that solution methods for combinatorial optimization be effectively parallelized and built to run not only on one machine but also on a large cluster.

Different combinatorial search methods have been developed for specific problem classes, including mixed integer programming (MIP), Boolean satisfiability (SAT), and constraint programming (CP). These methods represent (implicitly or explicitly) a complete enumeration of the solution space, usually in the form of a branching tree where the branches out of each node reflect variable assignments. The recursive nature of branching trees suggests that combinatorial search methods are amenable to efficient parallelization, since we may distribute sub-trees to different compute cores spread across multiple machines of a compute cluster. Yet, in practice this task has proved to be very challenging. For example, Gurobi, one of the leading commercial MIP solvers, achieves

H. Simonis (Ed.): CPAIOR 2014, LNCS 8451, pp. 351–367, 2014.
© Springer International Publishing Switzerland 2014

an average speedup factor of 1.7 on 5 machines (and only 1.8 on 25 machines) when compared to using only 1 machine [18]. Furthermore, during the 2011 SAT Competition, the best parallel SAT solvers obtained a average speedup factor of about 3 on 32 cores, which was achieved by employing an algorithm portfolio rather than a parallelized search [20]. In our experimentation, the winner of the parallel category of the 2013 SAT Competition also achieved a speedup of only about 3 on 32 cores. Constraint programming search appears to be more suitable for parallelization than search for MIP or SAT: different strategies, including a recursive application of search goals [24], work stealing [14], problem decomposition [25], and a dedicated parallel scheme based on limited discrepancy search [23] all exhibit good speedups (sometimes near-linear) of the CP search in certain settings, especially those involving infeasible instances or scenarios where evaluating search tree leaves is costlier than evaluating internal nodes. Yet, recent developments in CP have moved towards more constraint learning during search, for which efficient parallelization becomes increasingly more difficult.

In general, search schemes relying heavily on learning during search (such as learning new bounds, activities for search heuristics, cuts for MIP, nogoods for CP, and clauses for SAT) tend to be more difficult to efficiently parallelize. *It remains a challenge to design a robust parallelization scheme for solving combinatorial optimization problems* which must necessarily deal with bounds.

Recently, a branch-and-bound scheme based on *approximate decision diagrams* was introduced as a promising alternative to conventional methods (such as integer programming) for solving combinatorial optimization problems [5, 7]. In this paper, our goal is to study how this branch-and-bound search scheme can be effectively parallelized. The key observation is that relaxed decision diagrams can be used to partition the search space, since for a given layer in the diagram each path from the root to the terminal passes through a node in that layer. We can therefore *branch on nodes in the decision diagram* instead of branching on variable-value pairs, as is done in conventional search methods. Each of the subproblems induced by a node in the diagram is processed recursively, and the process continues until all nodes have been solved by an exact decision diagram or pruned due to reasoning based on bounds on the objective function.

When designing parallel algorithms geared towards dozens or perhaps hundreds of workers operating in parallel, the two major challenges are *i)* balancing the workload across the workers, and *ii)* limiting the communication cost between workers. In the context of combinatorial search and optimization, most of the current methods are based on either parallelizing the traditional tree search or using portfolio techniques that make each worker operate on the entire problem. The former approach makes load balancing difficult as the computational cost of solving similarly sized subproblems can be orders of magnitude different. The latter approach typically relies on extensive communication in order to avoid duplication of effort across workers.

In contrast, using decision diagrams as a starting point for parallelization offers several notable advantages. For instance, the associated branch-and-bound method applies relaxed and restricted diagrams that are obtained by limiting the size of the diagrams to a certain maximum value. The size can be controlled, for example, simply by limiting the maximum width of the diagrams. As the computation time for a (sub)problem is roughly proportional to the size of the diagram, by controlling the size we are able to control the computation time. In combination with the recursive nature of the framework, this makes it easier to obtain a balanced workload. Further, the communication between workers can be limited in a natural way by using both global and local pools of currently open subproblems and employing pruning based on shared bounds. Upon processing a subproblem, each worker generates several new ones. Instead of communicating all of these back to the global pool, the worker keeps several of them to itself and continues to process them. In addition, whenever a worker finds a new feasible solution, the corresponding bound is communicated immediately to the global pool as well as to other workers, enabling them to prune subproblems that cannot provide a better solution. This helps avoid unnecessary computational effort, especially in the presence of local pools.

Our scheme is implemented in X10 [13, 26, 28], which is a modern programming language designed specifically for building applications for multi-core and clustered systems. For example, Bloom et al. [11] recently introduced SatX10 as an efficient and generic framework for parallel SAT solving using X10. We refer to our proposed framework for parallel decision diagrams as DDX10. The use of X10 allows us to program parallelization and communication constructs using a high-level, type checked language, leaving the details of an efficient backend implementation for a variety of systems and communication hardware to the language compiler and run-time. Furthermore, X10 also provides a convenient parallel execution framework, allowing a single compiled executable to run as easily on one core as on a cluster of networked machines.

Our main contributions are as follows. First, we describe, at a conceptual level, a scheme for parallelization of a sequential branch-and-bound search based on approximate decision diagrams and discuss how this can be efficiently implemented in the X10 framework. Second, we provide an empirical evaluation on the maximum independent set problem, showing the potential of the proposed method. Third, we compare the performance of DDX10 with a state-of-the-art parallel MIP solver, IBM ILOG CPLEX 12.5.1 . Experimental results indicate that DDX10 can obtain much better speedups than parallel MIP, especially when more workers are available. The results also demonstrate that the parallelization scheme provides near-linear speedups up to 256 cores, even in a distributed setting where the cores are split across multiple machines.

The remainder of the paper is structured as follows. In Section 2 we provide a brief overview of the sequential branch-and-bound algorithm based on approximate binary decision diagrams, specifically in the context of the maximum independent set problem. We then, in Section 3, describe how the algorithm is well-suited for parallelization, and describe our framework. We report on experimental results in Section 4 and conclude in Section 5.

2 Review: Branch-and-Bound with Decision Diagrams

Binary decision diagrams (BDDs) were originally introduced to represent Boolean functions in the context of circuit design and formal verification [1, 12, 22]. More recently, BDDs, and more generally multi-valued decision diagrams (MDDs), have been successfully applied to represent the solution set to arbitrary discrete optimization problems, with applications in constraint programming [2, 6, 19], disjunctive scheduling [15], and general discrete optimization [4, 5, 7–10].

In this section, we briefly summarize the branch-and-bound search scheme based on decision diagrams proposed by Bergman [5] which was further extended by Bergman, Cire, van Hoeve, and Hooker [7]. It can be applied to any combinatorial optimization (and more generally discrete optimization) problem for which a decision diagram representation is available. For clarity, we discuss the application to the maximum independent set problem, although the presented techniques are generally applicable. (Bergman et al. [7] report results for the maximum independent set, maximum 2-SAT, and maximum cut problems.)

2.1 Maximum Independent Set Problem and BDDs

Given a graph $G = (V, E), V = \{1, \ldots, n\}$, an *independent set* I is a subset $I \subseteq V$ for which no two vertices in I are connected by an edge in E. Given a non-negative weight w_j for each vertex $j \in V$, the *maximum independent set problem* (MISP) asks for an independent set of G with maximum total weight. The MISP is equivalent to the maximum clique problem (in the complement graph) and finds application in areas ranging from data mining [17] to bioinformatics [16] and social network analysis [3].

We next describe how we can represent all independent sets of G using a binary decision diagram. To this end, we let $\mathcal{I}(G)$ be the family of independent sets in G, and $v^*(G)$ the value of a maximum independent set in G. Let the *neighborhood* of $j \in V$ be $N(j) = \{j' : (j, j') \in E\}$. In addition, for any subset $V' \subseteq V$, let $G[V']$ be the graph induced by V'.

For our purposes, a *binary decision diagram* (BDD) $B = (U, \ell, A, d)$ for a graph $G = (V, E)$ is a directed graph with nodes U and arcs A. The mapping $\ell : U \to \{1, \ldots, n+1\}$ associates a *layer* with each node in U and the mapping $d : A \to \{0, 1\}$ associates an *arc-domain* with each arc in A. For an arc $a = (u, u')$ in A with $\ell(u) = j$, we define its weight $w(a)$ as w_j if $d(a) = 1$ and 0 otherwise. We impose conditions that $\forall a = (u, u') \in A, \ell(u) < \ell(u')$ and that there exist two special nodes r, t (the *root* and *terminal*, respectively) which are the unique nodes with $\ell(r) = 1, \ell(t) = n + 1$. With these conditions, B is acyclic, and all maximal paths connect r to t. Let $B|_{u,u'}$ be the subgraph of B induced by the nodes that belong on some u, u' path in B.

Paths in B correspond to subsets of V as follows. Let $p = (a_1, \ldots, a_k)$ be any path in B with $a_i = (u_i, u'_i)$ for $i = 1, \ldots, k$. Denote by $V(p)$ the subset of vertices of V that are associated with the domains of the arcs in p, i.e. $V(p) = \{j : d(a_i) = 1, \ell(u_i) = j\} \subseteq V$. In this way, if $\mathcal{P}(B)$ is the set of all $r - t$ paths in B, then the family of subsets of V represented by B is given by $\mathrm{Sol}(B) =$

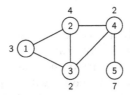

Fig. 1. Graph with vertex weights for the MISP

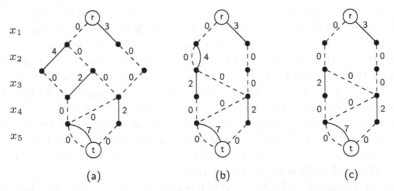

Fig. 2. Representing/approximating independent sets of the graph in Figure 1 with an exact BDD (a), relaxed BDD (b), and restricted BDD (c). Each layer in a BDD corresponds to binary decision x_i, where a dashed arc represents excluding vertex i, and a solid arc represents including vertex i.

$\cup_{p\in\mathcal{P}(B)}V(p)$. B is *exact* for G if $\text{Sol}(B) = \mathcal{I}(G)$ and is *relaxed* (resp. *restricted*) if $\text{Sol}(B) \supseteq \mathcal{I}(G)(resp.\text{Sol}(B) \subseteq \mathcal{I}(G))$. The weight $w(p)$ of a path $p \in \mathcal{P}(B)$ corresponds to the weight of the subset represented by p: $w(p) = \sum_{a\in p} w(a)$. We therefore have that the maximum path weight in B, $z^*(B)$, equals the size of the maximum weight set in $\text{Sol}(B)$. When B is exact, $z^*(B) = v^*(G)$, and when B is relaxed, resp. restricted, $z^*(B) \geq v^*(G), resp.z^*(B) \leq v^*(G)$.

To each $r - u$ path p we associate a *path-state* $\tilde{E}(p) \subseteq V$. $\tilde{E}(p)$ represents the set of vertices of V which correspond to the layers below node u that are not adjacent to any vertex in $V(p)$; i.e., $\tilde{E}(p) := \{i \geq \ell(u) \mid \forall j \in V(p) : (j, i) \notin E\}$. Path-states induce a *state* for each node u defined by the union of all path-states ending at u: i.e., $E(u) := \cup_{p\in\mathcal{P}(B|_{r,u})}\tilde{E}(p)$. Note that in an exact BDD, $\tilde{E}(p_1) = \tilde{E}(p_2)$ for $p_1, p_2 \in \mathcal{P}(B|_{r,u})$.

An example of an exact, relaxed, and restricted BDD for the graph in Figure 1 is given in Figure 2. The longest path length in the exact BDD (left-most diagram) is 11 and corresponds to vertex set $\{2, 5\}$. The longest path in the relaxed BDD (middle diagram) is 13, an upper bound on the optimal value, while the longest path value in the restricted BDD (right diagram) is 9, a lower bound on the objective value, and corresponds to vertex set $\{3, 5\}$.

2.2   BDD Construction

One technique for building BDDs is by top-down construction, which starts at the root node r, assigns the root state $E(r) = V$, and then creates the BDD layer-by-layer. Having constructed all nodes with $\ell(u) \leq j$, the algorithm builds layer $j + 1$ by examining the nodes with $\ell(u) = j$. If $j \notin E(u)$, $\ell(u)$ is increased by 1, pushing it to the next layer $j + 1$. Otherwise, two nodes u_0, u_1 with $\ell(u_0) = \ell(u_1) = j + 1$ are created along with arcs $a_k = (u, u_k), d(a_k) = k$ for $k = 0, 1$. $E(u_0) = E(u)\backslash\{j\}$ and $E(u_1) = E(u)\backslash(\{j\} \cup N(j))$. If two nodes w_1, w_2 are created for which $E(w_1) = E(w_2)$, they are merged by creating node w and redirected all arcs with tail w_1 or w_2 to w, deleting w_1 and w_2.

It has been shown [8] that the algorithm described above generates an exact BDD for the MISP, and that the states established by running the algorithm are equivalent to the definition of the state in the previous subsection. If an exact BDD can be constructed, it will exactly represent the family of independent sets, and therefore allow us to find the optimal value for the MISP by a single longest-path calculation. However, for most practically-sized problems, the BDD will grow exponentially large, requiring a modification of the algorithm that will create a relaxed/restricted BDD, from which we can extract upper/lower bounds, respectively, as described in the previous subsection.

To create a relaxed BDD, we forcibly merge nodes after each layer j is created if the number of nodes with $\ell(u) = j + 1$ exceeds a pre-set maximum allotted width W. We reassign the state of the node as the union of the states of the nodes that are merged, and continue the construction algorithm as before, merging nodes whenever the width exceeds W. This process creates a relaxed BDD [10]. To create a restricted BDD, if the number of nodes with $\ell(u) = j + 1$ exceeds W, we choose some nodes and delete them until the number of nodes in the layer equals W; see Bergman et al. [9]. This will ensure that a restricted BDD is created since we are never modifying any states – simply deleting nodes, and hence paths.

2.3   The Branch-and-Bound Algorithm

The branch-and-bound algorithm proceeds by branching on nodes in relaxed BDDs. Before describing the algorithm, we first define, recursively, *exact* versus *relaxed* nodes in relaxed BDDs. The root r is exact, and any other node u in a relaxed BDD B is exact if all nodes u' with arcs (u', u) directed to u are exact, and $\tilde{E}(p) = E(u)$, for all $p \in \mathcal{P}(B|_{r,u})$. A node is *relaxed* otherwise.

An *exact cut* C is a set of exact nodes whose removal disconnects r and t. The following theorem by Bergman [5] provides the basis for the algorithm:

Theorem 1 ([5]). *Let B be a relaxed BDD for a graph G and let C be an exact cut of B. Then,*

$$v^*(G) = \max_{u \in C} \{z^*(B|_{r,u}) + v^*(G[E(u)])\}.$$

Theorem 1 thereby establishes that, having constructed a relaxed BDD for G, we can branch on any exact cut C, and solve the problem defined by the graphs $G[E(u)]$ for every $u \in C$, adding the longest path to the nodes u to find the optimal value. Note that each node in the exact cut can therefore be processed individually by recording only the state of the node and the longest-path value up to that node in the relaxed BDD that created the node.

In general there are many choices for an exact cut to branch on. One possible cut is any layer prior to the first forced node merger during the top-down construction. Let $\tilde{\ell}$ be the first layer where we forcibly merged nodes together during the relaxation construction. For any $j < \tilde{\ell}$, each node must be exact, and any such layer (which contains all nodes with $\ell(u) = j$) will be an exact cut and can be used to create subproblems. Another possible cut is the *frontier cut* [5].

The algorithm branches on a set of partial solutions, as opposed to individual assignments of values to the problem variables, as is typically seen in algorithms designed to solve discrete optimization problems. This removes symmetry, and also allows the subproblems to be solved recursively, and individually. In principle, each node defines a subproblem which can be solved by *any* technique, but for the purpose of this paper, we will build restricted BDDs for primal heuristics, and relaxed BDDs for relaxation bounds, recursively defining subproblems.

3   Parallelizing BDD-Based Branch-and-Bound

The limited amount of information required for each BDD node makes the branch-and-bound algorithm naturally suitable for parallel processing. Once an exact cut C is computed for a relaxed BDD, the nodes $u \in C$ are independent and can be each processed in parallel. The information required to process a node $u \in C$ is its corresponding state, which is bounded by the number of vertices of G, $|V|$. After processing a node u, only the lower bound $v^*(G[E(u)])$ is needed to compute the optimal value, as shown in Theorem 1.

3.1   A Centralized Parallelization Scheme

There are many possible parallel strategies that can exploit this natural characteristic of the branch-and-bound algorithm for approximate decision diagrams. We propose here a centralized strategy defined as follows. A *master process* keeps a pool of BDD nodes to be processed, first initialized with a single node associated with the root state V. The master distributes the BDD nodes to a set of *workers*. Each worker receives a number of nodes, processes them by creating the corresponding relaxed and restricted BDDs, and either sends back to the master new nodes to explore (from an exact cut of their relaxed BDD) or sends to the master as well as all workers an improved lower bound from a restricted BDD. The workers also send the upper bound obtained from the relaxed BDD from which the nodes were extracted, which is then used by the master for potentially pruning the nodes according to the current best lower bound at the time these nodes are brought out from the global pool to be processed.

Even though conceptually simple, our centralized parallelization strategy involves communication between all workers and many choices that have a significant impact on performance. After discussing the challenge of effective parallelization, we explore some of these choices in the rest of this section.

3.2   The Challenge of Effective Parallelization

Clearly, a BDD constructed in parallel as described above can be very different in structure and overall size from a BDD constructed sequentially for the same problem instance. As a simple example, consider two nodes u_1 and u_2 in the exact cut C. By processing u_1 first, one could potentially improve the lower bound so much that u_2 can be pruned right away in the sequential case. In the parallel setting, however, while worker 1 processes u_1, worker 2 will be already wasting search effort on u_2, not knowing that u_2 could simply be pruned if it waited for worker 1 to finish processing u_1.

In general, *the order in which nodes are processed in the approximate BDD matters* — information passed on by nodes processed earlier can substantially alter the direction of search later. This is very clear in the context of combinatorial search for SAT, where dynamic variable activities and clauses learned from conflicts dramatically alter the behavior of subsequent search. Similarly, bounds in MIP and impacts in CP influence subsequent search.

Issues of this nature pose a challenge to effective parallelization of anything but brute force combinatorial search oblivious to the order in which the search space is explored. Such a search is, of course, trivial to parallelize. For most search methods of interest, however, a parallelization strategy that delicately balances independence of workers with timely sharing of information is often the key to success. As our experiments will demonstrate, our implementation, DDX10, achieves this balance to a large extent on both random and structured instances of the independent set problem. In particular, the overall size of parallel BDDs is not much larger than that of the corresponding sequential BDDs. In the remainder of this section, we discuss the various aspects of DDX10 that contribute to this desirable behavior.

3.3   Global and Local Pools

We refer to the pool of nodes kept by the master as the *global pool*. Each node in the global pool has two pieces of information: a state, which is necessary to build the relaxed and restricted BDDs, and the longest path value in the relaxed BDD that created that node, from the root to the node. All nodes sent to the master are first stored in the global pool and are then redistributed to the workers. Nodes with an upper bound that is no more than the best found lower bound at the time are pruned from the pool, as these can never provide a solution better than one already found.

In order to select which nodes to send to workers first, the global pool is implemented here using a data structure that mixes a priority queue and a stack. Initially, the global pool gives priority to nodes that have a larger upper bound, which intuitively are nodes with higher potential to yield better solutions. However, this search strategy simulates a best-first search and may result in an exponential number of nodes in the global queue that still need to be explored. To remedy this, the global pool switches to a *last-in, first-out* node selection strategy when its size reaches a particular value (denoted *maxPQueueLength*), adjusted according to the available memory on the machine where the master runs. This strategy resembles a stack-based depth-first search and limits the total amount of memory necessary to perform search.

Besides the global pool, workers also keep a *local pool* of nodes. The subproblems represented by the nodes are usually small, making it advantageous for workers to keep their own pool so as to reduce the overall communication to the master. The local pool is represented by a priority queue, selecting nodes with a larger upper bound first. After a relaxed BDD is created, a certain fraction of the nodes (with preference to those with a larger upper bound) in the exact cut is sent to the master, while the remaining fraction (denoted *fracToKeep*) of nodes are added to the local pool. The local pool size is also limited; when the pool reaches this maximum size (denoted *maxLocalPoolSize*), we stop adding more nodes to the local queue and start sending any newly created nodes directly to the master. When a worker's local pool becomes empty, it notifies the master that it is ready to receive new nodes.

3.4   Load Balancing

The global queue starts off with a single node corresponding to the root state V, which is assigned to an arbitrary worker which then applies a cut to produce more states and sends a fraction of them, as discussed above, back to the global queue. The size of the global pool thus starts to grow rapidly and one must choose how many nodes to send subsequently to other workers. Sending one node (the one with the highest priority) to a worker at a time would mimic the sequential case most closely. However, it would also result in the most number of communications between the master and the workers, which often results in a prohibitively large system overhead. On the other hand, sending too many nodes at once to a single worker runs the risk of starvation, i.e., the global queue becoming empty and other workers sitting idle waiting to receive new work.

Based on experimentation with representative instances, we propose the following parameterized scheme to dynamically decide how many nodes the master should send to a worker at any time. Here, we use the notation $[x]_\ell^u$ as a shorthand for $\min\{u, \max\{\ell, x\}\}$, that is, x capped to lie in the interval $[\ell, u]$.

$$nNodesToSend_{c,\bar{c},c^*}(s,q,w) = \left[\min\left\{\bar{c}s, c^*\frac{q}{w}\right\}\right]_c^\infty \qquad (1)$$

where s is a decaying running average of the number of nodes added to the global pool by workers after processing a node,[1] q is the current size of the global pool, w is the number of workers, and $c, \bar{c}$, and c^* are parametrization constants.

---

[1] When a cut C is applied upon processing a node, the value of s is updated as $s_{new} = rs_{old} + (1-r)|C|$, with $r = 0.5$ in the current implementation.

The intuition behind this choice is as follows. c is a flat lower limit (a relatively small number) on how many nodes are sent at a time irrespective of other factors. The inner minimum expression upper bounds the number of nodes to send to be no more than both (a constant times) the number of nodes the worker is in turn expected to return to the global queue upon processing each node and (a constant times) an even division of all current nodes in the queue into the number of workers. The first influences how fast the global queue grows while the second relates to fairness among workers and the possibility of starvation. Larger values of $c, \bar{c}$, and c^* reduce the number of times communication occurs between the master and workers, at the expense of moving further away from mimicking the sequential case.

Load balancing also involves appropriately setting the $fracToKeep$ value discussed earlier. We use the following scheme, parameterized by d and d^*:

$$fracToKeep_{d,d^*}(t) = [t/d^*]_d^1 \tag{2}$$

where t is the number of states received by the worker. In other words, the fraction of nodes to keep for the local queue is $1/d^*$ times the number of states received by the worker, capped to lie in the range $[d, 1]$.

3.5 DDX10: Implementing Parallelization Using X10

As mentioned earlier, X10 is a high-level parallel programming and execution framework. It supports parallelism natively and applications built with it can be compiled to run on various operating systems and communication hardware.

Similar to SatX10 [11], we capitalize on the fact that X10 can incorporate existing libraries written in C++ or Java. We start off with the sequential version of the BDD code base for MISP [7] and integrate it in X10, using the C++ backend. The integration involves adding hooks to the BDD class so that (a) the master can communicate a set of starting nodes to build approximate BDDs for, (b) each worker can communicate nodes (and corresponding upper bounds) of an exact cut back to the master, and (c) each worker can send updated lower bounds immediately to all other workers and the master so as to enable pruning.

The global pool for the master is implemented natively in X10 using a simple combination of a priority queue and a stack. The DDX10 framework itself (consisting mainly of the main DDSolver class in DDX10.x10 and the pool in StatePool.x10) is generic and not tied to MISP in any way. It can, in principle, work with any maximization or minimization problem for which states for a BDD (or even an MDD) can be appropriately defined.

4   Experimental Results

The MISP problem can be formulated and solved using several existing general purpose discrete optimization techniques. A MIP formulation is considered to be very effective and has been used previously to evaluate the sequential BDD approach [7]. Given the availability of parallel MIP solvers as a comparison point, we present two sets of experiments on the MISP problem: (1) we compare DDX10 with a MIP formulation solved using IBM ILOG CPLEX 12.5.1 on up to 32 cores, and (2) we show how DDX10 scales when going beyond 32 cores and employing up to 256 cores distributed across a cluster. We borrow the MIP encoding from Bergman et al. [7] and employ the built-in parallel branch-and-bound MIP search mechanism of CPLEX. The comparison with CPLEX is limited to 32 cores because this is the largest number of cores we have available on a single machine (note that CPLEX 12.5.1 does not support distributed execution). Since the current version of DDX10 is not deterministic, we run CPLEX also in its non-deterministic ('opportunistic') mode.

DDX10 is implemented using X10 2.3.1 [28] and compiled using the C++ backend with g++ 4.4.5.[2] For all experiments with DDX10, we used the following values of the parameters of the parallelization scheme as discussed in Section 3: $maxPQueueLength = 5.5 \times 10^9$ (determined based on the available memory on the machine storing the global queue), $maxLocalPoolSize = 1000, c = 10, \bar{c} = 1.0, c^* = 2.0, d = 0.1$ and $d^* = 100$. The maximum width W for the BDD generated at each subproblem was set to be the number of free variables (i.e., the number of active vertices) in the state of the BDD node that generated the subproblem. The type of exact cut used in the branch-and-bound algorithm for the experiments was the *frontier cut* [5]. These values and parameters were chosen based on experimentation on our cluster with a few representative instances, keeping in mind their overall impact on load balancing and pruning as discussed earlier.

4.1   DDX10 versus Parallel MIP

The comparison between DDX10 and IBM ILOG CPLEX 12.5.1 was conducted on 2.3 GHz AMD Opteron 6134 machines with 32 cores, 64 GB RAM, 512 KB L2 cache, and 12 MB L3 cache.

To draw meaningful conclusions about the scaling behavior of CPLEX vs. DDX10 as the number w of workers is increased, we start by selecting problem instances where both approaches exhibit comparable performance in the sequential setting. To this end, we generated random MISP instances as also used previously by Bergman et al. [7]. We report comparison on instances with 170 vertices and six graph densities $\rho = 0.19, 0.21, 0.23, 0.25, 0.27$, and 0.29. For each ρ, we generated five random graphs, obtaining a total of 30 problem instances. For each pair (ρ, w) with w being the number of workers, we aggregate the runtime over the five random graphs using the geometric mean.

Figure 3 summarizes the result of this comparison for $w = 1, 2, 4, 16$, and 32. As we see, CPLEX and DDX10 display comparable performance for $w = 1$ (the left-most data points). While the performance of CPLEX varies relatively little as a function of the graph density ρ, that of DDX10 varies more widely. As observed earlier by Bergman et al. [7] for the sequential case, BDD-based

---

[2] The current version of DDX10 may be downloaded from
http://www.andrew.cmu.edu/user/vanhoeve/mdd

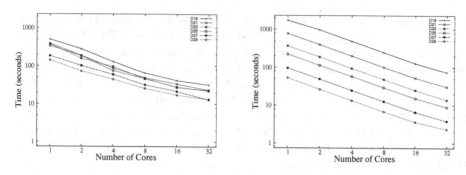

Fig. 3. Performance of CPLEX (left) and DDX10 (right), with one curve for each graph density ρ shown in the legend as a percentage. Both runtime (y-axis) and number of cores (x-axis) are in log-scale.

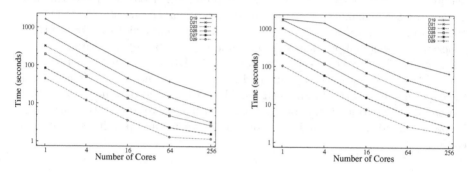

Fig. 4. Scaling behavior of DDX10 on MISP instances with 170 (left) and 190 (right) vertices, with one curve for each graph density ρ shown in the legend as a percentage. Both runtime (y-axis) and number of cores (x-axis) are in log-scale.

branch-and-bound performs better on higher density graphs than sparse graphs. Nevertheless, the performance of the two approaches when $w = 1$ is in a comparable range for the observation we want to make, which is the following: *DDX10 scales more consistently than CPLEX when invoked in parallel* and also retains its advantage on higher density graphs. For $\rho > 0.23$, DDX10 is clearly exploiting parallelism better than CPLEX. For example, for $\rho = 0.29$ and $w = 1$, DDX10 takes about 80 seconds to solve the instances while CPLEX needs about 100 seconds—a modest performance ratio of 1.25. This same performance ratio increases to 5.5 when both methods use $w = 32$ workers.

4.2   Parallel versus Sequential Decision Diagrams

The two experiments reported in this section were conducted on a larger cluster, with 13 of 3.8 GHz Power7 machines (CHRP IBM 9125-F2C) with 32 cores (4-way SMT for 128 hardware threads) and 128 GB of RAM. The machines are connected via a network that supports the PAMI message passing interface [21],

Table 1. Runtime (seconds) of DDX10 on DIMACS instances. Timeout = 1800

instance	n	density	1 core	4 cores	16 cores	64 cores	256 cores
hamming8-4.clq	256	0.36	25.24	7.08	2.33	1.32	0.68
brock200_4.clq	200	0.34	33.43	9.04	2.84	1.45	1.03
san400_0.7_1.clq	400	0.30	33.96	9.43	4.63	1.77	0.80
p_hat300-2.clq	300	0.51	34.36	9.17	2.74	1.69	0.79
san1000.clq	1000	0.50	40.02	12.06	7.15	2.15	9.09
p_hat1000-1.clq	1000	0.76	43.35	12.10	4.47	2.84	1.66
sanr400_0.5.clq	400	0.50	77.30	18.10	5.61	2.18	2.16
san200_0.9_2.clq	200	0.10	93.40	23.72	7.68	3.64	1.65
sanr200_0.7.clq	200	0.30	117.66	30.21	8.26	2.52	2.08
san400_0.7_2.clq	400	0.30	234.54	59.34	16.03	6.05	4.28
p_hat1500-1.clq	1500	0.75	379.63	100.3	29.09	10.62	25.18
brock200_1.clq	200	0.25	586.26	150.3	39.95	12.74	6.55
hamming8-2.clq	256	0.03	663.88	166.49	41.80	23.18	14.38
gen200_p0.9_55.clq	200	0.10	717.64	143.90	43.83	12.30	6.13
C125.9.clq	125	0.10	1100.91	277.07	70.74	19.53	8.07
san400_0.7_3.clq	400	0.30	–	709.03	184.84	54.62	136.47
p_hat500-2.clq	500	0.50	–	736.39	193.55	62.06	23.81
p_hat300-3.clq	300	0.26	–	–	1158.18	349.75	172.34
san400_0.9_1.clq	400	0.10	–	–	1386.42	345.66	125.27
san200_0.9_3.clq	200	0.10	–	–	–	487.11	170.08
gen200_p0.9_44.clq	200	0.10	–	–	–	1713.76	682.28
sanr400_0.7.clq	400	0.30	–	–	–	–	1366.98
p_hat700-2.clq	700	0.50	–	–	–	–	1405.46

although DDX10 can also be easily compiled to run using the usual network communication with TCP sockets. We used 24 workers on each machine, using as many machines as necessary to operate w workers in parallel.

Random Instances. The first experiment reuses the random MISP instances introduced in the previous section, with the addition of similar but harder instances on graphs with 190 vertices, resulting in 60 instances in total.

As Figure 4 shows, DDX10 scales near-linearly up to 64 cores and still very well up to 256 cores. The slight degradation in performance when going to 256 cores is more apparent for the higher density instances (lower curves in the plots), which do not have much room left for linear speedups as they need only a couple of seconds to be solved with 64 cores. For the harder instances (upper curves), the scaling is still satisfactory even if not linear. As noted earlier, coming anywhere close to near-linear speedups for complex combinatorial search and optimization methods has been remarkably hard for SAT and MIP. These results show that parallelization of BDD based branch-and-bound can be much more effective.

DIMACS Instances. The second experiment is on the DIMACS instances used by Bergman [5] and Bergman et al. [7], where it was demonstrated that sequential BDD-based branch-and-bound has complementary strengths compared

Table 2. Number of nodes in multiples of $1,000$ processed (#No) and pruned (#Pr) by DDX10 as a function of the number of cores. Same setup as in Table 1.

instance	1 core #No	1 core #Pr	4 cores #No	4 cores #Pr	16 cores #No	16 cores #Pr	64 cores #No	64 cores #Pr	256 cores #No	256 cores #Pr
hamming8-4.clq	43	0	42	0	40	0	32	0	41	0
brock200_4.clq	110	42	112	45	100	37	83	30	71	25
san400_0.7_1.clq	7	1	8	1	6	0	10	1	14	1
p_hat300-2.clq	80	31	74	27	45	11	46	7	65	12
san1000.clq	29	16	50	37	18	4	13	6	28	6
p_hat1000-1.clq	225	8	209	0	154	1	163	1	206	1
sanr400_0.5.clq	451	153	252	5	354	83	187	7	206	5
san200_0.9_2.clq	22	0	20	0	19	0	18	1	25	0
sanr200_0.7.clq	260	3	259	5	271	17	218	4	193	6
san400_0.7_2.clq	98	2	99	5	112	21	147	67	101	35
p_hat1500-1.clq	1586	380	1587	392	1511	402	962	224	1028	13
brock200_1.clq	1378	384	1389	393	1396	403	1321	393	998	249
hamming8-2.clq	45	0	49	0	49	0	47	0	80	0
gen200_p0.9_55.clq	287	88	180	6	286	90	213	58	217	71
C125.9.clq	1066	2	1068	0	1104	38	1052	13	959	19
san400_0.7_3.clq	–	–	2975	913	2969	916	2789	779	1761	42
p_hat500-2.clq	–	–	2896	710	3011	861	3635	1442	2243	342
p_hat300-3.clq	–	–	–	–	18032	4190	17638	3867	15852	2881
san400_0.9_1.clq	–	–	–	–	2288	238	2218	207	2338	422
san200_0.9_3.clq	–	–	–	–	–	–	9796	390	10302	872
gen200_p0.9_44.clq	–	–	–	–	–	–	43898	5148	45761	7446
sanr400_0.7.clq	–	–	–	–	–	–	–	–	135029	247
p_hat700-2.clq	–	–	–	–	–	–	–	–	89845	8054

to sequential CPLEX and outperforms the latter on several instances, often the ones with higher graph density ρ. We consider here the subset of instances that take at least 10 seconds (on our machines) to solve using sequential BDDs and omit any that cannot be solved within the time limit of 1800 seconds (even with 256 cores). The performance of DDX10 with $w = 1, 4, 16, 64$, and 256 is reported in Table 1, with rows sorted by hardness of instances.

These instances represent a wide range of graph size, density, and structure. As we see from the table, DDX10 is able to scale very well to 256 cores. Except for three instances, it is significantly faster on 256 cores than on 64 cores, despite the substantially larger communication overhead for workload distribution and bound sharing.

Table 2 reports the total number of nodes processed through the global queue, as well as the number of nodes pruned due to bounds communicated by the workers.[3] Somewhat surprisingly, the number of nodes processed does not increase by much compared to the sequential case, despite the fact that hundreds of workers

---

[3] Here we do not take into account the number of nodes added to local pools, which is usually a small fraction of the number of nodes processed by the global pool.

start processing nodes in parallel without waiting for potentially improved bounds which might have been obtained by processing nodes sequentially. Furthermore, the number of pruned nodes also stays steady as w grows, indicating that bounds communication is working effectively. This provides insight into the amiable scaling behavior of DDX10 and shows that it is able to retain sufficient global knowledge even when executed in a distributed fashion.

5   Summary and Future Work

We have introduced a parallelization scheme for a branch-and-bound search based on approximate binary decision diagrams. We implemented our approach using the X10 parallel programming and execution framework. Our application of the technique to the maximum independent set problem demonstrates that approximate BDD-based branch-and-bound can scale significantly better with increasing number of workers than a state-of-the-art commercial-strength solver. The results indicate that the solution technique is amenable to effective parallelization on hundreds of compute cores.

Besides extending DDX10 to different problem classes in addition to MISP, two interesting extensions of the underlying parallel framework would be to support *determinism* and *decentralized load balancing*. Determinism is important to many industrial users and therefore a desirable feature. Conveniently, X10 comes with native determinacy constructs called *clocks* that can be employed in order to ensure that DDX10 can also be executed deterministically. Second, while the centralized load balancing scheme proposed here scaled well to 256 cores, a central global queue is likely to become a bottleneck when extending to thousands of cores. Particularly attractive in the current setting is the recent work by Saraswat et al. [27] which provides a general determinate application framework, called the *Global Load-Balancing* (GLB) framework, available as a library in the latest release of X10. The GLB framework is responsible for automatically distributing the generated collection of tasks across all nodes, and detecting global termination. This can be used to design a distributed variant of DDX10 targeted for much larger compute clusters.

In summary, our work highlights how integrating branch-and-bound based on approximate decision diagrams into X10 allows one to leverage the features of a major parallel programming language in order to substantially improve our ability to solve hard combinatorial optimization problems by exploiting hundreds of compute cores in parallel.

Acknowledgements. Andre Cire and Willem-Jan van Hoeve were partially supported by NSF under grant CMMI-1130012 and a Google Research Award.

References

1. Akers, S.B.: Binary decision diagrams. IEEE Transactions on Computers 27, 509–516 (1978)

2. Andersen, H.R., Hadzic, T., Hooker, J.N., Tiedemann, P.: A Constraint Store Based on Multivalued Decision Diagrams. In: Bessière, C. (ed.) CP 2007. LNCS, vol. 4741, pp. 118–132. Springer, Heidelberg (2007)
3. Balasundaram, B., Butenko, S., Hicks, I.V.: Clique relaxations in social network analysis: The maximum k-plex problem. Operations Research 59(1), 133–142 (2011)
4. Behle, M.: Binary Decision Diagrams and Integer Programming. PhD thesis, Max Planck Institute for Computer Science (2007)
5. Bergman, D.: New Techniques for Discrete Optimization. PhD thesis, Tepper School of Business, Carnegie Mellon University (2013)
6. Bergman, D., Cire, A.A., van Hoeve, W.-J.: MDD Propagation for Sequence Constraints. JAIR (to appear)
7. Bergman, D., Cire, A.A., van Hoeve, W.-J., Hooker, J.N.: Discrete optimization with decision diagrams (2013) (under review)
8. Bergman, D., Cire, A.A., van Hoeve, W.-J., Hooker, J.N.: Optimization bounds from binary decision diagrams. INFORMS Journal on Computing (to appear)
9. Bergman, D., Cire, A.A., van Hoeve, W.-J., Yunes, T.: BDD-based heuristics for binary optimization. Journal of Heuristics (to appear)
10. Bergman, D., van Hoeve, W.-J., Hooker, J.N.: Manipulating MDD relaxations for combinatorial optimization. In: Achterberg, T., Beck, J.C. (eds.) CPAIOR 2011. LNCS, vol. 6697, pp. 20–35. Springer, Heidelberg (2011)
11. Bloom, B., Grove, D., Herta, B., Sabharwal, A., Samulowitz, H., Saraswat, V.: SatX10: A scalable plug & play parallel SAT framework. In: Cimatti, A., Sebastiani, R. (eds.) SAT 2012. LNCS, vol. 7317, pp. 463–468. Springer, Heidelberg (2012)
12. Bryant, R.E.: Graph-based algorithms for boolean function manipulation. IEEE Transactions on Computers C-35, 677–691 (1986)
13. Charles, P., Grothoff, C., Saraswat, V., Donawa, C., Kielstra, A., Ebcioglu, K., von Praun, C., Sarkar, V.: X10: an object-oriented approach to non-uniform cluster computing. In: OOPSLA 2005, San Diego, CA, USA, pp. 519–538 (2005)
14. Chu, G., Schulte, C., Stuckey, P.J.: Confidence-Based Work Stealing in Parallel Constraint Programming. In: Gent, I.P. (ed.) CP 2009. LNCS, vol. 5732, pp. 226–241. Springer, Heidelberg (2009)
15. Cire, A.A., van Hoeve, W.-J.: Multivalued decision diagrams for sequencing problems. Operations Research 61(6), 1411–1428 (2013)
16. Eblen, J.D., Phillips, C.A., Rogers, G.L., Langston, M.A.: The maximum clique enumeration problem: Algorithms, applications and implementations. In: Chen, J., Wang, J., Zelikovsky, A. (eds.) ISBRA 2011. LNCS, vol. 6674, pp. 306–319. Springer, Heidelberg (2011)
17. Edachery, J., Sen, A., Brandenburg, F.J.: Graph clustering using distance-k cliques. In: Kratochvíl, J. (ed.) GD 1999. LNCS, vol. 1731, pp. 98–106. Springer, Heidelberg (1999)
18. Gu, Z.: Gurobi Optimization - Gurobi Compute Server, Distributed Tuning Tool and Distributed Concurrent MIP Solver. In: INFORMS Annual Meeting (2013), http://www.gurobi.com/products/gurobi-compute-server/distributed-optimization
19. Hoda, S., van Hoeve, W.-J., Hooker, J.N.: A systematic approach to MDD-based constraint programming. In: Cohen, D. (ed.) CP 2010. LNCS, vol. 6308, pp. 266–280. Springer, Heidelberg (2010)
20. Järvisalo, M., Le Berre, D., Roussel, O., Simon, L.: The international SAT solver competitions. Artificial Intelligence Magazine (AI Magazine) 1(33), 89–94 (2012)

21. Kumar, S., Mamidala, A.R., Faraj, D., Smith, B., Blocksome, M., Cernohous, B., Miller, D., Parker, J., Ratterman, J., Heidelberger, P., Chen, D., Steinmacher-Burrow, B.: PAMI: A parallel active message interface for the Blue Gene/Q supercomputer. In: IPDPS-2012: 26th IEEE International Parallel & Distributed Processing Symposium, pp. 763–773 (2012)

22. Lee, C.Y.: Representation of switching circuits by binary-decision programs. Bell Systems Technical Journal 38, 985–999 (1959)

23. Moisan, T., Gaudreault, J., Quimper, C.-G.: Parallel Discrepancy-Based Search. In: Schulte, C. (ed.) CP 2013. LNCS, vol. 8124, pp. 30–46. Springer, Heidelberg (2013)

24. Perron, L.: Search Procedures and Parallelism in Constraint Programming. In: Jaffar, J. (ed.) CP 1999. LNCS, vol. 1713, pp. 346–361. Springer, Heidelberg (1999)

25. Régin, J.-C., Rezgui, M., Malapert, A.: Embarrassingly Parallel Search. In: Schulte, C. (ed.) CP 2013. LNCS, vol. 8124, pp. 596–610. Springer, Heidelberg (2013)

26. Saraswat, V., Bloom, B., Peshansky, I., Tardieu, O., Grove, D.: Report on the experimental language, X10. Technical report, IBM Research (2011)

27. Saraswat, V.A., Kambadur, P., Kodali, S., Grove, D., Krishnamoorthy, S.: Lifeline-based global load balancing. In: Proceedings of the 16th ACM Symposium on Principles and Practice of Parallel Programming, PPoPP 2011, pp. 201–212. ACM, New York (2011), http://doi.acm.org/10.1145/1941553.1941582 ISBN 978-1-4503-0119-0

28. X10. X10 programming language web site, http://x10-lang.org/ (January 2010)

A Portfolio Approach to Enumerating Minimal Correction Subsets for Satisfiability Problems

Yuri Malitsky[1], Barry O'Sullivan[1],
and Alessandro Previti[2], and Joao Marques-Silva[2]

[1] Insight Centre for Data Analytics, University College Cork, Ireland
{y.malitsky,b.osullivan}@4c.ucc.ie
[2] Complex and Adaptive Systems Laboratory, Dublin, Ireland
alessandro.previti@ucdconnect.ie, jpms@ucd.ie

Abstract. Even when it has been shown that no solution exists for a particular constraint satisfaction problem, one may still aim to restore consistency by relaxing the minimal number of constraints. In the context of a Boolean formula like SAT, such a relaxation is referred to as a Minimal Correction Subset (MCS). In the context of SAT, identifying MCSs for an instance is relevant in a wide range of applications, including MaxSAT solution approximation and Minimal Unsatisfiable Subset (MUS) enumeration. However, while there are a number of existing approaches to this problem, in this paper we demonstrate how performance can be significantly improved by employing algorithm portfolios. Yet, instead of applying the standard approach of selecting a single solver for the instance at hand, we present a new technique that within a predetermined timeout switches between enumeration algorithms multiple times. Through experimental study, this new approach is shown to outperform any possible optimal portfolio that solely relies on solvers that run uninterrupted for the allotted time.

1 Introduction

A set of constraints without a solution is referred to as being inconsistent, unsatisfiable, or over-constrained. Regardless of the terminology, in order to restore consistency some of the constraints have to be relaxed. To preserve as much information about the original problem as possible, in practice the goal is usually to do this with a minimal set of constraints. In the context of Boolean formulas, such a minimal set is referred to as the Minimal Correction Subset (MCS), and the importance of computing such a minimal relaxation is well-known, e.g. [9,15,5]. Furthermore, aside from their obvious benefits, MCSs for Boolean formulas also find other applications. As an example, the state-of-the-art in Minimal Unsatisfiable Subset (MUS) computation, complete enumeration relies upon the computation of the whole set of MCSs [12]. MCSs can also be used to compute minimal models [2,21]. On top of this, it is important to highlight that MCSs with the smallest size correspond to the solution of the Maximum Satisfiability (MaxSAT) problem.

H. Simonis (Ed.): CPAIOR 2014, LNCS 8451, pp. 368–376, 2014.

Defined formally, if $\mathcal{F}$ is a formula expressed as a set of clauses, then we define an MCS as follows:

Definition 1 (Minimal Correction Subset). $\mathcal{C} \subseteq \mathcal{F}$ *is a* Minimal Correction Subset *(MCS) iff* $\mathcal{F} \setminus \mathcal{C}$ *is satisfiable and* $\forall c \in \mathcal{C}, \mathcal{F} \setminus (\mathcal{C} \setminus \{c\})$ *is unsatisfiable.*

The study of over-constrained problems for these sets has been considered in a number of settings, e.g. [8], and include Reiter's seminal work in diagnosis [18]. Note that in this context MCSs and MUSs are referred to as *minimal diagnoses* and *minimal conflict sets* respectively. The computation of these sets has also been studied extensively in the area of constraints, e.g. [9,15,5]. In the last decade, various approaches have been proposed for computing MCSs. An intuitive approach for computing a single MCS was proposed in [1]. Here the idea being that, starting with the empty set, to add one constraint at a time and repeatedly calling a solver to check satisfiability. When the extended set of constraints became unsatisfiable, the constraint added last is guaranteed to be part of the MCS under construction. We will refer to this algorithm as the Basic Linear Search algorithm (bls). However, [12] showed that such an approach was inefficient in practice, and proposed an alternative based on MaxSAT. Additional research had also revealed how, in the context of model-based diagnosis, a divide-and-conquer strategy [15,5] could also be embraced for this task.

More recently, new techniques were introduced to enhance performance of existing algorithms [14]. In that same paper, a new algorithm was introduced where, inspired by the work of [3], it was shown how to outperform the alternatives for the computation of a single MCS. Results in that paper, show that the performance gains of the Virtual Best Solver (VBS) with respect to the other algorithms are quite significant. This meant that these algorithms complement each other and, as explicitly said by the authors, suggests the possibility of using portfolio solutions, one of the goals of this paper.

This paper shows how the algorithms presented in [14] can benefit from a portfolio approach. We show how the performance can be significantly improved for both discovering a single solution as well as the enumeration of all MCSs. In particular we show that instead of directly applying a portfolio approach to predict the next solver, it is far more beneficial to continuously switch among solvers within the allotted time.

2   A Portfolio Approach

Over the last few years, it has become increasingly evident that there is often no single solver that performs optimally on every instance [10,22,16,17], an observation strongly connected to the No Free Lunch Theorem. And in practice we often see that different solvers work noticeably better on certain instances than their counterparts. It is this observation that drives research into accurately identifying the most appropriate solver for a given problem instance. A task known in the community as Algorithm Selection.

Algorithm 1. Solver-based Nearest Neighbor for Algorithm Portfolios

---

1: **function** SNNAP-TRAIN(T, F, R)
2:      **for all** instances i in T **do**
3:          $\bar{R}_i \leftarrow Scaled(R_i)$
4:      **end for**
5:      **for all** solver j in the portfolio **do**
6:          $PM_j \leftarrow PredictionModel(T, F, \bar{R})$
7:      **end for**
8:      **return** PM
9: **end function**

1: **function** SNNAP-RUN($x, PM, T, R, \bar{R}, A, k$)
2:      $PR \leftarrow Predict(PM, x)$
3:      $dist \leftarrow CalculateDistance(PR, T, \bar{R})$
4:      $neighbors \leftarrow FindClosestInstances(dist, k)$
5:      $j \leftarrow FindBestSolver(neighbors, R)$
6:      **return** $A_j(x)$
7: **end function**

The premise is that there exist some detectable structural differences between the instances, and that by correctly discerning them, it is possible to predict the quality of a solvers' performance. There are of course many portfolio approaches that use a plethora of prediction techniques. One approach trains regression models to predict the performance of each solver, selecting the expected best one [19]. Alternatively, a ranking approach can suggest a preference order over all solvers [7]. Forests of trees have been used to distinguish between every pair of solvers in a portfolio, selecting the one voted upon most frequently [23]. Meanwhile, clustering can be used to group similar instances, and identifying a solver for each such grouping [13]. An overview of many of these approaches is presented in [11].

For the experiments detailed in this paper we utilize the Solver-based Nearest Neighbor for Algorithm Portfolios, or SNNAP [4]. In a recent study, this approach has been shown to work well in practice by making predictions not solely on the feature vector but by also considering the similarities in solver performances over the instances. This section first presents the SNNAP methodology in further detail and present the features which will be used to distinguish amongst instances. The section will then demonstrate the effectiveness of SNNAP in practice.

2.1   SNNAP

An acronym for Solver-based Nearest Neighbor for Algorithm Portfolios, this approach uses a two step procedure to select the best solver for an instance, the details of which are presented in Algorithm 1. In the training phase, provided are a collection of training instances T, a feature vector describing each instance F, and the performance of each algorithm in the portfolio on every instance R.

To avoid working with the actual performances, which can behave highly stochastically and take on a wide range of values, the values in R are normalized for each instance. Thus, for each instance the mean performance is set at 0 and the standard deviation to 1. Subsequently a random forest is trained for each solver to predict this normalized value when provided with the appropriate features.

During the testing phase, SNNAP utilizes the random forest to predict the effectiveness of each solver on the provided instance. Therefore, our trees predict how much better or worse a solver is than average on this instance, which helps avoid biasing predictions to instances that have a high range of possible outcomes. However, any prediction approach is likely to have some errors. Assuming, however, that the best solver would be among the top predictions, a binary vector is created where only the top n values are set to 1, while all others are set to 0. The distance between any two instances therefore becomes the Jaccard distance:

$$1 - \frac{|intersection((A_{a_1}, \ldots A_{a_n}), (A_{b_1}, \ldots A_{b_n}))|}{|union((A_{a_1}, \ldots A_{a_n}), (A_{b_1}, \ldots A_{b_n}))|}$$

Using this definition two instances that will prefer the exact same n solvers will have a distance of 0, while instances which prefer completely different solvers will have a distance of 1. Employing this distance metric, SNNAP finds the m nearest neighbors to the new instance, and uses the solver that performed best on those neighbors.

2.2   Features

The success or failure of Algorithm Selection largely depends on the ability to detect structural differences between the instances. In practice, these structural differences are defined by a vector of features. In the case of finding MCSs, we are dealing with Partial MaxSAT (PMS) instances. PMS instances are utilized rather than plain SAT in order to be able to consider SAT instances to which additional clauses have been added in order to avoid previously discovered MCSs. For this problem type, we introduce the following features:

Problem Size Features:
1-2 Number of variables and
   clauses in original formula:
   denoted v and c, respectively
3 Percentage of Soft Clauses
4-7 Soft Clause Weights: mean,
   stdev, min and max
8 Ratio of variables to clauses
Variable-Clause Graph Features:
9-13 Variable node degree statistics:
   mean, stdev, min, max, spread
14-18 Clause node degree statistics:
   mean, stdev, min, max, spread

Balance Features:
19-23 Positive to negative
   occurrences of each variable:
   mean, stdev, min, max, spread
24-28 Positive to negative
   literals in each clause:
   mean, stdev, min, max, spread
29-31 Fraction of unary, binary
   and ternary clauses
Proximity to Horn Formula:
32-36 Occurrences of a variable
   in a Horn clause: mean, stdev,
   min, max, spread
37 Fraction of Horn clauses

Table 1. Comparison of existing MCS solvers in terms of the time needed to find the first MCS

	PAR10	Average	# Not Solved
Best Single	587.13	98.08	8
SNNAP	127.88	66.75	1
VBS	11.58	11.58	0

All of these features can be computed from the problem specification and can therefore be computed within a few seconds. Above, by the Variable-Clause Graph, we refer to a bipartite graph where each node represents either a variable or a clause, with an edge existing between the two if that variable appears in the clause.

2.3 Numerical Results

Our first set of experiments focus on continuing the research proposed in [14], which targeted to find a single MCS as quickly as possible. Utilizing the same 265 instances and 15 solvers, we present the results in Table 1. The table shows the results after 10-fold cross validation in terms of average runtime, number of instances not solved, and the penalized average runtime. Here, the PAR10 means that if the instance was not solved within the allotted 1,800 seconds, it was recorded as having taken 18,000 seconds to complete. From these results we observe the best solver, a parameterization of efd, can on average find a solution in under 100 seconds. A portfolio approach like SNNAP can drop this average by nearly 35 seconds, and solve an additional 2% more instances. Yet the virtual best solver (VBS), the one which for every instance knows beforehand the fastest solver, can do significantly better still.

These results with finding a single MCS serve to confirm the findings presented in [14], showing how crucial portfolios are to this area. In practice, however, it is usually not enough to just find a single MCS. In order to find Minimal Unsatisfiable Subsets, for example, it is instead necessary to enumerate over all MCSs.

For this next round of experiments, we gather a collection of 180 SAT instances, which were randomly split into 100 training and 80 testing. These instances have been developed over the last two decades expressly for the enumeration task. Whereas when computing a single MCS one only needs a handful of calls to a SAT solver, enumeration requires millions of such calls. Thus the need for the two datasets to comprise different instances. Almost half of the instances come from DaimlerChrysler's Mercedes benchmarks [20], an extensively studied benchmark suite [12] from an automotive product configuration problem. The remaining instances are an assortment of small sized industrial instances, like the FVP-UNSAT/SSCALAR and UCLID hardware verification instances, CMDADD and DIMACS set of benchmarks. On these instances we ran the five MCS solvers contained in the MCSls tool [14]: bfd (Basic Fast-Diag), bls (Basic Linear Search), cld (Clause D based), efd (Enhanced FastDiag), and els (Enhanced Linear Search). The bls algorithm has already been presented in the introduction. The bfd algorithm computes MCSs by means of a

Table 2. Comparison of the 5 MCS finding solvers over 80 test instances. The table presents the total number of MCSs found over the test set in millions (M), and the average number of MCSs per instance in thousands (K).

	bfd	bls	cld	efd	els	SNNAP	VBS
Sum (M)	5.04	6.64	6.05	5.01	6.88	8.42	8.75
Average (K)	63.02	83.02	75.68	62.62	86.03	105.31	109.43

divide-and-conquer strategy. The cld algorithm, makes use of a clause (D clause) to lead the extension of an initial set of satisfied clauses. The two algorithms els and efd represent the enhanced version of bls and bfd respectively, that extend the basic algorithms with the new techniques presented in [14]. One reason for this choice of solvers, is that they have been previously proven to be among the best performing. Moreover, they are the only ones able to deal with Partial MaxSAT instances, which is crucial to the subsequent section of our experiments that need to restart solvers.

Each of the solvers were run uninterrupted for 1 hour and the number of found MCSs was recorded. Table 2 summarizes the results. What we observe is that although els is empirically the best, there is not much difference between the solvers. However, if we introduce SNNAP to predict which of the solvers to run, the performance improves dramatically. Furthermore, the table also presents the absolute best possible performance for an oracle solver that for each instance knows beforehand which of the 5 solvers will return the most number of MCSs. This virtual best solver (VBS) is the upper bound for any conceivable portfolio approach which relies solely on running a single solver for the allotted hour. What is encouraging is that SNNAP is able to almost match this performance.

3   A Dynamic Approach

Although algorithm selection can be shown to significantly improve the state-of-the-art in MCS solvers, it may not be the most effective approach to the problem at hand. At a first glance, it may seem advantageous to run a solver as long as possible, uninterrupted. However, as was shown with other NP-complete solvers like SAT, runtime can oftentimes be very heavy tailed [6]. This means that usually, by setting or trying just the right variables, commonly referred to as backdoor variables, the solver can finish very quickly. Yet, if something goes wrong, then the runtime can become essentially infinite. Therefore, it was found that if a solver is seemingly taking too long to solve an instance, it maybe beneficial to simply restart the search. Even though when looking for MCSs we are not interested in optimizing for time, the same concept of restarts can be exploited.

Furthermore, it can be assumed that solvers typically find many of the MCSs very quickly, with the rate of return decreasing as time progresses. Thus, as soon as a solver has found additional sets, the structure of the problem may change significantly enough that an alternate solving strategy may be preferred. Thus, instead of running the same solver for an hour, we propose to switch among solvers every 10 minutes. This particular shortened timeout was chosen empirically after observing that the available solvers were sure to find new MCSs in this time.

To avoid repeatedly finding the same MCSs, after each run the current instance is expanded with blocking clauses. It is important to point out here, that the restarts we employ here do not carry any information other than the previously found MCSs. Yet this setup allows us to investigate a portfolio approach that takes advantage of an ability to switch solvers multiple times.

Note that even though we are switching between solvers, the proposed approach is different from algorithm scheduling [10,16]. In these existing approaches, a schedule is determined which details the order in which solvers should run and for what duration. This schedule, however, is made once and then adhered to for the whole computation. What we propose is instead to make the decisions iteratively, once we have observed exactly how the problem has changed after the previous computations.

For the underlying approach to choose the next solver, we utilize SNNAP. However, we cannot simply train on performance of solvers on the original training instances. The portfolio will be faced with instances that are significantly altered through the addition of blocking clauses. We therefore expand the training set to include 300 additional instances. This dataset is formulated by registering all the MCSs found after an hour by any solver for a particular instance, and then creating instances that have a random subset of $\sim 25\%$, $\sim 50\%$, and $\sim 75\%$ of these MCSs. We say approximately, because to improve the coverage of possible instances, we allow for deviations of $\pm 5\%$. This, therefore, better simulates the type of instances a portfolio would encounter during its operation.

4   Evaluation

To observe the effects of restarting a solver every 10 minutes, Table 3 presents each solver with a "_r" suffix symbolizing a restart. We can see that although most solvers, like els, do not benefit from this change, others like cld, improve dramatically. Therefore we see that for just these pure solvers it is best to just leave them running. In the case of cld, however, the solver is known to occasionally return poor results, so we posit that restarts help stabilize its performance.

Subsequently, we experiment with the effects of just switching among solvers randomly after every 10 minutes. Table 3 presents the average over 3 seeds under Rand, and as expected, this does not lead to very good performance.

Finally we investigate a portfolio approach that takes advantage of an ability to switch solvers every 10 minutes. As we can determine from the table, this portfolio approach performs significantly better than any of the existing solvers, regular and restarted. Perhaps even more importantly, this portfolio also outperforms the upper bound of a portfolio that only runs solvers uninterrupted with a 1 hour timeout.

Interestingly for many of the test instances, the new portfolio solver switched between solvers after the first 10min. Subsequently, the same solver kept being used for each subsequent step. This suggests that even though we are able to

Table 3. Comparison of techniques that restart the solving process every 10 minutes. VBS-5 is the oracle solver which runs one of the 5 solvers for a full hour. Rand is the average of 3 seeds where after each 10 minute execution, a new solver is picked uniformly at random. Each of the 5 solvers were also run "*_r", where after 10 minutes it would be halted, and then restarted on an instance that had the previously found MCSs added. Portfolio is the solver that intelligently chooses the next solver to run and VBS is the approximated upper bound of a portfolio approach where solvers are restarted every 10 minutes.

	VBS-5	Rand	bfd_r	bls_r	cld_r	efd_r	els_r	Portfolio	VBS
Sum (M)	8.75	5.8	4.67	6.16	7.98	4.56	6.31	9.24	10.56
Average (K)	109.43	72.47	59.05	76.98	99.74	57.75	78.82	115.44	132.06

outperform the VBS of an uninterrupted solver, future research should investigate automatically varying the timeout of each run.

5   Conclusion

Enumerating all of the Minimal Conflict Subsets (MCS) is a prevalent problem in practice with many applications. In this paper we demonstrate that the state-of-the-art in this field can be greatly improved through the application of algorithm selection. In particular, we demonstrate a collection of features that can be used to distinguish between problem instances and suggest which is the best solver to apply. Although this approach led to a significant boost in the number of MCSs that could be found after running a solver uninterrupted with a 1 hour timeout, we also presented how these results can be further improved. Specifically, after running for a short time, the new approach would add the observed MCSs as blocking constraints, analyze the structure of the new instance, and then choose the most appropriate solver to run for the next short burst. The results of this methodology outperformed even the upper bound that could have been achieved by an optimal portfolio which left its solvers running uninterrupted. It is important to note that these improvements were there even with the overhead of repeatedly reading and writing the instance to file for each restart. A future algorithm that performs the proposed approach internally is likely to improve the results even further.

Acknowledgements. This work is partly supported by Science Foundation Ireland (SFI) Grant 10/IN.1/I3032, FP7 FET-Open Grant 284715, SFI PI grant BEACON (09/IN.1/I2618), FCT grants ATTEST (CMU-PT/ELE/0009/2009), POLARIS (PTDC/EIA-CCO/123051/2010), and by INESC-ID multiannual funding from the PIDDAC program funds. The Insight Centre for Data Analytics is supported by SFI Grant SFI/12/RC/2289.

References

1. Bailey, J., Stuckey, P.J.: Discovery of minimal unsatisfiable subsets of constraints using hitting set dualization. In: Practical Aspects of Declarative Languages, pp. 174–186 (2005)

2. Ben-Eliyahu, R., Dechter, R.: On computing minimal models. Ann. Math. Artif. Intell. 18(1), 3–27 (1996)
3. Birnbaum, E., Lozinskii, E.L.: Consistent subsets of inconsistent systems: structure and behaviour. J. Exp. Theor. Artif. Intell. 15(1), 25–46 (2003)
4. Collautti, M., Malitsky, Y., Mehta, D., O'Sullivan, B.: Snnap: Solver-based nearest neighbor for algorithm portfolios. In: Blockeel, H., Kersting, K., Nijssen, S., Železný, F. (eds.) ECML PKDD 2013, Part III. LNCS (LNAI), vol. 8190, pp. 435–450. Springer, Heidelberg (2013)
5. Felfernig, A., Schubert, M., Zehentner, C.: An efficient diagnosis algorithm for inconsistent constraint sets. AI EDAM 26(1), 53–62 (2012)
6. Gomes, C., Selman, B., Kautz, H.: Heavy-tailed phenomena in satisfiability and constraint satisfaction problems. J. of Automated Reasoning, 67–100 (2000)
7. Hurley, B., O'Sullivan, B.: Adaptation in a CBR-based solver portfolio for the satisfiability problem. In: Agudo, B.D., Watson, I. (eds.) ICCBR 2012. LNCS, vol. 7466, pp. 152–166. Springer, Heidelberg (2012)
8. Jampel, M., Freuder, E.C., Maher, M.J. (eds.): Over-Constrained Systems. Springer (1996)
9. Junker, U.: QUICKXPLAIN: Preferred explanations and relaxations for over-constrained problems. In: AAAI, pp. 167–172 (2004)
10. Kadioglu, S., Malitsky, Y., Sabharwal, A., Samulowitz, H., Sellmann, M.: Algorithm selection and scheduling. In: Lee, J. (ed.) CP 2011. LNCS, vol. 6876, pp. 454–469. Springer, Heidelberg (2011)
11. Kotthoff, L., Gent, I., Miguel, I.P.: An evaluation of machine learning in algorithm selection for search problems. AI Communications (2012)
12. Liffiton, M., Sakallah, K.: Algorithms for computing minimal unsatisfiable subsets of constraints. J. Autom. Reasoning 40(1), 1–33 (2008)
13. Malitsky, Y., Sellmann, M.: Instance-specific algorithm configuration as a method for non-model-based portfolio generation. In: Beldiceanu, N., Jussien, N., Pinson, É. (eds.) CPAIOR 2012. LNCS, vol. 7298, pp. 244–259. Springer, Heidelberg (2012)
14. Marques-Silva, J., Heras, F., Janota, M., Previti, A., Belov, A.: On computing minimal correction subsets. In: IJCAI (2013)
15. O'Callaghan, B., O'Sullivan, B., Freuder, E.C.: Generating corrective explanations for interactive constraint satisfaction. In: Principles and Practice of Constraint Programming, pp. 445–459 (2005)
16. O'Mahony, E., Hebrard, E., Holland, A., Nugent, C., O'Sullivan, B.: Using case-based reasoning in an algorithm portfolio for constraint solving. In: AICS (2008)
17. Pulina, L., Tacchella, A.: A self-adaptive multi-engine solver for quantified boolean formulas. Constraints 14(1), 80–116 (2009)
18. Reiter, R.: A theory of diagnosis from first principles. Artif. Intell. 32(1), 57–95 (1987)
19. Silverthorn, B., Miikkulainen, R.: Latent class models for algorithm portfolio methods. AAAI (2010)
20. Sinz, C., Kaiser, A., Küchlin, W.: Formal methods for the validation of automotive product configuration data. AI EDAM 17(1), 75–97 (2003)
21. Soh, T., Inoue, K.: Identifying necessary reactions in metabolic pathways by minimal model generation. In: ECAI, pp. 277–282 (2010)
22. Xu, L., Hutter, F., Hoos, H.H., Leyton-Brown, K.: Satzilla: Portfolio-based algorithm selection for SAT. CoRR (2011)
23. Xu, L., Hutter, F., Shen, J., Hoos, H.H., Leyton-Brown, K.: Satzilla2012: Improved algorithm selection based on cost-sensitive classification models. SAT Competition (2012)

Parallel Depth-Bounded Discrepancy Search

Thierry Moisan, Claude-Guy Quimper, and Jonathan Gaudreault

FORAC Research Consortium, Université Laval, Québec, Canada
Thierry.Moisan.1@ulaval.ca,
{Claude-Guy.Quimper,Jonathan.Gaudreault}@ift.ulaval.ca

Abstract. Search strategies such as Limited Discrepancy Search (LDS) and Depth-bounded Discrepancy Search (DDS) find solutions faster than a standard Depth-First Search (DFS) when provided with good value-selection heuristics. We propose a parallelization of DDS: Parallel Depth-bounded Discrepancy Search (PDDS). This parallel search strategy has the property to visit the nodes of the search tree in the same order as the centralized version of the algorithm. The algorithm creates an intrinsic load-balancing: pruning a branch of the search tree equally affects each worker's workload. This algorithm is based on the implicit assignment of leaves to workers which allows the workers to operate without communication during the search. We present a theoretical analysis of DDS and PDDS. We show that PDDS scales to multiple thousands of workers. We experiment on a massively parallel supercomputer to solve an industrial problem and improve over the best known solution.

1 Introduction

Parallelization has been of growing interest in recent years, including in the optimization community. de la Banda et al. [1] consider parallelization as one of the three main challenges in the future of optimization technologies. Search is at the core of optimization and constraint solvers. If one wants to parallelize a solver, it is natural to consider parallelizing search strategies.

Parallelization of constraint programming solver is a hard problem mainly due to communication between workers. When the number of workers is large, the time each worker spends communicating with the other workers often exceeds the time spent at solving the original problem.

We have recently seen good parallel algorithms without communication that are based on centralized algorithms. Régin et al. [2] split the problem into multiple subproblems. These subproblems are then given to workers that solve them using classic centralized algorithm.

Parallel Limited Discrepancy-based Search (PLDS[1]) [3] is based on Limited Discrepancy-based Search (LDS) [4]. LDS has a huge advantage over traditional search strategies such as Depth-First Search (DFS) when a good value ordering heuristic is used. The PLDS parallel version keeps this advantage by preserving

---

[1] In the original article this algorithm was named PDS. In this paper, we name it PLDS for clarity concerns.

H. Simonis (Ed.): CPAIOR 2014, LNCS 8451, pp. 377–393, 2014.

the node visit ordering of the centralized algorithm. Each leaf of the search tree is implicitly assigned to a worker. Every worker branches in the search tree while making sure there is at least one leaf in the subtree of the current node that is assigned to it. An important property of this approach is that, upon pruning the search tree, workload balance difference can be theoretically bounded. PLDS scales to thousands of workers.

In this paper, we show that the parallelization mechanism used by PLDS can also be used to parallelize other search strategies while keeping the same properties. We parallelize the Depth-bounded Discrepancy Search algorithm (DDS) [5] to obtain Parallel Depth-bounded Discrepancy Search (PDDS). Our motivation lies in the observation that in a centralized environment, DDS is generally more efficient than LDS when it is provided with good value ordering heuristics. We also show how the same parallelization mechanism can be applied to DFS which becomes Parallel Depth-First Search (PDFS). The theoretical analysis of PDFS will simplify the analysis of PDDS.

The rest of this paper is divided as follows. Section 2 describes the DDS algorithm and reviews related parallel computing work. Section 3 details the PDFS algorithm while section 4 details the PDDS algorithm. Section 5 presents a theoretical analysis of the algorithms. Finally, we experiment with an industrial problem coming from the wood-products industry in Section 6.

2   Literature Review

We review the related works by presenting different parallelization approaches. Then, we describe the original DDS algorithm that we parallelize in Section 4.

2.1   Shared Memory

It is possible to parallelize a search strategy by sharing, through a *shared memory* space, a list of open nodes, i.e. the visited nodes for which there are still values to branch on. Each worker can select an open node and process it until no more work can be done from that point. Then, the worker comes back to the pool of open nodes to obtain more work.

Perron [6] proposes a framework based on this idea. Good performances are often reported, as in [7] where a parallel Best First Search is implemented and evaluated up to 64 processors. However, this approach cannot easily scale up to thousands of processors due to communication overload.

2.2   Portfolios

Portfolio-based methods use a set of different solvers, parameters and/or search strategies. Workers are using different configurations to solve the exact same problem in parallel, increasing the probability of quickly finding a good solution. The approach can be improved by making use of randomized restarts [8] and nogoods learning [9].

Finding good alternative configurations for a specific problem can be a difficult problem by itself. Xu et al. [10] use machine learning to find appropriate SAT solver configurations to a new problem based on a set of learned examples.

2.3   Search Space Splitting and Work Stealing

Space splitting divides the search tree into small subtrees that are assigned to the workers. As it is unlikely that those subtrees are of equal size, a *work stealing* mechanism (see [11,12]) allows busy workers to share their workload with idle workers and therefore evenly balance the workload among all workers. In [13], Menouer et al. parallelize the constraint programming solver OR-Tools using a framework based on work-stealing.

Communication may cause issues when there are too many workers. At some point, the communication monopolizes the majority of the computing power. Reducing the amount of communication speeds up the search. For example, Xie and Davenport [14] allocate specific workers to coordinate the tasks, allowing more processors to be used before performance starts to decline.

Yun and Epstein [15] combined the use of portfolios with work-stealing. They start by launching a portfolio phase by making a choice of solver configuration. Then the search space is divided and work is distributed among the workers. During the search, information about the success (or lack thereof) is transmitted from the workers to the manager inducing a change in the future choices among the portfolio of solvers.

Recent work showed how to implicitly balance the workload while minimizing the communication during the search. Régin et al. [2] split the problem into a large number of subtrees. Some are quicker to explore than others, as pruning occurring during the search does not equally affect each part of the search tree. However, since a large number of subtrees is assigned to each worker, their workload tend to balance.

The exclusion of all communication during the search is also the solution we advocate in our previous work [3] where we introduced PLDS, a parallel version of Limited Discrepancy Search (LDS) [4]. The parallelization is done by implicitly assigning leaves of the search tree to workers. We showed how to test whether a worker has any work assigned to it in the subtree under the current node. This parallel algorithm has the property to keep the same node visit ordering as the sequential version. This is the approach we will generalize in this paper.

2.4   Depth-Bounded Discrepancy Search

Harvey and Ginsberg [4] introduce the concept of *discrepancy*. Each time a solver needs to assign a value to a variable, a value-ordering heuristic is used to select the value that will most likely lead to a solution. As a convention, when a binary search tree is represented graphically, the left branch under a node corresponds to the recommendation of the heuristic while the right one does not. Figure 1 shows such binary search tree. Therefore, each time the solver branches to the right in this tree, it goes against the heuristic recommendation. Such branching

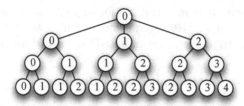

Fig. 1. A binary search tree with the number of discrepancies of each node

is called a *discrepancy*. Leaves on Figure 1 are labeled with the total number of discrepancies one must follow to go from the root of the tree to that leaf. Harvey and Ginsberg show that, when using a good value ordering heuristic, the expected quality of a leaf decreases as the number of discrepancies increases.

Limited discrepancy search (LDS) [4] is the first search strategy based on discrepancies. It visits the leaves of the tree in order of discrepancies. Improved Limited Discrepancy Search (ILDS) [16] is an improvement over LDS since it visits each leaf at most once (the original LDS has redundancy). There are other search strategies that take advantage of discrepancies such as Discrepancy-Bounded Depth First Search (DBDFS) [17] and Limited Discrepancy Beam Search (BULB) [18].

Depth-bounded Discrepancy Search (DDS) [5] makes the following assumption: it is more probable that the value ordering heuristic makes a mistake at the top of the search tree than at the bottom. A value-ordering heuristic can make better decisions lower in the search tree since it has more information about the problem. Hence, it is more likely that the heuristic makes a mistake at top the of the tree. Based on this assumption, if the search has to reconsider the choices it made, it better reconsider choices made at the top of the search tree rather than at the bottom.

Given a search tree of depth n, DDS performs $n + 1$ iterations. At iteration $k = 0$, DDS visits the leftmost leaf of the tree. At iteration k, for $1 \leq k \leq n$, DDS visits all the branches in the search tree above level $k - 1$. At level k, the algorithm visits all value assignments that do not respect the value ordering heuristic. Beyond level k, DDS visits all value assignments that respect the value ordering heuristic and therefore have no discrepancies. For example, for $k = 2$, DDS visits all nodes down to level 1, branches once on values that do not respect the heuristic, and then always branches on the leftmost child until it reaches a leaf.

Algorithms 1 and 2 are a generalization of the original DDS algorithm [5] for n-ary variables.

In the following description, we suppose that the variable ordering heuristic is deterministic and only depends on the states of the domains. Hence, under the same conditions, the algorithm will always make the same choices (otherwise, the variable ordering heuristics could cause the search strategy to visit multiple

---

Algorithm 1. DDS($[\text{dom}(X_1), \ldots, \text{dom}(X_n)]$)

> **for** $k = 0..n$ **do**
> > $s \leftarrow$ DDS-Probe($[\text{dom}(X_1), \ldots, \text{dom}(X_n)], k$)
> > **if** $s \neq \emptyset$ **then return** s
>
> **return** $\emptyset$

Algorithm 2. DDS-Probe($[\text{dom}(X_1), \ldots, \text{dom}(X_n)], k$)

> $Candidates \leftarrow \{X_i \mid |\text{dom}(X_i)| > 1\}$
> **if** $Candidates = \emptyset$ **then**
> > **if** $\text{dom}(X_1), \ldots, \text{dom}(X_n)$ *satisfies all the constraints* **then**
> > > **return** $\text{dom}(X_1), \ldots, \text{dom}(X_n)$
> >
> > **return** $\emptyset$
>
> Choose a variable $X_i \in Candidates$
> Let $v_0, \ldots, v_{|\text{dom}(X_i)|-1}$ be the values in $\text{dom}(X_i)$ sorted by the heuristic.
> **if** $k = 1$ **then** $\underline{d} \leftarrow 1$ **else** $\underline{d} \leftarrow 0$
> **if** $k = 0$ **then** $\overline{d} \leftarrow 0$ **else** $\overline{d} \leftarrow |\text{dom}(x_i)| - 1$
> **for** $d = \underline{d}..\overline{d}$ **do**
> > $s \leftarrow$ DDS-Probe($[\text{dom}(X_1), \ldots, \text{dom}(X_{i-1}), \{v_d\},$
> > $\qquad\qquad \text{dom}(X_{i+1}), \ldots, \text{dom}(X_n)], \max(0, k - 1)$)
> > **if** $s \neq \emptyset$ **then return** s
>
> **return** $\emptyset$

---

times some leaves and ignore other leaves). This supposition was also made in [4] and [5].

3 PDFS Algorithm

PDFS will simplify the theoretical analysis of DDS and PDDS for the following reason. At iteration k, DDS (PDDS) performs a DFS (PDFS) over the first $k-1$ variables of the problem.

To our knowledge, it is the first time that DFS and DDS are parallelized this way.

We parallelize DFS over ρ workers labeled from 0 to $\rho - 1$. Let s be a leaf of the search tree and $v(s)$ its order of visit in DFS, i.e. the first leaf visited by DFS has a visit order of 0, the second leaf visited by DFS has a visit order of 1, and so on. We implicitly assign a leaf s to worker $v(s) \bmod \rho$. Each worker is aware of its label and the total number of workers ρ. A worker w performs a standard DFS but only visits the nodes that have at least one leaf, among their descendants, whose assigned worker is w. Each worker needs to decide whether a node leads to a leaf of interest. This is done as follows.

Let a be the current node and left(a) its left child. The PDFS search keeps track of the worker $l(a)$ assigned to the leftmost leaf, in the subtree rooted at

a, to be visited in the current iteration of the centralized search strategy. In the case of PDFS, there is only one iteration but PDDS is run over multiple iterations. We necessarily have $l(a) = l(\text{left}(a))$ since both subtrees have the same leftmost leaf. The function C_{DFS} takes as input a node and returns the number of its descendants that are leaves to be visited in the current iteration of the centralized search strategy.

$$C_{\text{DFS}}([X_1, \ldots, X_n]) = \prod_{i=1}^{n} |\text{dom}(X_i)| \tag{1}$$

If all variable domains have cardinality δ, Equation 1 simplifies to Equation 2.

$$C_{\text{DFS}}([X_1, \ldots, X_n]) = \delta^n \tag{2}$$

The list of workers that needs to visit $\text{left}(a)$ is given by $l(a), (l(a) + 1) \bmod \rho$, $(l(a) + 2) \bmod \rho, \ldots, (l(a) + C_{\text{DFS}}(\text{left}(a)) - 1) \bmod \rho$. Consequently, the worker w only needs to visit the node $\text{left}(a)$ if it belongs to this list. This can be tested with the inequality $(w - l(\text{left}(a))) \bmod \rho < C_{\text{DFS}}(\text{left}(a))$. One can apply the same reasoning on the right child $\text{right}(a)$ knowing that that $l(\text{right}(a)) = (l(\text{left}(a)) + C_{\text{DFS}}(\text{left}(a))) \bmod \rho$.

Algorithm 3. PDFS($[\text{dom}(X_1), \ldots, \text{dom}(X_n)], l$)

$Candidates \leftarrow \{X_i \mid |\text{dom}(X_i)| > 1\}$
Choose a variable $X_i \in Candidates$
$z \leftarrow C_{\text{DFS}}(Candidates \setminus \{X_i\})$
for $v_d \in \text{dom}(X_i)$ **do**
> **if** $(currentProcessor - l) \bmod \rho < z$ **then**
> > $s \leftarrow$ PDFS($[\text{dom}(X_1), \ldots, \text{dom}(X_{i-1}), \{v_d\}, \text{dom}(X_{i+1}) \ldots, \text{dom}(X_n)], l$)
> > **if** $s \neq \emptyset$ **then return** s
>
> $l \leftarrow (l + z) \bmod \rho$

return $\emptyset$

Algorithm 3 describes PDFS. The first call to PDFS is done with the original variable domains and $l = 0$.

4 PDDS Algorithm

We show in this section how the same mechanism can be applied to DDS which becomes a Parallel Depth-bounded Discrepancy Search (PDDS). As in PLDS and PDFS, parallelization is done by assigning the leaves of the search tree to each worker in a round-robin fashion.

Algorithms 4 and 5 show how PDDS operates. Algorithm 4 visits all the leaves whose discrepancies appear within the first k variables. Algorithm 5 launches an iteration to visit all those leaves following a DFS search.

Algorithm 4. PDDS($[\mathrm{dom}(X_1), \ldots, \mathrm{dom}(X_n)]$)

$l \leftarrow 0$
for $k = 0..n$ **do**
    $Candidates \leftarrow \{X_i \mid |\mathrm{dom}(X_i)| > 1\}$
    $z \leftarrow C_{\mathrm{DDS}}(Candidates, k)$
    if $(currentProcessor - l) \bmod \rho < z$ **then**
        $s \leftarrow$ PDDS-Probe($[\mathrm{dom}(X_1), \ldots, \mathrm{dom}(X_n)], k, l$)
        if $s \neq \emptyset$ **then return** s
    $l \leftarrow (l + z) \bmod \rho$
return $\emptyset$

As for PDFS, the algorithm requires a function C_{DDS} that counts the number of leaves under the current node that should be visited during the current iteration of DDS. The next subsection shows how to implement the function C_{DDS}.

Algorithm 5. PDDS-Probe($[\mathrm{dom}(X_1), \ldots, \mathrm{dom}(X_n)]$, k, l)

$Candidates \leftarrow \{X_i \mid |\mathrm{dom}(X_i)| > 1\}$
if $Candidates = \emptyset$ **then**
    if $\mathrm{dom}(X_1), \ldots, \mathrm{dom}(X_n)$ *satisfies all the constraints* **then**
        return $\mathrm{dom}(X_1), \ldots, \mathrm{dom}(X_n)$
    return $\emptyset$
Choose a variable $X_i \in Candidates$
Let $v_0, \ldots, v_{|\mathrm{dom}(X_i)|-1}$ be the values in $\mathrm{dom}(X_i)$ sorted by the heuristic.
if $k = 1$ **then** $\underline{d} \leftarrow 1$ **else** $\underline{d} \leftarrow 0$
if $k = 0$ **then** $\overline{d} \leftarrow 0$ **else** $\overline{d} \leftarrow |\mathrm{dom}(X_i)| - 1$
for $d = \underline{d}..\overline{d}$ **do**
    $z \leftarrow C_{\mathrm{DDS}}(Candidates \setminus \{X_i\}, \max(0, k - 1))$
    if $(currentProcessor - l) \bmod \rho < z$ **then**
        $s \leftarrow$ PDDS-Probe($[\mathrm{dom}(X_1), \ldots, \mathrm{dom}(X_{i-1}), \{v_d\},$
                      $\mathrm{dom}(X_{i+1}), \ldots, \mathrm{dom}(X_n)], \max(0, k - 1), l$)
        if $s \neq \emptyset$ **then return** s
    $l \leftarrow (l + z) \bmod \rho$
return $\emptyset$

4.1 Counting Functions

We provide two functions that count the number of leaves in a subtree that have to be visited in the current iteration of the DDS. Both functions take as input the variables to be explored in this subtree and the number of levels k where discrepancies are allowed. Function 3 assumes that all variable domains have cardinality δ. Function 4 assumes that variables are selected in a static ordering. Without these assumptions, one would need to integrate the knowledge of the branching heuristic into the computation of the number of leaves. However, it is always possible to do a workaround and to extend the domains of all variables with dummy values to match the cardinality of the largest domain. The dummy values can be filtered out causing a slight workload imbalance among the processors as it will be discussed in Section 5.6.

If all variable domains have cardinality δ, then one can count the number of leaves as follows. At iteration k, DDS performs a DFS over a tree of height $k-1$. For each leaf of this tree, DDS explores the $\delta - 1$ solutions that cause one or more discrepancies to occur.

$$C_{\text{DDS}}([X_1,\ldots,X_n],k) = \begin{cases} 1 & \text{if } k = 0 \\ \delta^{k-1}(\delta-1) & \text{if } k > 0 \end{cases} \tag{3}$$

Interestingly, when all domains have the same size, the number of leaves depends only on the iteration number k and the cardinality of the domains δ but not on the number of variables.

We can also suppose a static variable ordering $X_1, X_2, \ldots, X_n$ which is used in every branch of the search. Under this assumption, variable domains can have different cardinalities.

$$C_{\text{DDS}}([X_1,\ldots,X_n],k) = \begin{cases} 1 & \text{if } k = 0 \\ |\operatorname{dom}(X_1)| - 1 & \text{if } k = 1 \\ (|\operatorname{dom}(X_k)| - 1) \prod_{i=1}^{k-1} |\operatorname{dom}(X_i)| & \text{if } k > 1 \end{cases} \tag{4}$$

If the variables do not have the same domain size and their ordering is not static, then the number of leaves in the search tree visited at iteration k depends on the variable ordering. The function C_{DDS} should be redefined according to the branching heuristic.

5   Analysis

This section provides an analysis of DFS, DDS, PDFS and PDDS. To compare these search strategies, we count the number of times each strategy visits a node while exploring an entire search tree of n binary variables.

5.1   Analysis of DFS

In a DFS, each node of the search tree is visited once. Since there are $2^{n+1} - 1$ nodes in a binary tree of height n, we define $\text{DFS}(n) = 2^{n+1} - 1$ to be the number of node visits in a complete DFS.

5.2 Analysis of DDS

Let n be the number of binary variables in a search tree and k the level of the last discrepancy for $k \leq n$. If the level of the last discrepancy is 0, then the search goes directly to the leftmost leaf of the subtree. Hence, the algorithm visits one node per variable left to instantiate, which is equal to n plus the root node which gives $n + 1$. Otherwise, the search does a DFS over the $k - 1$ first variables. For each of the 2^{k-1} leaves of this DFS, $n - k + 1$ nodes are visited down to the bottom of the search tree.

$$
\mathrm{DDS}(n, k) = \begin{cases} n + 1 & \text{if } k = 0 \\ \mathrm{DFS}(k - 1) + 2^{k-1}(n - k + 1) & \text{otherwise} \end{cases}
$$

$$
= \begin{cases} n + 1 & \text{if } k = 0 \\ 2^k - 1 + 2^{k-1}(n - k + 1) & \text{otherwise} \end{cases}
$$

The total number of node visits done by the DDS search strategy is given by the sum over all levels $k = 0..n$ in the search tree.

$$
\mathrm{DDS}(n) = \sum_{k=0}^{n} \mathrm{DDS}(n, k) \tag{5}
$$

$$
= 4 \cdot 2^n - n - 3 \tag{6}
$$

Surprisingly, this is the same number of node visits as a complete LDS search (the version proposed in [16]). The number of node visits of LDS was previously shown in [3].

5.3 Analysis of PDFS

We are interested in the number of node visits done by PDFS. To simplify the analysis, we suppose that the number of workers is a power of two: $\rho = 2^x$. If there are more workers than leaves ($\rho > 2^n$), then there are 2^n workers that each visits $n + 1$ nodes from the root to a leaf. The other $\rho - 2^n$ workers remain idle. If there are more leaves than workers ($\rho \leq 2^n$), then all nodes at level i, for $n - \log_2 \rho < i \leq n$, are visited by exactly 2^{n-i} workers, i.e. the 2^n leaves are visited by one worker each, the 2^{n-1} parents of the leaves are visited by 2 workers each, the 2^{n-2} grand-parents are visited by 4 workers each and so on until level $n - \log_2 \rho$ where all nodes are visited by all workers. All nodes in levels 0 to $n - \log_2 \rho$ are visited by all processors. The function $\mathrm{PDFS}(\rho, n)$ returns the number of node visits of PDFS with ρ workers in a tree of n binary variables.

$$
\mathrm{PDFS}(\rho, n) = \begin{cases} (n + 1)2^n & \text{if } 2^n < \rho \\ \rho \cdot \mathrm{DFS}(n - \log_2 \rho) + \sum_{i=n-\log_2 \rho+1}^{n} 2^i 2^{n-i} & \text{otherwise} \end{cases} \tag{7}
$$

$$
= \begin{cases} (n + 1)2^n & \text{if } 2^n < \rho \\ (2 + \log_2 \rho)2^n - \rho & \text{otherwise} \end{cases} \tag{8}
$$

This shows that as the number of workers grows, the computational power grows linearly while the number of node visits grows logarithmically until we reach the degenerate case where the workers outnumber the leaves of the tree.

5.4 Analysis of PDDS

An iteration of PDDS can be seen as a PDFS over $k - 1$ variables. For each of the 2^{k-1} leaves in this PDFS, PDDS completes the search by instantiating the remaining $n - k + 1$ variables. Let PDDS(ρ, n, k) be the number of node visits in iteration k of PDDS with ρ workers.

$$\text{PDDS}(\rho, n, k) = \begin{cases} n + 1 & \text{if } k = 0 \\ \text{PDFS}(\rho, k - 1) + 2^{k-1}(n - k + 1) & \text{otherwise} \end{cases} \quad (9)$$

which can be expanded to

$$\text{PDDS}(\rho, n, k) = \begin{cases} n + 1 & \text{if } k = 0 \\ (n + 1)2^{k-1} & \text{if } k > 0 \text{ and } 2^{k-1} \le \rho \quad (10) \\ (\log_2 \rho + n - k + 3)2^{k-1} - \rho & \text{otherwise} \end{cases}$$

We can further analyze the behavior of PDDS by summing the node visits over all the levels $k = 0..n$.

$$\text{PDDS}(\rho, n) = \sum_{k=0}^{n} \text{PDDS}(\rho, n, k) \quad (11)$$

$$= n + 1 + \sum_{k=1}^{\min(\log_2 \rho, n)} \text{PDDS}(\rho, n, k) + \sum_{k=\log_2 \rho + 1}^{n} \text{PDDS}(\rho, n, k) \quad (12)$$

$$= \begin{cases} (4 + \log_2 \rho)2^n - \rho(n - \log_2 \rho + 3) & \text{if } \rho \le 2^n \\ (n + 1)2^n & \text{otherwise} \end{cases} \quad (13)$$

In comparison, as reported in [3], equation (14) shows the number of node visits done by PLDS when searching a complete binary tree.

$$\text{PLDS}(\rho, n) = 2^n + 2^n \sum_{i=1}^{n} \sum_{k=0}^{i} \frac{1}{2^i} \min\left(\rho, \binom{i}{k}\right) \quad (14)$$

5.5 Speedup Analysis

The *speedup* is the ratio between the time for a single worker to accomplish a task over the time required for ρ workers to accomplish the same task. We measure the time in number of node visits while supposing that all nodes have an equal processing time. A single worker visits $\text{PDDS}(n, 1)$ nodes to explore an entire search tree of n binary variables while ρ workers each visits $\frac{\text{PDDS}(n,\rho)}{\min(\rho,2^n)}$ nodes to collectively explore the entire search tree. We therefore have a speedup of $\frac{\min(\rho,2^n)\text{PDDS}(n,1)}{\text{PDDS}(n,\rho)}$. A similar computation applies for PDFS.

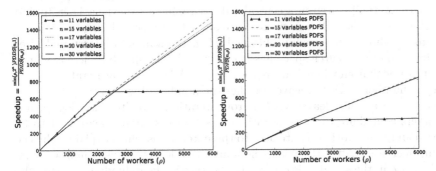

(a) Speedup of PDDS exploring a complete binary tree in function of the number of workers

(b) Speedup of PDFS exploring a complete binary tree in function of the number of workers

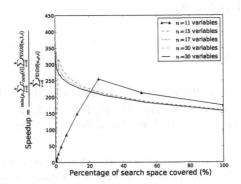

(c) Speedup of PDDS with $\rho = 512$ in function of the percentage of search space covered. Recalls $C_{DDS}(k)$ is the number of leaves processed at iteration k.

Fig. 2. Theoretical speedup of PDDS and PDFS algorithms

Figure 2a and Figure 2b show the speedup for PDDS and PDFS. For $n = 11$ variables, the speedup stops growing after 2048 workers. Beyond this point, there are more processors than leaves in the search tree. Since any additional worker is an idle worker, the speedup reaches a plateau. The number of variables affects the performance of PDDS, especially when there are few variables. However, as the number of variables grows, the effects become negligible.

One can see from Figure 2a and Figure 2b that the speedups in function of n for PDDS and PDFS are almost linear. In fact, while analyzing the functions $\mathrm{PDDS}(\rho, n)$ (Equation 10) and $\mathrm{PDFS}(\rho, n)$ (Equation 8), one sees that the most dominant term, 2^n, is multiplied by $\log_2 \rho$. This shows that the number of nodes to be visited logarithmically increases with the number of workers. However, the computation power increases linearly with ρ. It results in a speedup in $\Theta(\frac{\rho}{\log_2 \rho})$.

PDDS shows a greater speedup than PDFS when a complete search of the tree is performed. However, it is uncommon in practice to completely visit a search tree. Actually, even in a centralized environment we expect DDS to find a solution sooner than DFS as we better exploit the value ordering heuristic.

Figure 2c shows the speedup obtained when the search is interrupted after some fraction of the search space has been covered. The speedup increases until it reaches a peak from where it decreases. The peak is reached at iteration $\log_2 \rho$, when the number of visited leaves reaches the number of workers.

Since there are few leaves visited from iteration 0 to iteration $\log_2 \rho$, not all workers contribute to these iterations. As k grows, more leaves need to be explored and more workers contribute to these iterations which explains why the speedup increases. After the peak, there are more leaves to visit than workers. The decrease in the speedup is due to the increase in the redundancy among the workers. Indeed, at iteration k, the redundancy occurs when the workers visit the first $k - 1$ levels of the tree. The greater k is, the greater is the subtree in which the redundancy occurs.

5.6   Workload Analysis

Theorem 1. *Let n be the number of variables in the problem. If a branch is pruned from the search tree during the PDDS search, the number of leaves removed from the workload of each worker differs by at most n.*

Proof. If a branch of the tree is pruned, all the nodes under this branch are removed. Each leaf in the removed subtree are associated to a worker w and to an iteration k in which DDS visits the leaf. The leaves belonging to the same iteration are assigned to the workers in a round-robin fashion. Therefore, for the same iteration k, the workload among the workers differs by at most one. Since there are $n + 1$ iterations ($k = 0..n$), one concludes that the accumulated workload gap is bounded by $n + 1$.

However, the leaf visited at iteration 0 and the leaf visited at iteration 1 cannot be both filtered without filtering the whole tree. If the whole tree is filtered, then the workload between each worker is the same as there is no work to do. Otherwise, either iteration 0 or 1 does not create a workload difference of one. Hence, the maximum workload gap is bounded by n.   □

Theorem 1 shows the benefit of implicitly assigning leaves to workers in a round-robin fashion.

6 Experiments with Industrial Data

We carried out experiments with industrial data for an integrated planning and scheduling problem from the forest-products industry. Planning and scheduling lumber finishing operations is very challenging for the following reason: (1) the manufacturing operations lead to co-production (we simultaneously produce many different types of products from a single input) and (2) there are many alternative operations that can be used to transform a given raw product (each operation leads to a different basket of products). The result is that each operation contributes to partially fulfill many orders at the same time and each order can be fulfilled by many operations. The lumber finishing problem is fully described in [19] which provides a good heuristic to solve this problem. This heuristic inspired a search procedure (variable/value selection heuristics) [20] that allowed a constraint programming model to outperform standard mathematical programming. In [3], DFS, LDS and PLDS were compared using industrial data. LDS outperformed DFS, and PLDS allowed an impressive speedup and and solution quality that were never obtained before.

Using the same datasets and methodology as in [3], we compare DFS, LDS, PLDS, DDS, and PDDS. The search only considers integer variables. Once the values for these variables are known, the remaining continuous variables define a linear program that can be easily solved to optimality using the simplex method. Therefore, each time a valid assignment of the integer variables is obtained, we consider we have reached a leaf in the search tree and a linear program is solved to evaluate the value of this solution. The linear programs were solved using CPLEX version 12.5.

We used Colosse, a supercomputer with 7680 cores (dual, quad-core Intel Nehalem CPUs, 2.8 GHz with 24 GB RAM). Two Canadian lumber companies involved in the project provided the industrial instances.

Figures 3a to 3c show the objective value (minimizing backorder costs) according to computation time (maximum one hour) for 1, 512 and 4096 workers. DDS and PDDS with one worker showed the same performance. For this reason we omit the latter in the chart. The same comment applies to LDS (PLDS) and DFS (PDFS).

As expected, DDS outperformed LDS since we use a specialized value and variable ordering heuristic adapted to this problem. This shows that the assumption of DDS is true in this case: exploring discrepancies at the top of the search tree first leads to better solution faster. The centralized DDS even catches up PLDS running on 512 workers (see Figure 3b).

PDDS using 4096 workers obtains solutions of quality that was never reached before. The gap between the solutions obtained with PLDS (4096 workers) and PDDS (4096 workers) is considerable from an industrial point of view. PDDS has reduced the backorders by a ratio ranging between 68% and 85% when compared to DDS. Finally, if one needs a solution of a given quality, PDDS finds it with much less computation time than PLDS and PDFS.

Table 1 reports statistics computed during these experiments. The speedup is computed as the ratio of the number of leaves visited by multiple workers divided

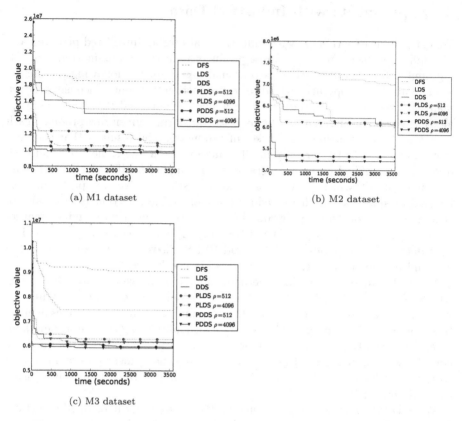

Fig. 3. Best objective value found depending on time for various datasets

by the number of leaves visited by one worker.[2] The true speedup measure based on wall-clock time is not used since it was not practical from an experimental point of view. For example, with dataset M1, DDS visits 615 leaves in one hour. The same leaves are visited in a few seconds with PDDS 1024 workers while 409 workers are idle. With the same dataset, PDDS 1024 workers visits 614885 leaves in 10 minutes which is equivalent to 110 days of work for a centralized DDS. Experiments with such high difference in task size would not lead to any significant results.

Even if the whole search tree is not visited, we wanted to measure the difference of workload between workers, in terms of visited leaves. Let χ_j be the number of leaves processed by worker j. Let $min(\chi)$ be the minimum value of χ_j for every $j \in 0, 1, \ldots, \rho - 1$ and $max(\chi)$ the maximum. Let $\overline{\chi}$ be the average

---

[2] The super-linear speedup obtained on instance M1 with $\rho = 512$ workers is explained by the uneven time required to solve the linear programs associated to each leaf. The average solving time is greater for the leaves both reached by DDS and PDDS than for the additional leaves visited by PDDS. Other instances do not show this behavior.

Table 1. Statistics of the industrial datasets experiments. The column $\overline{\chi}$ is the average number of leaves visited by each worker. The column σ_χ standard deviation of the number of leaves visited by each worker. The column $max(\chi) - min(\chi)$ is the maximum difference of processed leaves between workers.

dataset	ρ	speedup	$\overline{\chi}$	σ_χ	$max(\chi) - min(\chi)$
M1	512	517.9	622.04	1.94	12
M1	1024	1001.0	601.21	4.95	24
M1	2048	2008.2	603.06	4.68	24
M1	4096	4087.0	613.66	4.37	24
M2	512	475.0	756.11	2.68	14
M2	1024	945.4	752.41	2.44	14
M2	2048	1886.7	750.82	2.53	17
M2	4096	3732.8	742.72	2.76	18
M3	512	469.0	830.79	11.16	84
M3	1024	926.5	820.67	13.55	90
M3	2048	1844.2	816.75	11.8	85
M3	4096	3695.4	818.3	14.29	113

number of leaves visited by each worker. The standard deviation of the number of leaves visited by each worker is σ_χ. This measure shows that processors have visited roughly the same number of leaves.

7   Conclusion

We proposed a parallelization of DDS that we named Parallel Depth-bounded Discrepancy Search (PDDS). We theoretically showed that PDDS scales to unlimited number of workers until there are more workers than leaves in the search tree, thanks to the fact that there is no communication between the workers. When only a part of the tree is searched, as it is most common, the instances with more variables lead to a greater speedup.

We theoretically analyzed the numbers of node visits of DDS and PDDS. These numbers of node visits are used to analyze the theoretical speedup of PDDS. We showed that the number node visits of DDS is the same as LDS when visiting a complete search tree.

Finally we used an industrial problem from the forest-products industry to experiment with PDDS. We showed that PDDS consistently performs better than PLDS in our industrial context for which a good heuristic was provided. From an industrial point of view, the computation time is reduced and the solution quality is enhanced.

References

1. de la Banda, M.G., Stuckey, P.J., Van Hentenryck, P., Wallace, M.: The future of optimization technology. Constraints, 1–13 (2013)
2. Régin, J.-C., Rezgui, M., Malapert, A.: Embarrassingly parallel search. In: Schulte, C. (ed.) CP 2013. LNCS, vol. 8124, pp. 596–610. Springer, Heidelberg (2013)

3. Moisan, T., Gaudreault, J., Quimper, C.-G.: Parallel discrepancy-based search. In: Schulte, C. (ed.) CP 2013. LNCS, vol. 8124, pp. 30–46. Springer, Heidelberg (2013)
4. Harvey, W.D., Ginsberg, M.L.: Limited discrepancy search. In: Proceedings of the Fourteenth International Joint Conference on Artificial Intelligence (IJCAI 1995), pp. 607–613 (1995)
5. Walsh, T.: Depth-bounded discrepancy search. In: Proceedings of the Fifteenth International Joint Conference on Artificial Intelligence (IJCAI 1997), pp. 1388–1393 (1997)
6. Perron, L.: Search procedures and parallelism in constraint programming. In: Jaffar, J. (ed.) CP 1999. LNCS, vol. 1713, pp. 346–361. Springer, Heidelberg (1999)
7. Vidal, V., Bordeaux, L., Hamadi, Y.: Adaptive k-parallel best-first search: A simple but efficient algorithm for multi-core domain-independent planning. In: Proceedings of the Third International Symposium on Combinatorial Search, SOCS 2010 (2010)
8. Shylo, O.V., Middelkoop, T., Pardalos, P.M.: Restart strategies in optimization: Parallel and serial cases. Parallel Computing 37(1), 60–68 (2010)
9. Hamadi, Y., Sais, L.: ManySAT: a parallel SAT solver. Journal on Satisfiability, Boolean Modeling and Computation 6, 245–262 (2009)
10. Xu, L., Hutter, F., Hoos, H.H., Leyton-Brown, K.: Satzilla: Portfolio-based algorithm selection for sat. Journal of Artificial Intelligence Research (JAIR) 32, 565–606 (2008)
11. Michel, L., See, A., Van Hentenryck, P.: Transparent parallelization of constraint programming. INFORMS Journal on Computing 21, 363–382 (2009)
12. Chu, G., Schulte, C., Stuckey, P.J.: Confidence-based work stealing in parallel constraint programming. In: Gent, I.P. (ed.) CP 2009. LNCS, vol. 5732, pp. 226–241. Springer, Heidelberg (2009)
13. Menouer, T., Le Cun, B., Vander-Swalmen, P.: Partitioning methods to parallelize constraint programming solver using the parallel framework Bobpp. In: Nguyen, N.T., van Do, T., Thi, H.A. (eds.) ICCSAMA 2013. SCI, vol. 479, pp. 117–127. Springer, Heidelberg (2013)
14. Xie, F., Davenport, A.: Massively parallel constraint programming for supercomputers: Challenges and initial results. In: Lodi, A., Milano, M., Toth, P. (eds.) CPAIOR 2010. LNCS, vol. 6140, pp. 334–338. Springer, Heidelberg (2010)
15. Yun, X., Epstein, S.L.: A hybrid paradigm for adaptive parallel search. In: Milano, M. (ed.) CP 2012. LNCS, vol. 7514, pp. 720–734. Springer, Heidelberg (2012)
16. Korf, R.E.: Improved limited discrepancy search. In: Proceedings of the 30th National Conference on Artificial Intelligence and the 8th Innovative Applications of Artificial Intelligence Conference, vol. 1, pp. 286–291 (1996)
17. Beck, J.C., Perron, L.: Discrepancy-bounded depth first search. In: Proceedings of the Second International Workshop on Integration of AI and OR Techniques in Constraint Programming for Combinatorial Optimization Problems (CP-AI-OR 2000), pp. 8–10 (2000)
18. Furcy, D., Koenig, S.: Limited discrepancy beam search. In: Proceedings of the 17th International Joint Conference on Artificial Intelligence (IJCAI 2005), pp. 125–131 (2005)

19. Gaudreault, J., Forget, P., Frayret, J.M., Rousseau, A., Lemieux, S., D'Amours, S.: Distributed operations planning in the lumber supply chain: Models and coordination. International Journal of Industrial Engineering: Theory, Applications and Practice 17 (2010)
20. Gaudreault, J., Frayret, J.M., Rousseau, A., D'Amours, S.: Combined planning and scheduling in a divergent production system with co-production: A case study in the lumber industry. Computers and Operations Research 38, 1238–1250 (2011)

Self-splitting of Workload in Parallel Computation

Matteo Fischetti, Michele Monaci, and Domenico Salvagnin

DEI, University of Padova, Via Gradenigo 6/A, 35131 Padova, Italy
{matteo.fischetti,michele.monaci,domenico.salvagnin}@unipd.it

Abstract. Parallel computation requires splitting a job among a set of processing units called *workers*. The computation is generally performed by a set of one or more master workers that split the workload into chunks and distribute them to a set of slave workers. In this setting, communication among workers can be problematic and/or time consuming. Tree search algorithms are particularly suited for being applied in a parallel fashion, as different nodes can be processed by different workers in parallel. In this paper we propose a simple mechanism to convert a sequential tree-search code into a parallel one. In the new paradigm, called SelfSplit, each worker is able to autonomously determine, without any communication with the other workers, the job parts it has to process. Computational results are reported, showing that SelfSplit can achieve an almost linear speedup for hard Constraint Programming applications, even when 64 workers are considered.

1 Introduction

Parallel computation requires splitting a job among a set of workers. A commonly used parallelization paradigm is *MapReduce* [1]. According to the MapReduce paradigm, the overall computation is organized in two steps, and performed by two user-supplied operators, namely, map() and reduce(). The MapReduce framework is in charge of splitting the input data and dispatching it to an appropriate number of map workers, and also of the shuffling and sorting necessary to distribute the intermediate results to the appropriate reduce workers. The output of all reduce workers is finally merged. This scheme is very well suited for applications with a very large input that can be processed in parallel by a large number of mappers, while producing a manageable number of intermediate parts to be shuffled. However, the scheme may introduce a large overhead due to the need of heavy communication/synchronization between the map and reduce phases.

A different approach is based on work stealing [2,3,4]. The workload is initially distributed to the available workers. If the splitting turns out to be unbalanced, the workers that have already finished their processing *steal* part of the work from the busy ones. The process is periodically repeated in order to achieve a proper load balancing. Needless to say, this approach can require a significant amount of communication and synchronization among the workers.

H. Simonis (Ed.): CPAIOR 2014, LNCS 8451, pp. 394–404, 2014.
© Springer International Publishing Switzerland 2014

Tree search algorithms are particularly suited for being applied in a parallel fashion, as different nodes can be processed by different workers in parallel. However, traditional schemes can require an elaborate load balancing strategy, in which the set of active nodes is periodically distributed among the workers [5,6,7], in a work stealing fashion. Depending on the implementation, this may yield a deterministic or a nondeterministic algorithm, with the deterministic option being in general less efficient because of synchronization overhead. In any case, a non-negligible amount of communication and synchronization is needed among the workers, with negative effects on scalability [8,9].

Recently, strategies that try to overcome the traditional drawbacks of the work stealing approach within enumeration algorithms have been proposed. In particular, in [10] a master problem enumerates the partial solutions associated with a subset of the variables of the problem to solve, each of which will be later processed by a worker; the number of variables to consider is chosen in such a way to have significantly more subproblems than workers. All subproblems are put into a queue and distributed to workers as needed (usually, a subproblem is assigned to a given worker as soon as the worker is idle). In [11], a parallelization strategy for LDS [12] is presented, in which the leaves of the complete LDS tree are deterministically assigned to the workers, and each worker processes a subtree only if it contains a leaf assigned to it. These strategies share some similarities with our approach, although some important differences remain.

In the present paper we show how to modify a given deterministic (sequential) tree-search algorithm to let it run on a set of workers. The main features of the approach, that we call SelfSplit, are that

1. each worker works on the whole input data and is able to autonomously decide the parts it has to process;
2. almost no communication between the workers is required;
3. the resulting algorithm can be implemented to be deterministic;
4. in most cases, the modification only requires a few lines of codes.

The above features make SelfSplit very well suited for those applications in which encoding the input and the output of the problem requires a reasonably small amount of data (i.e., it can be handled efficiently by a single worker), whereas the execution of the job can produce a very large number of time-consuming job parts. This is indeed the case when using an enumerative method to solve an NP-hard problem. As such our method is well suited for, but not limited to, High Performance Computing (HPC) applications including Constraint Programming (CP) and Mixed Integer Programming (MIP), whereas approaches based on the MapReduce paradigm are more suited for Big Data applications.

The outline of the paper is as follows. Section 2 describes the basic self-splitting idea for tree search algorithms, along with possible variants aimed at improving load balancing, while Section 3 describes possible implementation strategies. Section 4 reports computational experiments of the application of the above technique within a CP solver. Finally, in Section 5 we draw some conclusions and outline future research directions.

2 SelfSplit Paradigm

SelfSplit addresses the parallelization of a given deterministic algorithm, called the *original algorithm* in what follows, that solves a given problem by breaking it into subproblems called *nodes*. In this paper we will only deal with original algorithms of enumeration type (branch-and-bound or alike), but other applications of SelfSplit are possible.

2.1 The Idea

Figure 1 illustrates our self-splitting method to parallelize an enumerative original algorithm.

a) Each worker reads the original input data and receives an additional input pair (k, K), where K is the total number of workers and $k \in \{1, \cdots, K\}$ identifies the current worker. The input is assumed to be of manageable size, so no parallelization is needed at this stage.

b) The same deterministic computation is initially performed, in parallel, by all workers. This initial part of the computation is called *sampling phase* and is illustrated in the figure by the fact that exactly the same enumeration tree is initially built by all workers. No communication at all is involved in this stage. It is assumed that the sampling phase is not a bottleneck in the

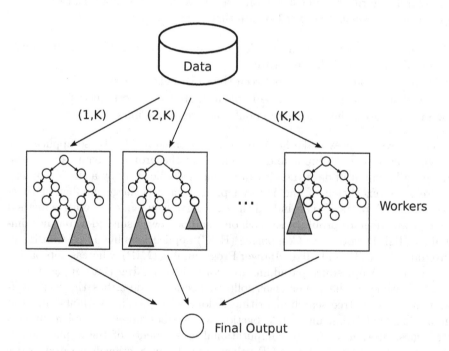

Fig. 1. Illustration of the SelfSplit paradigm

overall computation, so the fact that all workers perform redundant work introduces an acceptable overhead.

c) When enough open nodes have been generated, the sampling phase ends and each worker applies a deterministic rule to identify and solve the nodes that belong to it (gray subtrees in the figure), without any redundancy. No communication among workers is involved in this stage. It is assumed that processing the subtrees is the most time-consuming part of the algorithm, so the fact that all workers perform non-overlapping work is instrumental for the effectiveness of the self-splitting method.

d) When a worker ends its own job, it communicates its final output to a *merge worker* that process it as soon as it receives it. The merge worker can in fact be one of the K workers, say worker 1, that merges the output of the other workers after having completed its own job. We assume that output merging is not a bottleneck of the overall computation, as it happens, e.g., for enumerative algorithms where only the best solution found by each worker needs to be communicated.

Note that all steps but d) requires absolutely no communication among workers. SelfSplit is therefore very well suited for those computational environments where communication among workers is time consuming or unreliable as, e.g., in a large computational grid where the workers run in different geographical areas—a relevant context being cloud computing, or a constellation of mobile devices.

Though very desirable, the absence of communication implies the risk that workload is quite unbalanced, i.e., lucky and unlucky workers can finish their computation at very different points in time. To contrast this drawback, our recipe, as in [10], is to keep a significant number of open nodes for each worker after sampling, so as to increase the chances that workload is split in a fair way. More sophisticated rules can also be applied, as described later on.

2.2 Vanilla Algorithm

In its simplest version, our method modifies the original algorithm as follows:

1. Two integer parameters (k, K) are added to the original input: K denotes the number of workers, while k is an index that uniquely identifies the current worker ($1 \leq k \leq K$).
2. A global flag ON_SAMPLING is introduced and initialized to true. The flag becomes false when a given condition is met, e.g., when there are enough open nodes in the branch-and-bound tree. When the flag ON_SAMPLING is set to false we say that the *sampling phase* is over.
3. Each time a node n is created, it is deterministically assigned a *color* $c(n)$ which is a pseudo-random integer in $\{1, \cdots, K\}$ during the sampling phase, and $c(n) = k$ otherwise.

4. Whenever the modified algorithm is about to process a node n, condition

$$(\neg \text{ON\_SAMPLING}) \wedge (c(n) \neq k)$$

is evaluated. If the condition evaluates to true, node n is just discarded, as it corresponds to a subproblem assigned to a different worker; otherwise, the processing of node n continues as usual and no modified action takes place.

Each worker executes exactly the same algorithm, but receives a different input value for k. The above method ensures that each worker can autonomously and deterministically identify and skip the nodes that will be processed by other workers, and each node is covered by (at least) one worker. A similar strategy is exploited in [11], where however the details of the split are very much dependent of the LDS algorithm, and cannot be easily (and efficiently) generalized to an arbitrary tree search algorithm.

Load balancing is automatically obtained by the modified algorithm in a statistical sense: if the condition that triggers the end of the sampling phase is chosen appropriately, then the number of subproblems to distribute is significantly larger than the number of workers K, thus it is unlikely that a given worker will be assigned much more work to do than any other worker. Static decomposition and statistical load balancing were also at the base of the method proposed in [10], with some key differences:

- a master process is used to generate the subproblems to distribute. Enumeration in the master problem is more static than in SelfSplit, and some communication is needed to distribute the subproblems;
- open problems are dynamically assigned to idle workers as soon as they become available, which again requires some communication;
- the algorithm in [10] is non-deterministic because of dynamic scheduling.

SelfSplit is straightforward to implement if the original deterministic algorithm is sequential, and the random/hash function used to color a node is deterministic and identical for all workers. The algorithm can however be applied even if the original deterministic algorithm is itself parallel, provided that the pseudo-random coloring at Step 3 is done right after a synchronization point.

2.3   Paused-Node Queue Algorithm

A slightly more elaborate version, aimed at improving workload balancing among workers, can be devised using an auxiliary queue S of *paused nodes*. The modified algorithm reads as follows:

1. As before, two integer parameters (k, K) are added to the original input.
2. A paused-node queue S is introduced and initialized to empty.
3. Whenever the modified algorithm is about to process a node n, a boolean function NODE_PAUSE(n) is called: if NODE_PAUSE(n) is true, node n is moved into S and the next node is considered; otherwise the processing of node n continues as usual and no modified action takes place.

4. When there are no nodes left to process, the *sampling phase* ends. All nodes n in S, if any, are popped out and assigned a color $c(n)$ between 1 and K, according to a deterministic rule.

5. All nodes n whose color $c(n)$ is different from the input parameter k are just discarded. The remaining nodes are processed (in any order and possibly in a nondeterministic way) till completion.

Because it has access to all the nodes in S, the coloring phase at Step 4 has more chances to determine a balanced workload split among the workers than its "vanilla" counterpart", at the expense of a slightly more elaborate implementation.

3   Implementation Details

We will next give more details about the application of our method within an enumerative method for optimization problems. We will focus on the version exploiting the queue S of paused nodes. In this version, both the decision of moving a node into S as well as the color actually assigned to a node are based on an estimate of the computational difficulty of the node. The idea is to move a node in S if it is expected to be significantly easier than the root node (original problem), but not too easy as this would lead to an exceedingly time-consuming sampling phase.

Within NODE_PAUSE, a rough estimate of the difficulty of a node can be obtained by computing the logarithm of the cardinality of the Cartesian product of the current domains of the variables, to be compared with the same measure computed at the end of the root node—for problems involving binary variables only, this figure coincides with the number of free variables at the node. To cope with the intrinsic approximation involved in this estimate, the following adaptive scheme can be used to improve SelfSplit robustness. At the end of the sampling phase, if the number of nodes in S is considered too small for the number K of available workers, then the internal parameters of NODE_PAUSE are updated in order to make the move into the queue S less likely, and the sampling procedure is continued after putting the nodes in S back into the branch-and-bound queue—or the overall method is just restarted from scratch.

As to node coloring, in our implementation the color $c(n)$ associated with each node n in S is obtained in three steps: (1) compute a score estimating the difficulty of each node n, (2) sort the nodes by decreasing scores, and (3) assign a color c between 1 and K, in round-robin, so as to split node scores evenly among workers.

4   SelfSplit for Constraint Programming

We implemented SelfSplit within the CP solver Gecode 4.0 [13]. While the most natural option is to implement the scheme as a search engine, we opted for implementing self-splitting as a custom constraint propagator. This is only because we found the implementation much easier: implementing search engines is somewhat more involved, and requires some expertise in Gecode programming. In addition, our proof of concept implementation shows that the method can be implemented with very limited effort. Of course, different implementation strategies for different solvers can be devised.

Our global constraint, node_pause, is implemented as a generic n-ary propagator, that takes on input an array x of variables, a pointer to the queue S to store delayed nodes, a measure of node difficulty V_0 computed from the domains of the variables in x at the root node, and a threshold θ. Each time the propagate method of our propagator is called, the node-difficulty measure V is computed based on the local domains of the variables in x, and the resulting value is compared with V_0. If their ratio is greater than θ, then the local domains of the variables x are copied into a custom node class, which is stored in S, and the propagator returns a failure in order to kill the node. Otherwise, we just return. Note that this implementation is compatible with Gecode *copy and recomputation* backtracking model.

We next address the computation of the node-difficulty estimate within NODE_PAUSE, namely, the logarithm of the cardinality of Cartesian product of the variable domains. We provide an implementation both for arrays of integer variables and for arrays of set variables.

In the integer case, for each variable x_j in x we consider its current domain as a list of ranges $\{[l_j^k, u_j^k] | k \in K_j\}$, as implemented in Gecode. As such, the contribution of variable x_j to the difficulty measure can be computed as

$$\log_2 \sum_{k \in K_j} (u_j^k - l_j^k + 1)$$

which is a refinement of the simpler expression $\log_2(u_j - l_j + 1)$.

In the set case, Gecode approximates the domain of a variable x_j with three pieces of information:

i) a set of elements glb_j which is known to be contained in any feasible value for x_j, i.e., $glb_j \subseteq x_j$

ii) a set of elements lub_j which is known to contain any feasible value for x_j, i.e., $x_j \subseteq lub_j$

iii) bounds (m_j, M_j) on the cardinality of x, i.e., $m_j \leq |x_j| \leq M_j$.

With this encoding, we can compute the contribution of variable x_j as

$$\log_2 \sum_{i=m_j-|glb_j|}^{M_j-lub_j} \binom{|lub_j \setminus glb_j|}{i}$$

Note that the above expression can become expensive to compute, and prone to overflow for even modest values of $|lub_j \setminus glb_j|$. For this reason, if the resulting number is greater than 64 we use the valid upper bound $|lub_j \setminus glb_j|$.

The overall scheme is implemented as follows:

- the model to solve is coded into a C++ class (as usually done in Gecode) and the node_pause constraint is added to the model with the appropriate parameters. Note that the array x of variables that are considered for computing the measure of difficulty of the current subproblem is possibly a subset of the whole set of variables, chosen by exploiting knowledge about the model. We do not consider this an issue, since modeling a problem in Gecode requires some problem-specific coding in any case;
- the model (with the node_pause) propagator is completely enumerated (sampling phase);
- the nodes collected in S that survive the coloring phase are used to construct new models, copying the domains from the nodes, each of which is enumerated by Gecode. Note that the propagator node_pause is not used in this phase.

To avoid to have too few nodes in S after the sampling phase, we implemented the following simple adaptive mechanism, along the lines of the previous section. The boolean value returned by NODE_PAUSE(n) is true when the difference between the estimated difficulty of the root node and that of node n, computed as outlined above, is greater or equal to $\theta = 10$ (corresponding to a reduction of the cardinality of the Cartesian product of at least 1,024 times). When the sampling phase is over, if $|S| < 2000$ then θ is doubled and the overall method is just restarted.

We tested our implementation on several instances taken from the repository of modeling examples bundled with Gecode. Since we are interested in measuring the scalability of our method, we considered only instances which are either infeasible or in which we are required to find all feasible solutions (the parallel speedup for finding a first feasible solution can be completely uncorrelated to the number of workers, making the results hard to analyze). On some instances we added some form of symmetry breaking constraints in order to make the search for all solutions more efficient. We ran our method with number of workers $K \in \{1, 4, 16, 64\}$. Each worker is configured to use only a single thread: this is because a deterministic behavior is needed for correctness in the sampling phase, and it also makes the results more easily reproducible.

Detailed results are available in Table 1. According to the table, even on moderately easy instance requiring half a minute to solve, SelfSplit can achieve an almost linear speedup with up to 16 workers, and the speedup is still good for $K = 64$. On harder instances, the method scales almost linearly also with 64 workers. In all cases, the resulting algorithm is deterministic. Note that, despite the fact that our instances are very different, we used exactly the same parameter tuning on all of them, showing that the method is quite robust.

5   Conclusions and Future Work

We have presented SelfSplit, a new deterministic and (almost) communication free parallelization paradigm for tree search methods. The idea is that a given deterministic algorithm can easily be parallelized by letting each processing unit

Table 1. Measuring scalability with Gecode

| | time (s) | | speedup | |
instance	$K = 1$	$K = 4$	$K = 16$	$K = 64$
golomb_12	41.5	3.84	14.31	41.50
golomb_13	1195.8	4.00	15.67	57.49
golomb_14	19051.9	3.97	15.71	61.34
partition_16	30.0	3.75	13.64	46.15
partition_18	354.8	3.90	14.78	54.58
partition_20	4116.4	3.86	15.64	59.40
ortholatin_5	29.3	3.89	13.95	36.63
sports_10	98.7	3.91	14.51	44.86
hamming_7_4_10	32.3	3.85	14.04	40.38
hamming_7_3_6	2402.4	3.91	15.44	59.76

autonomously decide which are the parts of work it can skip as they will be performed by other units. This is achieved without any communication among the units, and only requires a same deterministic selection rule be applied by all units in the early part of the computation.

A main feature of the method is that exactly the same code is run independently by all units, without the need of any external coordination nor master-slave hierarchy—the work gently "splits itself" over the units. As a consequence, SelfSplit is very well suited for HPC on a computational grid (or cloud) where the processing units are geographically distributed and communication is expensive or unreliable, and task synchronization becomes a bottleneck of the overall computation.

Two different implementations of SelfSplit have been outlined, that only require minor changes of the deterministic algorithm to be parallelized. Computational results on a Constraint Programming implementation on top of an open-source solver show that an almost linear speedup can be achieved, even on 64 processing units, without any communication among them—besides the final merge of the solution(s) returned by each unit, and still with deterministic behavior.

A practically very important feature of SelfSplit is that it often requires just minor code changes. As an exercise, we took the Asymmetric TSP sequential codes in [14] (pure branch-and-bound) and in [15] (branch-and-cut code), which are optimized yet legacy FORTRAN codes. We parallelized them by using the SelfSplit approach, adding about 10 new lines of codes in the both cases, obtaining surprisingly good speedups.

Because of lack of communication among workers, SelfSplit can turn out to be not suitable for solvers that collect/learn important global information during the search, as this information can be crucial to reduce the search tree. Observe however that some solvers—notably, MIP branch-and-cut methods—do in fact collect their main global information (cuts, pseudocosts, incumbent, etc.)

in their early nodes, i.e., during sampling, thus all such information is automatically available to all workers. Therefore, performing the sampling phase redundantly in parallel by all workers has the advantage of sharing a potentially big amount of global information without communication—a distinguishing feature of SelfSplit. In any case, for those solvers a limited amount of communication (e.g., of the incumbent value) can be advisable.

Possible SelfSplit variants to be addressed in future work are outlined below:

a) SelfSplit can be run with just $K' \ll K$ workers, with input pairs $(1, K)$, $(2, K), \cdots, (K', K)$. In this case the overall procedure is heuristic in nature, meaning that some nodes will not be explored by any worker (namely, those with color $k = K' + 1, \cdots, K$). This setting is particularly attractive for the parallelization of heuristics for optimization/feasibility problem, as it guarantees that the solution spaces explored (exactly or heuristically) by the K' workers after sampling is non-overlapping—though their union does not necessarily cover the whole solution space.

b) The previous variant of running $K' \ll K$ workers can also be used to obtain a lower bound on the amount of computing time needed to solve the problem with K workers (just take the maximum computing time among the K' workers) as well as an estimate of the amount of computing time T_1 needed to solve the problem with the original (unmodified) algorithm by a single worker, e.g., through the simple formula

$$T_1 = T_s + K \cdot \overline{T}$$

where T_s is the sampling time and $\overline{T}$ is the average time spent by a worker after sampling.

c) SelfSplit can also be used to split the overall workload into K chunks to be solved at different points in time by a single (or few) worker(s), thus implementing a simple strategy to pause and resume the overall computation. This is also beneficial in case of failures, as it allows one to re-execute the affected chunks only.

d) A limited amount of communication may be introduced between the workers after the sampling and coloring phases. This communication is meant to exchange globally valid information, such as the primal bound in an enumerative scheme, which can be used to avoid unnecessary work. For example, if a feasibility problem is addressed, as soon as a worker finds the first feasible solution all the other workers can be interrupted as the overall problem is solved. Similarly, if the incumbent is periodically shared among the workers, each worker can be interrupted in case its own best bound is worse than the incumbent value. If the incumbent is not used in any other way, the search path followed by each workers is not affected by communication and each worker behaves deterministically till its own abort point.

e) Workers can be allowed to (periodically) communicate to deal with failures in the computational environment that require re-running a certain (k, K) pair.

f) After sampling, each worker can decide not to discard the nodes that have two or more colors $c_1, c_2, ..., c_m$, where $c_1 = k$ and the other colors $c_2, \cdots, c_m$ are selected according to some rules. In this case some redundant work is performed by the workers, e.g., with the aim of coping with failures in the computational environment. The final merge worker can stop the overall computation when all colors have been processed by some worker, even if other workers are still running or were aborted for whatever reason. Alternatively, two or more workers with the same (k, K) pair can be run, in parallel, making the event that all of them fail very unlikely, and still keeping the communication overhead negligible.

References

1. Dean, J., Ghemawat, S.: MapReduce: simplified data processing on large clusters. CACM 51(1), 107–113 (2008)
2. Grama, A., Kumar, V.: State of the art in parallel search techniques for discrete optimization problems. IEEE Trans. Knowl. Data Eng. 11(1), 28–35 (1999)
3. Michel, L., See, A., Hentenryck, P.V.: Transparent parallelization of constraint programming. INFORMS Journal on Computing 21(3), 363–382 (2009)
4. Chu, G., Schulte, C., Stuckey, P.J.: Confidence-based work stealing in parallel constraint programming. In: Gent, I.P. (ed.) CP 2009. LNCS, vol. 5732, pp. 226–241. Springer, Heidelberg (2009)
5. Bordeaux, L., Hamadi, Y., Samulowitz, H.: Experiments with massively parallel constraint solving. In: Boutilier, C. (ed.) IJCAI 2009, pp. 443–448 (2009)
6. Gent, I.P., Jefferson, C., Miguel, I., Moore, N.C., Nightingale, P., Prosser, P., Unsworth, C.: A preliminary review of literature on parallel constraint solving. In: Proceedings PMCS 2011 Workshop on Parallel Methods for Constraint Solving (2011)
7. Shinano, Y., Heinz, S., Vigerske, S., Winkler, M.: Fiberscip – a shared memory parallelization of scip. Technical report, ZIB (2013)
8. Koch, T., Ralphs, T.K., Shinano, Y.: Could we use a million cores to solve an integer program? Mathematical Methods of Operations Research 76(1), 67–93 (2012)
9. Achterberg, T., Wunderling, R.: Mixed integer programming: Analyzing 12 years of progress. In: Facets of Combinatorial Optimization, pp. 449–481 (2013)
10. Régin, J.C., Rezgui, M., Malapert, A.: Embarrassingly parallel search. In: Schulte, C. (ed.) CP 2013. LNCS, vol. 8124, pp. 596–610. Springer, Heidelberg (2013)
11. Moisan, T., Gaudreault, J., Quimper, C.-G.: Parallel discrepancy-based search. In: Schulte, C. (ed.) CP 2013. LNCS, vol. 8124, pp. 30–46. Springer, Heidelberg (2013)
12. Harvey, W.D., Ginsberg, M.L.: Limited discrepanc search. In: IJCAI 1995, pp. 607–615 (1995)
13. Gecode Team: Gecode: Generic constraint development environment (2012), Available at http://www.gecode.org
14. Fischetti, M., Toth, P.: An additive bounding procedure for the asymmetric travelling salesman problem. Mathematical Programming 53, 173–197 (1992)
15. Fischetti, M., Lodi, A., Toth, P.: Exact methods for the asymmetric traveling salesman problem. In: The traveling Salesman Problem and Its Variations, pp. 169–205. Springer, US (2004)

The Markov Transition Constraint

Michael Morin and Claude-Guy Quimper

Department of Computer Science and Software Engineering
Université Laval, Québec, Qc, Canada
Michael.Morin.3@ulaval.ca,
Claude-Guy.Quimper@ift.ulaval.ca

Abstract. We introduce a novel global Markov transition constraint (MTC) to model finite state homogeneous Markov chains. We present two algorithms to filter the variable domains representing the imprecise probability distributions over the state space of the chain. The first filtering algorithm is based on the fractional knapsack problem and the second filtering algorithm is based on linear programming. Both of our filtering algorithms compare favorably to the filtering performed by solvers when decomposing an MTC into arithmetic constraints. Cases where the fractional knapsack decomposition enforces bounds consistency are discussed whereas the linear programming filtering always perform bounds consistency. We use the MTC constraint to model and solve a problem of path planning under uncertainty.

1 Introduction

We introduce a novel global *markov transition constraint* (MTC) to model finite state space Markov processes with a finite number of steps T. Markov processes (also called Markov chains) are central to many applications. They are used in physics to model the motion of a target in a physical environment [22] and in computer science they are used in the Pagerank algorithm used by Google [20]. They are also used in economics and business science [8]. Markov processes even apply to arts for the generation of melodies [19] and lyrics [1]. They form the basis of decision making frameworks, such as Markov decision processes and hidden Markov decision processes which are fundamental to many artificial intelligence applications [16].

Let $\mathcal{N} = \{1, \ldots, N\}$ be a space of N states. States are mutually exclusive and jointly exhaustive, i.e., the process is in exactly one state of $\mathcal{N}$ at any time. Let $\mathbf{M}$ be the $N \times N$ *transition matrix* of the Markov process. That is, the probability of moving from state i to state j at any step $t \in \{1, \ldots, T\}$ is $0 \leq \mathbf{M}_{ij} \leq 1$ ($\forall i, j \in \mathcal{N}$), and the total probability of moving from state $i \in \mathcal{N}$ (given that the process is in state i) is 1, i.e., $\sum_{j \in \mathcal{N}} \mathbf{M}_{ij} = 1$ ($\forall i \in \mathcal{N}$). Let $\mathbf{x}^t = [x_1^t, \ldots, x_N^t]$ be the row vector of the probability distribution on the states at a time $t \in \{1, \ldots, T\}$ where x_i^t is the probability that the process is in state $i \in \mathcal{N}$ at step $t \in \{1, \ldots, T\}$. Given an initial distribution $\mathbf{x}^1$ over the states such that $\sum_{i \in \mathcal{N}} x_i^1 = 1$ and $0 \leq x_i^1 \leq 1$ for all $i \in \mathcal{N}$, the *Markov property* states that

$$\mathbf{x}^{t+1} = \mathbf{x}^t \mathbf{M}, \qquad\qquad \forall t \in \{2, \ldots, T\}. \tag{1}$$

H. Simonis (Ed.): CPAIOR 2014, LNCS 8451, pp. 405–421, 2014.
© Springer International Publishing Switzerland 2014

That is, the state at time $t + 1$ depends on the previous state only.[1] Given the distribution $\mathbf{x}^t$, the distribution after k steps from step t, $\mathbf{x}^{t+k}$, is computed as:

$$\mathbf{x}^{t+k} = \mathbf{x}^t \mathbf{M}^k, \qquad \forall t \in \{1, \ldots, T\}, k \in \{0, \ldots, T - t, \}. \qquad (2)$$

Example 1. Suppose that we wish to model the motion of a lost child between three stores (store 1, 2, and 3) of a mall with discrete time intervals of one minute. The child's behavior is as follows: s/he may spend time in the toys store (number 1); after one minute, her/his probability of leaving to the candy store (number 2) is $\frac{1}{8}$; from the candy store, s/he either returns to the toys store, stays there, or goes to the food market (number 3); whenever the child is in the food market, s/he directly returns to the candy store. The child's motion model is a transition matrix $\mathbf{M}$:

$$\mathbf{M} = \begin{bmatrix} \frac{7}{8} & \frac{1}{8} & 0 \\ \frac{1}{3} & \frac{1}{3} & \frac{1}{3} \\ 0 & 1 & 0 \end{bmatrix}. \qquad (3)$$

As a first experiment, suppose that the child starts in room 1. That is, the probability distribution at time 1 is $\mathbf{x}^1 = [1, 0, 0]$. We may, using equation (1), infer that the distribution over the child's location after a minute is $\mathbf{x}^2 = \left[\frac{7}{8}, \frac{1}{8}, 0 \right]$. It is as easy to know, using equation (2), the distribution over the child's location after a delay of $k > 1$ minutes.

As a second experiment, suppose that we have no information on the child's initial location, but that we want to get the best possible estimation over her/his location after 1 minute, i.e., at time $t = 2$. The probability distribution at time 1 is uncertain. The only thing we know is that $\mathbf{x}^1$ is a distribution: it sums up to 1 and the probabilities are in the interval $[0, 1]$. By relying on linear optimization techniques or on the fractional knapsack filtering algorithm we present in Section 4, we find that, the probabilities of locating the child in room 1, 2 or 3 are in the intervals $\left[0, \frac{7}{8} \right]$, $\left[\frac{1}{8}, 1 \right]$ and $\left[0, \frac{1}{3} \right]$ respectively; a solution that interval arithmetic alone is not able to provide. □

Clearly, filtering the uncertain probability distributions to the tightest possible intervals is not as easy in the second experiment as computing the probabilities in the first example. Uncertainty on probability distributions arises in modeling and problem solving due to external factors influencing the chain, imprecise data, and/or uncertain knowledge. For instance, in the search operation model presented in Section 6, the probability distributions are updated given the searcher's actions. There is no way, without knowing the entire searcher's path, to compute the exact probability distribution for each time step.

The introduction of a global constraint to model the evolution of the state space enables to intuitively model a Markov chain and leads to stronger filtering

---

[1] Even if some generalizations allow the current state to depend on $d \geq 1$ states (e.g., Markov chains of order d), we restrict ourselves to previous state dependencies only, i.e., to first-order Markov chains.

of the constrained variables when compared to a decomposition of the chain into elementary constraints.

The paper is organized as follows. We describe the MTC global constraint in Section 2. Section 3 discusses the literature related to similar constraints. In Section 4, we introduce two filtering algorithms: the first one based on linear optimization and the second one inspired from a fractional knapsack algorithm. We also discuss about the filtering of a decomposition into linear constraints. We identify cases where the different filtering techniques achieve bounds consistency. In Section 5, we conduct an empirical study of the filtering algorithms. Section 6 presents an application of the MTC constraint to a practical problem. We conclude in Section 7.

2   Modeling Markov Transition Processes as a Global Constraint

We define a new constraint that encodes a Markov transition process. As a convention, we write constrained variables in italic upper case and global constraints in small capitals. The MTC is defined as follows:

$$\text{MTC}([Y_1, \ldots, Y_N], [X_1, \ldots, X_N], \mathbf{M}) \Leftrightarrow \forall j \in \mathcal{N}, \sum_{i \in \mathcal{N}} X_i \mathbf{M}_{ij} = Y_j \wedge \sum_{i \in \mathcal{N}} X_i = 1.$$

The matrix $\mathbf{M}$ is a known Markovian transition matrix. The vectors $\mathbf{X} = [X_1, \ldots, X_N]$ and $\mathbf{Y} = [Y_1, \ldots, Y_N]$ are probability distributions over $\mathcal{N}$. The domains of the variables in vectors $\mathbf{X}$ and $\mathbf{Y}$ are:

$$\text{dom}(X_i) = \left[\underline{X}_i, \overline{X}_i\right], \qquad\qquad \forall i \in \mathcal{N}, \qquad (4)$$

where $\underline{X}_i$ and $\overline{X}_i$ are lower and upper bounds on variable X_i; and

$$\text{dom}(Y_j) = \left[\underline{Y}_j, \overline{Y}_j\right], \qquad\qquad \forall j \in \mathcal{N}, \qquad (5)$$

where $\underline{Y}_j$ and $\overline{Y}_j$ are lower and upper bounds on variable Y_j.

The constraint $\text{MTC}(\mathbf{Y}, \mathbf{X}, \mathbf{M})$ applies a transition matrix $\mathbf{M}$ to $\mathbf{X}$ and obtains $\mathbf{Y}$, i.e., $\text{MTC}(\mathbf{Y}, \mathbf{X}, \mathbf{M})$ states that $\mathbf{Y} = \mathbf{X}\mathbf{M}$. Multiple MTC constraints may be chained to compute a finite Markov chain of T steps:

$$\text{MTC}(\mathbf{X}^t, \mathbf{X}^{t-1}, \mathbf{M}), \qquad\qquad \forall t \{2, \ldots, T\}. \quad (6)$$

Other constraints may be added on the X_i variables to interact with the chain.

Since the constraint $\text{MTC}(\mathbf{Y}, \mathbf{X}, \mathbf{M})$ is satisfied only if $\mathbf{X}$ and $\mathbf{Y}$ represent probability distributions, filtering this constraint can both be seen as an application of the *theory of interval-probability* and as a linear optimization problem [12,24].

Definition 1 (Uncertain distribution). *We call a probability distribution an* uncertain distribution *if at least one of its probability is defined as an interval.*

For instance, the vector $[\text{dom}(X_1), \ldots, \text{dom}(X_n)]$ is a vector of intervals which is an uncertain distribution if and only if there exists at least an assignment to the variables X_i that sums up to 1.

3　Markov Constraints and Related Literature

Pachet and Roy [18] introduced the *elementary markov constraint* (EMC). The EMC is defined as follows[2]:

$$\text{EMC}(S, S', P_{S'}) \Leftrightarrow P_{S'} = \mathbf{M}_{SS'}. \tag{7}$$

That is, $\text{EMC}(S, S', P_{S'})$ states that the probability of moving from state S to state S' is $P_{S'}$. The domains of the state variables S and S' are subsets of $\mathcal{N}$. The domain of the probability variable is the set of conditional probabilities, computed during the generation of the model, of achieving state S' from any previous state: $\text{dom}(P_{S'}) = \{ p \mid (\exists i \in \mathcal{N} \mid p = \mathbf{M}_{iS'}) \}$.

Using multiple EMCs, the authors model *constrained Markov processes*, i.e., Markov processes with supplementary constraints on the generated sequence. Let $S_1, \ldots, S_T$ be the state variables of a sequence of a first-order Markov chain. Let $P_{S_2}, \ldots, P_{S_T}$ be the variables that represent the probabilities. The constrained sequence is modeled as chained EMCs:

$$\text{EMC}(S_t, S_{t+1}, P_{S_{t+1}}), \qquad\qquad \forall t \in \{1, \ldots, T-1\}. \tag{8}$$

The Markov property, enforced by the EMCs, is a cost function to optimize whereas supplementary constraints are used to steer the generation of the sequence. The approach is used on chords sequence and melody generation.

Following [18], Pachet et al. [19] show that when the scope of the supplementary constraints does not exceed the chain's order d, the constrained Markov process may be recompiled into a statistically equivalent unconstrained Markov process. The approach is illustrated on the melody generation problem. In [1], the authors apply constrained Markov processes to the generation of lyrics.

The *markov transition constraint* (MTC) we introduce differs from the *elementary markov constraint* (EMC) presented in [18,19]. The former aims at modeling the evolution of the state space by computing the distributions $\mathbf{x}^t$ defined for all $t \in \{1, \ldots, T\}$ while the latter aims at generating the sequence of states in a constrained fashion, i.e., a Markov sequence with supplementary constraints. Moreover, the MTC deals with interval-domain probabilities while the EMC deals with finite-domain probabilities.

Markov chains with uncertain data (imprecise Markov chains) are related to Markov constraints. Imprecise Markov chains are provided a credal set (or probability interval) for each possible transition whereas we deal with precise (singleton) transition probabilities and uncertain distributions. Further details on imprecise Markov chains may be found in [5,6].

4　Filtering the Markov Transition Constraint

A filtering algorithm for the MTC prunes the values from the domains of the variables X_i and Y_i that are inconsistent with the constraint.

---

[2] We restraint our EMC definition to first-order Markov chains even though the definition found in [18] is general.

Definition 2 (Interval support). *Let $C([X_1, \ldots, X_n])$ be a constraint of arity n. The assignment $[X_1, \ldots, X_n] = [x_1, \ldots x_n]$ is an interval support if and only if $C([x_1, \ldots x_n])$ is satisfied and the inequalities $\underline{X}_i \leq x_i \leq \overline{X}_i$ hold.*

Definition 3 (Bounds consistency). *A constraint $C([X_1, \ldots, X_n])$ is bounds consistent if and only if, for every variable X_i, there exists an interval support where the variable is assigned to the lower bound of its domain and a bounds support where the variable is assigned to the upper bound of its domain.*

A filtering algorithm *enforces bounds consistency* if and only if after being executed, no interval supports are eliminated and the constraint is bounds consistent. From now on, we simply say that an assignment has a *support* instead of an *interval support*.

We present three filtering algorithms to filter the variables X_i and Y_j ($\forall i, j \in \mathcal{N}$) subject to an MTC constraint. The first algorithm, denoted MTC-IA, uses the interval arithmetic [14] that is applied on a decomposition of the constraint. Following [24], the second algorithm, denoted MTC-LP, performs a linear optimization to achieve bounds consistency. The third algorithm, denoted MTC-FK, is a compromise between the two previous approaches. It relaxes the problem into a set of fractional knapsack constraints on which it enforces bounds consistency.

4.1 Filtering Using Interval Arithmetic (MTC-IA)

We decompose the global constraint $\text{MTC}(\mathbf{Y}, \mathbf{X}, \mathbf{M})$ into linear constraints as follows:

$$\sum_{i \in \mathcal{N}} X_i \mathbf{M}_{ij} = Y_j, \qquad\qquad \forall j \in \mathcal{N}, \quad (9)$$

$$\sum_{i \in \mathcal{N}} X_i = 1, \qquad\qquad (10)$$

$$\sum_{j \in \mathcal{N}} Y_j \mathbf{M}^{-1}{}_{ij} = X_i, \qquad\qquad \forall i \in \mathcal{N}, \quad (11)$$

$$\sum_{j \in \mathcal{N}} Y_j = 1. \qquad\qquad (12)$$

Constraints (9) and (10) follow from the definition of the MTC. Constraints (11) and (12) are implied constraints that enhance the filtering. We call MTC-IA the algorithm that uses the interval arithmetic to enforce bounds consistency on the constraints (9) to (12). This algorithm, that is already implemented in most constraint solvers, simply enforces bounds consistency on each constraint until a fixed point is reached. The implied constraints necessitate the inverse of the transition matrix $\mathbf{M}$. The inverse $\mathbf{M}^{-1}$ is computed during the generation of the model, prior to solving the problem. This pre-solving process is done in cubic time.

We discuss specific cases where interval arithmetic enforces bounds consistency on the constraint MTC. These cases require new definitions.

Definition 4 (Monomial matrix). *A matrix* **A** *is* monomial *if and only if it has one and only one non-null element in each column and each row.*

Proposition 1. *A monomial transition matrix* **M** *is a permutation matrix.*

Proof. The rows of **M** sums up to 1 by definition. Because **M** is monomial, we must have one element set to one in every row and all other elements are null which result into a binary matrix. Monomial binary matrices are permutation matrices.   □

Proposition 2. *The inverse of a monomial transition matrix* **M** *is a transition matrix.*

Proof. **M** is a permutation matrix. The inverse of a permutation matrix is its transpose which is also a permutation matrix.   □

Lemma 1. *If* **M** *is monomial, then enforcing bounds consistency on the linear constraints* (9) *and* (10) *enforces bounds consistency on* $\mathrm{MTC}(\mathbf{Y}, \mathbf{X}, \mathbf{M})$.

Proof. If **M** is monomial, it is a permutation matrix and the constraints (9) are binary equalities. Suppose that constraints (9) and (10) are bounds consistent. Let x be an interval support for (10) then $y = x\mathbf{M}$ is a permutation of x and, thanks to the equality constraints (9), we have $y_i \in \mathrm{dom}(Y_i)$. Consequently, the upper bound and lower bounds of the domains $\mathrm{dom}(X_i)$ have an interval support for $\mathrm{MTC}(\mathbf{Y}, \mathbf{X}, \mathbf{M})$. Let π be the permutation encoded by **M**. Since the bounds of $\mathrm{dom}(Y_{\pi(i)})$ are equal to the bounds of $\mathrm{dom}(X_i)$, the bounds of $\mathrm{dom}(Y_i)$ also have an interval support for $\mathrm{MTC}(\mathbf{Y}, \mathbf{X}, \mathbf{M})$.   □

Thanks to Proposition 2, Lemma 1 also holds when replacing constraint (9) by constraint (11) and/or constraint (10) by constraint (12).

4.2   Filtering Using Linear Programming (MTC-LP)

In this section, we describe our linear programming (LP) reformulation of the filtering problem for the MTC global constraint. Even though LP has not been applied (to our knowledge) to the filtering of global constraints encoding a Markov chain, there exist examples in the literature (e.g., [3] and [4]) where LP is used to filter global constraints. These successes justify the application of the idea to Markov chains.

The LP filtering algorithm solves two linear programs per variable: one for the lower bound and one for the upper bound. To obtain a lower bound on variable X_i, the LP minimizes X_i subject to the constraints (14) to (17).

$$\min X_i \tag{13}$$

subject to

$$\sum_{i \in \mathcal{N}} \mathbf{M}_{ij} X_i = Y_j, \qquad\qquad \forall j \in \mathcal{N}, \ (14)$$

$$\sum_{i \in \mathcal{N}} X_i = 1, \qquad\qquad (15)$$

$$\underline{X}_i \leq X_i \leq \overline{X}_i, \qquad\qquad \forall i \in \mathcal{N}, \ (16)$$

$$\underline{Y}_j \leq Y_j \leq \overline{Y}_j, \qquad\qquad \forall j \in \mathcal{N}. \ (17)$$

New bounds on $\overline{X}_i$, $\underline{X}_i$, and $\underline{Y}_i$ follow from a modification of the objective function. For each state $i \in \mathcal{N}$, we have the following LPs:

- $\overline{X}_i = \max X_i$ (resp. $\overline{Y}_i = \max Y_i$) subject to constraints (14) to (17);
- $\underline{X}_i = \min X_i$ (resp. $\underline{Y}_i = \min Y_i$) subject to constraints (14) to (17).

Each of the $4N$ linear programs may be solved using the simplex method [7]. We call this filtering technique based on linear optimization MTC-LP.

If an exact LP method is used (e.g., the simplex algorithm), we obtain optimal bounds on the domains of both X and Y.

Theorem 1. *MTC-LP enforces bounds consistency on* MTC$(\mathbf{Y}, \mathbf{X}, \mathbf{M})$.

Proof. The proof is a direct consequence of using an exact method for solving the linear programs. □

4.3 Filtering Using the Fractional Knapsack (MTC-FK)

The last filtering algorithm we present, denoted MTC-FK, is based on the fractional knapsack problem and improves on the filtering done by the interval arithmetic. It is inspired from the global knapsack constraint [9].

We consider, for some $l \in \mathcal{N}$, the pair of constraints $\sum_{i \in \mathcal{N}} X_i = 1$ and $\sum_{i \in \mathcal{N}} \mathbf{M}_{il} X_i = Y_l$. To compute an upper bound on the variable Y_l, one greedily assigns the largest possible values to the variables X_i that are multiplied by the greatest weights $\mathbf{M}_{il}$ while making sure that the constraint (10) is satisfied. On the other hand, to compute a lower bound on the variable Y_l, one needs to assign the largest values to the variables multiplied by the smallest weights.

Algorithm 1 propagates the constraints (9) to (12) as well as the knapsack constraints $(\sum_{i \in \mathcal{N}} X_i \mathbf{M}_{ij} = Y_j, \sum_{i \in \mathcal{N}} X_i = 1)$ and $(\sum_{j \in \mathcal{N}} Y_j \mathbf{M}^{-1}{}_{ij} = X_i, \sum_{j \in \mathcal{N}} Y_j = 1)$ until it reaches a precision of ϵ. Algorithm 2 computes a lower bound on Y_l (or X_l). Reversing the order of the iterations in the for loop makes Algorithm 2 compute an upper bound on Y_l (or X_l).

Lemma 2. *Let $\overline{m}_j$ and $\underline{m}_j$ be the greatest and smallest value in column j of the matrix $\mathbf{M}$. If $[\underline{m}_j, \overline{m}_j] \subseteq \mathrm{dom}\,(Y_j) \ \forall j \in \mathcal{N}$ then any distribution $x_1, \dots, x_N$ (i.e., any assignment to the variables of vector $\mathbf{X}$ that sums to one) has a support in* MTC$(\mathbf{Y}, \mathbf{X}, \mathbf{M})$.

Function MTC-FK($[X_1, \ldots, X_N], [Y_1, \ldots, Y_N], \mathbf{M}, \mathbf{M}^{-1}$)

> **Input**: A vector of variables that represents the current uncertain distribution over $\mathcal{N}$: $[X_1, \ldots, X_N]$; a vector of variables that represents the resulting uncertain distribution: $[Y_1, \ldots, Y_N]$; a transition matrix and its inverse: $\mathbf{M}$ and $\mathbf{M}^{-1}$.
>
> **Output**: The vectors of probability variables with filtered domain: $[X_1, \ldots, X_N]$, and $[Y_1, \ldots, Y_N]$.
>
> **repeat**
> > **for** $i \in \mathcal{N}$ **do** $x_i^{\text{old}} \leftarrow \overline{X}_i - \underline{X}_i$;
> > **for** $i \in \mathcal{N}$ **do** $y_i^{\text{old}} \leftarrow \overline{Y}_i - \underline{Y}_i$;
> > Enforce bounds consistency on constraints (9) to (12);
> > **foreach** $l \in \mathcal{N}$ **do**
> > > $\underline{Y}_l \leftarrow$ Fk-FilterLowerBound ($[X_1, \ldots, X_N], Y_l, [\mathbf{M}_{1l}, \ldots, \mathbf{M}_{Nl}]$);
> > > $\overline{Y}_l \leftarrow$ Fk-FilterUpperBound ($[X_1, \ldots, X_N], Y_l, [\mathbf{M}_{1l}, \ldots, \mathbf{M}_{Nl}]$);
> > > $\underline{X}_l \leftarrow$ Fk-FilterLowerBound ($[Y_1, \ldots, Y_N], X_l, [\mathbf{M}_{1l}^{-1}, \ldots, \mathbf{M}_{Nl}^{-1}]$);
> > > $\overline{X}_l \leftarrow$ Fk-FilterUpperBound ($[Y_1, \ldots, Y_N], X_l, [\mathbf{M}_{1l}^{-1}, \ldots, \mathbf{M}_{Nl}^{-1}]$);
> >
> **until** $\sqrt{\sum_{i \in \mathcal{N}} (x_i^{\text{old}} - \overline{X}_i + \underline{X}_i)^2} + \sqrt{\sum_{i \in \mathcal{N}} (y_i^{\text{old}} - \overline{Y}_i + \underline{Y}_i)^2} \leq \epsilon$;
> **return** $[X_1, \ldots, X_N], [Y_1, \ldots, Y_N]$;

Algorithm 1. The MTC-FK filtering algorithm

Proof. Let $\mathbf{M}^j$ be the jth column of $\mathbf{M}$. Since the components of $\mathbf{X}$ sums to one and that none are negative, the scalar product of $\mathbf{X}$ and $\mathbf{M}^j$ is a convex combination of the elements in $\mathbf{M}^j$. Consequently, the result lies in the convex hull of $\mathbf{M}^j$ and it cannot be greater nor smaller than any element in $\mathbf{M}^j$. □

Lemma 3. *Let $\overline{m}_j$ and $\underline{m}_j$ be the greatest and smallest value in column j of the matrix $\mathbf{M}$. If $[\underline{m}_j, \overline{m}_j] \subseteq \text{dom}(Y_j) \; \forall j \in \mathcal{N}$ then MTC-FK enforces bounds consistency on* $\text{MTC}(\mathbf{Y}, \mathbf{X}, \mathbf{M})$.

Proof. The algorithm MTC-FK enforces bounds consistency on the constraint $\sum_{i \in \mathcal{N}} X_i = 1$ so that the lower bound and upper bounds of the domains of X_i can each form an assignment of the variables X_i that sums to one. From Lemma 2, any assignment that sums to 1 can be extended to a support of $\text{MTC}(\mathbf{Y}, \mathbf{X}, \mathbf{M})$. We now need to prove that the variables Y_i are fully pruned. If a bound of $\text{dom}(Y_i)$ is modified, this bound was computed by the Algorithm 2 which constructed a valid support for the constraint. If a bound of $\text{dom}(Y_i)$ is not filtered, then either Algorithm 2 computed the same bound (in which case, it has a support) or it computed a larger one. However, the second case cannot occur since, as seen in Lemma 2, the scalar product of any distribution with a column of $\mathbf{M}$ leads to a value in $[\underline{m}_j, \overline{m}_j] \subseteq \text{dom}(Y_j)$. □

Lemma 3 is particularly useful at the beginning of the search in a problem where the domains of the variables Y_i are the intervals $[0, 1]$.

Function FK-FilterLowerBound $([U_1, \ldots, U_N], \underline{V}_l, [t_1, \ldots, t_N])$

   Input: A vector of variables that represents an uncertain distribution over $\mathcal{N}$: $[U_1, \ldots, U_N]$; a lower bound on the variable that represents the probability that the process is in state l after applying transitions: $\underline{V}_l$; a vector of the transition probabilities to state l: $[t_1, \ldots, t_N]$.

   Output: A new lower bound on state l probability: $\underline{V}_l$.

   for $i \in \mathcal{N}$ **do** $u_i \leftarrow \underline{U}_i$;

   $\lambda \leftarrow 1 - \sum_{l \in \mathcal{N}} \underline{U}_l$;

   for $k \in \mathcal{N}$ *in non-decreasing order of* t_k **do**

      $\delta \leftarrow \min\left(\lambda, \overline{U}_k - \underline{U}_k\right)$;

      $u_k \leftarrow u_k + \delta$;

      $\lambda \leftarrow \lambda - \delta$;

      if $\lambda = 0$ **then** break;

   return $\max\left(\sum_{i \in \mathcal{N}} u_i t_i, \underline{V}_l\right)$;

Algorithm 2. The FK-FilterLowerBound lower bound filtering algorithm

Theorem 2. *The consistency achieved by each algorithm satisfies MTC-IA $\prec$ MTC-FK $\prec$ MTC-LP.*

Proof. We first prove MTC-IA $\preceq$ MTC-FK $\preceq$ MTC-LP. The MTC-FK algorithm filters the same constraints as MTC-IA but the knapsack constraints consider pairs of constraints which offer a filtering that is not weaker. The algorithm MTC-LP achieves bounds consistency which is optimal. We now show two examples which prove that MTC-IA $\neq$ MTC-FK, and that MTC-FK $\neq$ MTC-LP. Let $N = 3$. Let $\mathrm{dom}(X_1) = [.3, 1]$, and $\mathrm{dom}(X_2) = \mathrm{dom}(X_3) = [0, 1]$. Let $\mathrm{dom}(Y_1) = \mathrm{dom}(Y_2) = \mathrm{dom}(Y_3) = [0, 1]$. Let the transition matrix be

$$
\mathbf{M} = \begin{bmatrix} 0 & .4 & .6 \\ .3 & .4 & .3 \\ .4 & .6 & 0 \end{bmatrix}. \tag{18}
$$

By Lemma 2, MTC-FK enforces bounds consistency whereas MTC-IA does not. In fact, MTC-FK sets $\overline{Y}_1 = 0.28$ whereas MTC-IA sets $\overline{Y}_1 = 0.49$. Suppose that $\mathrm{dom}(Y_1) = [.1, 1]$. By Theorem 1, MTC-LP enforces bounds consistency which is not the case of MTC-FK. In fact, MTC-LP sets $\overline{X}_1 = 0.75$ while MTC-FK sets $\overline{X}_1 = 0.9$. □

5 Empirical Experiments and Discussion

We generated random domains and transition matrices to compare the three filtering algorithms and see how much their filtering differ. For MTC-IA, we present the results with and without the implied constraints (11) and (12). The MTC-LP enforces bounds consistency on all instances providing the optimal bounds. Let $\underline{x}^{old}$, $\overline{x}^{old}$, $\underline{y}^{old}$ and $\overline{y}^{old}$ be the initial bounds of the filtering problem. Let $\underline{x}^*$, $\overline{x}^*$, $\underline{y}^*$ and $\overline{y}^*$ be the optimal bounds found by MTC-LP. Let $\underline{x}$, $\overline{x}$, $\underline{y}$ and $\overline{y}$ be the bounds found by a given filtering method. We define the *proportion of optimality* as the ratio of the sum of the distances traveled, in the domain space, by a filtering method to the sum of the distances traveled by MTC-LP:

$$Ind_p = \frac{\left\|\underline{x} - \underline{x}^{old}\right\| + \left\|\overline{x}^{old} - \overline{x}\right\| + \left\|\underline{y} - \underline{y}^{old}\right\| + \left\|\overline{y}^{old} - \overline{y}\right\|}{\left\|\underline{x}^* - \underline{x}^{old}\right\| + \left\|\overline{x}^{old} - \overline{x}^*\right\| + \left\|\underline{y}^* - \underline{y}^{old}\right\| + \left\|\overline{y}^{old} - \overline{y}^*\right\|}. \tag{19}$$

We generated five sets of filtering problem instances: a transition matrix $\mathbf{M}$ along with random bounds on the variables of vectors $\mathbf{X}$ and $\mathbf{Y}$. The randomly generated bounds are feasible, i.e., the constraint MTC is satisfiable. Table 1 summarizes the sets' characteristics. The first set (*random*) contains randomly generated transition matrices. The second and the third sets (*star* and *plus grids*) contain transition matrices that represents square grids, i.e., states are located on a square grid and are only connected to their neighbors. *Plus grids* are grids where a state is linked to its North, South, West and East neighbors. *Star grids* also include diagonals (NW, NE, SW, SE). Let ρ be the conditional probability that the process stays in state i when it is in state i for any $i \in \mathcal{N}$, i.e., the *probability of stationarity* $\rho = \mathbf{M}_{ii}$ ($\forall i \in \mathcal{N}$). Let $\deg(i)$ be the degree of cell i (loops included) in the adjacency matrix of the grid. The transition matrix of a grid instance is defined as:

$$\mathbf{M}_{ij} = \begin{cases} \frac{1-\rho}{\deg(i)-1} & \text{if } i \neq j; \\ \rho & \text{if } i = j. \end{cases} \tag{20}$$

We chose $\rho \in \{.2, .4, .6, .8\}$. The fourth set (*zero-one-Y*) contains randomly generated transition matrices, but the domains of the variables of vector $\mathbf{Y}$ are $[0, 1]$. That is, the information about the future (i.e., the uncertain distribution of vector $\mathbf{Y}$) comes from the present (i.e., the uncertain distribution of vector $\mathbf{X}$). This is the usual way to process Markov chains in Markov processes. The fifth set (*zero-one-X*) models the converse case where the information about the present process' state comes from the future. Our benchmark library includes non-singular matrices only. We generated 10 different instances for each pair of state space size N and probability of stationarity ρ in each set for a total of 3690 instances.

Figures 1 and 2 show scatter plots comparing the proportion of optimality achieved by different filtering algorithms. The higher the proportion of optimality is on a given axis, the better the filtering algorithm performs (a value of 1 represents a bounds consistent domain as obtained by the MTC-LP algorithm). The dotted line is a visual frontier between the two compared algorithms' performance. A dot representing an instance for which both algorithms produce the same filtering lies on this visual frontier. A dot for which the algorithm on the x-axis (y-axis) performs better than the algorithm on the y-axis (x-axis) lies on the right (left) hand-side of the frontier. The algorithm with the highest density

Table 1. The characteristics of the instance sets

Name	N	$\mathbf{M}$	$\rho = \mathbf{M}_{ii}$	Set size
Random	$\{2, \ldots, 100\}$	Random	Random	990
Star grids	$\{4, 9 \ldots, 100\}$	Star grids	$\{.2, .4, .6, .8\}$	360
Plus grids	$\{4, 9, \ldots, 100\}$	Plus grids	$\{.2, .4, .6, .8\}$	360
Zero-one-Y	$\{2, \ldots, 100\}$	Random	Random	990
Zero-one-X	$\{2, \ldots, 100\}$	Random	Random	990

of dots on its side of the frontier tends to achieve the best overall performance. Darker blue shades are used for instances with a larger state space size (N); lighter blue shades are used for instances with a smaller N.

As shown on Figure 1, MTC-IA (x-axis) outperforms MTC-IA without implied constraints (y-axis) on all instance sets. While the performance of the two algorithms is similar on some random instances (Figure 1a), the importance of implied constraints is clear as all the dots fall on the right hand side of the frontier. The scatter plots of grid instance sets (Figures 1b and 1c) favor MTC-IA. Zero-one-Y instances represent forward in time inferences using $\mathbf{M}$ (Figure 1d). Zero-one-X instances represent backward in time inferences using $\mathbf{M}^{-1}$ (Figure 1e). On these instances, we see the benefits of the interaction between a set of elementary constraints: the implied constraints enable the algorithm to further filter the domains backward in time whenever knowledge on the future is acquired by forward filtering. The difficulty of backward inferences is shown by the fact that the distribution of the results on the Zero-one-Y instances (forward inference) is closer to 1 when compared to the distribution of the results on the Zero-one-X instances (backward inference). This is partly due to the negative values in the inverse of most transition matrix $\mathbf{M}$.

As shown on Figure 2, MTC-FK (x-axis) outperforms MTC-IA (y-axis) on all instance sets. The performance of both algorithms is close on the Random set (Figure 2a), but still, MTC-FK performs better. The same tendency appears for grids (Figures 2b and 2c). MTC-FK enforced bounds consistency on all Zero-one-Y instances (a result of Lemma 3) whereas this is not the case for MTC-IA. Forward inference is easier than backward inference for both algorithms: the distribution of the optimality results is closer to 1 on the Zero-one-Y set than on the Zero-one-X set.

6   An Application to Path Planning under Uncertainty

We illustrate the MTC constraint usage on an *optimal search path* (OSP) problem [15]. A searcher moves on a graph $G_A = (\mathcal{V}(G_A), \mathcal{E}(G_A))$, within T time steps, in order to find a lost object. In absence of search, the object would simply move from vertex to vertex according to a transition matrix $\mathbf{M}$. The searcher, however, influences the chain evolution by searching the vertices and by

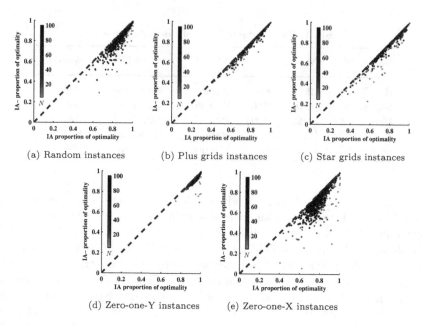

(a) Random instances (b) Plus grids instances (c) Star grids instances

(d) Zero-one-Y instances (e) Zero-one-X instances

Fig. 1. Proportion of optimality achieved by MTC-IA (IA) when compared to MTC-IA without implied constraints (IA-)

removing the object as soon as it detects it: a special "removed" state that represents the searcher's hands is added to the state space.

The probability that a vertex r contains the object at a time t is called the *probability of containment*, or $poc_t(r)$. The initial distribution (poc_1) is known a priori. The distribution $\mathbf{x}_t$ over the states is

$$\mathbf{x}_t = [poc_t(1), \ldots, poc_t(N), cos_{t-1}], \tag{21}$$

where cos_t is the searcher's *cumulative overall probability of success* linked to the removed state at time t ($cos_0 = 0$). When the searcher is located in a given vertex r at time t, i.e., $y_t = r$, s/he searches that vertex. Her/his known probability of detection (pod) is conditional to the object's presence. The local success of the searcher in r at time t is defined as:

$$pos_t(r) = \begin{cases} poc_t(r) \times pod & \text{if } r = y_t, \\ 0 & \text{if } r \neq y_t. \end{cases} \tag{22}$$

We split the chain to introduce a distribution $\widehat{\mathbf{x}}^t$ that represents the search and to apply the object's motion:

$$\widehat{\mathbf{x}}^t = [poc_t(1) - pos_t(1), \ldots, poc_t(N) - pos_t(N), cos_t], \tag{23}$$

$$\mathbf{x}^{t+1} = \widehat{\mathbf{x}}^t \begin{bmatrix} \mathbf{M} & 0 \\ 0 & 1 \end{bmatrix}. \tag{24}$$

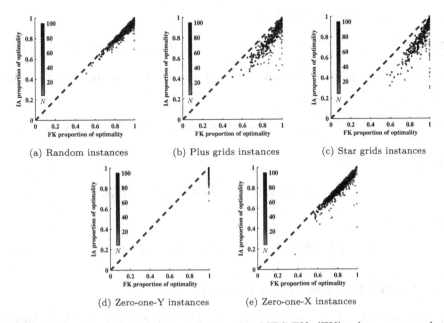

(a) Random instances   (b) Plus grids instances   (c) Star grids instances

(d) Zero-one-Y instances   (e) Zero-one-X instances

Fig. 2. Proportion of optimality achieved by MTC-FK (FK) when compared to MTC-IA (IA)

The searcher's goal is to maximize cos_T. The OSP problem is known to be $\mathcal{NP}$-hard even for stationary objects [23]. Recent examples of OSP-related researches with a single searcher may be found in the works of [2,11,21].

This leads to a CP model with four sets of interval-domain probability variables: the variables COS_t that represent the probability of finding the object up to time t, and the variables $POC_t(r)$, $POC_t^{\text{Searched}}(r)$, and $POS_t(r)$ that represent the probabilities in the split Markov chain ($\forall t \in \{1, \ldots, T\}, r \in \mathcal{V}(G_A)$). A set of finite-domain variables $PATH_t$ models the searcher's path. The domains of each path variable is a subset of vertices, i.e., $\text{dom}(PATH_t) \subseteq \mathcal{V}(G_A)$ ($\forall t \in \{1, \ldots, T\}$). $POC_1(r) = poc_1(r)$, and $PATH_0 = y_0$ are known data. The probability of finding (removing) the object up to time t is:

$$COS_t = \sum_{1 \leq t' \leq t} \max_{r \in \mathcal{V}(G_A)} POS_{t'}(r). \tag{25}$$

The objective is to maximize COS_T subject to the graph edges constraints:

$$(PATH_{t-1}, PATH_t) \in \mathcal{E}(G_A), \qquad \forall t \in \{1, \ldots, T\}; \tag{26}$$

the probability of success along the path of the searcher:

$$PATH_t = r \implies POS_t(r) = POC_t(r)pod, \qquad \forall t \in \{1, \ldots, T\}, \forall r \in \mathcal{V}(G_A); \tag{27}$$

$$PATH_t \neq r \implies POS_t(r) = 0, \qquad \forall t \in \{1, \ldots, T\}, \forall r \in \mathcal{V}(G_A); \tag{28}$$

the effect of searching on the chain:

$$POC_t^{\text{Searched}}(r) = POC_t(r) - POS_t(r), \qquad \forall t \in \{1, \ldots, T\}, \forall r \in \mathcal{V}(G_A); \quad (29)$$

and the application of the Markovian transition matrix of the object on the split chain:

$$\text{MTC}\left(\mathbf{X}^{t+1}, \widehat{\mathbf{X}}^t, \begin{bmatrix} \mathbf{M} & 0 \\ 0 & 1 \end{bmatrix}\right), \qquad \forall t \in \{1, \ldots, T-1\}, \forall r \in \mathcal{V}(G_A), \quad (30)$$

where

$$\mathbf{X}^{t+1} = [POC_{t+1}(1), \ldots, POC_{t+1}(N), COS_t], \quad (31)$$

and

$$\widehat{\mathbf{X}}^t = \left[POC_t^{\text{Searched}}(1), \ldots, POC_t^{\text{Searched}}(N), COS_t\right]. \quad (32)$$

6.1   Applied Experiment and Discussion

We implemented two Markov chain-based models for the OSP using Choco Solver 2.1.3 [10], a state-of-the-art CP solver. The first, OSP-IA-, uses a standard elementary arithmetic constraints decomposition in (30). The second, OSP-FK, uses fractional knapsack filtering in (30). We kept the model as close as possible to the one presented in [15] while adding the necessary variables to model the search with MTCs. The goal of the experiment is to show the benefits of the filtering achieved by the MTC when compared to elementary arithmetic constraint decomposition. All implementations of the application are done using the Java programming language, the Apache Commons Math [13] library, and the Java Universal Network/Graph (JUNG) 2.0.1 framework [17]. The total detection value selection heuristic developed for the OSP problem [15] is used.

The accessibility graphs (G_A) are the following: G^+, a 11×11 plus grid, G^* a 11×11 star grid, and the University Laval tunnels map G^L [15]. We allowed a total of $T = 17$ time steps. The probability of detection are $pod(r) \in \{0.3, 0.6, 0.9\}$ $(\forall r \in \mathcal{V}(G_A))$. The assumed Markovian object's motion model of the objective, i.e., its transition matrix $\mathbf{M}$, is based on equation (20). The probability of stationary of the object in $\mathbf{M}$ are $\rho \in \{0.3, 0.6, 0.9\}$. We allowed a total number of 5,000,000 backtracks and a time limit of 20 minutes. We tested with different epsilon values and retained $\epsilon = 0.3$ as it is sufficient in practice for most search operations. All tests were run on an Intel(R) Core(TM) i7-2600 CPU with 4 GB of RAM.

Table 2 compares standard filtering (OSP-IA-) to the fractional knapsack filtering (OSP-FK). We chose three performance criteria: the objective value (COS), the time consumed to the last incumbent solution, and the total number of backtracks to obtain the last incumbent solution. The objective value and the time to last incumbent tell us which model achieves the highest performance on a given problem instance independently of its filtering performance.

Table 2. Results on OSP problem instances with $\epsilon = 10^{-3}$; underlined values are better

pod(r)	ρ	Objective (COS)		Time[†] (s)		Backtracks	
		IA-	FK	IA-	FK	IA-	FK
G^L with $T = 17$							
0.3	0.3	0.519	0.519	1	1	1948	<u>1</u>
	0.6	0.578	0.578	<u>1</u>	2	1880	<u>2</u>
	0.9	0.738	0.738	<u>1</u>	1	1550	<u>11</u>
0.6	0.3	0.735	0.735	1	1	1980	<u>10</u>
	0.6	0.758	0.758	1	1	1851	<u>9</u>
	0.9	0.845	0.845	1	1	1475	<u>35</u>
0.9	0.3	0.810	0.810	1	1	1911	<u>23</u>
	0.6	0.835	0.835	1	1	1832	<u>35</u>
	0.9	0.890	0.890	1	1	1396	<u>61</u>
G^+ with $T = 17$							
0.3	0.3	0.106	0.106	<u>4</u>	7	7205	<u>1</u>
	0.6	0.165	0.165	<u>3</u>	5	5343	<u>0</u>
	0.9	0.442	0.442	<u>3</u>	5	2236	<u>20</u>
0.6	0.3	0.189	0.189	<u>4</u>	7	7044	<u>2</u>
	0.6	0.298	0.298	<u>3</u>	5	5241	<u>0</u>
	0.9	0.551	0.551	<u>3</u>	5	2325	<u>43</u>
0.9	0.3	0.259	0.259	<u>4</u>	7	6946	<u>2</u>
	0.6	0.329	0.329	<u>3</u>	5	5130	<u>0</u>
	0.9	0.623	0.623	<u>3</u>	5	2152	<u>73</u>
G^* with $T = 17$							
0.3	0.3	0.131	0.131	<u>5</u>	9	7757	<u>1</u>
	0.6	0.219	0.219	<u>4</u>	7	6444	<u>0</u>
	0.9	0.584	0.584	<u>3</u>	6	3324	<u>35</u>
0.6	0.3	0.226	0.226	<u>5</u>	10	7514	<u>6</u>
	0.6	0.314	<u>0.348</u>	<u>4</u>	7	6334	<u>0</u>
	0.9	0.686	0.686	<u>3</u>	6	3344	<u>58</u>
0.9	0.3	<u>0.304</u>	0.301	<u>5</u>	9	7209	<u>9</u>
	0.6	0.381	0.381	<u>4</u>	7	6065	<u>2</u>
	0.9	0.734	0.734	<u>3</u>	6	2945	<u>54</u>

† The time to last incumbent solution.

The total number of backtracks to last incumbent and the time consumed tell us whether the model filtering performance is good or not. For each criterion, underline results belongs to the best performing techniques. We do not underline ties. Overall, OSP-FK outperformed OSP-IA- by achieving a higher or equal objective values while maintaining a lower number of backtracks on most instances. Even though we demonstrated that the theoretical consistency achieved by MTC-FK is better than the one achieved by MTC-IA (with and without implied constraints), some cases may occurs where MTC-IA (with or without implied constraints) outperforms MTC-FK on applications. This may be due to the interaction between a heuristic (e.g., the total detection heuristic) and the solving process. Efficient heuristics lead to smaller search trees removing some filtering needs. As demonstrated by our positive results, thorough filtering of the variables is beneficial in most cases. The MTC-FK filtering algorithm achieves high quality solutions soon in the solving process while greatly reducing the total number of backtracks when compared to a model with elementary constraints.

7   Conclusion

We introduced MTC, a global constraint that models Markov chains. We discussed cases where elementary arithmetic constraints enforce bounds consistency. We proved that elementary arithmetic constraints, even with implied constraints, are insufficient to enforce bounds consistency in all cases and discussed a linear programming algorithm that always performs bounds consistency. However, a constraint solver needs to run the filtering algorithms on an exponential number of nodes, and linear optimization turns out to be an expensive filtering solution. Thus, we provided, as a trade-off, a filtering algorithm based on the fractional knapsack problem. The method is proved to achieve bounds consistency when the transition matrix is monomial and in the case of forward reasoning. Finally, we successfully applied the MTC global constraint on the OSP problem, a path planning problem under uncertainty involving Markov transitions.

References

1. Barbieri, G., Pachet, F., Roy, P., Degli Esposti, M.: Markov constraints for generating lyrics with style. In: Proceedings of 20th Biennial European Conference on Artificial Intelligence. IOS Press (2012)
2. Berger, J., Lo, N., Noel, M.: Exact solution for search-and-rescue path planning. International Journal of Computer and Communication Engineering 2(3) (2013)
3. Bessière, C., Hebrard, E., Hnich, B., Kiziltan, Z., Walsh, T.: Filtering algorithms for the NValue constraint. In: Barták, R., Milano, M. (eds.) CPAIOR 2005. LNCS, vol. 3524, pp. 79–93. Springer, Heidelberg (2005)
4. Bessière, C., Hebrard, E., Hnich, B., Kiziltan, Z., Walsh, T.: Filtering algorithms for the NValue constraint. Constraints 11(4), 271–293 (2006)
5. Blane, H., Den Hertog, D.: On markov chains with uncertain data. Available at SSRN 1138144 (2008)
6. De Cooman, G., Hermans, F., Quaeghebeur, E.: Imprecise markov chains and their limit behavior. Probability in The Engineering and Informational Sciences 23(4), 597 (2009)
7. Garfinkel, R.S., Nemhauser, G.L.: Integer programming, vol. 4. Wiley, New York (1972)
8. Hamilton, J.: A new approach to the economic analysis of nonstationary time series and the business cycle. In: Econometrica: Journal of the Econometric Society, pp. 357–384 (1989)
9. Katriel, I., Sellmann, M., Upfal, E., Van Hentenryck, P.: Propagating knapsack constraints in sublinear time. In: Proceedings of the National Conference on Artificial Intelligence (AAAI-2007), pp. 231–236 (2007)
10. Laburthe, F., Jussien, N.: Choco Solver Documentation (2012), http://www.emn.fr/z-info/choco-solver/
11. Lau, H., Huang, S., Dissanayake, G.: Discounted mean bound for the optimal searcher path problem with non-uniform travel times. European Journal of Operational Research 190(2), 383–397 (2008)
12. Luis, M., Campos, D., Huete, J., Moral, S.: Probability intervals: a tool for uncertain reasoning. International Journal of Uncertainty, Fuzziness and Knowledge-Based Systems 2(02), 167–196 (1994)

13. Apache Commons: Commons math: The apache commons mathematics library (2010) (accessed: October 2013)
14. Moore, R.E.: Interval analysis, vol. 2. Prentice-Hall Englewood Cliffs (1966)
15. Morin, M., Papillon, A.-P., Abi-Zeid, I., Laviolette, F., Quimper, C.-G.: Constraint programming for path planning with uncertainty. In: Milano, M. (ed.) CP 2012. LNCS, vol. 7514, pp. 988–1003. Springer, Heidelberg (2012)
16. Norvig, P., Russell, S.J.: Artificial Intelligence: A Modern Approach, 3rd edn. Prentice-Hall (2013)
17. O'Madadhain, J., Fisher, D., Nelson, T., White, S., Boey, Y.: Jung: Java universal network/graph framework (2010), http://jung.sourceforge.net
18. Pachet, F., Roy, P.: Markov constraints: steerable generation of markov sequences. Constraints 16(2), 148–172 (2011)
19. Pachet, F., Roy, P., Barbieri, G.: Finite-length markov processes with constraints. In: Proceedings of the Twenty-Second International Joint Conference on Artificial Intelligence-Volume Volume One. AAAI Press, pp. 635–642 (2011)
20. Page, L., Brin, S., Motwani, R., Winograd, T.: The pagerank citation ranking: bringing order to the web. Technical report, Stanford InfoLab (1999)
21. Sato, H., Royset, J.: Path optimization for the resource-constrained searcher. Naval Research Logitics 57(5), 422–440 (2010)
22. Stone, L.: Theory of Optimal Search. Academic Press, New York (2004)
23. Trummel, K., Weisinger, J.: The complexity of the optimal searcher path problem. Operations Research 34(2), 324–327 (1986)
24. Weichselberger, K.: The theory of interval-probability as a unifying concept for uncertainty. International Journal of Approximate Reasoning 24(2), 149–170 (2000)

New Lower Bounds on the Number of Vehicles for the Vehicle Routing Problem with Time Windows

Sohaib Afifi, Rym Nesrine Guibadj, and Aziz Moukrim

Université de Technologie Compiègne
Laboratoire Heudiasyc, UMR 7253 CNRS, 60205 Compiègne, France
{sohaib.afifi,rym-nesrine.guibadj,aziz.moukrim}@hds.utc.fr

Abstract. The Vehicle Routing Problem with Time Windows (VRPTW) consists in determining the routing plan of vehicles with identical capacity in order to supply the demands of a set of customers with predefined time windows. This complex multi-constrained problem has been widely studied due to its industrial, economic and environmental implications. In this work, we are interested in defining the number of vehicles needed to visit all the customers. This objective is very important to evaluate the fixed costs for operating the fleet. In this paper, we provide an analysis of several lower bounds based on incompatibility between customers and on vehicle capacity constraints. We also develop an adaptation of Energetic Reasoning algorithm for VRPTW with a limited fleet. The proposed approach focuses on some time-intervals and exploits time constraints, incompatibility graph and bin packing models in order to obtain new valid lower bounds for the fleet size. Experiments conducted on the standard benchmarks show that our algorithms outperform the classical lower bound techniques and give the minimum number of vehicles for 339 out of 468 instances.

Keywords: vehicle routing, time windows, lower bounds, energetic reasoning.

1 Introduction

In today's business world, transportation costs become a major share of the total logistic expenses of companies. That is why many companies try to improve their transportation by using rational manners and effective tools. The objective of these problems is to make a vehicle scheduling strategy in order to minimize the number of routes and the corresponding total travel distance or cost. In the literature such problems are referred to as *routing problems.*

The vehicle routing problem with time windows (VRPTW) [10] is among the most studied variants of routing problems due its wide range of applications. Common examples are newspaper delivery, beverage and food delivery, commercial and industrial waste collection [13]. In VRPTW, a set of customers must

H. Simonis (Ed.): CPAIOR 2014, LNCS 8451, pp. 422–437, 2014.

be served by a fleet of vehicles located in a single depot. A quantity of goods should be delivered to each customer whose service takes an amount of time. Each customer is associated with a time window that represents the interval of time when the customer is available to receive the service. This means that if the vehicle arrives too soon, it should wait until the opening of the time window to serve the customer while too late arrival is not allowed. Since deliveries cannot be split, a customer must be served by a single vehicle. All vehicles are identical and have a maximum capacity Q. The aim is to plan the minimal number of routes starting and ending in a unique depot in order to serve all the customers while respecting all the time windows and capacity constraints.

VRPTW was first introduced by Solomon [25]. Both exact and heuristic algorithms have been proposed to solve VRPTW. Most of the exact methods focus on the variant of the problem where the number of available vehicles is not fixed. A review on the exact methods up to 2002 is reported in [7]. Kallehauge in [17] gave a detailed analysis of existing formulations. More recently, Baldacci et al. [3] reviewed mathematical formulations, relaxations and recent exact methods. They reported the computational comparison between the methods proposed in [15], [8] and [2] that are considered as the most effective exact methods in the literature. These approaches have significantly improved the quality of the lower bounds for instances with up to 100 customers. The key factor of their success is the effective combination between the set partitioning formulation and the column generation based algorithms.

Since, VRPTW is an NP-Hard problem [21], the computational times for exact methods can be very high, even for instances with a moderate size. This has been the motivation for some researches to focus on approximate methods. It is worth pointing out that the literature concerning VRPTW is split according to the objective considered. While exact methods usually minimize the total traveled distance, most heuristics consider a hierarchical objective which first minimizes the number of vehicles used and then the total distance. Thus, a solution that employs fewer vehicles is always better than a one using more, even if its total traveled distance is worse. A good survey of heuristic methods is reported in the papers of Bräysy and Gendreau [5] [6]. Among the best performing heuristics are the hybrid genetic algorithm of [16], the column generation heuristic of [1] and the memetic algorithm of [20]. A new optimization framework was later developed by Ursani et al. [27] for the distance minimization objective only. This framework is an iterative procedure between optimization and deterioration phases and uses a genetic algorithm as an optimization methodology. In the recent paper of Vidal et al. [28], a hybrid genetic solver is developed to deal with a large class of time-constrained vehicle routing problem. A third stream of research focuses on solving VRPTW as a multi-objective problem in which both vehicles and cost are considered depending on the needs of the user [26] [24].

The goal of this paper is to use scheduling methods via Energetic Reasoning in order to develop new lower bounding procedures for VRPTW. This is mainly based on constraint propagation concept. The objective is to reduce the computational effort by removing some values from the variables of the problem because a given subset of the constraints cannot be satisfied. The remained of the paper is organized as follows. Section 2 briefly describes the problem. In Sections 3 and 4, the detailed description of the proposed lower bound methods is given and in Section 5 the results of a computational study are reported. Finally, Section 6 provides some concluding remarks.

2   Problem Formulation

In the following, we present a mixed integer formulation for VRPTW. The problem is modeled using an oriented graph $G = (V^+, E)$, where $V^+ = \{0, 1, 2, ..., n\}$ is the vertex set representing the set of customers $V = \{1, 2, ..., n\}$ and the depot 0. $E = \{(i, j) : i \neq j, i, j \in V^+\}$ is the edge set. The capacities of all vehicles are equal and are denoted by Q. A demand q_i, a service time s_i and a time window $[e_i, l_i]$ are associated to each vertex $i \in V$. Vehicle v cannot arrive later than l_i and if it arrives earlier than e_i, it must wait before the service can start. Each edge $(i, j) \in E$ is associated with a travel cost $\delta_{i,j}$ which satisfies the triangle inequality. Each vehicle must start and finish its tour at the depot. Each customer must be served within a predefined time window and assigned to exactly one vehicle. The total size of deliveries for customers assigned to the same vehicle must not exceed the vehicle capacity Q and the travel cost/time $C(R)$ of each tour R must not exceed l_0 which is the latest possible arrival time to the depot.

The model involves three types of variables: the binary routing variables $x_{ij} \in \{0, 1\}$ $(i, j \in V^+)$, the scheduling variables $w_i \geq 0$ $(i \in V)$ and the vehicle load variables y_i $(i \in V)$. The routing variables x_{ij} is one if a vehicle traverses the arc $(i, j) \in E$. The scheduling variable w_i denotes the time the vehicle arrives at customer $i \in V$. y_i denotes the vehicle load at departure from customer i. The formulation is as follows:

$$\min \sum_{i \in V} x_{0i} \tag{1}$$

subject to:

$$\sum_{j \in V^+} x_{ij} = 1 \qquad\qquad \forall i \in V \tag{2}$$

$$\sum_{j \in V^+} x_{ij} - \sum_{j \in V^+} x_{ji} = 0 \qquad\qquad \forall i \in V^+ \tag{3}$$

$$w_j \geq w_i + x_{ij}(\max(\delta_{i,j} + s_i, e_j - l_i)) - (1 - x_{ij})(l_i - e_j) \qquad \forall i, j \in V \tag{4}$$

$$e_i \leq w_i \leq l_i \qquad\qquad \forall i \in V \tag{5}$$

$$y_j \geq y_i + q_j - (1 - x_{ij})Q \qquad\qquad \forall i, j \in V \tag{6}$$

$$q_i \leq y_i \leq Q \qquad\qquad \forall i \in V \tag{7}$$

$$x_{ij} \in \{0, 1\} \qquad\qquad \forall i, j \in V^+ \tag{8}$$

The objective function (1) is to minimize the total number of vehicles used to serve all the customers. Constraints (2) and (3) define the routing network and the constraints (4) and (5) guarantee the connectivity of each tour and ensure that the time windows are respected. We assume that the time windows are adjusted such that $e_i = \max(e_i, \delta_{0,i})$ and $l_i = \min(l_i, l_0 - (\delta_{i,0} + s_i))$ $\forall i \in V$. Constraints (6) and (7) ensure that the vehicle's capacity is not exceeded. Also, constraints (6) eliminate subtours in a manner similar to (4). Finally, (8) are integral constraints.

3   Classical Lower Bounding Techniques

There were only few attempts to propose lower bounds for VRPTW when the objective is to minimize the number of vehicles. To the best of our knowledge, the most competitive results are currently offered by Kontoravdis and Bard [19]. In this section, we briefly review the main features of their lower bounding heuristics.

3.1   A Lower Bound Based on Incompatibilities between Customers

The first lower bound is deduced from the incompatibility constraints. Let i and j be two customers. If there is no feasible route containing i and j then they define an incompatible pair denoted by $i||j$. Such a situation occurs if one of the following conditions is verified:

1. Customers i and j cannot be in the same route due to their time window constraints:
   $(e_i + s_i + \delta_{i,j} > l_j) \wedge (e_j + s_j + \delta_{j,i} > l_i) \Rightarrow i||j.$
2. The travel cost of any tour with i and j exceeds the cost limit l_0:
   $(C(R_1) > l_0) \wedge (C(R_2) > l_0)$ where $R_1 = (0, i, j, 0) \wedge R_2 = (0, j, i, 0) \Rightarrow i||j.$
3. The sum of the demands is greater than the vehicle capacity:
   $q_i + q_j > Q \Rightarrow i||j.$

Using these conditions, we build the graph of incompatibilities between customers defined as: $G_{inc}^V = (V, E_V)$ where $E_V = \{(i, j) \in V \times V : i||j\}$. Based on this graph the minimum number of routes to be used, denoted LB_{Clique}, is equal to the size of the maximum clique extracted from G_{inc}^V.

3.2   A Lower Bound Based on Vehicle Capacity Constraints

The second bound is based on a relaxation of time window constraints. When considering only the capacity constraints, VRPTW can be reduced to a Bin Packing Problem (BPP). Each vehicle is considered as a bin with fixed size Q and each customer demand as an item with size q_i that should be put in a bin. Any lower bound $LB_{Capacity}$ on the number of bins required to pack all the items is considered as a valid lower bound for VRPTW.

3.3   A Lower Bound Based on the Amount of Needed Travel Time

This lower bound consists of calculating the minimum number of bins LB_{BP} of capacity l_0 to pack $n + m$ items. The size θ_i of an item $i, 1 \leq i \leq n$, represents the necessary amount of time that a vehicle needs to serve customer i and to travel to its closest neighbor. This time is defined by:

$$\theta_i \leftarrow \min_{j \in V^+} \{\max(\delta_{i,j} + s_i, e_j - l_i)\} \tag{9}$$

The sizes of the other m items correspond to the m least travel times from the depot to the first served customers where $m = \max(LB_{Clique}, LB_{Capacity})$.

4   New Lower Bounds Inspired from Energetic Reasoning

In this section, we first present a brief overview of Energetic Reasoning, then we discuss its adaptation to VRPTW.

4.1   Energetic Reasoning

Energetic Reasoning (ER) is one of the most powerful propagation algorithms. It has been originally developed by Erschler et al. [9] for Cumulative Scheduling Problems (CuSP). The idea is to propose a smart way to simultaneously consider time and resource constraints in a unique reasoning. In this context, the energy is generally defined by multiplying the time duration by the resource quantity of a given time interval. Considering the quantities of energy supplied by the resources and consumed by the tasks within given intervals, the energetic approach aims to develop satisfiability tests to ensure that a given schedule is feasible. Since its inception, Energetic Reasoning has gained popularity and has been used for solving more complex scheduling problems [23].

In order to keep the same notation used for vehicle routing problem, we describe the CuSP as follows. We consider a set V of n activities to be scheduled on a resource of quantity m. Each activity i has a release time e_i, a latest start time l_i and a processing time s_i. Moreover, the activity i requires a constant amount b_i of resource throughout its processing. We will deal here only with the case where $b_i = 1, \forall i \in V$. This is equivalent to the problem of scheduling n activities on m identical parallel machines. For ease of presentation, we denote this problems as PMSP.

Given a time interval $[t_1, t_2]$, with $t_1 < t_2$, the part of an activity i that must be processed between t_1 and t_2 is called *work* of i in the time interval $[t_1, t_2]$. To compute this *work*, the activities are either *left-shifted* or *right-shifted* on their time window, which means that, they can start either at their release date e_i, or at their latest start time l_i. Thus, the work of an activity i over $[t_1, t_2]$ is equal to the minimum between its *left work* and its *right work*. For convenience, the left work, the right work and the work of an activity i over $[t_1, t_2]$ are denoted respectively $W_{left}(i,t_1,t_2)$, $W_{right}(i,t_1,t_2)$ and $W(i,t_1,t_2)$. They are formally defined as follows:

$$W_{left}(i,t_1,t_2) = \min\{ t_2 - t_1, s_i, \max(0, e_i + s_i - t_1)\} \qquad (10)$$

$$W_{right}(i,t_1,t_2) = \min\{ t_2 - t_1, s_i, \max(0, t_2 - l_i)\} \qquad (11)$$

$$W(i,t_1,t_2) = \min(W_{left}(i,t_1,t_2), W_{right}(i,t_1,t_2)) \qquad (12)$$

Finally, we define the total work over $[t1,t2]$ as the sum of the works of all the activities $W(t_1,t_2) = \sum_{i=1}^{i=n} W(i,t_1,t_2)$ and the available energy in the considered interval as $E(t_1,t_2) = m*(t_2 - t_1)$. If the total work is greater than the available energy then no feasible solution exists.

Proposition 1. *satisfiability test*
if $\exists[t_1,t_2]$, $W(t_1,t_2) > E(t_1,t_2)$ then the instance is infeasible.

Note that one crucial point to apply efficiently Energetic Reasoning is to determine the relevant time-intervals on which it may be useful to check feasibility conditions. Baptiste et al. [4] have proved that the only relevant time intervals $[t_1,t_2]$ that need to be considered are those where $t_1 \in T_1$ and $t_2 \in T_2$ such as $t_1 < t_2$, $T_1 = \{e_i, i \in V\} \cup \{l_i, i \in V\} \cup \{e_i + s_i, i \in V\}$ and $T_2 = \{l_i + s_i, i \in V\} \cup \{e_i + s_i, i \in V\} \cup \{l_i, i \in V\}$. Therefore, the satisfiability test algorithm runs in $O(n^3)$. The detailed steps are summarized in Algorithm 1.

Algorithm 1. satisfiability test of Energetic Reasoning

Data: I : PMSP instance;
1 **begin**
2   initialization;
3    $T_1 = \{e_i, i \in V\} \cup \{l_i, i \in V\} \cup \{e_i + s_i, i \in V\}$;
4    $T_2 = \{l_i + s_i, i \in V\} \cup \{e_i + s_i, i \in V\} \cup \{l_i, i \in V\}$;
5   **foreach** $t_1 \in T_1$ **do**
6   **foreach** $t_2 \in T_2$ *such as* $t_1 < t_2$ **do**
7          $W \leftarrow 0$;
8   **foreach** $i \in V$ **do**
9             $W \leftarrow W + W(i,t_1,t_2)$;
10   **if** $W > m*(t_2 - t_1)$ **then**
11   Infeasible instance ;

4.2 From VRPTW to PMSP

Our approach is to relax a VRPTW instance, where a limited number of vehicles is given, in order to obtain a PMSP instance. Once the transformation is performed, we apply the same satisfiability test on the relaxed m-VRPTW instance, using Algorithm 1. Starting from a trivial value $m = \max(LB_{Clique}, LB_{Capacity})$, feasibility tests are carried out to detect an infeasibility (if in at least one of the time intervals, the minimum number of vehicles found exceeds m). If an infeasibility is detected, then $m + 1$ is a valid lower bound. The process is iterated until no infeasibility is detected.

A trivial relaxation of an m-VRPTW instance can be done by ignoring travel times, customer demands and vehicle capacities. We obtain a PMSP where the vehicles are considered as m identical parallel machines, the number of activities is equal to the number of customers n and each activity i has to be processed for s_i units of time by only one machine. The processing of activity i cannot be started before its release date e_i or after its lastest start time l_i.

In vehicle routing problems, travel times are not negligible compared to the service times. Ignoring the travel time would undervalue the energy consumed. Therefore, few adjustments could be performed and Energetic Reasoning becomes inefficient. Better results are obtained by considering the time that a vehicle needs to travel in order to visit each customer.

First, the travel time $\delta_{i,j}$ between the customers i and j is updated to eliminate the waiting time at the customer j.

$$\delta_{i,j} \leftarrow \max(\delta_{i,j}, e_j - (l_i + s_i)) \quad \forall i \in V \; \forall j \in V^+ \tag{13}$$

Then, the number of potential successors of customer i is reduced. This is performed by eliminating the transition $\delta_{i,j}$ if j cannot be served after i due to its time window:

$$if(e_i + s_i + \delta_{i,j} > l_j) \quad then \quad \delta_{i,j} \leftarrow \infty \quad \forall i \in V^+ \; \forall j \in V^+ \setminus \{i\} \tag{14}$$

Before giving the detail of our travel evaluation procedure, we note by I' the instance derived from the m-VRPTW instance I. We associate I' with a graph $G' = (V', E')$ after performing the following transformations:

1. We introduce m artificial departure vertices V_d and m artificial arrival vertices V_a. Then, we define the set $V' = V \cup V_d \cup V_a$ with $V = \{1, ..., n\}$, $V_d = \{n + 1, ..., n + m\}$ and $V_a = \{n + m + 1, ..., n + 2 \times m\}$.
2. The set of arcs is defined by $E' = E \cup \{(i, j) : i \neq j, i \in V \cup V_d, j \in V \cup V_a\}$.
3. The distance matrix $\Delta = (\delta_{i,j})$ is extended to $\Delta' = (\delta'_{i,j})$ which is associated to E' such as:

$$\delta'_{i,j} = \begin{cases} \delta_{i,j} & (i, j \in V), \\ \delta_{0,j} & (i \in V_d, j \in V), \\ \delta_{i,0} & (i \in V, j \in V_a), \\ \infty & (i \in V', j \in V_d) \; or \; (i \in V_a, j \in V') \end{cases} \tag{15}$$

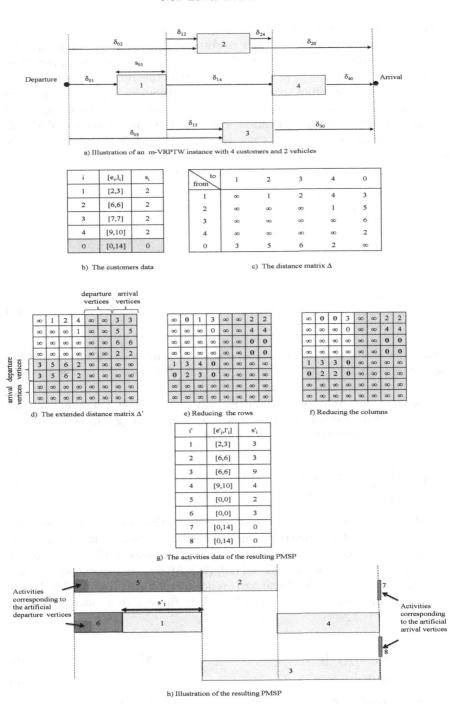

Fig. 1. An example of m-VRPTW instance relaxed to PMSP instance

The set of vertices V' in G' denotes the $n + 2 \times m$ activities assigned to I'. The artificial departure activities corresponding to V_d have a time window equal to $[0,0]$ and durations equal to the m smallest travel time from the depot to customers $\{\delta_{0,1}^{min}, ..., \delta_{0,m}^{min}\}$. This supposes that the vehicles must leave the depot immediately in order to visit the m first customers. The artificial arrival activities corresponding to V_a have the largest possible time window $[0, l_0]$ and no processing time. For the remaining activities, the range of the possible start dates is equal to the customer's time window $[e_i, l_i]$, while the processing time is equal to the sum of the service time s_i with the minimal travel time that the vehicle will necessary perform to reach the next customer.

$$[e_i', l_i'] = \begin{cases} [0,0] & i \in V_d, \\ [0, l_0] & i \in V_a, \\ [e_i, l_i] & i \in V \end{cases} \tag{16}$$

$$s_i' = \begin{cases} \delta_{0,i-n}^{min} & i \in V_d, \\ 0 & i \in V_a, \\ s_i + min_{j \in V'}\{\delta_{i,j}'\} & i \in V \end{cases} \tag{17}$$

Each row i , $i \in V'$ of the extended matrix Δ' is updated by subtracting the smallest element from the remaining ones (18). This means that the minimal travel time after serving customer i is subtracted from the total distance of any solution since every solution must include only one customer from this row. This process is called *reducing the rows*. It was introduced by [22] in order to solve the well known Traveling Salesman Problem.

$$\delta_{i,j}' = \begin{cases} \delta_{i,j}' - min_{j \in V'}\{\delta_{i,j}'\} & \forall i \in V \, j \in V', \\ max(0, \delta_{i,j}' - \delta_{0,i-n}^{min}) & \forall i \in V_d \, j \in V' \end{cases} \tag{18}$$

Next, we apply the same argument to the resulting matrix, by considering the minimal travel time to arrive from any customer j to customer i (19). This time is added at the beginning of activity i. For this reason, the bounds of the corresponding time window are shifted (20) (21). After these reducing operations, the matrix Δ' contains at least one zero in each row and each column. The Figure 1 illustrates the relaxation of an m-VRPTW instance with 4 customers and 2 vehicles.

$$s_i' \leftarrow s_i' + min_{j \in V'}\{\delta_{j,i}'\} \qquad \forall i \in V' \tag{19}$$
$$e_i' \leftarrow max(0, e_i' - min_{j \in V'}\{\delta_{j,i}'\}) \qquad \forall i \in V' \tag{20}$$
$$l_i' \leftarrow max(0, l_i' - min_{j \in V'}\{\delta_{j,i}'\}) \qquad \forall i \in V' \tag{21}$$

According to the evaluation procedure of travel times, we distinguish two possible lower bounds $LB_{ER_{eval1}}$ and $LB_{ER_{eval2}}$. The former is obtained if the travel times to the successors are considered before the remaining travel times from the predecessors whereas the latter is obtained by reversing the order of the considered travels. LB_{ER} denotes the maximum between $LB_{ER_{eval1}}$ and $LB_{ER_{eval2}}$.

4.3 Bin-Packing Lower Bounds and Energetic Reasoning

We extend Energetic Reasoning, using the Bin-Packing Problem with Conflicts (BPPC), to get tighter lower bounds for VRPTW. In each time-interval $[t_1, t_2]$, we compute the mandatory parts of activities and then we deduce an associated bin-packing instance. The decision version of BPPC that we use can be formulated as follows: given a set of items with different weights and a graph where the vertices represent the items and the edges represent the conflicts between the pairs of items; is there a packing of these items in less than m bins with a capacity T ?

We now state the link between a necessary condition for the existence of m-VRPTW solution and the existence of BPPC solution. Let $I'(V', G_{inc}, m)$ denote a relaxed instance of m-VRPTW where V' is the set of activities, m the number of available vehicles and G_{inc} the graph of incompatibilities between activities. Let $[t_1, t_2]$ be a time-interval, we assume that the corresponding mandatory parts of activities have been computed: $W(i, t_1, t_2), \forall i \in V'$. Then, $BPPC(I', G_{inc}, t_1, t_2)$ denotes the packing instance which is associated to the scheduling instance I' in the time-interval $[t_1, t_2]$. $BPPC(I', G_{inc}, t_1, t_2)$ is made of m bins and n' items of size $W_i = W(i, t_1, t_2), \forall i \in \{1, ..., n\}$. The size of the bin is equal to the length of the time-interval $T = t_2 - t_1$. Then, deciding whether all mandatory parts of the activities can be scheduled within $[t_1, t_2]$ in I' is equivalent to determine for $BPPC(I', G_{inc}, t_1, t_2)$ if all items can be packed into the available bins.

Property 1. If there exists a time-interval $[t_1, t_2]$, such that $BPPC(I', G_{inc}, t_1, t_2)$ has no solution, then there is no solution to the initial problem $I'(V, G_{inc}, m)$.

Since this lower bound is based on NP-hard relaxation of m-VRPTW, it is naturally much time consuming than LB_{ER}. Therefore, we did not lunch it on all the previously defined time intervals. An interval $[t_1, t_2]$ is selected if its conflict sub graph density is greater than or equal 80% or if the ratio of activities works to the available energy is close to 1 i.e. $W(t_1, t_2)/[m * (t_2 - t_1)] > 0.9$.

As stated in Section 4.2, Energetic Reasoning uses two procedures to determine the processing time of activities. The new obtained lower bound LB_{ERBPPC} represents the maximum between LB_{ERBPPC_eval1} and LB_{ERBPPC_eval2}. In the same way and by ignoring the conflict constraints, we can obtain a quicker lower bound LB_{ERBPP}.

The example in Figure 2 illustrates the contribution of bin packing lower bounds in the improvement of Energetic Reasoning results. We consider a VRPTW instance with 8 customers defined by their time windows and service times. We suppose that the vehicle capacity is large enough to satisfy all customers demands and

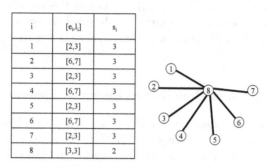

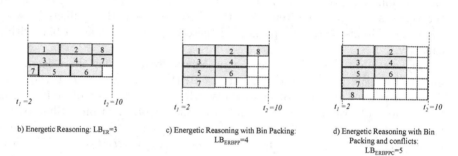

a) m-VRPTW instance with its associated incompatibility graph

b) Energetic Reasoning: LB_{ER}=3

c) Energetic Reasoning with Bin Packing: LB_{ERBPP}=4

d) Energetic Reasoning with Bin Packing and conflicts: LB_{ERBPPC}=5

Fig. 2. Illustration of Energetic Reasoning lower bounds

that customer 8 cannot be served with any other customer. The results obtained by $LB_{Capacity}$ and LB_{Clique} are equal to 1 and 2 respectively. When analyzing the interval $[t_1, t_2]$, the Energetic Reasoning LB_{ER} gives 3. This result is improved by applying Bin Packing lower bound and taking into account the conflicts between customers ($LB_{ERBPP} = 4$ and $LB_{ERBPPC} = 5$).

5   Numerical Results

We tested our algorithms on the well known instances of Solomon [25], Gehring and Homberger [11]. The benchmark comprises 6 sets ($R1$, $C1$, $RC1$, $R2$, $C2$, $RC2$). Each data set contains $25, 50, 100, 200, 400, 600, 800$ and 1000 customers who have specific euclidean coordinates. Customers' locations are determined using a random uniform distribution for the problem sets $R1$ and $R2$, but are restricted to be within clusters for the sets $C1$ and $C2$. Sets $RC1$ and $RC2$ have a combination of clustered and randomly placed customers. Sets $R1$, $C1$ and $RC1$ have a short scheduling horizon with tight time windows, while $R2$, $C2$ and $RC2$ are based on wide time windows. Our algorithms are coded in C++ and all experiments were conducted on an Intel(R) Core(TM) 2 Duo 2.93GHz.

Table 1. Average lower bound results and CPU times for Solomon instances

Data Set	n	Classical						New						BestLB	BestUB	AvgGAP
		Clique		Capacity		BP		ER		ERBPP		ERBPPC				
		LB	CPU	LB	CPU	LB	CPU	LB	CPU	LB	CPU	LB	CPU			
C1	25	1.89	0	3	0	2	0	3	0	3	0	3	0	3	3	0
C2	25	1	0	1	0	1	0	1.13	0	1.13	0	1.13	0	1.13	1.13	0
R1	25	3.58	0	2	0	3.25	0	3.92	0	3.92	0.01	4	0.26	4	4.75	0.75
R2	25	1	0	1	0	1	0	1.09	0	1.09	0	1.09	0.04	1.09	1.27	0.18
RC1	25	2.75	0	3	0	2.25	0	3.13	0	3.13	0	3.13	0.03	3.13	3.25	0.13
RC2	25	1	0	1	0	1	0	1	0	1	0.01	1	0.08	1	1.5	0.5
C1	50	3	0	5	0	4	0	5	0	5	0	5	0	5	5	0
C2	50	1	0	2	0	2	0	2	0	2	0	2	0	2	2	0
R1	50	5.25	0	4	0	5.17	0	6.33	0.02	6.33	0.09	6.42	76.41	6.42	7.42	1
R2	50	1	0	1	0	1.18	0	1.27	0.01	1.27	0.06	1.27	0.23	1.27	2	0.73
RC1	50	4.25	0	5	0	4.13	0	5.25	0.01	5.25	0.05	5.38	89.88	5.38	6.5	1.13
RC2	50	1.13	0	1	0	1	0	1.13	0.01	1.13	0.05	1.13	0.28	1.13	2	0.88
C1	100	4.89	0	10	0	8	0	10	0	10	0	10	0	10	10	0
C2	100	1.38	0	3	0	3	0	3	0	3	0	3	0	3	3	0
R1	100	7.92	0	8	0	8.33	0	10.42	0.11	10.42	0.45	10.42	364	10.42	11.92	1.5
R2	100	1.18	0	2	0	2	0	2.09	0.07	2.09	0.31	2.09	27.2	2.09	2.73	0.64
RC1	100	6.38	0	9	0	7.63	0	9.63	0.1	9.63	0.45	9.63	99.52	9.63	11.5	1.88
RC2	100	1.13	0	2	0	2	0	2.13	0.11	2.13	0.44	2.13	33.16	2.13	3.25	1.13

Table 2. Average lower bound results and CPU times for Gehring and Homberger instances

Data Set	n	Classical						New						BestLB	BestUB	AvgGAP
		Clique		Capacity		BP		ER		ERBPP		ERBPPC				
		LB	CPU	LB	CPU	LB	CPU	LB	CPU	LB	CPU	LB	CPU			
C1	200	8.8	0	18	0	15	0	18.3	0.41	18.3	1.82	18.3	429.41	18.3	18.9	0.6
C2	200	2.1	0	6	0	6	0	6	0	6	0	6	0	6	6	0
R1	200	10.4	0	18	0	7.3	0	18.2	0	18.2	0	18.2	0	18.2	18.2	0
R2	200	1.6	0	4	0	2	0	4	0	4	0	4	0	4	4	0
RC1	200	6.8	0	18	0	6.5	0	18	0	18	0	18	0	18	18	0
RC2	200	1.9	0	4	0	2	0	4	0.15	4	0.75	4	140.56	4	4.3	0.3
C1	400	16.9	0.01	36	0.01	26.5	0.01	36.5	2.8	36.5	12.47	36.5	2880.12	36.5	37.6	1.1
C2	400	2.5	0.01	11	0.01	11	0.01	11.2	2.6	11.2	12.31	11.2	2880.09	11.2	11.6	0.4
R1	400	17.9	0.01	36	0.01	11.7	0.01	36.4	0	36.4	0	36.4	0	36.4	36.4	0
R2	400	2.4	0.01	8	0.01	2.7	0.01	8	0	8	0	8	0	8	8	0
RC1	400	12.8	0.01	36	0.01	10.8	0.01	36	0	36	0	36	0	36	36	0
RC2	400	3.1	0.01	8	0.01	2.8	0.01	8	1.11	8	5.55	8	1440.04	8	8.4	0.4
C1	600	24.2	0.03	56	0.02	40.4	0.02	56.4	5.35	56.4	31.25	56.4	2160.32	56.4	57.2	0.8
C2	600	4.3	0.02	17	0.01	15.9	0.02	17.1	5.92	17.1	28.72	17.1	2160.31	17.1	17.4	0.3
R1	600	28.1	0.03	54	0.01	13.9	0.02	54.5	0	54.5	0	54.5	0	54.5	54.5	0
R2	600	4.3	0.02	11	0.01	3.4	0.01	11	0	11	0	11	0	11	11	0
RC1	600	19.7	0.04	55	0.02	12.2	0.02	55	0	55	0	55	0	55	55	0
RC2	600	4.7	0.02	8	0.01	2.9	0.02	11	3.63	11	18.38	11	1440.16	11	11.4	0.4
C1	800	34.4	0.08	72	0.05	49.3	0.05	72.8	18.36	72.8	97.45	72.8	4322.5	72.8	75	2.2
C2	800	5.7	0.05	22	0.04	21	0.04	22.2	48.55	22.2	228.25	22.2	9722.18	22.2	23.3	1.1
R1	800	35.3	0.06	72	0.05	15.8	0.05	72.8	0	72.8	0	72.8	0	72.8	72.8	0
R2	800	5.3	0.05	15	0.04	3.5	0.04	15	0	15	0	15	0	15	15	0
RC1	800	27.5	0.41	72	0.05	14.9	0.05	72	0	72	0	72	0	72	72	0
RC2	800	6.6	0.05	15	0.04	3.5	0.04	15	8.96	15	45.7	15	2160.52	15	15.4	0.4
C1	1000	44.6	0.34	90	0.09	58	0.09	91	35.22	91	185.64	91	4324.8	91	93.9	2.9
C2	1000	7.9	0.11	28	0.09	25.1	0.07	28.1	67.81	28.1	284.21	28.1	6484.59	28.1	28.8	0.7
R1	1000	44.5	0.13	91	0.09	19.5	0.08	91.9	0	91.9	0	91.9	0	91.9	91.9	0
R2	1000	6.5	0.09	19	0.08	4	0.07	19	0	19	0	19	0	19	19	0
RC1	1000	31.5	0.79	90	0.08	17.7	0.08	90	0	90	0	90	0	90	90	0
RC2	1000	7.8	0.1	18	0.08	4.1	0.07	18	8.21	18	40.32	18	1080.57	18	18.2	0.2

Finding a clique with the greatest cardinality involves the use of an exact method with exponential worst case performance. Nevertheless, our experiments on the standard benchmarks show that the maximum clique can be identified in a fraction of a second using the exact method described in [18]. For the Bin Packing Problem, we use the heuristic algorithm developed by [14] to get good lower bounds in a reasonable computational times. When conflicts are considered in solving Bin Packing, we apply the approach proposed in [12]. For performance purpose, we launch this algorithm with a time out of 3 hours.

Table 1 and Table 2 compare the performance of our Energetic Reasoning bounds: LB_{ER}, LB_{ERBPP} and LB_{ERBPPC} to the elementary bounds present in the literature: LB_{Clique}, $LB_{Capacity}$ and LB_{BP}. The column $BestUB$ represents the overall best-published upper bounds. The maximum of the lower bounds is reported in column $BestLB$. In $AvgGAP$, we present the average gap between $BestUB$ and $BestLB$.

In general, the proposed techniques give the minimum number of vehicles of 339 instances among the 468 instances tested and give near optimal solution for the rest. The average performance of $LB_{Capacity}$ is consistently better than LB_{Clique}, but LB_{Clique} outperforms $LB_{Capcaity}$ in 5 instances in $C1$, 25 instances in $R1$, 4 instances in $RC1$ and 1 instance in $RC2$ by a margin of 128. This is due to the structure of the data sets which does not favor time and capacity incompatible pairs. On the other hand, the three new lower bounds: LB_{ER}, LB_{ERBPP} and LB_{ERBPPC} produced consistent results across all data sets. Compared to the classical lower bound techniques: LB_{Clique}, $LB_{Capacity}$ and LB_{BP}, they give better bounds for 23 instances.

When Energetic Reasoning is combined to BPP (LB_{ERBPP}) and BPPC (LB_{ERBPPC}), the results outperform the bounds produced by LB_{ER} in 3 instances. This is due to the fact that the incompatibilities are considered at each examined time interval. These results confirm that the association of ER and BPPC is very efficient for VRPTW. We believe that ERBPPC will clearly outperform ER on highly constrained data set with more incompatible pairs. To conclude, the overall performance of the new lower bounding procedures has been encouraging. The use of Energetic Reasoning improves many lower bounds and gives good results for both capacity constrained problems and time constrained problems.

6   Conclusion

In this paper, we introduced several combinatorial optimization methods which can be used to get lower bounds for the Vehicle Routing Problem with Time Windows (VRPTW). Investigating the concept of Energetic Reasoning, we were able to propose new lower bounding techniques based on the transformation of m-VRPTW instance to PMSP. The numerical results confirm the contribution brought by the new proposed techniques. With a very fast computing time, we were able to provide the exact number or a reasonable approximation of the minimum number of vehicles required to visit all the customers. This suggests that our lower bounds techniques can quickly produce a good estimation of the fleet size. A challenging area for future research is to develop an exact method using the proposed lower bound procedures.

Acknowledgments. This work is partially supported by the Regional Council of Picardie and the European Regional Development Fund (ERDF), under PRIMA project. It is also partially supported by the National Agency for Research (project ATHENA, Reference ANR-13-BS02-0006-01).

Bibliography

[1] Alvarenga, G.B., Mateus, G.R., De Tomi, G.: A genetic and set partitioning two-phase approach for the vehicle routing problem with time windows. Computers & Operations Research 34(6), 1561–1584 (2007)

[2] Baldacci, R., Mingozzi, A., Roberti, R.: New route relaxation and pricing strategies for the vehicle routing problem. Operations Research 59(5), 1269–1283 (2011)

[3] Baldacci, R., Mingozzi, A., Roberti, R.: Recent exact algorithms for solving the vehicle routing problem under capacity and time window constraints. European Journal of Operational Research 218(1), 1–6 (2012)

[4] Baptiste, P., Le Pape, C., Nuijten, W.: Satisfiability tests and time-bound adjustments for cumulative scheduling problems. Annals of Operations Research 92, 305–333 (1999)

[5] Bräysy, O., Gendreau, M.: Vehicle routing problem with time windows, part i: Route construction and local search algorithms. Transportation Science 39(1), 104–118 (2005)

[6] Bräysy, O., Gendreau, M.: Vehicle routing problem with time windows, part ii: Metaheuristics. Transportation Science 39(1), 119–139 (2005)

[7] Cordeau, J.F., Desaulniers, G., Desrosiers, J., Solomon, M.M., Soumis, F.: The vehicle routing problem. ch. VRP with Time Windows, pp. 157–193. Society for Industrial and Applied Mathematics, Philadelphia (2001)

[8] Desaulniers, G., Lessard, F., Hadjar, A.: Tabu search, partial elementarity, and generalized k-path inequalities for the vehicle routing problem with time windows. Transportation Science 42(3), 387–404 (2008)

[9] Erschler, J., Lopez, P., Thuriot, C.: Raisonnement temporel sous contraintes de ressources et problèmes d'ordonnancement. Revue d'Intelligence Artificielle 5(3), 7–36 (1991)

[10] Fisher, M.L., Jörnsten, K.O., Madsen, O.B.: Vehicle routing with time windows: Two optimization algorithms. Operations Research 45(3), 488–492 (1997)

[11] Gehring, H., Homberger, J.: A parallel hybrid evolutionary metaheuristic for the vehicle routing problem with time windows. In: Proceedings of EUROGEN 1999, vol. 2, pp. 57–64 (1999)

[12] Gendreau, M., Laporte, G., Semet, F.: Heuristics and lower bounds for the bin packing problem with conflicts. Computers & Operations Research 31(3), 347–358 (2004)

[13] Golden, B.L., Assad, A.A., Wasil, E.A.: Routing vehicles in the real world: Applications in the solid waste, beverage, food, dairy, and newspaper industries. In: The Vehicle Routing Problem, pp. 245–286. SIAM Monographs on Discrete Mathematics and Applications, Philadelphia (2002)

[14] Haouari, M., Gharbi, A.: Fast lifting procedures for the bin packing problem. Discrete Optimization 2(3), 201–218 (2005)

[15] Jepsen, M., Petersen, B., Spoorendonk, S., Pisinger, D.: Subset-row inequalities applied to the vehicle-routing problem with time windows. Operations Research 56(2), 497–511 (2008)

[16] Jung, S., Moon, B.R.: A hybrid genetic algorithm for the vehicle routing problem with time windows. In: GECCO 2002: Proceedings of the Genetic and Evolutionary Computation Conference, July 9-13, pp. 1309–1316. Morgan Kaufmann, New York (2002)

[17] Kallehauge, B.: Formulations and exact algorithms for the vehicle routing problem with time windows. Computers & Operations Research 35(7), 2307–2330 (2008)

[18] Konc, J., Janezic, D.: An improved branch and bound algorithm for the maximum clique problem. Proteins 58, 569–590 (2007)

[19] Kontoravdis, G., Bard, J.F.: A grasp for the vehicle routing problem with time windows. ORSA Journal on Computing 7(1), 10–23 (1995)

[20] Labadi, N., Prins, C., Reghioui, M.: A memetic algorithm for the vehicle routing problem with time windows. Rairo-Operations Research 42, 415–431 (2008)

[21] Lenstra, J.K., Kan, A.: Complexity of vehicle routing and scheduling problems. Networks 11(2), 221–227 (1981)

[22] Little, J.D.C., Murty, K.G., Sweeney, D.W., Karel, C.: An algorithm for the traveling salesman problem. Operations Research 11(6), 972–989 (1963)

[23] Néron, E., Baptiste, P., Gupta, J.N.: Solving hybrid flow shop problem using energetic reasoning and global operations. Omega 29(6), 501–511 (2001)

[24] Ombuki, B., Ross, B.J., Hanshar, F.: Multi-objective genetic algorithms for vehicle routing problem with time windows. Applied Intelligence 24, 17–30 (2006)

[25] Solomon, M.M.: Algorithms for the vehicle routing and scheduling problems with time window constraints. Operations Research 35(2), 254–265 (1987)

[26] Tan, K.C., Chew, Y., Lee, L.: A hybrid multiobjective evolutionary algorithm for solving vehicle routing problem with time windows. Computational Optimization and Applications 34(1), 115–151 (2006)

[27] Ursani, Z., Essam, D., Cornforth, D., Stocker, R.: Localized genetic algorithm for vehicle routing problem with time windows. Applied Soft Computing 11(8), 5375–5390 (2011)

[28] Vidal, T., Crainic, T.G., Gendreau, M., Prins, C.: A hybrid genetic algorithm with adaptive diversity management for a large class of vehicle routing problems with time-windows. Computers & Operations Research 40(1), 475–489 (2013)

Constrained Clustering Using Column Generation

Behrouz Babaki[1], Tias Guns[1], and Siegfried Nijssen[1,2]

[1] Department of Computer Science, KU Leuven, Belgium
{firstname.lastname}@cs.kuleuven.be
[2] LIACS, Universiteit Leiden, The Netherlands

Abstract. In recent years, it has been realized that many problems in data mining can be seen as pure optimisation problems. In this work, we investigate the problem of constraint-based clustering from an optimisation point of view. The use of constraints in clustering is a recent development and allows to encode prior beliefs about desirable clusters. This paper proposes a new solution for minimum-sum-of-squares clustering under constraints, where the constraints considered are must-link constraints, cannot-link constraints and anti-monotone constraints on individual clusters. Contrary to most earlier approaches, it is exact and provides a fundamental approach for including these constraints. The proposed approach uses column generation in an integer linear programming setting. The key insight is that these constraints can be pushed into a branch-and-bound algorithm used for generating new columns. Experimental results show the feasibility of the approach and the promise of the branch-and-bound algorithm that solves the subproblem directly.

1 Introduction

One of the core problems studied in the data mining and machine learning literature is that of *clustering*. Given a database of examples, the clustering task involves identifying groups of similar examples; such groups are for instance indicative for patients with similar clinical observations, customers with similar purchase behaviour, or website visitors with similar click behaviour.

While the clustering problem is common in the data mining literature, it is only recently realized in the data mining community that this problem is closely related to problems studied in the optimization literature, and hence that open problems in clustering may be solved using generic optimization tools. In this paper, we study one such open problem: *Optimal Constrained Minimum Sum-of-Squares Clustering* (MSSC). We show that a generic optimization strategy can be used to address this problem.

Many types of clustering problems are known in the literature; however, MSS clustering is arguably one of the most popular clustering settings. In MSS clustering, the task is to find a clustering in which each example is put into *exactly* one cluster. Clusters do not overlap and together they cover all the available data. Clusters should be chosen such that points within a cluster have small sum-of-squared distances.

H. Simonis (Ed.): CPAIOR 2014, LNCS 8451, pp. 438–454, 2014.
© Springer International Publishing Switzerland 2014

The popularity of the MSS clustering setting is partially due to the *k-means* algorithm. K-means is a heuristic algorithm which quickly converges to a local minimum and is included in most data mining toolkits. Even though successful, basic *k*-means has several disadvantages. One is its randomized nature: each run of the algorithm may yield a different clustering. Another is its lacking ability to take into account prior knowledge of a user.

There are many types of prior knowledge that a user may have. A common perspective is to formalize prior knowledge in terms of *constraints* on the clusters one wishes to find [6], where the most popular constraints are *must-link* and *cannot-link* constraints. A must-link constraint enforces that examples that are known to be related, are part of the same cluster. A cannot-link constraint, on the other hand, enforces that examples that are not related are not part of the same cluster. These constraints are popular as many other constraints can be transformed into must-link and cannot-link constraints [9]. The maximum cluster diameter constraint, for instance, requires that each cluster must have a diameter of at least distance α; hence, any two points that are further than α apart cannot link together. The minimum cluster separation constraint requires that clusters must be separated by at least distance β; hence, any two points that are less than β apart must link together.

Other clustering settings that can be seen as constraint-based clustering problems are problems in which clusters need to have a minimum or maximum size, or where one is looking for alternative clusterings [15].

An important question is how to find a clustering that satisfies constraints. Here, most algorithms in the data mining literature take a heuristic approach. Arguably, the most well-known example is the *COP-k-means* algorithm [23], which modifies the *k-means* algorithm to deal with must-link and cannot-link constraints. Unfortunately, even though the algorithm is fast, it may not find a solution that satisfies all constraints even if such a solution exists [9]. In itself, this is not surprising as the problem is known to be NP hard and hence a polynomial solution is not likely to exist [2,9]. As a result, the problem of how to solve the MSSC problem under constraints is still open.

This paper addresses this challenge and develops a generic approach that can find an optimal solution to constrained MSSC problems. While we will focus on must-link and cannot-link constraints, the approach allows for the inclusion of several other constraints as well; we will show that the approach works for all constraints that are *anti-monotone*.

Our approach builds on earlier work that showed the feasibility of *unconstrained* optimal MSS clustering [14,4] by using *column generation* in an *integer linear programming* setting. The column generation process is here responsible for identifying candidate clusters that can be put into a clustering. We will show that most clustering constraints can be dealt with by *pushing* the constraints in a branch-and-bound algorithm for column generation.

This paper is organized as follows. Section 2 introduces MSS clustering and MSS clustering under constraints. Section 3 gives an overview of how to find a solution using a column generation process, building on the earlier work of [14,4].

In Section 4 we introduce a branch-and-bound approach for generating columns under constraints. Section 5 discusses practical considerations in the implementation of this algorithm. Section 6 provides experiments, Section 7 discusses related work and Section 8 concludes.

2   MSSC

Assumed given is a dataset D with n data points. Each example in the dataset is a point p in an m-dimensional space and is represented by a vector with m values. One cluster is defined as a set of data points $C \subseteq D$. A clustering consists of k such clusters, and corresponds to a partitioning of the data into k groups. The number of clusters k is typically given upfront by the user. In MSS clustering, the clusters in a clustering are usually non-overlapping, that is, each data point belongs to exactly one cluster.

Given a cluster $C \subseteq D$, the cluster center or *centroid* is the mean of the data points that belong to that cluster:

$$z_C = \text{mean}(C) = \frac{\sum_{p \in C} p}{|C|} \tag{1}$$

The quality of a clustering can be measured in many different ways. In MSS clustering, the quality of a clustering is measured using the sum of squared distances between each point in a cluster and the centroid of the cluster: $SSC(\mathcal{C}) = \sum_{C \in \mathcal{C}} \sum_{p \in C} d^2(p, z_C)$, where $d(\cdot, \cdot)$ calculates the distance between two points, for example, the Euclidean distance.

Note that for a cluster C, the sum of squared distances to its centroid equals the sum of all pairwise distances between the points of that cluster, divided by the size of the cluster:

$$\sum_{p \in C} d^2(p, z_C) = \frac{\sum_{p_1, p_2 \in C} d^2(p_1, p_2)}{|C|} \tag{2}$$

For simplicity of notation, when we write $p_1, p_2 \in C$ we assume that every pair of two points in C is included in the sum exactly once. To summarize the MSSC problem, a mathematical programming formulation is given in Table 1.

The best known clustering algorithm that uses sum of squared distances is the k-means algorithm. It is an approximate algorithm that starts with an initial random clustering and iteratively minimizes the sum-of-squares using the following two steps: 1) add each data point to the cluster with closest cluster centre; 2) compute the new cluster centre of the resulting clusters. These two steps are iterated until convergence, that is, the cluster centres do not change any more. This procedure can get stuck in local minima and it is not uncommon that two different runs (e.g. with different initial clusters) produce different clusterings.

Constraints. The most well-known constraints are must-link and cannot-link constraints. Let ML and CL be subsets of $D \times D$. Then a cluster C satisfies a must-link constraint $(p_1, p_2) \in ML$ iff $|\{p_1, p_2\} \cap C| \neq 1$; it satisfies a cannot-link constraint $(p_1, p_2) \in CL$ iff $|\{p_1, p_2\} \cap C| \leq 1$.

Note that both constraints can be evaluated on the individual clusters in a clustering. This is a key observation for our work.

In a seminal paper by Wagstaff et al [23], the COP-k-means algorithm is proposed. COP-k-means is an extention of the k-means algorithm towards must-link and cannot-link constraints. It modifies the k-means algorithm by not assigning each point to its closest cluster centre, but rather to the closest centre that satisfies all constraints. If no such centre exists, the algorithm terminates. An alternative approach is to continue running the algorithm, even though the final solution might then not satisfy all constraints; in any case, the algorithm is not guaranteed to find a solution even if there exists one.

Many other constraints are possible. We will not give a complete overview here (see Section 7 and [6]). For this work it is however important to observe that many problems can be formalized using constraints that are *anti-monotone*. We call a boolean constraint $\varphi(C)$ on a cluster C of data points *anti-monotone* iff $\varphi(C)$ implies $\varphi(C')$ for all $C' \subseteq C$. The cannot-link constraint is anti-monotone: if a cluster C satisfies a cannot-link constraint, every subset also satisfies this constraint. There are many other anti-monotone constraints:

- a maximum cluster size constraint on clusters $|C| \leq \theta$, which can be used to avoid that one cluster dominates a clustering;
- a maximum overlap constraint $|C \cap X| \leq \theta$, which can be used to avoid that any cluster found is too similar to a given set of points X; this generalizes the cannot-link constraint;
- a minimum difference constraint $|C \backslash X| \leq \theta$, which requires a certain similarity to cluster X;
- a soft cannot-link constraint, which requires that the number of pairs of points in a cluster that have a cannot-link constraint among them is bounded;
- conjunctions or disjunctions of anti-monotone constraints.

A conjunction of anti-monotone constraints can for instance be used to find an alternative clustering: starting from a clustering $\mathcal{C}$, we can enforce that in a new clustering every cluster is different from all clusters in the earlier clustering.

Must-link constraints are an example of constraints that are not anti-monotone. In the following sections, we will show how to solve the MSS problem under a combination of anti-monotone constraints and must-link constraints, by adapting a state-of-the-art unconstrained optimal clustering algorithm. A feature of the algorithm is that it exploits the anti-monotonicity of cluster constraints.

3   Column Generation Framework

In this section we give a brief overview of an ILP formulation of MSSC and a column generation method for solving it, based on the (unconstrained) MSSC column generation framework of Aloise et al. [4]. The next section will introduce our proposed approach for taking constraints into account.

Table 1. MSS clustering   **Table 2.** An ILP model for MSS clustering

$$\underset{\mathcal{C}}{\text{minimize}} \quad \sum_{C \in \mathcal{C}} \sum_{p \in C} d^2(p, z_C), \qquad (3)$$

$$\text{s.t.}$$

$$C_1 \cap C_2 = \emptyset \quad \forall C_1, C_2 \in \mathcal{C} \qquad (4)$$

$$\left| \bigcup_{C \in \mathcal{C}} C \right| = n \qquad (5)$$

$$|\mathcal{C}| = k \qquad (6)$$

$$\underset{x}{\text{minimize}} \quad \sum_{t \in T} c_t x_t, \qquad (7)$$

$$\text{s.t.}$$

$$\sum_{t \in T} x_t a_{it} = 1 \quad \forall i \in \{1, \dots, n\} \qquad (8)$$

$$\sum_{t \in T} x_t = k \qquad (9)$$

$$x_t \in \{0, 1\} \qquad \forall t \in T \qquad (10)$$

An ILP formulation of MSSC. Given a dataset with n data points, the number of possible clusters is 2^n. In principle, we can hence reformulate the clustering problem using a Boolean n by 2^n matrix A that represents all possible clusters: each column is a cluster where $a_{it} = 1$ if data point p_i is in cluster t and $a_{it} = 0$ otherwise. We define the *cost* of a cluster (column) as the sum of squared distances of the points in the cluster to its mean: $c_t = \sum_{i=1}^{n} d^2(p_i, z_t) a_{it}$.

The problem in equations 3-6 can then be formulated as an Integer Linear Program as in Table 2 [14], where $T = \{1, \dots, 2^n\}$ denotes all possible clusters. Equation 7 corresponds to the SSC criterion. Equation 8 states that each data point must be covered exactly once. Hence it enforces both that the clusters are not overlapping and that all points are covered. Equation 9 finally ensures that exactly k clusters are found. Note that the k-means (and COP-k-means) algorithm can return empty clusters and hence less than k clusters in some occasions. This can not arise in the above formulation.

For even moderate sizes of n the number of clusters will be too large to solve the above ILP by first materializing A. However, we can use a column generation approach in which the master problem (Eq. 7-10) is *restricted* to a smaller set $T' \subseteq T$ and columns (clusters) are incrementally added until the optimal solution is provably found.

Column Generation. iterates between solving the restricted master problem and adding one or multiple columns. A column is a candidate for being added to the restricted master problem if adding it can improve the objective function. If no such column can be found, one is certain that the optimal solution of the restricted master problem is also the optimal solution of the full master problem. Whether a column can improve on the objective can be derived from the dual.

The dual of the master problem (Table 2) is given in Table 3. Here λ_i indicates a dual value corresponding to the constraint in equation 8 and σ a dual value corresponding to equation 9. One column in the master problem corresponds to one constraint in the dual (Equation 12).

Table 3. Dual of the optimization problem

Table 4. Model with stabilization included ($N = \{1, \ldots, n\}$)

$$\underset{\lambda, \sigma}{\text{maximize}} \quad -k\sigma + \sum_{i=1}^{n} \lambda_i \qquad (11)$$

s.t.

$$-\sigma + \sum_{i=1}^{n} a_{it}\lambda_i \leq c_t \quad \forall t \in T \qquad (12)$$

$$\lambda_i \geq 0 \qquad \forall i \in \{1, \ldots, n\} \qquad (13)$$

$$\sigma \geq 0 \qquad (14)$$

$$\underset{x}{\text{minimize}} \quad \sum_{t \in T} c_t x_t + \sum_{i=1}^{n} \theta_i y_i, \qquad (15)$$

s.t.

$$\sum_{t \in T} x_t a_{it} + y_i = 1 \quad \forall i \in N \qquad (16)$$

$$\sum_{t \in T} x_t = k \qquad (17)$$

$$x_t \in \{0, 1\} \qquad \forall t \in T \qquad (18)$$

$$-\mu \leq y_i \leq \mu \qquad \forall i \in N \qquad (19)$$

Given values for λ and σ, obtained by solving a restricted master problem, we need to determine whether there are columns for which $\sigma - \sum_{i=1}^{n} a_{it}\lambda_i + c_t < 0$, that is, whether there are columns with a *negative reduced cost*. If no such column can be found, the current solution is optimal.

Finding a column with negative reduced cost is called *pricing*. While a pricing routine can return any column with a negative reduced cost, one typically searches for the smallest one; hence we are interested in finding:

$$\underset{t \in T}{\arg\min} \quad \sigma - \sum_{i=1}^{n} a_{it}\lambda_i + c_t. \qquad (20)$$

Solving this pricing problem is not trivial, given the large number of columns. The details of solving the pricing subproblem will be discussed in more detail in Section 4.

When solving the restricted master problem, it is possible that it has no feasible solution. In this case, Farkas' Lemma [22] can be used to add columns that gradually move the solutions of the restricted master problems closer to the feasible region, or to prove infeasibility of the master problem. This Farkas pricing is similar to the regular pricing explained above. In this case, the problem to optimize is:

$$\underset{t \in T}{\arg\min} \quad \sigma' - \sum_{i=1}^{n} a_{it}\lambda_i' \qquad (21)$$

where σ' and λ' are the dual Farkas values. Note that this is the same problem as the regular pricing problem above, with the exception that the cost c_t of the cluster does not need to be taken into account.

4   Column Generation with Constraints

Given the earlier observations, one can see that enforcing constraints on clusters C amounts to removing from the cluster matrix A all clusters that do not satisfy these constraints. In a column generation scheme, this means that it is sufficient to add these constraints to the subproblem solver; they do not need to be added to the master problem. The rest of this section explains our proposed branch-and-bound method for solving the (constrained) subproblem.

4.1   Subproblem Solving

Essentially, in each iteration of the column generation process we need to solve a constrained minimisation problem. The objective function to minimize is given by equation 20 (equation 21 in case of infeasibility). By removing the constant σ and using equation 2, we can rewrite the objective as:

$$\arg\min_{t \in T} \sum_{i=1}^{n} d^2(p_i, z_t)a_{it} + \sigma - \sum_{i=1}^{n} a_{it}\lambda_i \tag{22}$$

$$= \arg\min_{t \in T} \frac{\sum_{i=1}^{n} \sum_{j=i+1}^{n} d^2(p_i, p_j)a_{it}a_{jt}}{\sum_{i=1}^{n} a_{it}} - \sum_{i=1}^{n} a_{it}\lambda_i \tag{23}$$

Let us represent the cluster $t \in T$ and its corresponding column $a_{\cdot t}$ as a set X. We define $d(X) = \sum_{i,j \in X} d^2(p_i, p_j)$, where every pair is only considered once in the sum, $d(X, Y) = \sum_{i \in X, j \in Y} d^2(p_i, p_j)$ and $\lambda(X) = \sum_{i \in X} \lambda_i$. We can now rephrase our problem as that we wish to search for a cluster X:

$$\arg\min_{X} \frac{d(X)}{|X|} - \lambda(X) \tag{24}$$

and such that all constraints on clusters are satisfied.

Blocks. A first simple observation is that the must-link constraints are transitive and hence the must-link relation is an equivalence relation. We will refer to the equivalence classes as *blocks*. We can rephrase our optimization problem as an optimization problem over the blocks. Let $X = [p_i]_{ML}$ denote the block that point $p_i \in D$ belongs to (a point can never belong to two blocks) and let $D/ML = \{[p_i]_{ML} \mid p_i \in D\}$ denote the blocks in the data. We are looking for a subset of the blocks $\bar{X} \subseteq D/ML$ such that the following criterion is minimized:

$$f(\bar{X}) = \left(\sum_{X \in \bar{X}} d(X) + \sum_{X,Y \in \bar{X}} d(X,Y) \right) \Big/ \sum_{X \in \bar{X}} |X| - \sum_{X \in \bar{X}} \lambda(X). \tag{25}$$

Note that we can precompute the terms $d(X)$, $d(X,Y)$, $|X|$ and $\lambda(X)$ for all $X, Y \in D/ML$. Note furthermore that if $ML = \emptyset$ then $\forall X \in \bar{X} : |X| = 1$ and this formula is identical to the one without constraints.

In addition, the choice of $\bar{X}$ has to satisfy the cannot-link constraint: for no two $X, Y \in \bar{X}$ it may be the case that $i \in X, j \in Y, (i, j) \in CL$.

Algorithm 1. Branch-and-bound(Set: $\bar{X}$, Set: $\bar{C}$)

$\bar{X}$ is the current set of blocks under consideration, $\bar{C}$ the possible extensions to $\bar{X}$.

1: $\bar{C} :=$ reduce-candidates($\bar{X}, \bar{C}$)
2: **if** not prunable($\bar{X}, \bar{C}$) **then**
3:    Store $\bar{C}$ in a stack
4:    Process $\bar{X}$ as candidate cluster
5:    **while** $\bar{C}$ is not empty **do**
6:       $C := \bar{C}$.pop ()
7:       Branch-and-bound ($\bar{X} \cup \{C\}$, $\bar{C}$)
8:    **end while**
9: **end if**

Algorithm. We propose to use a branch-and-bound algorithm to solve this problem. This algorithm performs a set-enumeration and is given in Algorithm 1 (initialized with Branch-and-bound($\{\}$, D/ML)). It uses newly developed pruning strategies to make the search feasible and is easily extended to include a wide range of constraints. In order to prune candidates, we either remove some candidates from consideration (line 1) or discard a branch of the search tree using bounds on the objective function (line 2).

The removal of candidates in line 1 corresponds to propagation in a constraint programming setting [8]. However, we will show that the proposed bound used in line 2 is not valid in the presence of arbitrary constraints and hence cannot be used in general.

4.2 Reducing the Number of Candidates

We employ three strategies to reduce the set of candidates in line 1 of Algorithm 1:

Cannot-Link Constraints. The cannot-link constraint is easily taken into account: when there is a cannot-link constraint between a block in $\bar{C}$ and a block in $\bar{X}$, the block is removed from $\bar{C}$.

Anti-monotone constraints other than cannot-link constraints are easily included as well: if a set $\bar{X} \cup \{C\}$ does not satisfy an anti-monotone constraint, the candidate C can be removed in line 1.

Block Compatibility. Assume that we have a block $C_1 \in \bar{X}$ and a block $C_2 \in \bar{C}$ and the following holds:

$$\frac{d(C_1) + d(C_2) + d(C_1, C_2)}{|C_1| + |C_2|} - \lambda(C_1) - \lambda(C_2) > 0,$$

then any cluster $\bar{X}'$ we could build that includes both C_1 and C_2 can be improved by removing both C_1 and C_2:

$$f(\bar{X}) = \sum_{p_i \in \cup \bar{X}} d(p_i, z_{\cup \bar{X}})^2 - \sum_{X \in \bar{X}} \lambda(X) \geq$$

$$\sum_{p_i \in \cup \bar{X} \setminus \{C_1, C_2\}} d(p_i, z_{\cup \bar{X} \setminus \{C_1, C_2\}})^2 + \sum_{p_i \in C_1 \cup C_2} d(p_i, z_{C_1 \cup C_2})^2 - \sum_{X \in \bar{X}} \lambda(X) \geq$$

$$\sum_{p_i \in \cup \bar{X} \setminus \{C_1, C_2\}} d(p_i, z_{\cup \bar{X} \setminus \{C_1, C_2\}})^2 + - \sum_{X \in \bar{X} \setminus \{C_1, C_2\}} \lambda(X). \quad (26)$$

Note that this argument is only valid in the presence of anti-monotone constraints in combination with must-link constraints. We refer to this test as a *compatibility test*. When a block in $\bar{C}$ is incompatible with a block in $\bar{X}$, the block is removed from $\bar{C}$.

4.3 Pruning Using a Bound on the Objective Function

For the remaining set of candidates, a more elaborate test is carried out to determine whether to continue the search (line 2). This test consists of calculating a bound on achievable solutions and comparing it with the best solution found so far. A key feature of this bound is that it can be calculated efficiently.

The key idea is as follows. Let $\bar{X}'$ be a set that is found below a set $\bar{X}$ in the search tree, that is, $\bar{X}' \subseteq \bar{C} \cup \bar{X}$. We can write its quality as follows:

$$\underbrace{(d(\cup \bar{X}) +}_{\text{old}} \underbrace{\sum_{X \in \bar{X}' \setminus \bar{X}} \beta(\bar{X}, X)}_{\text{(1) between old and new}} + \underbrace{\sum_{X, Y \in \bar{X}' \setminus \bar{X}} d(X, Y))}_{\text{(2) between new blocks}} / \underbrace{\sum_{X \in \bar{X}'} |X|}_{\text{(3) sizes}} - \underbrace{\sum_{X \in \bar{X}'} \lambda(X)}_{\text{(4) lambdas}},$$

where $\beta(\bar{X}, X) = d(X) + \sum_{Y \in \bar{X}} d(X, Y)$.

Essentially, we need to have a bound on the best $\bar{X}'$. An important first concern is that we do not know the size of the best $\bar{X}'$ and hence we do not know term (3). We simplify this problem by iterating over all cluster sizes $\sum_{X \in \bar{X}} |X| \leq s \leq \sum_{X \in \bar{X}} |X| + \sum_{C \in \bar{C}} |C|$ and calculating a bound on the quality assuming the best cluster has size s, i.e., we calculate a bound on the above formula assuming part (3) is iteratively fixed. The overall bound is the best bound among all the sizes considered.

Calculating a lower bound for a fixed value s of (3) requires a lower bound on (1) and (2), and an upper bound on (4). We discuss each in turn. A lower bound on part (1) for a given size s is obtained as follows:

- sort all $C \in \bar{C}$ increasing in their $\beta(\bar{X}, C)/|C|$ values, yielding order $C_1, \ldots, C_m$;
- determine the largest value k such that $\sum_{i=1}^{k} |C_i| \leq s$;
- determine $\sum_{i=1}^{k} \beta(\bar{X}, C_i)$ as bound.

The argument for this is as follows. All additional points that are selected by the algorithm above in $C_1, \ldots, C_k$ are characterized by the $\beta(\bar{X}, C)/|C|$ value their corresponding block has. If we sum these characteristic values over all points, the result is $\sum_{i=1}^{k} \beta(\bar{X}, C_i)$. Choosing the lowest possible characteristic values is a lower bound as the sum of characteristic values of the points in the optimum $\bar{X}^*$, and hence also the value $\sum_{X \in \bar{X}^* \setminus \bar{X}} \beta(\bar{X}, X)$, can never be better.

A similar algorithm can be used to determine an upper bound for term (4):

- sort all $C \in \bar{C}$ decreasing in their $\lambda(C)/|C|$ values, yielding order $C_1, \ldots, C_m$;
- determine the smallest value k such that $\sum_{i=1}^{k} |C_i| \geq s$;
- determine $\sum_{i=1}^{k} \lambda(C_i)$ as bound.

A simple lower bound on term (2) is that it is always higher than zero. Calculating a good bound is hard, as we essentially need to solve an edge-weighted clique problem.

While the overall bound obtained is not very tight, also because term (1) and term (4) are sorted independently, it has important computational advantages. First, we can sort the $\lambda(C)/|C|$ and $\beta(\bar{X}, C)/|C|$ values before iterating over potential sizes; hence, we can avoid doing this repeatedly for each size s. Second, we do not need to consider all sizes s indicated earlier. If we consider the sorted ranges of λ and β values, there are ranges of sizes in which the bound does not change; the bound only changes when either a lambda value changes or a β value changes. It hence suffices to consider $2|\bar{C}|$ different sizes for s. Finally, we can maintain the bounds incrementally.

As a result, the overall bound over all sizes s can be calculated in $O(|\bar{C}| \log |\bar{C}|)$ time. As furthermore all required counts can be maintained incrementally in $O(|\bar{C}|)$ time, the overall time spent in one call of the Branch-and-bound algorithm (excluding recursive calls) is $O(|\bar{C}| \log |\bar{C}|)$; in other words, the complexity of the algorithm is *not* dependent on the number of points in the data, but *only* on the number of blocks that the must-link constraints identify in it.

5   Practical Considerations

The column generation approach, in combination with the branch-and-bound algorithm, provides a fundamental approach for finding optimal solutions under constraints. However, several practical considerations are of importance when implementing the column generation approach.

5.1   Initialisation

Initially, there are no columns in the restricted master problem. This means that Farkas pricing needs to be performed until a feasible solution is found, which can be time consuming. However, assuming a heuristic solver such as COP-k-means finds a solution, one can initialize the restricted master problem with this known (sub-optimal) solution. This avoids the need for Farkas pricing, provides a number of good initial columns (cuts to the dual problem) as well as an upper bound for the master problem.

5.2   Branching

Integer linear programs are typically solved by solving a number of LP relaxations and using branching to enforce integrality. So far, we have described how we employ the column-generation method for solving the LP relaxations. In theory, if the solution to the linear program is fractional any type of branching can be used. In previous work [14] a Ryan-Foster branching scheme was employed. In this scheme, in the restricted problem two columns are determined that have a corresponding fractional value and that cover the same data point (p_1). Branching will enforce that in subsequent problems only one of these two columns can cover that point. Observe that no two columns cover exactly the same data points and hence they must differ in at least one data point (p_2). We can now branch by enforcing that in one branch points p_1 and p_2 are in the same cluster and that in the other branch p_1 and p_2 are not in the same cluster.

This type of branching naturally fits our approach as it corresponds to adding a must-link or cannot-link constraint. Compared to [14], the proposed approach can hence handle both constrained and unconstrained cases in the same principled manner.

5.3   Slow Convergence

Many large-scale column generation approaches suffer from slow convergence. Similar to [14], we also observed degeneracy in our experiments: even when given the optimal solution, a large number of column generation iterations is required before the optimality is proved. We implemented a dual stabilisation scheme similar to the one of [14]: adding a linear penalisation to the dual objective corresponds to adding a *perturbation* variable to each of the constraints in equation 8 and adding them to the objective function, given in Table 4. Here y_i are the perturbation variables, $+/-\mu$ its bounds and θ_i its coefficients in the objective function. The θ_i form a stabilisation centre in the dual that will penalize duals that are too far from it. A good choice for θ is the dual λ values from the best known solution so far. The value of μ has to be progressively decreased until 0. At this point, all perturbation variables are 0 and the problem is identical to the original restricted master problem.

We employ a scheme where the θ_i are given an equal initial value and μ is set to 0.99. Each time an optimal solution to the perturbed restricted master problem is found, the θ_i values are changed to the duals of that optimal solution and μ is divided by 2^ℓ where ℓ is a counter of the number of such updates.

6   Experiments

Data was obtained from the UCI machine learning repository [5]. Table 5 lists the properties of the datasets.

We used the open-source SCIP [1] system as column generation framework. The branch-and-bound pricer is written in C++. Source code is available at

Table 5. Description of datasets

name	# points	dimensions	# labels
Iris	150	4	3
Wine	178	13	3
Soybean	47	35	4

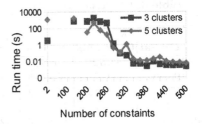

Fig 1. Run times on the Iris data set

http://dtai.cs.kuleuven.be/CP4IM/cccg/. All experiments were run on quad-core Intel 64 bit computers with 16GB of RAM running Ubuntu Linux 12.04.3.

Constraints were generated according to the common methodology of [23]: two data points are repeatedly sampled randomly from labelled data; if they have the same label a ML constraint is generated, otherwise a CL constraint. This is repeated until the required number of constraints is generated. The code for generating these constraints and for the COP-k-means algorithm were obtained from http://www.cs.ucdavis.edu/~davidson/constrained-clustering/ .

It is common practice to run (COP-)k-means multiple times to avoid that it is stuck in a local minimum. For each setting, we ran COP-k-means 500 times. The implementation obtained continues until convergence and is not guaranteed to satisfy all constraints. We will report on the number of runs that satisfy all constraint (COP sat). Only when at least one solution is found that satisfies all constraints will we report on its quality (COP max).

We initialized our column generation method with the *best* solution found by COP-k-means. Best is here defined by the clustering with the largest number of clusters satisfying all constraints. Among these clusterings, the one with the lowest MSS is selected. Note that in case COP-k-means did not find a solution satisfying all constraints, our column generation method started with the *best* infeasible solution. The stabilisation parameter μ was set to 0.99. Initial perturbation values θ_i can be set to any value; the update mechanism is explained in Section 3. In case a feasible solution is at hand, a good initial value for θ_i can be obtained from bounds on the dual variables. These bounds are calculated as in [14], and we used the lower bounds of the dual variables to initialize the corresponding θ_i.

The branch-and-bound method for solving the subproblem maintains a list of all clusters that improve the bound during search (including the final best one). All these clusters are added as columns to the restricted master problem.

Results. We compare the result of our column generation approach to that of repeated runs of COP-k-means. Our column generation approach is initialized as explained above, and a time-out of 30 minutes is used.

Table 6 shows the quality of the results for the Iris dataset, once for $k = 3$ (the true number of class labels, left) and once for $k = 5$ (right); Figure 1 gives an impression for the amount of run time it took to calculate these results.

Table 6. Clustering with '#c' constraints, Iris dataset. *optimality proven

	k=3				k=5			
#c	COP sat	COP max	CG best	#c	COP sat	COP max	CG best	impr.
2	100.00%	90.3725	90.3725	2	100%	46.5616	46.5616	0%
60	100.00%	83.6675	83.6675	60	100%	53.399	53.399	0%
100	37.20%	87.2082	87.2082	100	100%	57.3827	57.3804*	0.004%
140	0%	-	87.8750*	140	100%	63.1699	62.2115*	1.5%
200	0%	-	89.1496*	200	100%	71.1401	69.3154*	2.56%
240	0%	-	85.2477*	240	100%	72.7078	69.9776*	3.76%
300	0%	-	89.3868*	300	83.6%	82.0819	81.9792*	0.13%
340	31.40%	89.3868	89.3868*	340	100%	85.9036	82.9945*	3.39%
400	0%	-	89.3868*	400	100%	84.0495	84.0357*	0.02%
440	31.00%	88.6409	88.6409*	440	100%	82.6373	82.6373*	0%
500	0%	-	89.3868*	500	100%	85.8908	85.8719*	0.02%

A first observation is that in case of $k = 3$, and a low number of clusters, COP-k-means easily finds clusterings that satisfy the constraints (indicated by "COP sat"). For higher numbers of constraints, COP-k-means encounters more problems finding clusterings satisfying all constraints. In multiple cases none of the 500 runs finds a clustering satisfying all constraints. When we increase the number of clusters to $k = 5$, the constrained clustering problem becomes easier [9]; as a consequence, COP-k-means can find satisfying solutions more easily. Even when COP-k-means can not find a solution, our method finds acceptable clusterings; even optimal ones are found for higher numbers of constraints. The case of 140 constraints is an exception. For $k = 5$ and higher numbers of constraints, our method can find the optimal constrained clustering.

Table 7 shows the results for the bigger Wine dataset. This dataset is much harder, both for COP-k-means and for the column generation approach. In case of $k = 3$, the true number of class labels, COP-k-means is again rarely able to find a solution satisfying all constraints. The CG approach is able to find solutions for some cases, but can not prove them optimal within the time-out. In case of $k = 5$ the problem becomes easier, as was the case on Iris. We can see that the CG approach can sometimes greatly improve the best solution found in 500 COP-k-means runs, even without being able to prove its optimality.

Table 8 shows results on the Soybean dataset, a smaller dataset of higher dimensionality; its true number of labels is 4. Observe that for $k = 3$ and 80 constraints, CG is able to **prove** that this problem is infeasible. The heuristic COP-k-means simply does not find a solution, as happens for 40 and 60 constraints. We further note that in contrast to $k = 4$, for $k = 5$ COP-k-means is often not able to find the optimal solution.

Table 7. Clustering with '#c' constraints, Wine dataset. *optimality proven

k=3				k=5				
#c	COP sat	max	CG best	#c	COP sat	COP max	CG best	impr.
240	0%	-	4860250*	240	100%	4021090	3327908*	17.24%
300	0%	-	5133144*	300	0%	-	4077296*	+
340	0%	-	5214981*	340	16.6%	4659910	4329603*	7.09%
380	0%	-	5220299*	380	66.6%	4729860	4450036*	5.92%
420	0%	-	5232632*	420	59.6%	4740180	4537678*	4.27%
460	0%	-	5232632*	460	94.2%	4819200	4540041*	5.79%
500	0%	-	5232632*	500	15%	4922560	4684355*	4.84%

Table 8. Soybean, different k and number of clusters (#c); GC quality gap = difference between best solution quality of cop-kmeans and the solution of CG, INF = infeasible

	k=3		k=4		k=5	
# cons	COP sat.	CG quality gap	COP sat.	CG quality gap	COP sat.	CG quality gap
2	100.00%	0	100.00%	0	100.00%	0.00%
10	100.00%	0	100.00%	0*	100.00%	4.56%*
20	100.00%	0*	100.00%	0.12%*	100.00%	0.29%*
40	0.00%	339*	100.00%	0*	100.00%	1.25%*
60	0.00%	418*	52.60%	0*	81.20%	0.24%*
80	0.00%	INF	74.00%	0*	27.00%	0.38%*

7   Related Work

We build on a column generation approach first described in [14] and improved in [4]. This earlier work only studies *unconstrained* clustering settings. We show that with modifications it can also be used in the presence of constraints. The main necessary modification is in the subproblem solver. We use a branch-and-bound approach that directly solves the subproblem and can be used in the presence of any constraint that is anti-monotone.

A feature of the first approach [14] is that it uses a heuristic *Variable Neighbourhood Search* method to solve a subproblem, and only when a solution can not be found in this way an exact method is used. The exact method uses Dinkelbach's lemma [13] to solve equation 23 through a series of unconstrained quadratic 0-1 problems. The latter are solved using a heuristic VNS combined with an exact branch-and-bound algorithm for verifying the stopping criterion of the Dinkelbach method.

This method is improved in [4]. One of these improvements is the introduction of a compatibility test. We adapted this test for use in the presence of must-link constraints.

Other exact methods for MSSC are branch and bound methods [18,12,7], a cutting plane algorithm that starts from the observation that MSSC is a concave optimisation problem [24], dynamic programming [16,20] and a branch-and-cut semi-definite programming algorithm [3]. These methods do not consider the addition of extra constraints.

Exact methods for constrained-based clustering have been studied before. Typical is that they do not use MSS as optimisation criterion, but rather a function that is linear or quadratic. Saglam et al. [21] use an integer linear programming approach for minimizing the maximum cluster diameter. More recently, constraint programming has been used for solving constrained clustering tasks [8]. A range of constraints is supported including instance-level constraints, size of cluster constraints and constraints on the separation between clusters and maximum diameter of a cluster. As objective function the (non-normalized) sum of squared distances between clusters or maximum diameter is supported.

A large class of clustering methods are those that evaluate the quality of a cluster based on a cut-value. Also in such methods the use of column generation has been proposed [17]. The inclusion of constraints in this method may be a topic for further research.

Exact methods are also used as part of approximate constraint-based clustering methods. Demiriz et al. [11] propose to modify k-means such that the assignment step, where points are assigned to their nearest feasible cluster, corresponds to solving an LP. Constraints on minimum cluster size can be taken into account, as well as instance level constraints. Davidson et al. studied the use of SAT solvers, also using diameter as optimization criterion [10]. Müller and Kramer [19] use integer linear programming to solve constrained clustering tasks where a fixed number of candidate clusters is given upfront. The problem consists of selecting the right subset of clusters, which can be compared to solving one iteration of the restricted master problem. They investigate a number of different optimisation criterion, as well as constraints at the clustering level, such as the maximum amount of overlap between clusters or logical formula over entire clusters. These methods are not guaranteed to find globally optimal solutions.

8   Conclusions

We proposed a column generation strategy for solving the constrained MSS clustering problem. The main novelty is a branch and bound algorithm that directly solves the subproblem. Experiments showed its promise: in cases where the COP-k-means algorithm is not able to find a solution satisfying all constraints even in 500 runs, CG could find solutions and in several cases even prove their optimality.

Several open questions remain. Degeneracy was not a main concern in this study, however we observe that with the simple stabilisation scheme described in section 5 the master problem still converges very slowly. It is worth investigating if advanced stabilisation techniques work better [14]. Furthermore, the pruning strategy in the branch-and-bound algorithm could be improved and the branch-and-bound could be expanded to deal with additional constraints.

Acknowledgments. This work was supported by the European Commission under the project "Inductive Constraint Programming" contract number FP7-284715 and by the Research Foundation–Flanders by means of two Postdoc grants.

References

1. Achterberg, T.: SCIP: solving constraint integer programs. Mathematical Programming Computation 1(1), 1–41 (2009)
2. Aloise, D., Deshpande, A., Hansen, P., Popat, P.: NP-hardness of euclidean sum-of-squares clustering. Machine Learning 75(2), 245–248 (2009)
3. Aloise, D., Hansen, P.: A branch-and-cut SDP-based algorithm for minimum sum-of-squares clustering. Pesquisa Operacional 29, 503–516 (2009)
4. Aloise, D., Hansen, P., Liberti, L.: An improved column generation algorithm for minimum sum-of-squares clustering. Mathematical Programming 131(1-2), 195–220 (2012)
5. Bache, K., Lichman, M.: UCI machine learning repository (2013)
6. Basu, S., Davidson, I., Wagstaff, K.: Constrained Clustering: Advances in Algorithms, Theory, and Applications. Chapman & Hall/CRC Press (2008)
7. Brusco, M.J., Stahl, S.: Minimum within-cluster sums of squares partitioning. In: Branch-and-Bound Applications in Combinatorial Data Analysis. Springer (2005)
8. Dao, T.-B.-H., Duong, K.-C., Vrain, C.: A declarative framework for constrained clustering. In: Blockeel, H., Kersting, K., Nijssen, S., Železný, F. (eds.) ECML PKDD 2013, Part III. LNCS, vol. 8190, pp. 419–434. Springer, Heidelberg (2013)
9. Davidson, I., Ravi, S.S.: The complexity of non-hierarchical clustering with instance and cluster level constraints. Data Min. Knowl. Discov. 14(1), 25–61 (2007)
10. Davidson, I., Ravi, S.S., Shamis, L.: A sat-based framework for efficient constrained clustering. In: SDM, pp. 94–105 (2010)
11. Demiriz, A., Bennett, K., Bradley, P.: Using assignment constraints to avoid empty clusters in k-means clustering. In: Constrained Clustering: Algorithms, Applications and Theory. Chapman & Hall/CRC (2008)
12. Diehr, G.: Evaluation of a branch and bound algorithm for clustering. SIAM Journal on Scientific and Statistical Computing 6(2), 268–284 (1985)
13. Dinkelbach, W.: On nonlinear fractional programming. Management Science 13(7), 492–498 (1967)
14. du Merle, O., Hansen, P., Jaumard, B., Mladenovic, N.: An interior point algorithm for minimum sum-of-squares clustering. SIAM J. Sci. Comput. 21(4), 1485–1505 (1999)
15. Gondek, D., Hofmann, T.: Non-redundant data clustering. In: ICDM, pp. 75–82 (2004)
16. Jensen, R.E.: A dynamic programming algorithm for cluster analysis. Operations Research 17(6), 1034–1057 (1969)
17. Johnson, E.L., Mehrotra, A., Nemhauser, G.L.: Min-cut clustering. Mathematical Programming 62(1-3), 133–151 (1993)
18. Koontz, W.L.G., Narendra, P.M., Fukunaga, K.: A branch and bound clustering algorithm. IEEE Trans. Comput. 24(9), 908–915 (1975)
19. Mueller, M., Kramer, S.: Integer linear programming models for constrained clustering. In: Pfahringer, B., Holmes, G., Hoffmann, A. (eds.) DS 2010. LNCS, vol. 6332, pp. 159–173. Springer, Heidelberg (2010)

20. Os, B., Meulman, J.: Improving dynamic programming strategies for partitioning. Journal of Classification 21(2), 207–230 (2004)
21. Saglam, B., Salman, F.S., Sayin, S., Türkay, M.: A mixed-integer programming approach to the clustering problem with an application in customer segmentation. European Journal of Operational Research 173(3), 866–879 (2006)
22. Schrijver, A.: Combinatorial Optimization – Polyhedra and Efficiency. Springer (2003)
23. Wagstaff, K., Cardie, C.: Clustering with instance-level constraints. In: ICML, pp. 1103–1110 (2000)
24. Xia, Y., Peng, J.: A cutting algorithm for the minimum sum-of-squared error clustering. In: SDM (2005)

A Constraint Programming-Based Column Generation Approach for Operating Room Planning and Scheduling

Seyed Hossein Hashemi Doulabi[1,3], Louis-Martin Rousseau[1,3], and Gilles Pesant[2,3]

[1] Department of Mathematics and Industrial Engineering, École Polytechnique de Montréal,
Montréal, Canada
[2] Department of Computer and Software Engineering, École Polytechnique de Montréal,
Montréal, Canada
[3] Centre interuniversitaire de recherche sur les réseaux d'entreprise, la logistique et le transport
(CIRRELT), Montréal, Canada
{hashemi.doulabi,louis-martin.rousseau,gilles.pesant}@polymtl.ca

Abstract. The Operating Room Planning and Scheduling Problem combines the assignment of surgeries to operating rooms and their scheduling over a short-term planning horizon while respecting constraints such as maximum daily working hours of surgeons and due dates of surgeries. We formulate the problem within a constraint programming-based column generation framework. Computational results demonstrate that compared with a compact formulation the proposed algorithm is robust in finding good solutions.

Keywords: operating room planning and scheduling, column generation, constraint programming.

1   Introduction

The most important goal of hospital managers is to make sure that medical resources such as operating rooms, which are the main source of revenue for hospitals, are utilized as efficiently as possible. Ad hoc planning and scheduling of operating rooms can lead to a high rate of cancellations which decreases patient satisfaction or leads to the inefficient use of many medical resources including staff and equipment. Operating theatre management is a very complex task since several groups such as surgeons, nurses, anesthesiologists, patients and operating room managers with different preferences and aversions play a role in this context. As a result, satisfying all constraints at a reasonable cost, and obtaining a solution which satisfies the objectives of all groups is very complicated or unattainable [1].

Generally operating theatre management has been studied as a three-stage process in the literature. The last stage is usually named elective case scheduling and is considered as the operational level of the operating theatre management which can be divided itself into the following two steps: 1) *Operating room planning*: each surgery is assigned to a particular operating room and a day in the planning horizon (one week). This step is also called *Advance scheduling.* 2) *Operating room scheduling*: surgeries are scheduled to be performed in particular periods in the day or the sequence of

H. Simonis (Ed.): CPAIOR 2014, LNCS 8451, pp. 455–463, 2014.
© Springer International Publishing Switzerland 2014

surgeries is determined in operating rooms. Some researchers have referred to this step as *Allocation scheduling*.

The focus of this paper is on the integration of operating room planning and scheduling. In the literature, integrated advance and allocation scheduling have rarely been addressed by researchers due to the intrinsic complexity of the synchronization [2-9]. A few papers devise some two-step methods to solve the planning and scheduling problems sequentially rather than an exact method capable of finding the optimal solution in a unified manner [2-5]. Exact approaches in the literature are some mathematical formulations based on four-index binary variables x_{ikdt} which take value 1 if surgery i starts at time t in operating room k on day d [6-9].

In this paper a new mathematical programming model is presented to solve the integrated operating room planning and scheduling problem. To solve the developed model, a column generation approach is proposed which applies a constraint programming model to solve subproblems. The rest of this paper is organized as follows. Definition of the problem and assumptions are explained in Section 2. The master problem and subproblem formulations are presented in Section 3. In Section 4, a competitive compact formulation of the problem is formulated based on the aforementioned four-index binary variables. Computational results are reported in Section 5 and some conclusions are drawn in Section 6.

2   Problem Definition and Assumptions

An integrated operating room planning and scheduling problem is defined over a planning horizon which is usually a week. On each day, there are some operating rooms with different availability times. All available surgeries can be categorized into two sets: 1) Mandatory surgeries whose due dates are within the current planning horizon and must be performed over this period. 2) Optional surgeries whose due dates are out of the planning horizon and can be postponed to a later period. The problem is considered in an open scheduling environment, i.e. the capacity of each operating room on each day can be used by different surgeons.

The other assumptions in this problem are as follows: 1) Durations of surgeries are assumed to be deterministic. 2) Availabilities of surgeons are known regarding their preferences. It is assumed that each surgeon has determined his maximum daily surgery time in the planning horizon. 3) A surgeon can be physically available in at most one operating room at a time on a given day. This constraint is referred to as coloring constraint in the literature. 4) Surgeries are generally categorized into two groups of infectious and non-infectious surgeries. In an operating room, for switching from an infectious surgery to a non-infectious one a mandatory cleaning activity must be performed. However, no cleaning is required between two infectious or two non-infectious surgeries. 5) Enough beds are available in recovery rooms and this resource does not happen to be a bottleneck. 6) Eight hours of regular time are available in each operating room. The objective function is to maximize the sum of scheduled surgery time in operating rooms over the planning horizon. 7) Available operating rooms on each day are identical from the viewpoint of available time and available equipment.

3   The Proposed Column Generation Approach

We define binary variable x_j which takes value 1 if plan j is accepted, 0 otherwise. A plan $j \in J$ shows a schedule for an operating room on a particular day. The following sets are defined to formulate the problem; Ω_1 and Ω_2 are the sets of mandatory and optional surgeries respectively. There is a due date for each mandatory surgery within the planning horizon (Note that $\Omega = \Omega_1 \cup \Omega_2$). The set of surgeons is denoted by L and for each $l \in L$, Ω_l stands for the set of surgeries corresponding to surgeon l. The set of days is represented by D. On each day $d \in D$, $|K_d|$ identical operating rooms are available. J_i and J_d represent the set of all feasible plans which include surgery i and the set of all feasible plans on day d respectively. T_d denotes the set of time slots on day d. In fact the available time in each operating room is discretized to a set of time slots. The parameters of the model are defined as follows; B_j is the sum of durations of surgeries scheduled in plan j. t_i and e_i show the duration and the due date of surgery i respectively. The surgeon corresponding to surgery i is represented by S_i. A_l^d denotes the maximum time that surgeon l prefers to work on day d. It is also assumed that R_d time slots are available in each operating room on day d. Some other auxiliary parameters are also defined as follows; a_{jt}^i is set to 1 if surgical case i starts at time t regarding plan j. a'_{ij} is equal to 1 if surgical case i is included in plan j. b_j^d is set to 1 if plan j is scheduled on day d. These auxiliary parameters are set to 0 if the stated condition is not respected.

3.1   Master Problem Formulation

Using the above definitions the master problem can be formulated as follows:

$$Max \sum_{j \in J} B_j x_j \tag{1}$$

Subject to:

$$\sum_{j \in J_i} x_j = 1 \qquad\qquad \forall i \in \Omega_1 \tag{2}$$

$$\sum_{j \in J_i} x_j \leq 1 \qquad\qquad \forall i \in \Omega_2 \tag{3}$$

$$\sum_{j \in J_d} x_j \leq |K_d| \qquad\qquad \forall d \in D \tag{4}$$

$$\sum_{j \in J_d} \sum_{i \in \Omega_l} t_i a'_{ij} x_j \leq A_l^d \qquad\qquad \forall l \in L, \forall d \in D \tag{5}$$

$$\sum_{j \in J_d} \sum_{i \in \Omega_l} \sum_{t' \in T_d : t' < t < t' + t_i} a_{it'}^i x_j \leq 1 \qquad \forall d \in D, \forall l \in L, \forall t \in T_d \tag{6}$$

$$x_j \in \{0,1\} \qquad\qquad \forall j \in J \tag{7}$$

In this formulation Constraints (2)-(3) state that some surgeries must be scheduled within the planning horizon and others are optional. Note that due dates of mandatory surgeries will be respected in subproblems. Constraint (4) states that on each day, the number of selected schedules must be equal to or less than the number of available operating rooms in the day. Each surgeon has some preferences about maximum working hours during the planning horizon which is presented by Constraint (5). Constraint (6) prevents from consideration simultaneous operations for a surgeon in different operating rooms (coloring constraint). Constraint (7) declares that variables of the model are binary. The presented master problem is inspired from the model presented by Fei et. al [10]. The difference between this model and the one presented in [10] is that we have included the surgeon overlapping constraint (Constraint 6). The presented objective function is also different from [10]. Moreover the definition of a binary variable in this model determines the schedule of an operating room in a day while in [10] the binary variable only determines the assigned surgeries to the operating room not the detailed schedule. In fact the model presented in [10] is just a planning model and cannot prevent assigning more than one surgery to a surgeon at a time. Therefore it is possible that the generated plans do not lead to feasible schedules. This difference makes the structure of the subproblem of this formulation totally different from that of [10].

3.2   Subproblem Formulation

In this section a constraint programming model is presented to formulate the subproblem. The reason for applying constraint programming is that we plan to extend the presented column generation to a branch-and-price-and-cut in the future, and the flexibility of constraint programming will make it easier to handle the effect of adding cuts and branching rules.

Assuming $\pi_i^{(2)}, \pi_i^{(3)}, \pi_d^{(4)}$, $\pi_{ld}^{(5)}$ and $\pi_{ldt}^{(6)}$ as the dual variables corresponding to Constraints (2)-(6), the reduced cost of variable x_j can be computed by the following formula:

$$\sigma_j = B_j - \sum_{d \in D} \sum_{i \in \Omega} b_j^d a_{ij}' \pi_i^{(2,3)} - \sum_{d \in D} b_j^d \pi_d^{(4)} - \sum_{d \in D} \sum_{l \in L} \sum_{i \in \Omega_l} t_i b_j^d a_{ij}' \pi_{ld}^{(5)} -$$
$$\sum_{d \in D} \sum_{l \in L} \sum_{i \in \Omega_l} \sum_{t \in T} \sum_{t' \in T_d : t' \le t < t' + t_i} a_{jt'}^i b_j^d \pi_{ldt}^{(6)}$$

In the subproblem of column generation we have to maximize this reduced cost expression because we are looking for the column with the most positive reduced cost. We can decompose the subproblem of the column generation over index d as follows:

$$Max \left\{ B_j - \sum_{i \in \Omega} a_{ij}' \pi_i^{(2,3)} - \pi_d^{(4)} - \sum_{l \in L} \sum_{i \in \Omega_l} t_i a_{ij}' \pi_{ld}^{(5)} - \right.$$
$$\left. \sum_{l \in L} \sum_{i \in \Omega_l} \sum_{t \in T} \sum_{t' \in T_d : t' \le t < t' + t_i} a_{jt'}^i \pi_{ldt}^{(6)} \right\}$$

Since in each operating room only a small number of surgeries (n) can be performed on a given day, we adopt a position-based model and define the variables of the constraint programming model in the subproblem as follows: W_p shows the index of surgery to be operated in position p in the pattern on day d. ST_p shows the start time of the surgery assigned to position p in the pattern on day d. Based on these variables, constraints in the subproblem can be presented as follows:

$$W_1 \in \{i \in \Omega \mid e_i \geq d\} \tag{8}$$

$$W_p \in \{i \in \Omega \mid e_i \geq d\} \cup \{0\} \qquad \forall p \in \{2, \dots, n\} \tag{9}$$

$$ST_p \in \{0, 1, \dots, |T_d|\} \qquad \forall p \in \{1, \dots, n\} \tag{10}$$

$$\text{If } (W_p = 0) \text{ then } (W_{p+1} = 0) \qquad \forall p \in \{2, \dots, n-1\} \tag{11}$$

$$ST_{p+1} \geq ST_p + t_{[W_p]} + CL_{[W_p][W_{p+1}]} \qquad \forall p \in \{1, \dots, n-1\} \tag{12}$$

$$\text{gcc}(W, \{0\} \cup \Omega, [0, \dots, 0], [n-1, 1, \dots, 1]) \tag{13}$$

$$\sum_{p=1}^{n} \left(S_{[W_p]} == l \right) t_{[W_p]} \leq A_l^d \qquad \forall l \in L \tag{14}$$

$$\text{If } W_p = 0 \text{ then } ST_p = ST_{p-1} + t_{[W_{p-1}]} \qquad \forall p \in \{2, \dots, n\} \tag{15}$$

$$t_{[0]} = 0, \; S_{[0]} = 0 \tag{16}$$

$$CL_{[0][i]} = 0, \; CL_{[i][0]} = 0 \qquad \forall i \in \Omega \cup \{0\} \tag{17}$$

Constraint (8) states that the first position in the pattern must be assigned to one of the surgeries in set Ω whose due date is on or after day d. Constraint (9) shows that other positions can be either assigned to a surgery in set Ω or they can be left empty by taking 0. In fact, n is an upper bound on the number of assigned surgeries to an operating room and a pattern in a day can include fewer surgeries. Constraint (10) declares the domain of start time variables. Constraint (11) states that if a position is left empty the next available positions must be empty. It means that positions must be filled one after the other and there cannot be any empty position between two assigned surgeries. It should be noted that having a hole in the schedule is possible since the start time of surgeries are determined by variables ST_p and Constraint (11) only breaks the symmetry of the model by preventing it from obtaining the same solution through different assignment of surgeries to the available positions.

Constraint (12) ensures that start time of a surgery in a position is greater than the finish time of previously scheduled surgeries plus the duration of a possible cleaning activity. In this constraint, CL is a data matrix which includes possible cleaning times between surgeries. The entry in row i and column j of this matrix is equal to the duration of the cleaning activity, if surgery i is infectious and surgery j is non-infectious. Each surgery can be scheduled at most once in a pattern. This constraint is formulated by the global cardinality constraint in (13). The upper bound value for $\{0\}$ is $n-1$ in order to prevent from generating an empty plan since empty plans have a coefficient of 0 in the objective function and can never be generated in the subproblem. The upper bound values for other values in Ω is 1. The lower bound values for all members of $\Omega \cup \{0\}$ are zero. In fact this constraint means that at most $n-1$ positions can be left empty and each surgery can be assigned at most once in a schedule. Constraint (14) prevents from generating columns which do not satisfy the maximum possible working hours of surgeons. Constraint (15) forces the start time of an empty position to be equal to the finish time of the previous position. Constraint (16) states that when

W_p equals 0, $t_{[W_p]}$ and $S_{[W_p]}$ will be equal to zero. Therefore, this fact together with Constraints (12) and (15) ensure that when W_p equals 0 the start time of this position and also subsequent positions will be fixed to the start time of the last position with a surgery from Ω. Similarly, Constraint (17) forces the entries in row 0 or column 0 of data matrix CL to be equal to 0. In summary, the constraint programming model to be solved as the subproblem is as follows:

$$Max \left\{ \sum_{p=1}^{n} t_{[W_p]} - \sum_{p=1}^{n} \pi_{[W_p]}^{(2,3)} - \sum_{p=1}^{n} \left(t_{[W_p]} \pi_{d[S_{[W_p]}]}^{(5)} \right) - \sum_{p=1}^{n} \sum_{t \in T_d} \left(t < \right. \right.$$
$$\left. \left(ST_p + t_{[W_p]} \right) \ \wedge \ t \geq ST_p \right) \pi_{d[S_{[W_p]}]t}^{(6)} - \pi_d^{(4)} \right\}$$

Subject to: Constraints (9)-(18)

In the objective function, $\pi_{[W_p]}^{(2,3)}$, $\pi_{d[S_{[W_p]}]}^{(5)}$ and $\pi_{d[S_{[W_p]}]t}^{(6)}$ are element constraints which will be zero if W_p equals 0. Each iteration of the column generation starts by solving the relaxed master problem. Then using the dual values taken from the relaxed master problem, the subproblems will be solved consecutively. During constraint programming search, solutions with a positive reduced cost will be added as new columns to the master problem. If at least one column is generated the next iteration of column generation will start, otherwise, the column generation stops. At the end of the column generation, the objective value obtained by solving the LP relaxation of the master problem is an upper bound for the problem and by solving an integer programming model of the master problem restricted to the generated columns a feasible lower bound of the problem can be obtained. It is worth mentioning that modeling the subproblem as a constrained scheduling problem is not likely to be efficient because only a small fraction of surgeries are scheduled in a subproblem on a given day. Surgeries are all considered to be optional activities, leading to very weak filtering.

4   A Compact Formulation

We consider the compact formulation of our problem obtained by enhancing the formulation presented in [6-8] which is based on binary variables x_{ikdt}. Such a variable takes value 1 if the surgery is scheduled on day d in operating room k at time t and 0 otherwise. Based on these variables, constraints equivalent to (2), (3), (5) and (6) can be formulated as presented in [8]. A constraint to prevent overlapping of surgeries in operating rooms can be also formulated based on these variables as presented in [8]. However, to formulate the additional cleaning time due to switching from infectious to non-infectious surgeries two new binary variables y_{ikd} and z_{ijkd} are required to be defined. y_{ikd} takes value 1 if surgery i is scheduled on day d in operating room k. Variable z_{ijkd} is defined for those pairs of surgeries i and j where one of them is infectious and the other one is non-infectious and also the due dates of both surgeries must be on or after day d. This variable takes 1 if surgeries i and j are scheduled on day d in operating room k and surgery i is scheduled to start before surgery j and 0 otherwise. Based on these variables the following constraints must be included in the model. The set of feasible pairs of surgeries for variable z_{ijkd} is denoted by $Pairs$.

$$y_{ikd} = \sum_{t \in T} x_{ikdt} \qquad \forall d \in D, \forall k \in K_d, \forall i \in \Omega: e_i \geq d \qquad (18)$$

$$z_{ijkd} \leq y_{ikd}, \ z_{jikd} \leq y_{jkd} \qquad \forall d \in D, \forall k \in K_d, \forall (i,j) \in Pairs \qquad (19)$$

$$z_{ijkd} + z_{jikd} \leq 1 \qquad \forall d \in D, \forall k \in K_d, \forall (i,j) \in Pairs \ (i < j) \qquad (20)$$

$$z_{ijkd} + z_{jikd} \geq y_{ikd} + y_{jkd} - 1 \qquad \forall d \in D, \forall k \in K_d, \forall (i,j) \in Pairs \ (i < j) \qquad (21)$$

$$\sum_{t \in T}(tx_{ikdt}) + t_i + CT_{ij} \leq \atop \sum_{t \in T}(tx_{jkdt}) + M(1 - z_{ijkd}) \qquad \forall d \in D, \forall k \in K_d, \forall (i,j) \in Pairs \qquad (22)$$

Constraint (18) links variable y_{ikd} to the model. Constraints (20)-(21) forces $z_{ijkd} + z_{jikd}$ to be equal to 1, if surgeries i and j are scheduled in operating room k on day d. In Constraint (22), CT_{ij} is equal to the additional cleaning time if surgery i is infectious and surgery j is non-infectious and 0 otherwise. In this constraint, M is a big constant and if z_{ijkd} takes 1 the start time of surgery j will be forced to be after the finish time of surgery i plus the additional cleaning time.

5   Computational Experiments

To solve the master problem and subproblem IBM ILOG CPLEX Optimization Studio V12.4 is used which includes CPLEX and CP Optimizer to solve linear programming and constraint programming models respectively. Experiments were run on a computer with 2 processors Intel Xeon X5675, 3.07 gigahertz and 96 gigabyte of RAM. To conduct computational experiments we generated a set of instances following the method that Fei et. al [10] suggested. Durations of surgeries are uniformly generated from interval [2 hours, 4 hours] where time is discretized based on five-minute units. Six operating rooms are assumed to be available on each day and the available time in each operating room is eight hours. 50% of surgeries are randomly picked to be infectious. The duration of mandatory cleaning activities is set to 30 minutes. A constructive heuristic is applied to generate the initial solution of the column generation. This heuristic fills operating rooms consecutively starting from the first operating room in the first day to the last one in the last day. Surgeries which have not been already scheduled are sorted based on their due dates.

The computational results are shown in Table 1. In this table the first and the second columns present the instance number and the number of surgeries. The next column presents the objective value of the initial solution that we provide for column generation as initial columns. Four columns under "Column Generation" show the computational results obtained from solving the instances by the proposed column generation method. The Columns "Best Sol." and "UB" present the lower and upper bound values. Upper bound values are the objective value of the relaxed master after convergence of the column generation and lower bound values have been obtained by solving the master problem restricted to the generated columns. The next column, "Imp. (%)", presents the percentage of improvement over the initial solution which is calculated by 100*(Best Sol- Initial Obj)/ Initial Obj. The total time consumed to get the upper and lower bounds for each instance is presented under Column "Time". Information on lower and upper bound values, percent of initial solution improvement and computational time of the compact formulation presented in Section 4, are shown in the next four columns. All computational times are in seconds. As shown in the table the column generation method has converged within 45 minutes for all instances.

Computational results show that the compact formulation finds optimal solutions for small instances with 40 surgeries, but it is not efficient in solving larger instances and cannot improve the initial solution in many cases. However, the column generation is robust in improving the initial solution in all but one instance. Moreover, we can see that the upper bound values obtained by the column generation have been dominated by those of the compact formulation. This implies that there is a possibility to improve the column generation by embedding it in a branch-and-price framework.

Table 1. Comparison of the developed column generation and a compact formulation

Ins. #	$|\Omega|$	Initial Obj.	Column Generation				Compact Formulation			
			Best Sol.	UB	Imp. (%)	Time	Best Sol.	UB	Imp. (%)	Time
1	40	1263	1342	1359	6.25	12	1342	1342	6.25	505
2	40	1365	1365	1383	0.00	8	1365	1365	0.00	36
3	40	1302	1420	1446	9.06	9	1431	1431	9.91	349
4	40	1257	1378	1512	9.63	10	1420	1420	12.9	210
5	40	1417	1457	1457	2.82	6	1457	1457	2.82	54
6	80	1874	2060	2285	9.93	296	-	2200	-	2700
7	80	1895	2146	2415	13.2	755	-	-	-	2700
8	80	1873	2059	2324	9.93	883	-	2222	-	2700
9	80	1837	2057	2203	11.9	191	1942	2174	5.72	2700
10	80	1991	2183	2308	9.64	403	1905	2252	-4.32	2700
11	120	1966	2126	2398	8.14	592	-	2228	-	2700
12	120	2039	2194	2421	7.60	2496	-	2368	-	2700
13	120	2074	2176	2432	4.92	911	-	2385	-	2700
14	120	2011	2180	2432	8.40	693	-	2388	-	2700
15	120	1978	2213	2432	11.8	1682	-	-	-	2700

6   Summary and Conclusion

We developed an efficient column generation method for the Operating Room Planning and Scheduling Problem. The proposed column generation uses a constraint programming model to solve the subproblem. Based on the formulations in the literature, a compact formulation is also presented as a competitive algorithm. Computational experiments demonstrate that the proposed column generation is robust in finding good solutions compared with the compact formulation. As future research, studying the enhanced versions of the problem with more realistic constraints such as capacity constraints in recovery rooms is very interesting. Including the preferences of operating rooms staff such as anesthesiologists and nurses can also be another research direction.

References

1. Guerriero, F., Guido, R.: Operational research in the management of the operating theatre: A survey. Health Care Management Science 14(1), 89–114 (2011)
2. Jebali, A., Hadj Alouane, A.B., Ladet, P.: Operating rooms scheduling. . International Journal of Production Economics 99(1), 52–62 (2006)
3. Guinet, A., Chaabane, S.: Operating theatre planning. . International Journal of Production Economics 85(1), 69–81 (2003)
4. Fei, H., Meskens, N., Chu, C.: An operating theatre planning and scheduling problem in the case of a block scheduling strategy. In: 2006 International Conference on Service Systems and Service Management, vol. 1, pp. 422–428. IEEE (2006)
5. Fei, H., Meskens, N., Chu, C.: A planning and scheduling problem for an operating theatre using an open scheduling strategy. Computers & Industrial Engineering 58(2), 221–230 (2010)
6. Roland, B., Di Martinelly, C., Riane, F.: Operating theatre optimization: A resource-constrained based solving approach. In: 2006 International Conference on Service Systems and Service Management, vol. 1, pp. 443–448. IEEE (2006)
7. Roland, B., Di Martinelly, C., Riane, F., Pochet, Y.: Scheduling an operating theatre under human resource constraints. Computers & Industrial Engineering 58(2), 212–220 (2010)
8. Marques, I., Captivo, M.E., Pato, M.V.: An integer programming approach to elective surgery scheduling. OR Spectrum 34(2), 407–427 (2012)
9. Vijayakumar, B., Parikh, P.J., Scott, R., Barnes, A.,Gallimore, J.: A dual bin-packing approach to scheduling surgical cases at a publicly-funded hospital. European Journal of Operational Research 224(3), 583–591 (2012)
10. Fei, H., Chu, C., Meskens, N.: Solving a tactical operating room planning problem by a column-generation-based heuristic procedure with four criteria. Annals of Operations Research 166(1), 91–108 (2009)

Dynamic Controllability and Dispatchability Relationships

Paul Morris

NASA Ames Research Center
Moffett Field, CA 94035, U.S.A.

Abstract. An important issue for temporal planners is the ability to handle temporal uncertainty. Recent papers have addressed the question of how to tell whether a temporal network is *Dynamically Controllable*, i.e., whether the temporal requirements are feasible in the light of uncertain durations of some processes. We present a fast algorithm for Dynamic Controllability. We also note a correspondence between the reduction steps in the algorithm and the operations involved in converting the projections to dispatchable form. This has implications for the complexity for sparse networks.

1 Introduction

Many Constraint-Based Planning systems (e.g. [1]) use Simple Temporal Networks (STNs) to test the consistency of partial plans encountered during the search process. These systems produce *flexible* plans where every solution to the final Simple Temporal Network provides an acceptable schedule. The flexibility is useful because it provides scope to respond to unanticipated contingencies during execution, for example where some activity takes longer than expected. However, since the uncertainty is not modeled, there is no guarantee that the flexibility will be sufficient to manage a particular contingency.

Many applications, however, involve a specific type of temporal uncertainty where the duration of certain processes or the timing of exogenous events is not under the control of the agent using the plan. In these cases, the values for the variables that are under the agent's control may need to be chosen so that they do not constrain uncontrollable events whose outcomes are still in the future. This is the *controllability* problem. By formalizing this notion of temporal uncertainty, it is possible to provide guarantees about the sufficiency of the flexibility.

In [2], several notions of controllability are defined, including *Dynamic Controllability* (DC). Roughly speaking, a network is dynamically controllable if there is an execution strategy that satisfies the constraints and depends only on knowing the outcomes of uncontrollable events up to the present time.

The fastest known algorithm for computing Dynamic Controllability (DC) is the $O(N^4)$ algorithm of [3] (N is number of nodes). That paper introduces a structural characterization of DC in terms of the absence of a particular type of cycle, called a *semi-reducible negative cycle*. This is analogous to the result

H. Simonis (Ed.): CPAIOR 2014, LNCS 8451, pp. 464–479, 2014.

characterizing consistency of ordinary STNs in terms of the absence of negative cycles in the distance graph. Other properties of semi-reducible negative cycles have been studied by Hunsberger [4], who corrected a flaw in the formal definition of DC. An excellent tutorial on Dynamic Controllability is available online [5].

In this paper, we exploit recursive structure within the semi-reducible paths and present a new algorithm that runs in $O(N^3)$ time. We also consider the relationship to the dispatchability of the projections.

Other authors (e.g. [6,7]) have pursued incremental algorithms where the Dynamic Controllability property is rechecked after the addition of a new edge to a network that has already been shown to be Dynamically Controllable. This corresponds to a common situation in temporal planning where edges are added incrementally to resolve flaws in the plan. These algorithms have been shown empirically to have $O(N^3)$ complexity for each increment on a suite of randomly generated problems. We do not address incrementality in this paper. Additional work has studied related concepts in wider contexts, e.g. [8,9].

2   Background

This background section defines the basic formalism of Dynamic Controllability, following [3,4].

A Simple Temporal Network (STN) [10] is a graph in which the edges are annotated with upper and lower numerical bounds. The nodes in the graph represent temporal events or *timepoints*, while the edges correspond to constraints on the durations between the events. Each STN is associated with a *distance graph* derived from the upper and lower bound constraints. An STN is consistent if and only if the distance graph does not contain a negative cycle. This can be determined by a single-source shortest path propagation such as in the Bellman-Ford algorithm [11] (faster than Floyd-Warshall for sparse graphs, which are common in practical problems). To avoid confusion with edges in the distance graph, we will refer to edges in the STN as *links*.

A Simple Temporal Network With Uncertainty (STNU) is similar to an STN except the links are divided into two classes, *requirement links* and *contingent links*. Requirement links are temporal constraints that the agent must satisfy, like the links in an ordinary STN. Contingent links may be thought of as representing causal processes of uncertain duration, or periods from a reference time to exogenous events; their finish timepoints, called *contingent timepoints*, are controlled by Nature, subject to the limits imposed by the bounds on the contingent links. All other timepoints, called *executable timepoints*, are controlled by the agent, whose goal is to satisfy the bounds on the requirement links. We assume the durations of contingent links vary independently, so a control procedure must consider every combination of such durations. Each contingent link is required to have non-negative (finite) upper and lower bounds, with the lower bound strictly less than the upper. We assume contingent links do not share finish points. (Networks with coincident contingent finishing points cannot be Dynamically Controllable.)

Choosing one of the allowed durations for each contingent link may be thought of as reducing the STNU to an ordinary STN. Thus, an STNU determines a family of STNs corresponding to the different allowed durations; these are called *projections* of the STNU.

Given an STNU with N as the set of nodes, a *schedule* T is a mapping

$$T : N \to \Re$$

where $T(x)$ is called the *time* of timepoint x. A schedule is said to be *consistent* if it satisfies all the link constraints.

The *history* of a specific time t with respect to a schedule T, denoted by $T\{\prec t\}$, specifies the durations of all contingent links that have finished up to and including time t.

Hunsberger [4] corrected a flaw in the original definition of Dynamic Controllability by defining history in terms of a specific time rather than a timepoint; we follow that approach here and in the definition of dynamic strategy below. However, we also follow the variation of including the present that was introduced in [3]. The latter issue is whether the agent can react instaneously to an observation to execute a new timepoint, or requires an infinitesimal amount of time to react. Both of these are mathematical idealizations: a realistic reaction might take a finite amount of time, which could be modeled by a separate link or folded into the contingent process being observed. The instantaneous idealization choice leads to a cleaner mathematical formulation and simpler algorithms.

An *execution strategy* S is a mapping

$$S : \mathcal{P} \to \mathcal{T}$$

where $\mathcal{P}$ is the set of projections and $\mathcal{T}$ is the set of schedules. An execution strategy S is *viable* if $S(p)$, henceforth written S_p, is consistent with p for each projection p.

An STNU is *Dynamically Controllable* if there is a *dynamic* execution strategy, that is, a viable execution strategy S such that

$$S_{p1}\{\prec t\} = S_{p2}\{\prec t\} \Rightarrow S_{p1}(x) = S_{p2}(x)$$

for each executable timepoint x and projections $p1$ and $p2$, where $t = S_{p1}(x)$ [4]. Thus, a Dynamic execution strategy assigns a time to each executable timepoint that may depend on the outcomes of contingent links in the past (or present), but not on those in the future. This corresponds to requiring that only information available from observation may be used in determining the schedule. We will use *dynamic strategy* in the following for a (viable) Dynamic execution strategy.

3   Previous Algorithms

In [12], an algorithm for Dynamic Controllability was presented that runs in pseudo-polynomial time. The algorithm analyzes triangles of links and possibly tightens some constraints in a way that makes explicit the limitations to the execution strategies due to the presence of contingent links.

Some of the tightenings involve a novel temporal constraint called a *wait*. Given a contingent link AB and another link AC, the $<B, t>$ annotation on AC indicates that execution of the timepoint C is not allowed to proceed until after either B has occurred or t units of time have elapsed since A occurred. More precisely, it corresponds to the constraint $C - A \geq \min(B - A, t)$. Thus, a wait is a ternary constraint involving A, B, and C. Note that a wait reduces to a binary constraint in any projection, since there the value $B - A$ is fixed.

The tightenings in the original algorithm, called *reductions*, were expressed in terms of rules that were applied to the STNU graph. We now review developments in [13,3,4], which re-express the reductions in a more mathematically concise form, using a derived graph.

An ordinary STN has an alternative representation as a *distance graph* [10]. Similarly, there is an analogous representation for an STNU called the *labeled distance graph* [13]. This is actually a multigraph (which allows multiple edges between two nodes), but we refer to it as a graph for simplicity. In the labeled distance graph, each requirement link $A \xrightarrow{[x,y]} B$ is replaced by two edges $A \xrightarrow{y} B$ and $A \xleftarrow{-x} B$, just as in an STN. For a contingent link $A \xRightarrow{[x,y]} B$, we have the same two edges $A \xrightarrow{y} B$ and $A \xleftarrow{-x} B$, but we also have two additional edges of the form $A \xrightarrow{b:x} B$ and $A \xleftarrow{B:-y} B$. These are called *labeled edges* because of the additional "b:" and "B:" annotations indicating the contingent timepoint B with which they are associated. Note especially the reversal in the roles of x and y in the labeled edges. We refer to $A \xleftarrow{B:-y} B$ and $A \xrightarrow{b:x} B$ as *upper-case* and *lower-case* edges, respectively. Observe that the upper-case labeled weight B:-y gives the value the edge would have in a projection where the contingent link takes on its maximum value, whereas the lower-case labeled weight corresponds to the contingent link minimum value.

There is also a representation for a $A \xrightarrow{<B, t>} C$ wait constraint in the labeled distance graph. This corresponds to a single edge $A \xleftarrow{B:-t} C$. Note the analogy to a lower bound. This weight is consistent with the lower bound that would occur in a projection where the contingent link has its maximum value.

We can now represent the tightenings in terms of the labeled distance graph. As in [3], we present a version of the rules that assumes the agent can react instantaneously to observed events.

The reductions from the classic algorithm are replaced by what is essentially a single reduction with different flavors, together with a label removal rule:

(UPPER-CASE REDUCTION)

$$A \xleftarrow{B:x} C \xleftarrow{y} D \quad \text{adds} \quad A \xleftarrow{B:(x+y)} D$$

(LOWER-CASE REDUCTION) If $x < 0$,

$$A \xleftarrow{x} C \xleftarrow{c:y} D \quad \text{adds} \quad A \xleftarrow{x+y} D$$

(CROSS-CASE REDUCTION) If $x < 0$, $B \neq C$,

$A \xleftarrow{B:x} C \xleftarrow{c:y} D$ adds $A \xleftarrow{B:(x+y)} D$

(NO-CASE REDUCTION)

$A \xleftarrow{x} C \xleftarrow{y} D$ adds $A \xleftarrow{x+y} D$

(LABEL REMOVAL REDUCTION) If $z \geq -x$,

$B \xleftarrow{b:x} A \xleftarrow{B:z} C$ adds $A \xleftarrow{z} C$

With this reformulation, the "Case" (first four) reductions can all be seen as forms of composition of edges, with the labels being used to modulate when those compositions are allowed to occur. In light of this, the *unlabeled distance*[1] of a path in the labeled distance graph is defined to be the sum of edge weights in the path, ignoring any labels. Thus, the reductions preserve the unlabeled distance.

Morris [3] observes that a duration-uncertain contingent link $A \xrightarrow{[x,y]} B$ can be decomposed $A \xrightarrow{[x,x]} C \xrightarrow{[0,y-x]} B$ into a duration-certain part $A \xrightarrow{[x,x]} C$ and a "pure" duration-uncertain part $C \xrightarrow{[0,y-x]} B$. If this is done for all contingent links, the STNU is said to be in *Normal Form*. In that case, the contingent link lower bounds all become zero, so the LABEL REMOVAL reduction assumes a simpler form as follows.

(LABEL REMOVAL) If $x \geq 0$,

$A \xleftarrow{B:x} C$ adds $A \xleftarrow{x} C$

We will assume in the remainder of the paper that STNU networks are in Normal Form since this simplifies the analysis and algorithms without loss of generality.

3.1   Path Transformations

The key to speeding up the determination of Dynamic Controllability is to perform the reductions in an organized way. This in turn is facilitated by considering the relationship of paths to Dynamic Controllability. To this end, Morris [3] defines a concept of *semi-reducible path*, which we review here.

An ordinary STN is consistent if and only if its distance graph does not contain a negative cycle. There is a related characterization of DC in terms of negative cycles in the labeled distance graph. This involves a notion of path transformation.

Consider a path $\mathcal{P}$ that contains a subpath $\mathcal{Q}$ between two points A and B and suppose $\mathcal{Q}$ matches the left side of a reduction. Then applying the reduction

---

[1] Terminology from [4]. Called *reduced distance* in [3], which is somewhat misleading.

to Q yields a new edge e between A to B. Consider the path $\mathcal{P}'$ obtained from $\mathcal{P}$ by replacing Q by e. We may regard $\mathcal{P}$ as being *transformed* into $\mathcal{P}'$ by the reduction. Note that $\mathcal{P}'$ has the same unlabeled distance as $\mathcal{P}$ since the reductions preserve unlabeled distance.

Definition 1. *A path is* **reducible** *if it can be transformed into a single edge by a sequence of reductions. A path is* **semi-reducible** *if it can be transformed into a path without lower-case edges by a sequence of reductions.*

The property of semi-reducible can be directly characterized in structural terms. The following notation is useful. We write $e < e'$ in $\mathcal{P}$ if e is an earlier edge than e' in $\mathcal{P}$, where $\mathcal{P}$ is a path in the labeled distance graph. If A and B are nodes in the path, we write $\mathcal{D}_{\mathcal{P}}(A, B)$ for the unlabeled distance from A to B in $\mathcal{P}$. We denote the start and end nodes of an edge e by $\text{start}(e)$ and $\text{end}(e)$, respectively.

Definition 2. *Suppose e is a lower-case edge in $\mathcal{P}$ and e' is some other edge such that $e < e'$ in $\mathcal{P}$. The edge e' is a* **drop edge** *for e in $\mathcal{P}$ if $\mathcal{D}_{\mathcal{P}}(\text{end}(e), \text{end}(e')) < 0$. The edge e' is a* **moat edge** *for e in $\mathcal{P}$ if it is a drop edge and there is no other drop edge e'' such that $e'' < e'$ in $\mathcal{P}$. In this case, we call the subpath of $\mathcal{P}$ from $\text{end}(e)$ to $\text{end}(e')$ the* **extension** *of e in $\mathcal{P}$. We say the moat edge is* **unusable** *if e and e' have labels that come from the same contingent link; otherwise it is* **usable**.

Thus, a drop edge is where the path following e becomes negative, and a moat edge is a closest drop edge. An unusable moat edge will have a label that is the upper-case version of the label on the lower-case edge.[2]

The extension subpath $\mathcal{P}$ turns out to have a very useful property called the *prefix/postfix property*, which says that every nonempty proper prefix of $\mathcal{P}$ has non-negative unlabeled distance and every nonempty proper postfix of $\mathcal{P}$ has negative unlabeled distance. The *Nesting Lemma* [3] says if two prefix/postfix paths have a non-empty intersection, then one of the paths is contained in the other. This means that if a path has two subpaths with the property, then the subpaths are either nested or disjoint.

The main results of [3] provided a characterization of Dynamic Controllability in terms of the structure of the labeled distance graph, as follows.

Theorem 1. *A path $\mathcal{P}$ is semi-reducible if and only if every lower-case edge in $\mathcal{P}$ has a usable moat edge in $\mathcal{P}$.*

Theorem 2. *An STNU is Dynamically Controllable if and only if it does not have a semi-reducible negative cycle.*

The labeled distance graph can be used to calculate distances between nodes in a similar manner to an ordinary STN distance graph, provided the restrictions imposed by the labels are respected. The approach in [3] calculates distances forward from each contingent timepoint looking for a usable moat edge to reduce

---

[2] We will see later the reductions may be viewed as performing composition of edges in projections, and these edges belong to incompatible projections.

away the associated lower-case edge. For the innermost nested extensions, this can be done in a single pass. This produces new edges that bypass these extensions, which decrements the level of nesting. Each pass has a complexity bound of $O(N^3)$ for the distance calculation. It was shown that the depth of nesting can be linearly bounded leading to a linear cutoff and an overall $O(N^4)$ complexity.

4   Cubic Algorithm

We now present a cubic algorithm for Dynamic Controllability using the same formal framework as [3], but with a different organization of the computation. Note that a moat edge must have negative unlabeled distance; thus, it must be either a negative ordinary edge, or a negative upper case labeled edge. Before leaving NASA, Nicola Muscettola [14, Personal Communication] proposed the following key ideas, which he anticipated would lead to a cubic algorithm. However, to the best of our knowledge, such an algorithm has not been published. We have formulated one based on these ideas and include it here. The key ideas are:

- Calculate distance *backwards* from potential moat edges. (That is, calculate distance *to* rather than distance *from*.)
- Calculate the distance over non-negative edges using Dijkstra's algorithm [11].
- If a new negative edge is encountered, invoke a recursive call before continuing the distance calculation.
- A recursive cycle indicates the network is not Dynamically Controllable.

The backward distance calculation implicitly uses the reduction rules and may add new non-negative edges. The following example illustrates the approach (parentheses added for readability):

$$[(E \xrightarrow{4} B \xrightarrow{B:-2} A) \xrightarrow{b:0} B \xrightarrow{1} D \xrightarrow{D:-3} C] \xrightarrow{d:0} (D \xrightarrow{3} B \xrightarrow{B:-2} A) \xrightarrow{b:0} B \xrightarrow{-2} E$$

Consider a backward distance calculation starting at E. This will invoke a recursive call when it gets to A, which will add a $D \xrightarrow{1} A$ edge. (It also adds a $E \xrightarrow{2} A$ edge, which we ignore for now.) The top-level call will then continue using the $D \xrightarrow{1} A$ edge until it gets to C, where it causes another recursive call. When the call at C gets to A, it will use the already added $E \xrightarrow{2} A$ edge and leave behind a new $E \xrightarrow{0} C$ edge, at which point the top-level call resumes and encounters a recursive cycle.

An issue of special note is that Dijkstra's algorithm is normally restricted to graphs with non-negative edges whereas in our case the initial edges connected to the Dijkstra source may be negative. However, it is easy to see (e.g., discussion in [15]) that the algorithm is still valid provided that (1) the only negative edges are the initial edges and (2) the propagation does not compute a negative distance to the source.[3] If the propagation computes a negative distance to the source, this will be detected as a recursive cycle.

The goal of the computation is to discover semi-reducible negative cycles. It turns out that a restricted form of the reduction rules is sufficient to make this discovery because of the negativity. In particular, the Case reductions can all be restricted to $x < 0$ and $y \geq 0$. Application of the rules will stop when all the edges have the same sign, which must be negative since the whole cycle has negative unlabeled distance. (A rule cannot fail because of the label restrictions; if it did, the original cycle would contain an unusable moat edge and would not be semi-reducible.) A semi-reducible negative cycle is thus transformed by the rules to a cycle of all-negative edges, similar to a result of [7].

The restricted reduction rules are equivalent to the BackPropagate-Tighten rules used in the Incremental Dynamic Controllability work (e.g. [6,7]).[4] We will refer to the restricted reduction rules as *Plus-Minus* reductions since they involve a non-negative edge followed by a negative edge. We will also regard LABEL REMOVAL as being implicitly applied wherever it is applicable.

Before presenting the detailed algorithm (Fig. 1), we make a modification to the STNU to simplify the exposition. (An implementation could behave as if this modification is made without actually changing the data structures.) The modification separates the start nodes of contingent links from other contingent links and from the targets of ordinary negative edges. Thus,

$$B \Longleftarrow A \Longrightarrow C \quad \text{becomes} \quad B \Longleftarrow A \xrightarrow{[0,0]} A' \Longrightarrow C$$

$$B \Longrightarrow A \Longrightarrow C \quad \text{becomes} \quad B \Longrightarrow A \xrightarrow{[0,0]} A' \Longrightarrow C$$

$$B \xrightarrow{-x} A \Longrightarrow C \quad \text{becomes} \quad B \xrightarrow{-x} A \xrightarrow{[0,0]} A' \Longrightarrow C$$

It is easy to see that the resulting STNU is equivalent to the original one. (Recall that we allow instantaneous reactions.) The modification keeps distance calculations involving different (and no) labels separate and adds at most $O(K)$ nodes and edges, where K is the number of contingent links.

The algorithm is summarized in Fig. 1. We have used indentation instead of begin-end to set off blocks of code. The `continue` statement, as in Java, skips to the next turn of the immediately enclosing loop. A negative node is a node that is the target of some negative edge. There is a separate distance function for the distance to each source, but we have abbreviated distance(x,source) as distance(x) to avoid clutter.

This is essentially a distance-limited version of Dijkstra's algorithm except for lines 00-01 and 19-21, which deal with the recursive aspect, lines 02-03, which

---

[3] Consider the proof of correctness [11] of the usual algorithm. This relies on the fact that the distance to head nodes of the priority queue cannot be superseded by paths from later nodes, which start at a greater distance and are over non-negative edges. In our case, after processing the initial node, this will still be true for subsequent head nodes because paths from later nodes will use non-initial edges.

[4] They have more rules because of multiple choices of focus edge, and because they make a distinction between direct and derived upper-case edges.

```
Boolean procedure determineDC()
  for each negative node n do
    if DCbackprop(n) = false
      return false;
  return true;
end procedure


Boolean procedure DCbackprop(source)
00 if ancestor call with same source
01   return false;
02 if prior terminated call with source
03   return true;
04 distance(source) = 0;
05 for each node x other than source do
06   distance(x) = infinity;
07 PriorityQueue queue = empty;
08 for each e1 in InEdges(source) do
09   Node n1 = start(e1);
10   distance(n1) = weight(e1);
11   insert n1 in queue;
12 while queue not empty do
13   pop Node u from queue;
14   if distance(u) >= 0
15     Edge e' = new Edge(u, source);
16     weight(e') = distance(u);
17     add e' to graph;
18     continue;
19   if u is negative node
20     if DCbackprop(u) = false
21       return false;
22   for each e in InEdges(u) do
23     if weight(e) < 0
24       continue;
25     if e is unsuitable
26       continue;
27     Node v = start(e);
28     new = distance(u) + weight(e);
29     if new < distance(v)
20       distance(v) = new;
35       insert v into queue;
36 return true;
end procedure
```

Fig. 1. Cubic Algorithm

short-circuit later calls with the same source, lines 08-11, which unroll the initial propagation, lines 14-18, which add non-negative edges to the graph, and the unsuitability condition in lines 25-26, which occurs if the source edge is unusable for e (from the same contingent link).[5] The distance limitation occurs at the first non-negative value reached along a path (where a non-negative edge is added).

Notice that if the e in line 22 is a non-negative suitable edge, then since distance(u) is negative, the Plus-Minus reductions will apply. The derived distance in line 28 will be that of either an ordinary or an upper-case edge. The added e' edge in line 17 is ordinary (by virtue of LABEL REMOVAL if necessary).

The whole algorithm terminates if the same source node is repeated in the recursion; thus, an infinite recursion is prevented. We will show that this condition occurs if and only if the STNU has a semi-reducible negative cycle. Thus, the algorithm does not require a subsidiary consistency check.

The early termination conditions in lines 00-03, which prevent infinite recursion and multiple calls with the same source, can be detected by marking schemes. Thus, the algorithm involves at most N (number of nodes in the network) non-trivial calls to DCbackprop, each of which (not including the recursive call) has complexity $O(N^2)$ if a Fibonacci heap is used for the priority queue, giving $O(N^3)$ in all. At most $O(N^2)$ edges are added to the graph; this cost is absorbed by the $O(N^3)$ overall complexity. The early termination calls to DCbackprop have $O(1)$ cost and come from superior calls or from determineDC. The former may be absorbed into the cost of line 20, while there are at most N of the latter. We now turn to the task of proving correctness.

Theorem 3. *The DCbackprop procedure encounters a recursive repetition if and only if the STNU is not Dynamically Controllable.*

Proof. Suppose first there is a recursive repetition. Note that if DCbackprop calls itself recursively, then there is a (backwards) negative path from the source of the superior call to that of the inferior. Since the distance calculations involve applications of the Plus-Minus reductions, that implies there is a reducible negative path from the first source to the second. Thus, a recursive repetition involves a cycle stitched together from reducible negative paths, which is a semi-reducible negative cycle.

Suppose conversely that the STNU is not Dynamically Controllable. Then it must have some semi-reducible negative cycle C. The intuition behind the proof is that the negative segments in C will either be bypassed, or will pile up against each other. Since they cannot all be bypassed, this will result in a recursive cycle. For the argument, it is convenient to temporarily remove lines 00-01 of the algorithm. In that case, a recursive cycle will result in an infinite recursion rather than returning false and we can talk about termination instead of what value is returned.

---

[5] It is useful to think of the distance calculation as taking place in the projection where any initial contingent link takes on its maximum duration and every other contingent link has its minimum. An unsuitable edge does not belong to that projection.

Note that for every negative node A in $\mathcal{C}$, DCbackprop(A) will be called eventually, either as a recursive call or as a top-level call from determineDC. By line 17, the execution of DCbackprop(A) may add a non-negative edge BA from some other node B in $\mathcal{C}$ to A. We will call this a *cross-edge*. Since the Dijkstra algorithm computes shortest paths, we have weight(BA) $\leq \mathcal{D}_{\mathcal{C}}(\text{B, A})$.

If there is no infinite recursion, then every call to DCbackprop must terminate. Our strategy will be to show that every terminating call to DCbackprop(A) for some A in $\mathcal{C}$ will add at least one cross-edge. These will then bypass all the negative edges in $\mathcal{C}$, which contradicts the fact that $\mathcal{C}$ is a negative cycle.

First assume all the non-negative edges in $\mathcal{C}$ are ordinary edges. (Lower-case edges add a slight complication that we will address in due course.) Consider the very first termination of a DCbackprop(A) call for A in $\mathcal{C}$. Note that the execution cannot have included a recursive call; otherwise the recursive call would have terminated first. The backward Dijkstra propagation must reach the predecessor A$'$ of A in $\mathcal{C}$. This cannot be a negative node, since otherwise it would cause a recursive call. If the distance to A$'$ is non-negative, then DCbackprop(A) will add a non-negative A$'$A edge. Otherwise the propagation must reach predecessor A$''$ of A$'$ since A$'$ is not a negative node. (Thus, A$'$A is a non-negative edge.) The propagation will continue to predecessors until a non-negative distance is reached. This must happen eventually. (Otherwise the propagation would continue all the way back to A and cause a recursive loop.) Then the execution of DCbackprop(A) will add a cross-edge before it terminates. For the inductive step, the argument is similar, except any recursive calls that have already terminated will have left behind cross-edges, and the propagation will be over those rather than the edges in the original cycle.

Now consider the case where $\mathcal{C}$ contains lower-case edges. The difficulty here is that subpaths of $\mathcal{C}$ are not necessarily shortest. Since the cross-edges resulting from the Dijkstra calculation are shortest paths, the shortenings could result in a closer moat edge for a lower-case edge. However, by Theorem 3 of [3], we can assume without loss of generality that $\mathcal{C}$ is *breach-free*. (A lower-case edge in the cycle has a *breach* if its extension contains an upper-case edge from the same contingent link.) In that case, the closer moat edge would still be usable.

Thus, we have shown every terminating call to DCbackprop leaves behind a cross-edge, which results in a contradiction. It follows there is some non-terminating call, i.e., an infinite recursion. When we put back lines 00-01, the recursive cycle is trapped, and results instead in a determination that the STNU is not Dynamically Controllable.   □

The algorithm as presented only adds non-negative edges, which are the only ones needed for the Dijkstra calculations. However, derived negative edges are implicit in the distance calculation. Thus, when distance(u) is negative in line 14 of the algorithm, we could infer and save a negative edge or wait condition. We will call the algorithm that does this determineDC+. Results in the next section show that the network resulting from determineDC+ is suitable for execution.

5   Dispatchability

Previous papers (e.g. [6]) have noted a relationship between Dynamic Controllability and dispatchability, but have not investigated it formally. In this section,
we clarify the relationship between dispatchability and Dynamic Controllability and explore how dispatchability applies to an STNU. We start by relating
dispatchability of an STNU to the better-understood dispatchability of its STN
projections. Since wait edges may be needed for this property to hold, we consider an *extended* STNU (ESTNU) that may include wait edges. Recall that wait
edges reduce to ordinary edges in a projection.

A *dispatching execution* [16] is one that respects direct precedence constraints,
and propagates execution times only to neighboring nodes, but otherwise is free
to execute a timepoint at any time within its propagated bounds. An STN is *dispatchable* if a dispatching execution is guaranteed to succeed. It is shown in [16]
that every consistent STN can be reformulated into an equivalent *minimum
dispatchable network*. The reformulation procedure first constructs the AllPairs
network and then eliminates *dominated* edges that are not needed for dispatchability. A fast version [17] of the algorithm computes distances from one node at
a time and uses that to determine which edges from that node are dominated.

We can extend these notions to an ESTNU by essentially pretending that
contingent timepoints are executed by the agent and propagating the observed
time. For an ESTNU dispatching execution, we require the free choices are made
only for executable timepoints, and respect the waits and precedences. We cannot directly mandate that the contingent timepoints respect the precedences;
however (see proof of next result), this is indirectly achieved if the projections
of the ESTNU are dispatchable. Note that such a strategy does not depend on
future events. This leads to the following.

Definition 3. *An ESTNU is dispatchable if every projection is dispatchable.*

Theorem 4. *A dispatchable ESTNU is Dynamically Controllable.*

Proof. Suppose all the projections are dispatchable. We show that an ESTNU
dispatching execution will satisfy precedence constraints for the contingent time-
points. Suppose othewise and let $A \overset{[x,y]}{\Longrightarrow} B \overset{-z}{\longrightarrow} C$ be the subnetwork where a
precedence is violated for the first time in some projection. Consider the state
of the execution after A and strictly before B, but within $z/2$ units of time prior
to B. This is a dispatching execution in the STN sense since no precedence has
been violated yet. However, the constraints in the projection now force C into
the past although it has not been executed yet, which implies the projection is
not dispatchable [17] contrary to our assumption.[6]

Thus, the ESTNU dispatching strategy restricted to each projection is a dispatching execution. If it fails, the projection in which it fails, and hence the
ESTNU, are not dispatchable.   □

---

[6] As an example, given an STNU $A \overset{[2,4]}{\Longrightarrow} B \overset{-1}{\longrightarrow} C$, the minimum dispatchable network
for the minimum-duration projection includes $A \overset{1}{\longrightarrow} C$, which makes C precede B.

Since dispatchability is itself a desirable property, this suggests the objective of transforming an STNU into an ESTNU such that each projection is dispatchable, preferably in minimum dispatchable form [16]. It is natural to consider if the fast algorithm discussed in [17], or some variant, can be adapted for this purpose. As it turns out, the determinDC+ algorithm may itself be viewed as such a variant, although it does not achieve minimum form. First we prove some basic facts about dispatchability of an STN in preparation for considering dispatchability of STNU projections.

Recall that a path has the prefix/postfix property if every nonempty proper prefix is non-negative and every nonempty proper postfix is negative. It turns out that prefix/postfix paths in an STN are related to edges in a minimum dispatchable network (MDN) for the STN. By results in [16,17], an MDN edge is either undominated or is one (which may be arbitrarily chosen) of a group of concurrent mutually dominating edges that are not strictly dominated.

Theorem 5. *Every consistent STN has a minimum dispatchable network such that if AB is an edge in that MDN, then there is a shortest path $\mathcal{P}$ from A to B in the STN such that $\mathcal{P}$ has the prefix/postfix property.*

Proof. First suppose AB is an undominated MDN edge. Let $\mathcal{P}$ be any shortest path from A to B in the STN and let C be any node strictly between A and B in $\mathcal{P}$. Consider the case where AB is negative. Then the prefix AC must be non-negative; otherwise AB would be lower-dominated [16]. It follows that the postfix CB is negative. On the other hand, if AB is non-negative, then CB must be negative (otherwise AB would be upper-dominated), and then AC is non-negative.

For the mutual dominance case, we restrict the MDN edge choice. Among a group of mutually dominating edges that are not strictly dominated, we choose an MDN edge AB such that a shortest path $\mathcal{P}$ from A to B does not contain a shortest path for another edge in the group. We can then use an argument similar to the undominated case because if AB is dominated by AC or CB, it would be strictly dominated.   □

This leads to a sufficient condition for dispatchability.

Definition 4. *An STN is* prefix/postfix complete *(PP complete) if whenever the distance graph has a shortest path from A to B with the prefix/postfix property, there is also a direct edge from A to B whose weight is equal to the shortest path distance.*

Theorem 6. *A consistent STN that is PP complete is dispatchable.*

Proof. By Theorem 5, the STN contains all the edges of one of its MDNs; thus, it is dispatchable.   □

Prefix/postfix paths have a well-behaved structure. Suppose a path $\mathcal{P}$ has the prefix/postfix property. Clearly if it has more than one edge, then the first edge must be non-negative and the last edge must be negative. Now consider a negative edge e in the interior of $\mathcal{P}$. The proper prefix of $\mathcal{P}$ that ends with e will be non-negative; thus, there must be some closest e' to e such that the subpath Q from e' to e is non-negative. It is not difficult to see that Q will also have the prefix/postfix property. We will call the subpath Q the *train* of e in $\mathcal{P}$. We get a similar train for every negative edge within $\mathcal{P}$. Since they all have the prefix/postfix property, the trains must be nested or disjoint. (This result is similar to the Nesting Lemma for extensions of lower-case edges.) Note that an innermost nested train consists of a negative edge preceded by all non-negative edges, and (like all trains) has non-negative total distance.

For an ESTNU, it turns out that we only need to ensure PP completeness for a subset of the projections. The *AllMin* projection is where every contingent link takes on its minimum duration. In an *AllMinButOne* projection, one of the contingent links takes on its maximum duration, and every other contingent link takes on its minimum duration. We will call these the *basic projections* of an ESTNU.

Theorem 7. *Given an ESTNU, if the basic projections are PP complete, then every projection is PP complete.*

Proof. Suppose the basic projections are PP complete, and consider a prefix/postfix shortest path $\mathcal{P}$ in one of the projections. The proof is by induction on the depth of nesting of the trains in $\mathcal{P}$. Recall that an innermost train must consist of a negative edge preceded by all non-negative edges. It is not hard to see that such a path has its minimum distance in one of the basic projections. The subpath corresponding to the train will still have the prefix/postfix property there. (The postfixes are not larger, and the proper prefixes have non-negative edges.) By PP-completeness of that basic projection, there must be an edge in the ESTNU that bypasses the subpath. Its distance cannot be less than the subpath since $\mathcal{P}$ is a shortest path; thus, they have the same distance. This reduces the depth of nesting. If the depth is zero, the same reasoning can be used to infer a bypass edge for the whole path.   □

This gives us some insight into the functioning of the determineDC+ algorithm. The nested trains cause recursive calls of DCBackprop. If DCBackprop(A) is called where AB is a contingent link, then the algorithm may be viewed as adding PP-completeness edges for the AllMinButOne projection for AB. If instead, the call is where A is the target of ordinary negative edges, then it is adding PP-completeness edges for the AllMin projection. The recursive calls ensure that non-negative edges are added in the order corresponding to the nesting. In summary, the algorithm is extending the STNU so that it is prefix/postfix complete with respect to the basic projections, and thus is dispatchable.

6   Closing Remarks

Note that in contrast to the fast MDN algorithm [17], all the edges in the original network are kept by determineDC+. The algorithm may also add unneeded dominated edges in addition to the non-dominated ones that it needs. The number of added edges is significant because the complexity of a Dijkstra computation is sensitive to the number of edges. This suggests a potential avenue for future improvement.

A class of STNs is said to be *sparse* if the number of edges E is a fixed multiple of the number of nodes N (i.e., E scales as $O(N)$). The cost of one Dijkstra call is $O(E + N \log N)$, which is $O(N \log N)$ for a sparse network. Networks encountered in practical problems tend to be sparse. Typically for an STN, if the original network is sparse, the minimum dispatchable network is also sparse [17],[7] since it essentially contains the same information in a concise form. It is reasonable to think the same might be true for an STNU if only non-dominated edges are added. Thus, if the algorithm could be improved to not add any dominated edges, then the complexity might in practice be comparable to that of the Fast Dispatchability algorithm, i.e., $O(N^2 \log N)$ for sparse networks. The issue essentially is to prune unneeded edge additions from each recursive call before the higher-level calls use them.

It might seem the ideal solution would be to adapt the fast minimum dispatchability algorithm [17] (FMDA) directly. For an STNU the distance calculation could be backwards and invoke recursive calls at negative nodes. However, the adaption would also need to handle, or sidestep, the contraction of rigid components in FMDA, which may be complicated by the fact that a contingent link is itself a rigid constraint in a projection. Besides that, there is the question of how to adapt the reweighting approach of FMDA (which makes possible a Dijkstra computation where the original weights may be negative), so that it works for all the basic projections. Also, to be worth it, the adaptions would need to fit within the existing FMDA cost. These are challenges for future research.

Acknowledgment. This paper owes a profound debt to the many insights of Nicola Muscettola, which include the key ideas underpinning the cubic algorithm. The current author is responsible for the formal treatment and proofs, and the results relating Dispatchability and Dynamic Controllability.

References

1. Muscettola, N., Nayak, P., Pell, B., Williams, B.: Remote agent: to boldly go where no AI system has gone before. Artificial Intelligence 103(1-2), 5–48 (1998)
2. Vidal, T., Fargier, H.: Handling contingency in temporal constraint networks: from consistency to controllabilities. JETAI 11, 23–45 (1999)
3. Morris, P.: A structural characterization of temporal dynamic controllability. In: Benhamou, F. (ed.) CP 2006. LNCS, vol. 4204, pp. 375–389. Springer, Heidelberg (2006)
4. Hunsberger, L.: Magic loops in simple temporal networks with uncertainty - exploiting structure to speed up dynamic controllability checking. In: ICAART, vol. (2), pp. 157–170 (2013)

---

[7] Artificially constructed exceptions are unlikely to occur in practice.

5. Hunsberger, L.: Tutorial on dynamic controllability (2013),
   http://icaps13.icaps-conference.org/wp-content/uploads/2013/06/
   hunsberger.pdf
6. Shah, J.A., Stedl, J., Williams, B.C., Robertson, P.: A fast incremental algorithm
   for maintaining dispatchability of partially controllable plans. In: Boddy, M.S.,
   Fox, M., Thiébaux, S. (eds.) ICAPS, pp. 296–303. AAAI (2007)
7. Nilsson, M., Kvarnström, J., Doherty, P.: Incremental Dynamic Controllability
   Revisited. In: Proceedings of the 23rd International Conference on Automated
   Planning and Scheduling (ICAPS). AAAI Press (2013)
8. Rossi, F., Venable, K.B., Yorke-Smith, N.: Uncertainty in soft temporal constraint
   problems: A general framework and controllability algorithms for the fuzzy case.
   Journal of Artificial Intelligence Research 27, 617–674 (2006)
9. Tsamardinos, I., Pollack, M.E.: Efficient solution techniques for disjunctive temporal reasoning problems. Artificial Intelligence 151, 43–89 (2003)
10. Dechter, R., Meiri, I., Pearl, J.: Temporal constraint networks. Artificial Intelli-
    gence 49, 61–95 (1991)
11. Cormen, T., Leiserson, C., Rivest, R.: Introduction to Algorithms. MIT Press,
    Cambridge (1990)
12. Morris, P., Muscettola, N., Vidal, T.: Dynamic control of plans with temporal
    uncertainty. In: Proc. of IJCAI 2001 (2001)
13. Morris, P., Muscettola, N.: Dynamic controllability revisited. In: Proc. of AAAI
    2005 (2005)
14. Muscettola, N.: Personal Communication (2006)
15. Web (2010),
    http://stackoverflow.com/questions/3833500/
    dijkstras-algorithm-with-negative-edges-on-a-directed-graph
16. Muscettola, N., Morris, P., Tsamardinos, I.: Reformulating temporal plans for ef-
    ficient execution. In: Proc. of Sixth Int. Conf. on Principles of Knowledge Repre-
    sentation and Reasoning (KR 1998) (1998)
17. Tsamardinos, I., Muscettola, N., Morris, P.: Fast transformation of temporal plans
    for efficient execution. In: Proc. of Fifteenth Nat. Conf. on Artificial Intelligence
    (AAAI 1998) (1998)

Author Index

Printed in the United States
By Bookmasters